A Picture is Worth a Thousand Tables

Andreas Krause · Michael O'Connell
Editors

A Picture is Worth a Thousand Tables

Graphics in Life Sciences

Editors
Andreas Krause
Actelion Pharmaceuticals Ltd
Department of Clinical Pharmacology
Modeling and Simulation
Allschwil
Switzerland

Michael O'Connell
Tibco Software Inc.
New York, NY
USA

ISBN 978-1-4614-5328-4 ISBN 978-1-4614-5329-1 (eBook)
DOI 10.1007/978-1-4614-5329-1
Springer New York Heidelberg Dordrecht London

Library of Congress Control Number: 2012950035

© Springer Science+Business Media New York 2012
This work is subject to copyright. All rights are reserved by the Publisher, whether the whole or part of the material is concerned, specifically the rights of translation, reprinting, reuse of illustrations, recitation, broadcasting, reproduction on microfilms or in any other physical way, and transmission or information storage and retrieval, electronic adaptation, computer software, or by similar or dissimilar methodology now known or hereafter developed. Exempted from this legal reservation are brief excerpts in connection with reviews or scholarly analysis or material supplied specifically for the purpose of being entered and executed on a computer system, for exclusive use by the purchaser of the work. Duplication of this publication or parts thereof is permitted only under the provisions of the Copyright Law of the Publisher's location, in its current version, and permission for use must always be obtained from Springer. Permissions for use may be obtained through RightsLink at the Copyright Clearance Center. Violations are liable to prosecution under the respective Copyright Law.
The use of general descriptive names, registered names, trademarks, service marks, etc. in this publication does not imply, even in the absence of a specific statement, that such names are exempt from the relevant protective laws and regulations and therefore free for general use.
While the advice and information in this book are believed to be true and accurate at the date of publication, neither the authors nor the editors nor the publisher can accept any legal responsibility for any errors or omissions that may be made. The publisher makes no warranty, express or implied, with respect to the material contained herein.

Printed on acid-free paper

Springer is part of Springer Science+Business Media (www.springer.com)

Foreword

The graphical display of data in medical research is a key component of communicating one's findings, yet sadly over the years little attention is paid to how to make such graphics truly effective and non-distortive.

For statisticians, clinicians, or other scientists there is virtually no training in the art and science of successful and reliable graphics, so the end result is that we all tend to muddle by as best we can. We resort to certain stereotypes of graph, and displays that distort research findings, e.g., by exaggerating true effects or not conveying statistical uncertainty, are quite common.

Hence, this book is an exciting contribution in explaining in a practical setting how graphics can be used best in medical research. The breadth of topics is particularly suited to the whole spectrum of pharmaceutical research, and the editors are to be congratulated on gathering such a well-informed and insightful set of experts to tackle each topic. Also noteworthy is that the book is an attractive read, illustrating the value of color for better perception of graphics. Whether it is for in-house and personal insights to data, publications in scientific journals, or conveying findings in regulatory submissions, this book's content carries a wealth of ideas that both statisticians and the broader realm of research scientists need to avidly absorb. Indeed, besides aiming at statisticians, modelers, and quantitative people the book's content should be appreciated by MDs, pharmacologists, managers, and decision makers, i.e., the key recipients of data analyses and their visualization. The hope is that in future we can all do a better job in utilizing graphics to convey the essence of our research findings.

London, UK   Stuart Pocock

Preface

Graphics are an essential means of communication for humans, from prehistoric cave painting to modern computer graphics. A graphic can often summarize and visualize information much more efficiently than tables or words.

A difference between verbal and graphical communication is that verbal communication is largely direct, the sender and the recipient of the message can exchange information. If necessary, in multiple iterations.

Graphics on the other hand are frequently an asynchronous means of communication: the creator of the graphic is not present to elucidate on what is being shown, to help understanding and facilitate correct interpretation. The graph has to speak for itself, deliver a clear message, and avoid misinterpretation.

Nowadays, graphs are often created to extract and reveal the information contained in a data set, reminding of the fact that having collected data is not equal to having generated knowledge. The knowledge must be distilled from the data and displayed in ways that make interpretation easy, accurate, and correct.

The life sciences and the pharmaceutical industry generate a wealth of data. A clinical study that aims to quantify the effects of a newly tested drug in humans, collects data from many domains: demographics, dosing, pharmacokinetics and pharmacodynamics, laboratory measurements, adverse events, ECG and vital signs, concomitant medications, medical history, and more; at many different time points in many different subjects, healthy volunteers, and treated patients. Data regarding the operations of the clinical trial, e.g., sites and investigators, are also voluminous, as are data generated in chemistry and biology research, and preclinical studies leading to the clinical trials. Other functional areas in pharmaceutical companies are also highly data-driven, e.g., post-marketing areas such as epidemiology and sales and marketing.

Clinical trial data are analyzed in various ways: by examining the raw data listings, aggregating the data in summary statistics, in tables and graphics. Tables, listings, and figures are used to summarize the data in clinical study reports and submissions to regulatory agencies at the conclusion of the clinical trials. They are also used to review the study data during clinical trials.

Wikipedia defines Graphics as follows:

Graphics (from Greek γραφικός graphikos) are visual presentations on some surface, such as a wall, canvas, computer screen, paper, or stone; to brand, inform, illustrate, or entertain. Examples are photographs, drawings, Line Art, graphs, diagrams, typography, numbers, symbols, geometric designs, maps, engineering drawings, or other images. Graphics often combine text, illustration, and color. Graphic design may consist of the deliberate selection, creation, or arrangement of typography alone, as in a brochure, flier, poster, Web site, or book without any other element. Clarity or effective communication may be the objective, association with other cultural elements may be sought, or merely, the creation of a distinctive style.
Graphics can be functional or artistic. [...]
Wikipedia entry on Graphics (Jan 23, 2012)

The Wikipedia entry states that graphics can be functional or artistic. The philosophy of this book is that efficient graphics are both functional and artistic. It takes a good amount of creativity on how to display the data, in order for the information in the data to be visualized effectively to the reader of the graph.

Creativity requires inspiration. This book aims at providing that inspiration. We have assembled a group of life science graphics experts who have described graphics principles, techniques, and case studies to provide inspiration and for you, to incorporate graphical best practices into your work.

Many of the visualization principles that hold for life science data are also encountered in daily life. For example, the map you use to orient yourself on the London Underground ("The Tube") has only limited similarity to the actual geographical positions of train tracks and stops. The simplification to largely horizontal, vertical, and diagonal lines improves the interpretation and the usefulness enormously.[1]

The London underground map does not only simplify the layout by reducing the geographical layout to straight lines, but it also actually distorts the geography substantially: if you take the underground—the Piccadilly line—from the city to Heathrow airport, you will probably be surprised by how quickly the inner city stations follow one another and then by how long it takes to arrive at Heathrow airport. If the tube map would be scaled according to geography, the inner city part, the part most used by travelers, would shrink and become illegible. So leave early for the outer stations!

Garland (1994) gives a historical account of the development of the tube "map," highlighting the interactions that improved its usefulness to the user of the London tube system. The book highlights prominently that the use of the word map is actually inappropriate because the geographical reality is not—or not accurately—mapped onto paper ("Though not strictly speaking a map, this term is almost universally used by people when referring to the London Underground *Diagram* [...]," p. 2, italics added).

---

[1] The simplification was taken too far in 2009 when it was suggested that the River Thames should not be shown on the tub map any longer (BBC London, "Thames reunited with tube map", http://news.bbc.co.uk/local/london/hi/people_and_places/newsid_8259000/8259435.stm).

	Relative survival rate, % (SE)			
	5 years	10 years	15 years	20 years
Cancer site				
Oral cavity and pharynx	56·7 (1·3)	44·2 (1·4)	37·5 (1·6)	33·0 (1·8)
Oesophagus	14·2 (1·4)	7·9 (1·3)	7·7 (1·6)	5·4 (2·0)
Stomach	23·8 (1·3)	19·4 (1·4)	19·0 (1·7)	14·9 (1·9)
Colon	61·7 (0·8)	55·4 (1·0)	53·9 (1·2)	52·3 (1·6)
Rectum	62·6 (1·2)	55·2 (1·4)	51·8 (1·8)	49·2 (2·3)
Liver and intrahepatic bile duct	7·5 (1·1)	5·8 (1·2)	6·3 (1·5)	7·6 (2·0)
Pancreas	4·0 (0·5)	3·0 (0·5)	2·7 (0·6)	2·7 (0·8)
Larynx	68·8 (2·1)	56·7 (2·5)	45·8 (2·8)	37·8 (3·1)
Lung and bronchus	15·0 (0·4)	10·6 (0·4)	8·1 (0·4)	6·5 (0·4)
Melanomas	89·0 (0·8)	86·7 (1·1)	83·5 (1·5)	82·8 (1·9)
Breast	86·4 (0·4)	78·3 (0·6)	71·3 (0·7)	65·0 (1·0)
Cervix uteri	70·5 (1·6)	64·1 (1·8)	62·8 (2·1)	60·0 (2·4)
Corpus uteri and uterus, NOS	84·3 (1·0)	83·2 (1·3)	80·8 (1·7)	79·2 (2·0)
Ovary	55·0 (1·3)	49·3 (1·6)	49·9 (1·9)	49·6 (2·4)
Prostate	98·8 (0·4)	95·2 (0·9)	87·1 (1·7)	81·1 (3·0)
Testis	94·7 (1·1)	94·0 (1·3)	91·1 (1·8)	88·2 (2·3)
Urinary bladder	82·1 (1·0)	76·2 (1·4)	70·3 (1·9)	67·9 (2·4)
Kidney and renal pelvis	61·8 (1·3)	54·4 (1·6)	49·8 (2·0)	47·3 (2·6)
Brain and other nervous system	32·0 (1·4)	29·2 (1·5)	27·6 (1·6)	26·1 (1·9)
Thyroid	96·0 (0·8)	95·8 (1·2)	94·0 (1·6)	95·4 (2·1)
Hodgkin's disease	85·1 (1·7)	79·8 (2·0)	73·8 (2·4)	67·1 (2·8)
Non-Hodgkin lymphomas	57·8 (1·0)	46·3 (1·2)	38·3 (1·4)	34·3 (1·7)
Multiple myeloma	29·5 (1·6)	12·7 (1·5)	7·0 (1·3)	4·8 (1·5)
Leukaemias	42·5 (1·2)	32·4 (1·3)	29·7 (1·5)	26·2 (1·7)

Rates derived from SEER 1973–98 database (both sexes, all ethnic groups).[12]
NOS=not otherwise specified.

Fig. 1 Cancer survival rates by cancer site and time from Brenner (2002)

Another fine example of highlighting information in data with graphics comes from Edward Tufte. Tufte has made a long and distinguished career by clearly displaying information from data. His book, "The Visual Display of Quantitative Information" was named one of the top 100 books of the twentieth century by Amazon.

In this particular example, Tufte used a table of relative cancer survival rates published in a landmark paper (Brenner, The Lancet, 2002) and showed how this display of data can be improved.

Figure 1 shows the original table. Each row gives the survival statistics for a particular type of cancer at 5, 10, 15, and 20 years, showing the relative survival rates and the associated standard errors. Figure 2 shows a simplified version: the font size of the standard errors is shrunk since they are arguably less important than the survival rates, and the rows are sorted by survival rate, the most likely measure of interest to the reader. The legibility of the information inside the data is improved considerably (Tufte 2007, p. 174).

Estimates of relative survival rates, by cancer site

	% survival rates and standard errors						
	5 year		10 year		15 year		20 year
Prostate	98.8	0.4	95.2	0.9	87.1	1.7	81.1 3.0
Thyroid	96.0	0.8	95.8	1.2	94.0	1.6	95.4 2.1
Testis	94.7	1.1	94.0	1.3	91.1	1.8	88.2 2.3
Melanomas	89.0	0.8	86.7	1.1	83.5	1.5	82.8 1.9
Breast	86.4	0.4	78.3	0.6	71.3	0.7	65.0 1.0
Hodgkin's disease	85.1	1.7	79.8	2.0	73.8	2.4	67.1 2.8
Corpus uteri, uterus	84.3	1.0	83.2	1.3	80.8	1.7	79.2 2.0
Urinary, bladder	82.1	1.0	76.2	1.4	70.3	1.9	67.9 2.4
Cervix, uteri	70.5	1.6	64.1	1.8	62.8	2.1	60.0 2.4
Larynx	68.8	2.1	56.7	2.5	45.8	2.8	37.8 3.1
Rectum	62.6	1.2	55.2	1.4	51.8	1.8	49.2 2.3
Kidney, renal pelvis	61.8	1.3	54.4	1.6	49.8	2.0	47.3 2.6
Colon	61.7	0.8	55.4	1.0	53.9	1.2	52.3 1.6
Non-Hodgkin's	57.8	1.0	46.3	1.2	38.3	1.4	34.3 1.7
Oral cavity, pharynx	56.7	1.3	44.2	1.4	37.5	1.6	33.0 1.8
Ovary	55.0	1.3	49.3	1.6	49.9	1.9	49.6 2.4
Leukemia	42.5	1.2	32.4	1.3	29.7	1.5	26.2 1.7
Brain, nervous system	32.0	1.4	29.2	1.5	27.6	1.6	26.1 1.9
Multiple myeloma	29.5	1.6	12.7	1.5	7.0	1.3	4.8 1.5
Stomach	23.8	1.3	19.4	1.4	19.0	1.7	14.9 1.9
Lung and bronchus	15.0	0.4	10.6	0.4	8.1	0.4	6.5 0.4
Esophagus	14.2	1.4	7.9	1.3	7.7	1.6	5.4 2.0
Liver, bile duct	7.5	1.1	5.8	1.2	6.3	1.5	7.6 2.0

Fig. 2 Cancer survival rates by cancer site and time, adaptation by Tufte

Subsequently, Tufte took these survival rates and created a semi-graphic: Fig. 3 shows the same data but now the row and column structure has been broken up: vertical distances approximately represent numerical differences in survival rates instead of just another row. Interconnecting the values that correspond to the same type of cancer visually reveals trends in time: a negative slope indicates a decrease in survival rate with time.

The left-hand side graph shows an artistically adapted version: for better legibility, the vertical dimension—the invisible y-axis—corresponds only roughly to the numerical values of the 5 year survival statistics. The cancer types are sorted vertically by 5 year survival statistics, and the slopes of the subsequent survival statistics of the particular cancer type correspond to the decrease in survival. A negative slope corresponds to a decline in survival over time. Note how the ease of interpretation is markedly improved over the original table (Tufte 2006, p. 176, originally published on the Internet in 2003).

Preface

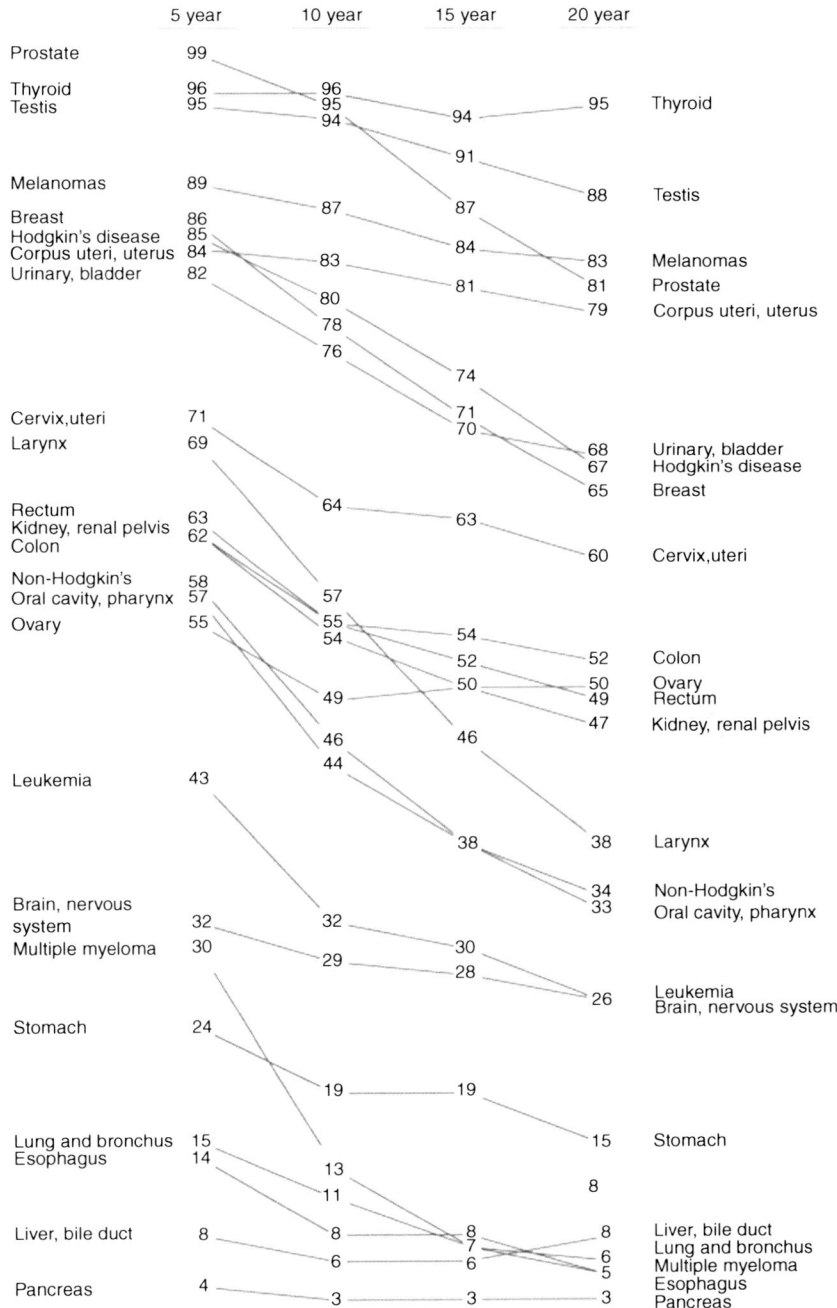

Fig. 3 Cancer survival rates by cancer site and time displayed as semi-graphic

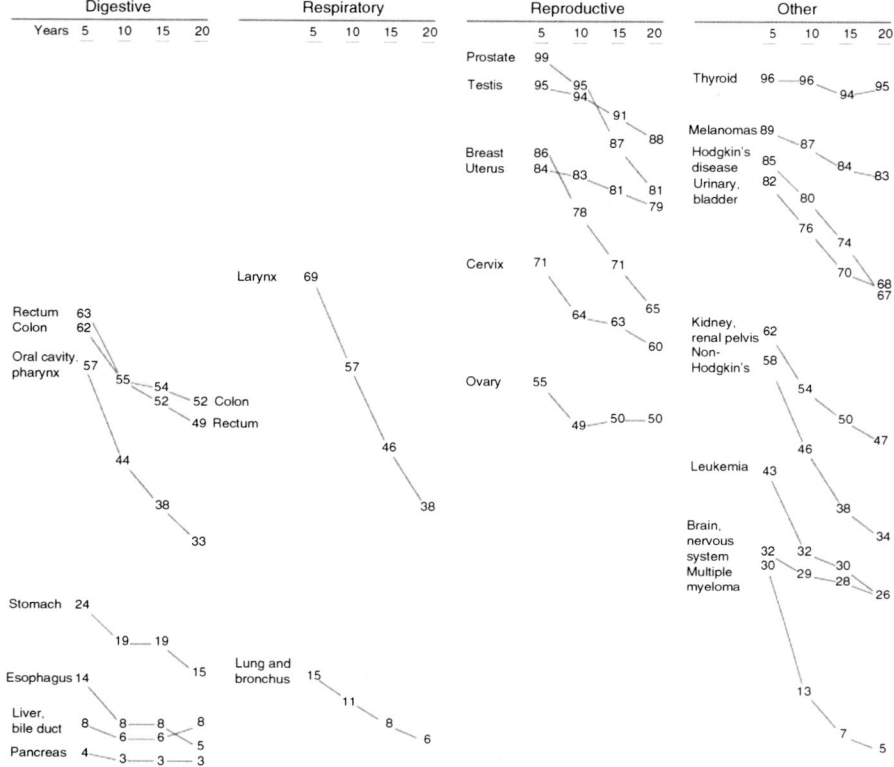

Fig. 4 Cancer survival rates by cancer site and time as semi-graphic, stratified by locus

Nash (2006) in a Web discussion of the Tufte graphs suggested further modifications, suggesting that grouping on cancer types allows for intersecting lines without substantial loss of legibility due to the lower number of intersecting lines (Fig. 4).

Nash collaborated with Tufte on the choice of line weights and text alignments and—most importantly—the placing of the names of the cancer types that are repeated on the right-hand side for better visualization.

These examples show that graphs can be art—creative, informative, and beautiful to the eye. They also demonstrate clearly the importance of extracting, distilling, and presenting efficiently the information contained in the data.

Even though no one disputes the importance of good graphics for efficient data analysis, they are still not as widely used as they could be. Part of the reason might be that graphs can often be improved once the data are seen and understood. However, the life sciences industry works under time pressure, and analysis programs producing tables, listings, and figures are often on the critical path in the development of new treatment therapies. Thus, the graphs are programmed before seeing the data; for example, before database lock and release of the treatment code

Preface   xiii

("unblinding"). Once the data are available, the results must be available quickly. The graphs are created and delivered without a close looking at the data in the interest of time. That conflict between time pressure and thorough analysis can be critical in that information is missed. But of course, it will probably not even get noticed.

Help is on its way: as data volumes explode in all industries, investment and research in analysis, visualization, business intelligence, and data discovery software is ever-increasing. In the life sciences, the most popular software systems for graphical exploration include R and S-PLUS™, Spotfire™, and SAS™. Microsoft Excel™ seems to still be the most widely used graphical analysis software.

While modern software is necessary in enabling graphical analysis it is not sufficient; and sound graphics principles are required for optimal discovery and communication of information in data. Some of the aforementioned software applications enable sound principles to be followed and others make this difficult!

The landmark books by Tufte (1990, 1997, 2001, 2006) and Few (2004, 2009) establish general principles for good visualizations. Cleveland (1993, 1994) establishes the techniques from a statistical point of view, including the groundbreaking conditional multivariate Trellis display (Becker et al., 1996), codified now in R through the lattice package (Sarkar, 2008). Tufte can be viewed as somewhat of a minimalist and artist, while Cleveland's methods brought strength to the scientific community in the quest to discover and distill information from multivariate data.

Color was researched by Brewer et al. (2002), resulting in the groundbreaking ColorBrewer and corresponding R package on CRAN.

CTSpedia (CTSpedia, 2012) is a knowledge base for clinical and translational research. Among the topics covered are statistical graphics. The Web page gives general advice and tries to organize the types of graphics by the question and the data at hand. It guides through a variety of graphical tools and suggests standards for good graphics.

Visualization and data discovery has been the topic of several TED lectures (TED, 2012), with prime example being Hans Rosling's presentation of interactive graphics analyzing the world's poverty data (Rosling, 2006). The accompanying software, GapMinder (Rosling, 2009), is now available for free download (Gapminder, 2012). It is based on Google's Motion Chart (Google Chart Tools), another free software tool. Similarly, IBM offers the creation and discussion of data sets and analyses online with its Many Eyes initiative (IBM Many Eyes, 2012).

There are several other recent graphics references that are relevant to graphical analysis in the life sciences. Robbins (2004) introduces general principles by showing hands-on examples; Unwin et al (2006) focus on visualization of large data sets. Leckart (2010) provides an interesting update on medication package inserts in the popular magazine Wired. The Jung + Wenig (2010) art design company is the creator of one of the redesigned package inserts. One of their designs is shown in Fig. 5, illustrating that "boring" laboratory results that mostly contain only a large amount of text can be made so much more visually interesting and informative.

The aim of our book is to provide a standard reference for life science graphics, leveraging the pioneering concepts described by Tufte, Cleveland, Few, and others; in a modern framework and with life science context.

PATIENT NAME	BIRTH DATE	PATIENT ID NO	GENDER	COLLECTED 11	07	2010	11:40 a.m.
Grant, Boyd	12.09.1976	9131-10-1/11Jks	Male	RECEIVED 11	07	2010	1:53 p.m.
				ORDERED BY Dr. Michaels

ALIGN HERE | SEND TO: **QUEST DIAGNOSTICS NI**
Attn: Dona Little/Send Outs
33608 Ortega highway
San Juan Capistrand. CA 92690-6130

About this Test

This test measures the amount of a substance called prostate-specific antigen, or PSA, in the blood. The prostate gland releases more of this antigen as you age, but PSA levels can also rise due to an inflammation of the prostate or prostate cancer. As such, PSA is used as a tool for screening for cancer.

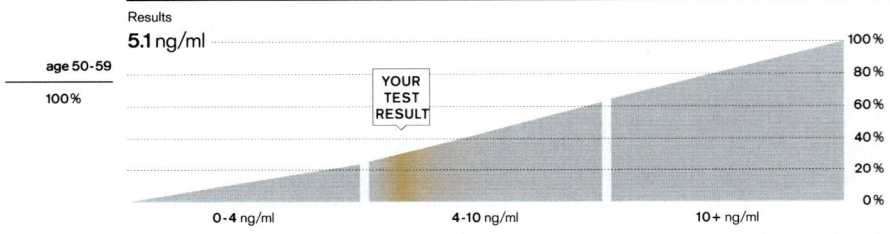

What do your results mean?

Patients with levels between 4.0 and 10.0 ng/ml have a 25 percent risk of prostate cancer according to the American Cancer Society; men with a PSA greater than 10.0 ng/ml have a 67 percent risk of prostate cancer.

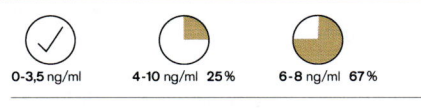

But this test is not definitive. Up to 20 percent of patients in your age range (55-69) will experience at least one false-positive. And 65-75 percent of men with an elevated PSA level who are subsequently biopsied are NEVER diagnosed with prostate cancer.

Additional Perspective
Risk of being diagnosed with prostate cancer

 1 in 6 all ages

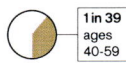 1 in 39 ages 40-59

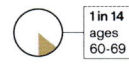 1 in 14 ages 60-69

What now?

1   Talk to your physician about alternative reasons for an elevated PSA
benign prostate enlargement | inflammation | infection | age | race

2   Talk to your physician about additional tests, including:
digital rectal exam
biopsy (may cause harmful side effects, including bleeding and infection)

Survival Rate
for patients diagnosed with prostate cancer

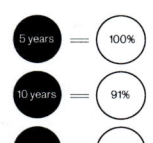

Fig. 5   Blood work results redesigned

Preface

We hope that the book serves as a source of inspiration—not only to the creators of scientific graphics, e.g., statisticians and programmers, but in particular to consumers of graphics and key decision makers, physicians, and managers in the life sciences industry.

We will have achieved our mission if business users and decision makers ask statisticians and information technology functional areas for more and better graphs of all their data. As better graphics are used throughout the industry, everyone in the life science ecosystem benefits, including the general population receiving better information and medications through better decision making.

Book Structure

The book is divided into parts. The part "General Principles and Reviews of Graphics" motivates the topic, reviews and establishes general principles, and suggests structured and innovative uses of graphics and also tables.

The part "Preclinical and Early Clinical Development" contains chapters that illustrate a variety of applications for PK/PD, biomarker, and genetic data.

The largest part, "Clinical Trial Graphics," contains chapters on efficacy and safety in therapeutic areas such as oncology and respiratory disease, safety reviews for cardiac and general safety, meta-analysis, and dose–response visualization. The principles are generally applicable to a large variety of applications.

The part on "Operations, Marketing, and Post-Approval Graphics" shows illustrative case studies in clinical trials data management, exploratory visualization of medical safety in observational studies, and post-approval visualizations.

Book Web Site

Being able to use computer programs right away will help getting up to speed quickly. For this purpose, this book has a companion Web site:
   http://www.elmo.ch/doc/life-science-graphics/
   The Web site contains computer programs for download and further information about the book.

References

Becker RA, Cleveland WS, Shyu MJ (1996) The visual design and control of trellis display. J Comput Stat Graph 5:123–155

Brenner H (2002) Long-term survival rates of cancer patients achieved by the end of the 20th century: a period analysis. Lancet 360:1131–1135

Brewer CA, Hatchard GW, Harrower MA (2003) ColorBrewer in print: a catalog of color schemes for maps. cartography and geographic information. Science 30(1):5–32. http://www.personal.psu.edu/cab38/ColorBrewer/ColorBrewer_instructions.html. Accessed 11 Feb 2012

Cleveland WS (1993): Visualizing data. Hobart Press

Cleveland WS (1994): The elements of graphing data. Hobart Press, 2nd edition

CTSpedia: a knowledge base for clinical and translational research. http://www.ctspedia.org/. Accessed 11 Feb 2012

Few S (2004) Show me the numbers: designing tables and graphs to enlighten. Analytics Press

Few S (2009) Now you see it: simple visualization techniques for quantitative analysis. Analytics Press

Gapminder home page: http://www.gapminder.org/. Accessed 11 Feb 2012

Garland K (1994) Mr. Beck's underground map: a history. Capital Transport Publishing, London

Google chart tools. http://code.google.com/apis/chart/interactive/docs/gallery/motionchart.html. Accessed 1 April 2012

IBM Many Eyes. http://www-958.ibm.com/software/data/cognos/manyeyes/. Accessed 11 Feb 2012

Jung C, Wenig T (2010) Redesign of the prostate laboratory results. Personal communication. http://www.jungundwenig.com/. Accessed 21 May 2012

Leckart S (2010) The blood test gets a makeover. Wired Magazine. http://www.wired.com/magazine/2010/11/ff_bloodwork/all/1. Accessed 1 April 2012

Nash D (2006) Response to cancer survival rates: tables, graphics, PP. http://www.edwardtufte.com/bboard/q-and-a-fetch-msg?msg_id=0000Jr. Accessed 10 April 2012

Robbins N (2004) Creating more effective graphs. Wiley Rosling H: Hans Rosling shows the best stats you've ever seen. Presentation at the TED conference 2006, Monterey CA. http://www.ted.com/talks/hans_rosling_shows_the_best_stats_you_ve_ever_seen.html. Accessed 11 Feb 2012

Rosling H, Johansson C (2009) Gapminder: liberating the x-axis from the burden of time. Statistical Computing and Graphics Newsletter 20 (1):4–7 http://stat-computing.org/newsletter/issues/scgn-20-1.pdf. Accessed 11 Feb 2012

Sarkar D (2008) Lattice: multivariate data visualization with R. Springer

TED (2012): Ideas worth spreading. http://www.ted.com/. Accessed 1 April 2012

Tufte ER (1990): Envisioning information. Graphics Press

Tufte ER (1997) Visual explanations: images and quantities, evidence and narrative. Graphics Press

Tufte ER (2001) The visual display of quantitative information. Graphics Press, 2nd edition

Tufte ER (2006) Beautiful evidence. Graphics Press

Unwin A, Theus M, Hofmann H (2006) Graphics of large datasets: visualizing a million. Springer

Wainer H (2007) Graphic discovery: a trout in the mil and other visual adventures. Princeton University Press, 2nd edition

Wikipedia: The Free Encyclopedia. Wikimedia Foundation Inc. Updated 27 Feb 2012, 11:27 UTC. Encyclopedia on-line. Available from http://en.wikipedia.org/wiki/Graphics. Internet. Retrieved 29 Feb 2012

Allschwil, Switzerland	Andreas Krause
New York, NY, USA	Michael O'Connell

Acknowledgments

Over the past 10 years, we have been involved in many sessions and tutorials on life sciences graphics in a wide variety of public forums including the Joint Statistical Meetings (JSM), the American Conference on Pharmacometrics (ACOP), the Drug Information Association (DIA) Annual Meeting, the BASS Conference, the Deming Conference, the Graybill Conference, the Midwest Biopharmaceutical Workshop and the Association of American Pharmaceutical Scientists (AAPS) Annual Meeting. The AAPS Sunrise School workshop (at 7 am!) was organized by Stacey Tannenbaum. It was Stacey who suggested the title "A picture is worth a thousand tables." So thank you, Stacey, for the inspiration!

All of these sessions have included our collaborators and colleagues, many of whom are featured in this book. A big hats-off to all the contributors to this book. The community that all of us have created has grown dramatically over this period and spawned other initiatives that have progressed the use of graphics across the life science industry.

Finally, we gratefully acknowledge the highly professional and efficient collaboration with Springer-Verlag throughout the entire process. Our special thanks go to Carolyn Honour, Renata Hutter, and John Kimmel in this regard; and to Springer for the terrific job they do with publications in the life sciences.

Contents

Part I General Principles and Reviews of Graphics

1. **Concepts and Principles of Clinical Data Graphics** 3
   Andreas Krause

2. **Graphics for Exploratory Analysis and Data Discovery in the Life Sciences** 23
   Michael O'Connell, Ian Cook, Difei Luo, and Josh Patel

3. **Grables: Visual Displays That Combine the Best Attributes of Graphs and Tables** 41
   Thomas E. Bradstreet

4. **The Use of Figures in Epidemiological Publications: A Survey of Current Practice and Consequent Recommendations** 71
   Elisabeth Wreford Andersen and Stuart J. Pocock

Part II Preclinical and Early Clinical Development

5. **Use of Graphics for Studies with Small Sample Sizes: A Simulated Case Study of an Early-Phase Asset for Treatment of Type 2 Diabetes Mellitus** 87
   Denise Shortino, Ann Walker, and Andrew Miskell

6. **Exploring Pharmacokinetic and Pharmacodynamic Data** 99
   Charles Roosen, Richard Pugh, and Andrew Nicholls

7. **(Interactive) Graphics for Biomarker Assessment** 117
   Michael Merz

8. **Graphical Displays for Biomarker Data** 139
   Manuela Zucknick, Thomas Hielscher, Martin Sill, and Axel Benner

Part III Clinical Trial Graphics

9 Statistical Graphics in Clinical Oncology 173
Kye Gilder

10 Efficient and Effective Review of Clinical Trial Safety Data Using Interactive Graphs and Tables 199
Harry Southworth

11 Visualizing Dose–Response When the Signal to Noise Ratio Is Low: The Bronchodilatory Response in Chronic Obstructive Pulmonary Disease 217
Michael Looby and Didier Renard

12 Statistical Graphics in Late Stage Drug Development 241
Julia Wang and Surya Mohanty

13 Graphical Data Exploration in QT Model Building and Cardiovascular Drug Safety 255
Ihab G. Girgis and Surya Mohanty

14 Data Visualization at the Individual Patient Level 273
Matthew Austin and Alicia Zhang

15 Graphics for Meta-Analysis 295
Peter W. Lane, Judith Anzures-Cabrera, Steff Lewis, and Jeffrey Tomlinson

16 Visualization of QT Data for Thorough QT Study Analysis and Review 309
Christoffer W. Tornøe

17 Graphics for Safety Analysis 325
Peter W. Lane and Ohad Amit

18 Cardiac Safety 343
Richard J. Anziano

Part IV Operations, Marketing, and Post-Approval Graphics

19 Data Visualization for Clinical Trials Data Management and Operations 359
Ted Snyder

20 Post-approval Uses of Clinical Data, Phase IV Data, and Sales and Marketing Data Visualizations 373
Sam Weerahandi, Birol Emir, and Ed Whalen

**21 Using Exploratory Visualization in the Analysis
of Medical Product Safety in Observational Healthcare Data**............ 391
Patrick Ryan

About the Editors... 415

Index.. 417

Contributors

Ohad Amit Quantitative Sciences, GlaxoSmithKine, Collegeville, PA, USA

Elisabeth Wreford Andersen London School of Hygiene and Tropical Medicine, London, UK

Technical University of Denmark, Department of Informatics, Data Analysis, Lyngby, Denmark

Richard J. Anziano Primary Care Statistics, Pfizer Inc., Groton, CT, USA

Judith Anzures-Cabrera Roche, Welwyn Garden City, UK

Matthew Austin One Amgen Center Drive, Global Biostatistical Science, Amgen Inc., Thousand Oaks, CA, USA

Axel Benner Division of Biostatistics, German Cancer Research Center (DKFZ), Im Neuenheimer Feld, Heidelberg, Germany

Thomas E. Bradstreet Experimental Medicine Statistics, Merck Research Labs, North Wales, PA, USA

Ian Cook Tibco Software Inc., Chapel Hill, NC, USA

Birol Emir Pfizer Inc., New York, NY, USA

Kye Gilder Biostatistics Department, NuVasive, Inc., San Diego, CA, USA

Ihab G. Girgis Johnson and Johnson Pharmaceutical Research and Development, LLC, Advanced Modeling and Simulation Group, Raritan, NJ, USA

Thomas Hielscher Division of Biostatistics, German Cancer Research Center (DKFZ), Heidelberg, Germany

Andreas Krause Department of Clinical Pharmacology, Modeling and Simulation, Actelion Pharmaceuticals Ltd, Allschwil, Switzerland

Peter W. Lane Quantitative Sciences, GlaxoSmithKine, Stevenage, UK

Steff Lewis Centre for Population Health Sciences, University of Edinburgh, Edinburgh, UK

Michael Looby Modeling and Simulation, Novartis Pharma AG, Basel, Switzerland

Difei Luo Tibco Software Inc., London, UK

Michael Merz Novartis Institutes for BioMedical Research, Translational Sciences, Basel, Switzerland

Andrew Miskell GlaxoSmithKline, Clinical Statistics, Durham, NC, USA

Surya Mohanty Johnson and Johnson Pharmaceutical Research and Development, LLC, Advanced Modeling and Simulation Group, LLC, Raritan, NJ, USA

Andrew Nicholls Mango Solutions Ltd, Chippenham, UK

Michael O'Connell Tibco Software Inc., New York, NY, USA

Josh Patel Tibco Software Inc., Chapel Hill, NC, USA

Stuart J. Pocock London School of Hygiene and Tropical Medicine, London, UK

Richard Pugh Mango Solutions Ltd, Chippenham, UK

Didier Renard Modeling and Simulation, Novartis Pharma AG, Basel, Switzerland

Charles Roosen Mango Solutions AG, Basel, Switzerland

Patrick Ryan Epidemiology Analytics, Janssen Research and Development, Titusville, NJ, USA

Martin Sill Division of Biostatistics, German Cancer Research Center, Heidelberg, Germany

Ted Snyder Clinical Informatics, Infinity Pharmaceuticals, Cambridge, MA, USA

Denise Shortino GlaxoSmithKline, Clinical Statistics, Durham, NC, USA

Harry Southworth AstraZeneca, Macclesfiled, United Kingdom

Jeffrey Tomlinson Roche, Welwyn Garden City, UK

Christoffer W. Tornøe Quantitative Clinical Pharmacology, Novo Nordis, Søborg, Denmark

Ann Walker GlaxoSmithKline, Clinical Statistics, Durham, NC, USA

Julia Wang Janssen Pharmaceutical Companies of Johnson and Johnson, Raritan, NJ, USA

Sam Weerahandi Pfizer Inc., New York, NY, USA

Ed Whalen Pfizer Inc., New York, NY, USA

Alicia Zhang One Amgen Center Drive, Global Biostatistical Science, Amgen Inc., Thousand Oaks, CA, USA

Manuela Zucknick Division of Biostatistics, German Cancer Research Center, Heidelberg, Germany

Part I
General Principles and Reviews of Graphics

Chapter 1
Concepts and Principles of Clinical Data Graphics

Andreas Krause

Abstract Graphics are an essential tool for detecting and analyzing structure in data, developing statistical models, presenting results, and communicating about data and results. Drug development is still largely based on tables and listings, and graphics are an underutilized resource. Processing the data with suitable graphics can make the data analysis substantially more efficient. Principles are introduced and illustrated with corresponding figures of clinical data.

The principles of good graphics in the sense of Cleveland and Tufte's works are introduced. A further step is taken by establishing these and further principles for graphics of clinical data as generated from patient data in drug development.

Graphical setups such as different graph types and formats and elements such as axes, symbols, lines, legends, and colors are discussed. The principles are followed by applications to clinical data by examples, covering pharmacological, efficacy, and safety data.

Following some basic principles enables the provision of graphs to the clinical team that makes use of the power of graphics: looking at the data, understanding its structure, and learning efficiently about the information contained in the data.

1.1 Introduction

Statistical graphics are "information graphics in the field of statistics used to visualize quantitative data" (Wikipedia) on ("Statistical graphics", 2012). Good statistical graphics can provide a convincing means of communicating the underlying message that is present in the data.

A. Krause (✉)
Department of Clinical Pharmacology, Modeling and Simulation, Actelion Pharmaceuticals Ltd
Gewerbestr. 16, 4123 Allschwil, Switzerland
e-mail: Andreas.Krause@actelion.com

Table 1.1 Safety data tabulated: occurrence of events by category and treatment

Category	Event	Active	Placebo	Absolute difference	Relative difference (%)
Nervous system	Headache	45	23	22	96
Gastrointestinal	Diarrhea	23	15	8	53
Nervous system	Dizziness	11	17	−6	−35
Gastrointestinal	Nausea	7	1	6	600
Gastrointestinal	Vomiting	5	2	3	150
Gastrointestinal	Dyspepsia	11	7	4	57
Gastrointestinal	Abdominal pain	8	7	1	14

If used well, graphics can condense complex information into a simple and easily interpretable display. Patterns that might have stayed undetected can—with suitable graphics—easily be spotted by the human eye. If used badly, graphics can hide relevant information or even mislead to produce wrong interpretations. Not using graphics increases the risk of missing a signal in the data, an efficacy or safety signal or a signal that the model fitted to the data might not be appropriate.

According to Cleveland (1993), graphical statistical methods can have four objectives: exploration of the content of a data set, finding structure in data, checking assumptions in statistical models, and communication of results of an analysis.

This paper introduces some of the basic principles of good graphics as defined and illustrated in the works of Tufte (2002, 2007) and Cleveland (1993). For general purposes, Robbins (2005) outlines basic concepts to a wide audience, Heiberger and Holland (2004) discuss displays for statistic analysis, and Unwin et al. (2006) focus on large data sets. A further step is taken by establishing these principles for visualization of clinical data: data that are generated in drug development by measuring drug effects (efficacy and safety) in healthy subjects and patients.

Consider the following example: a study was conducted in which patients were treated with either an experimental drug or a placebo in a double-blind manner. Safety events were collected to assess if the active drug's safety differs from the placebo group's safety.

Table 1.1 displays the raw data whereas Fig. 1.1 gives the same information in a graphical format. The graph's rows are ordered such that the largest relative differences between active drug and placebo are shown on top and the smallest differences at the bottom. Arguably it is substantially easier to detect the most prevalent safety signals in the graph.

This paper first establishes some general principles for good visualizations before proceeding toward particular applications in clinical drug development. After establishing and illustrating some principles, I will cover graphical elements such as axes, symbols, lines, legends, and colors and then move on to applications including comparisons, categorical covariate analysis, change from baseline graphs, and graphs with outliers as well as higher-dimensional graphs. Some illustrative graphs are going to incorporate the principles, serving as inspiration for further application.

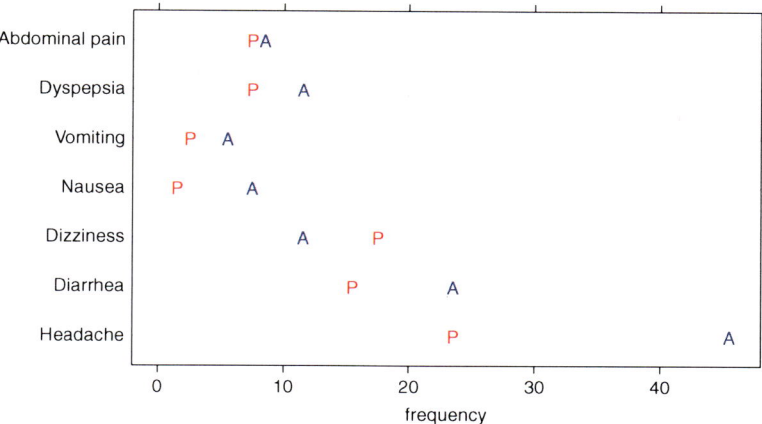

Fig. 1.1 Safety data graphed: occurrence of events by category and treatment. Symbols denote treatments active ("A") and placebo ("P")

1.2 Principles

General aims in the display of data include the effective communication of the information contained in the data in a non-distorting manner. Key elements comprise the accuracy, ease, and speed of the interpretation.

The purpose of a graph is not to have a graph but to reveal to the reader of such graph the information contained in the data. Thus, the interpretation is in the eye of the beholder. This is in particular relevant if the creator of a graph (a programmer, a statistician, a modeler) has a background different from the reader (a medical doctor, a marketing expert). This can imply that graphs need to be adapted to the background of the reader, be it a statistician, a modeler, a medical doctor, or a marketing expert.

Tufte establishes the principle of maximizing the "data to ink ratio": too much ink or a large number of pixels per observation indicates that the visualization might be improved whereas little ink or few pixels per data point indicate an effective visualization.

An example of a low data to ink ratio is a bar chart: a single number is represented by a filled area that consumes a large space on the display. In the extreme, one can question if just a few numbers are not better displayed by a table. An example for a high data to ink ratio is a scatterplot that shows all the data and each observation is represented by a single pixel or symbol. Frequently one can choose between various displays, and the full data set often allows the best judgment about the data structure.

When the data consist of thousands of observations, aggregation can not be possibly avoided. Unwin et al. (2006) provide examples of displays particularly suited for large data sets.

1.2.1 A Graph is a Model

A statistical model aims at extracting the pattern from the data. In this respect a graph is not different. The statistical model and the graph transform the data into coefficients and patterns, respectively, to facilitate the recognition of the information in the data.

The S Language as implemented in S-PLUS (S-PLUS 2005) and R (R Development Core Team 2009) introduces a language that makes the similarity apparent. To calculate a linear model that regresses y on x in a data set named study101, the command is

```
lm(y~x, data=study101)
```

To create a graph of the same data, the command is identical except that a plotting routine is called:

```
xyplot(y~x, data=study101)
```

Note that the syntax suggests that the y-variable is a function of the x-variable, in the linear model as well as in the graph.

Researchers in visualization but also in related fields such as user interfaces for everyday's appliances (Norman 2002) have noted for long that additional elements in visualization that do not contribute to the reading and interpretation might actually be misleading. In particular, as noted by Tufte (2002), color shading and three-dimensional illusions frequently lead to misreading of graphs.

Generally, it is well known that some patterns are easier to recognize and interpret accurately than others for the human eye. Judging relative lengths and comparing values lined up along a common axis (such as in a scatterplot) are easier tasks than judging angles, slopes, and areas (as, for example, in pie charts).

On a finer scale, Cleveland (1993) notes that all elements that do not belong directly to the information should be placed outside the central graph, including tick marks and labels. Thus, tick marks and axis labels that stretch into the graph are to be avoided. If possible, legends shall also play a minor role (we will address the particular topic later).

Generally, the message is that a graph requires some thought about the aims of the graph as well as the perception of it by the reader.

1.2.2 Encouraging Comparison

Cleveland (1993, 1998) introduced the principle of Trellis™ graphics in S-PLUS (see also Becker et al. 1996). In R, the concept is labeled lattice graphics for trademark reasons. SAS version 9.2 recently introduced the lattice layout in SAS/GRAPH in the SGPANEL procedure (SAS Institute Inc. 2009).

The concept makes heavy use of the fact that once a reader is used to a particular graphics type, it becomes very easy to interpret a whole series of such graphs for matters of comparison.

We illustrate the principle with Fig. 1.2. The data comprise drug exposure of patients in four different dose groups, pre and post an intervention. Four dose groups

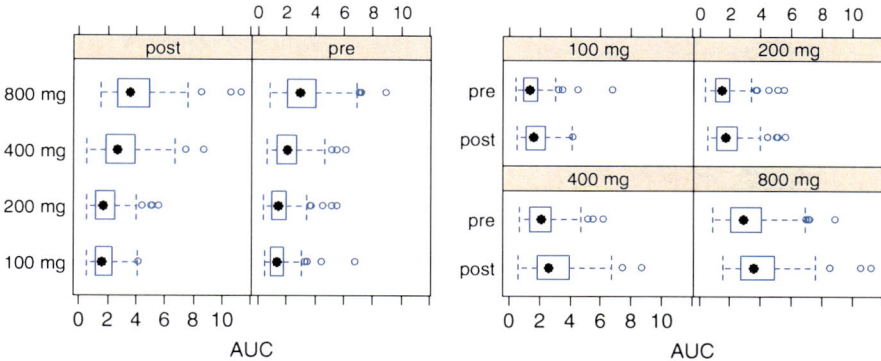

Fig. 1.2 Exposure to a drug for different doses, pre and post an intervention. *Left*: one panel for pre and post event. *Right*: one panel per dose group

at two time points (pre and post) yield eight box-and-whisker plots. The arrangement of the eight box-and-whisker plots facilitates particular comparisons. The left-hand side graph enables comparing the exposure in the four dose groups pre intervention (left panel) and in the four dose groups post intervention (right panel). However, to compare the exposures for a particular dose group pre and post intervention, the graph on the right is much more suitable: it arranges the same eight box-and-whisker plots such that the pairs of dose groups are adjacent to one another, facilitating visual comparison.

As a consequence, axis ranges are identical and each panel contains corresponding elements, facilitating interpretation.

1.3 Graphical Elements

A graph can and typically does contain a large number of elements: axes with ranges, tick marks, labels, points and symbols, lines, legends, colors, and more. They are typically predefined with default values and settings, but one needs to be aware that it can help to deliberately choose particular settings.

1.3.1 Choosing the Axes

The choice of axes of a graph is a central decision. Using the analogy of a statistical model, most readers of graphs intuitively interpret the *y*-axis variable as the dependent variable and the *x*-axis as independent variable. In other words, *y* is a function of *x*. For clinical applications, blood pressure might be a function of age. Thus, blood pressure should be shown on the *y*-axis and age on the *x*-axis. Similarly, body weight is influenced by sex and not vice versa.

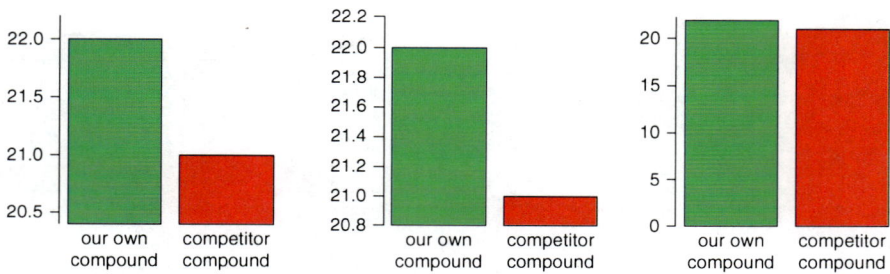

Fig. 1.3 Comparison of two values using different y-axis limits

1.3.2 Axis Ranges

The range of the axis, mostly the y-axis, can heavily influence a graph. Figure 1.3 shows three graphs, each representing the same two values (21 and 22). Thus, the choice of the y-axis range is fairly arbitrary. The figures show the different visual impressions of the same data and the choice of colors is suggestive (green="good", red="bad").

Using relative values poses a further challenge: the choice of where to place the reference. Relative values will be discussed with an example of visualizing change from baseline.

1.3.3 Point Symbols

In a scatterplot, an observation can be marked by a single pixel, an open circle, a closed circle, or, for data in groups, symbols (squares, triangles, circles) or letters indicating the groups.

A single pixel is mostly too small to be visible and filled circles or symbols introduce visual distortions if the data set is so large that the symbols overlay. Possibly the best choice for plotting symbols is open circles since if they overlay, the circles intersect and it is still visible that there are multiple observations.

Symbols such as squares, triangles, and circles to indicate groups (for example treatment groups) require an additional step: the symbols need to be mapped to the groups such that the graph needs a legend and the reader needs to memorize the legend or look it up frequently if there are many groups. Furthermore, a legend will probably use up space that can otherwise be used for the graph. Symbols that are easy to map, for example, "P" for placebo, "A" for active or "H" for high dose, and "L" for low dose, are easy to remember and require only a single look at the legend. The introductory Fig. 1.1 used "P" and "A" for active and placebo and can do without a legend since the mapping is obvious.

1.3.4 Lines

Lines are used to interconnect observations. Since they connect the first point to the second and the second to the third, lines inherently define an *order*. There are cases where the order can be visualized by means of lines, for example if drug effect is graphed against drug concentration: interconnecting the observations can show delays in (change of) drug effect when drug concentration changes (the graph type is known as hysteresis plot). In many other cases, for example when races or sexes are interconnected, it does not make sense and actually misleads the interpretation by suggesting an order that is not present.

1.3.5 Legends

Legends are required if a representation such as a plot symbol or a color is used for different groups in the data. However, the additional burden is that it is not intuitively clear which group a triangle or an orange line represents. The reader of the graph is forced to move the eyes from the graph to the legend and back many times, disabling to focus on the graph. In addition, the legend requires valuable space that can otherwise be used for the graph.

Thus, intuitive grouping symbols or colors should be used if possible. In particular if the groups have an inherent order, one might be able to map that order to another order in colors, symbols, line types, or line widths.

For example, some colors carry intuitive connotation: red is a warning color (or "bad") and green is "good", yellow or orange is in between. Response to treatment might thus be color-coded with response to treatment colored in green, progressive disease in red, and stable disease in orange or yellow. In a black and white graph, line widths can range from small (for progressive disease) to thick (for clinical response).

Alternatively, characters might be used that have an intuitive connection to the groups. Examples would be letters such as H, M, L for high, medium, and low dose.

If a legend is included in a graph, it is worth a second thought if the reading can be made easier or if the graph could even be created without a legend. Consider, for example, adding the legend to the figure caption to gain space for graphing the data (See Fig. 1.1).

1.3.6 Colors

Colors are a much disputed topic in reporting of clinical trial results. It is still widespread standard to use only black and white or grayscale figures in clinical study reports. The reasons given are frequently that the graph must be interpretable if printed on a black and white printer.

At times, such arguments severely reduce the usefulness of a graph because instead of intuitive colors one is forced to use much less intuitive coding of groups, typically triangles, squares, and circles. If color is used it can help to provide an intuitive understanding of graphs and a sensible choice of colors can help avoiding legends. If color does not facilitate interpretation it might be counterproductive, in particular if redundant elements such as shading, gradients, or textures are used.

Several researchers have discussed the topic of color schemes and choice of colors, including Zeileis et al. (2007, 2009), Lumley (2006), and Ihaka (2003). It seems that the researcher in drug development must separate presentations and internal reports—that may use color in graphs—from submission documents and publications that still require grayscale graphs for the time being. Some journals are moving toward online supplements that provide color figures while the print publication is still in black and white.

1.4 Particular Applications

This section deals with particular applications in graphing clinical data. It moves from individual elements to entire graphs of clinical data, illustrating the principles established before.

1.4.1 Comparisons: Compare Like with Like

Frequently, two variables containing observations that correspond to one another are compared: baseline value versus after treatment, observed versus model-predicted, and so on. There are several pitfalls in such comparisons, and knowing about the effects of perception can improve these graphs substantially.

Figure 1.4 shows three graphs of the same data set. The difference between the two graphs on the left is simply that the axes are swapped. Note the difference in perception, for example with respect to the two predicted values just below and above 70.

Where do these differences in perception come from? Inspecting the graphs closer shows that there are several differences: the axis ranges differ numerically and the x-axis is stretched compared to the y-axis: 10 pixels on the y-axis represent a larger range than 10 pixels on the x-axis. As a consequence, the line of identity, $y=x$, has a slope larger or smaller than 1 (in terms of pixel slope).

The principle should thus be that axis ranges shall be identical and the same number of units of measurement must correspond to the same number of pixels in x and y. This leads to the conclusion that a comparison graph must be square, naturally

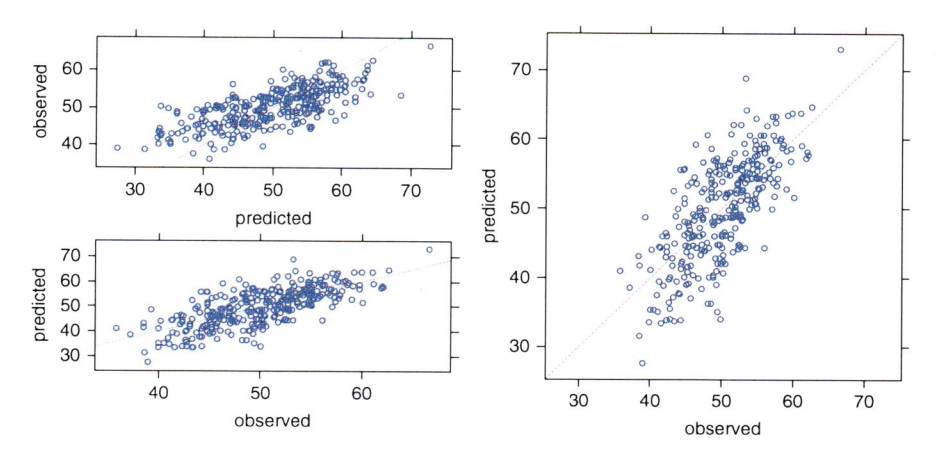

1 Concepts and Principles of Clinical Data Graphics

Fig. 1.4 Comparing pairs of data. *Left*: observed value versus predicted value (*top*) and vice versa (*bottom*) with aspect ratios not equal to 1. *Right*: same graph with an aspect ratio of 1. The *line* represents the line of identity, $y = x$

forcing the line of identity to have a 45° slope from corner to corner. The third graph in Fig. 1.4 implements these principles.

1.4.2 Categorical Variables

Showing categorical data in a graph poses a particular challenge: if there is no natural order in the groups, one must define the order such that the data can be shown along axes that possess an order. Thinking carefully about such ordering can substantially facilitate the reading of the graph.

Figure 1.5 shows different ordering of the event frequency. Note that the relative difference between the two groups is another variable that is used for sorting the categories but not plotted. Risk difference might be more suitable for small numbers of observations, avoiding division by 0 and misleading values due to low event rates.

1.4.3 Crossover Studies

Sometimes, the same patient undergoes different treatments (such as two treatments to be compared or the same treatment in fed and fasted status). Naturally, one can either compare the data observed in the same patients to one another or the data in the same group. In the resulting graph, Fig. 1.6, the left-hand side graph serves well for comparing the two states of the patients, whereas the right-hand side graph

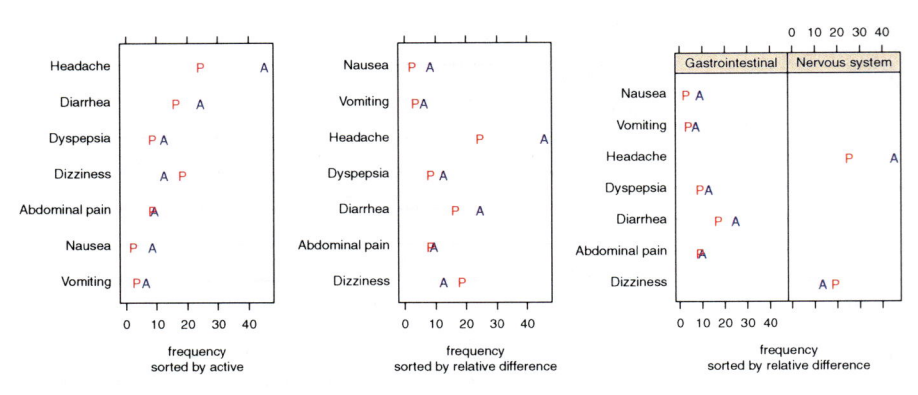

Fig. 1.5 Frequency of adverse events for placebo and active treatment in different sorting orders

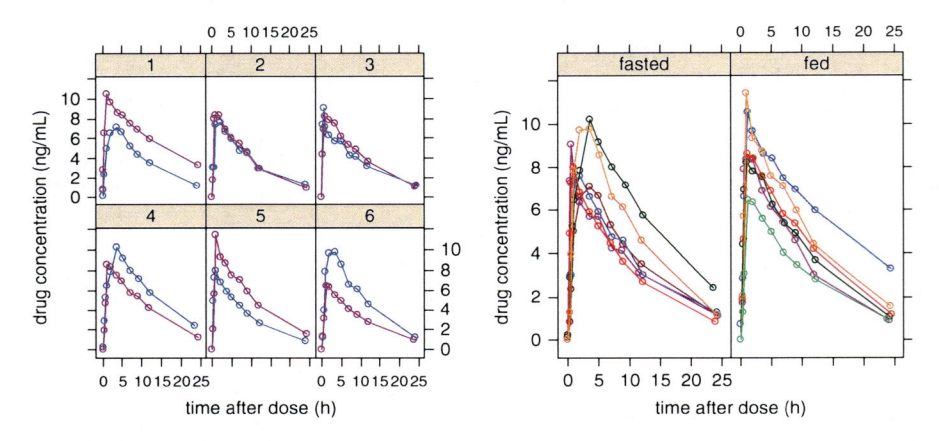

Fig. 1.6 Crossover data. *Left*: categorized by subject ID. *Right*: categorized by food status

serves better to compare profiles within each group. Bock (2001) discusses a variety of graphs for single patient results in crossover studies.

1.4.4 Change from Baseline

Changes such as the change from baseline in blood pressure or some other clinical measure are often shown as simple profile: one polygon per individual. However, even simple displays like these can be created in different ways, making the intended perception more or less easy. Following the idea of Trellis™/lattice displays, there should be a common axis for all groups in the data. Furthermore, changes of continuous variables can be either positive (the value is above the baseline value) or negative (the value is below the baseline value). The choice of the y-axis range influences the perception of the change pattern as shown in Fig. 1.7.

On the left, the data are simply shown with default plot options, in particular a y-axis range that is determined by the data range. On the right, the common y-axis

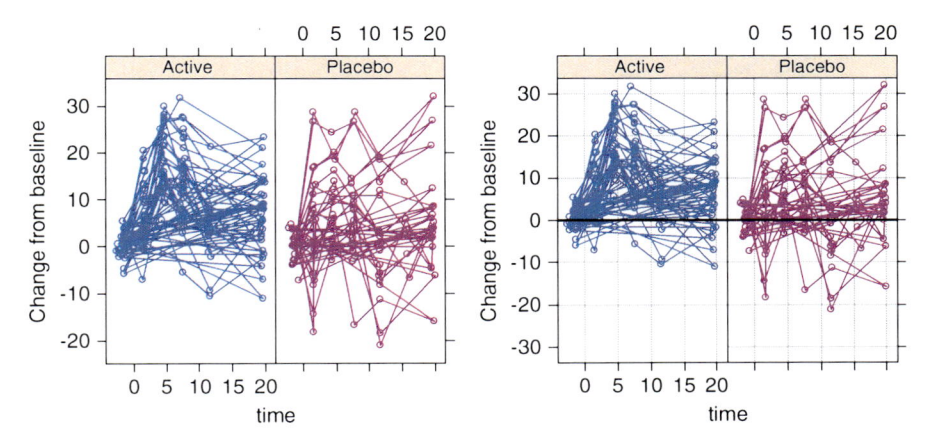

1 Concepts and Principles of Clinical Data Graphics 13

Fig. 1.7 Change from baseline. The *right graph* improves the default graph by use of a symmetric *y*-axis and auxiliary lines to facilitate reading

is maintained but two elements were modified to facilitate judgment about the change from baseline: the *y*-axis range is symmetric around 0 such that the point of no change, 0, comes to lie in the middle of the graph. All data in the upper half correspond to positive change from baseline, all data in the lower half correspond to negative change from baseline. The direction of the change can thus be read off without even looking at the *y*-axis. The human eye is even able to roughly judge whether there is more data with positive or negative change from baseline.

Further enhancements include a horizontal line at the point of no change from baseline and vertical light gray lines to facilitate reading the numerical values off the *x*-axis.

1.4.5 Extreme Values

Extreme values or outliers pose challenges to many techniques. Simple summary statistics are heavily influenced, i.e., the mean and the standard deviation change substantially by inclusion or omission of the extreme values. Model-based analyses can yield very different results and graphs can be distorted because of visual effects.

Figure 1.8 illustrates several problems: 99 of the 100 observations occupy about 1/10 of the area of the graph. 9/10 of the area of the graph is "occupied" by a single observation in the lower right corner. That single observation prevents seeing any structure in the bulk of the data simply because the space allocated is too small. Furthermore, a linear regression line is added (thick red line) and the "fit" is heavily influenced by that single extreme value. If that value is excluded from the data set, the fit corresponds to the thin red line.

Possible solutions for such situations include use of logarithmic scales, truncation, and placement of the extreme value at the truncation limits, or even omission of the single value.

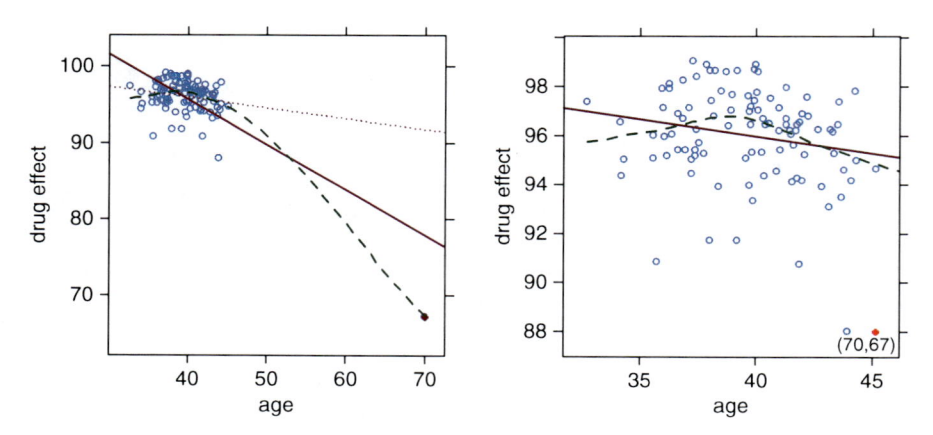

Fig. 1.8 A single extreme value distorts the visualization. *Left*: all data. *Right*: extreme value truncated with values indicated (70, 67)

1.4.6 Higher Dimensions

Using more than two dimensions on a screen or a paper implies the creation of an illusion. An imaginary position of the observer is defined, the projection of the three-dimensional graph into two dimensions is derived, and the graph is shown in two dimensions. Such manipulations can create highly irritating or even misleading results. The misleading enhancement of more than two dimensions in graphs has been noted as early as 1951 by Haemer (1951).

Consider the simple example of a three-dimensional bar chart in Fig. 1.9 (Krause 2009, Robbins 2009). The four bars on the left correspond to four numbers, 10, 20, 30, and 40. However, looking at the horizontal lines at the rear wall of the bars and the y-axis on the left, the bars are all clearly below the lines that correspond to their values. The graph is correct perspective-wise. Just the reader is misled because one tends to forget about the third dimension when looking at "simple" bars.

The most important part of this figure is the distance between a bar and the rear wall. Since the surface that the bars are placed on is tilted toward the reader, the bar must be projected onto the rear wall to read off the values correctly. The graph on the right-hand side shows how to read off the value: by "shoving" the bar against the rear wall—toward the back and up the slope!

Figure 1.10 shows an even more extreme graph: the bars have become cylinders placed on a tilted surface. Can one judge the values or compare the heights of two bars? Not to mention that the order of the data plays a role: smaller bars can easily disappear behind larger ones. The graph on the right shows exactly the same data, possibly less "fancy" but much easier to interpret.

The settings of both figures are the default settings in a popular spreadsheet package. Previous versions allowed adding the raw data as a table to the 3D display, suggesting that the graph is not of much use.

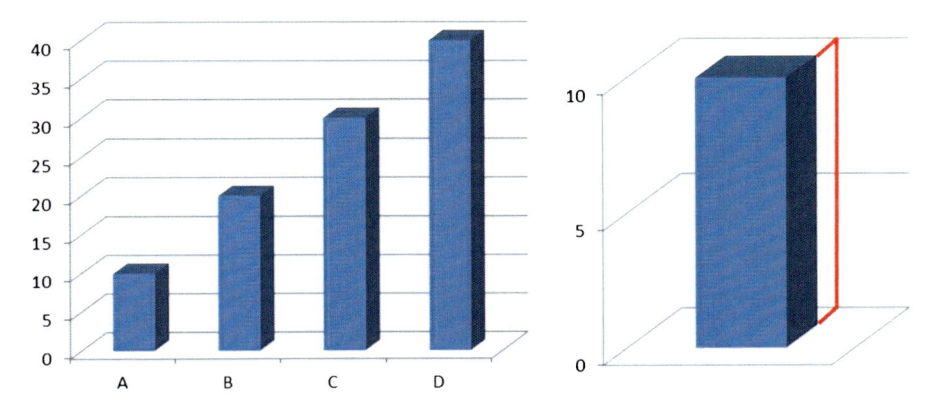

Fig. 1.9 A three-dimensional bar chart. *Right*: reading of a value

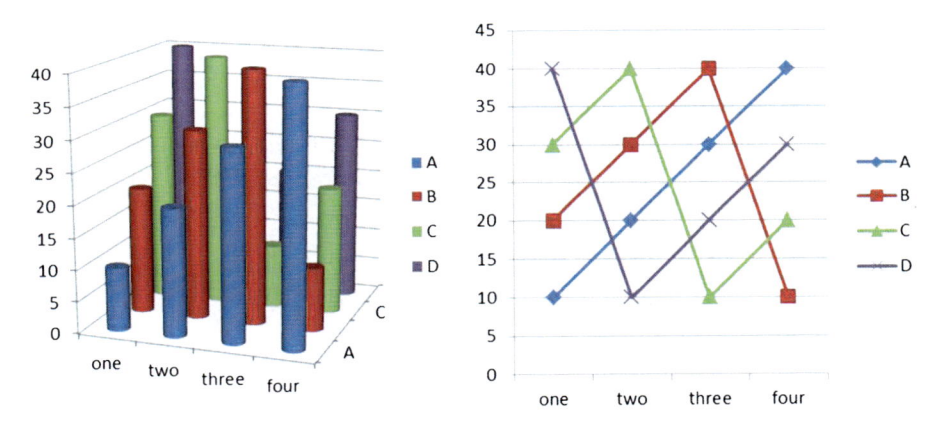

Fig. 1.10 A three-dimensional matrix arrangement of three-dimensional columns (*left*) and an alternative two-dimensional display of the same data (*right*)

The criticism does not necessarily apply to all three-dimensional displays. For example, if there is a correlation between two variables that influence a third, a surface can be of help to illustrate the structure. Figure 1.11 shows how the amount of air a healthy male person can exhale in 1 s (FEV_1) is a function of age and height (following the formula by Morris et al. 1971). The particular values of age and height are not relevant such that the *x*- and *y*-axes are omitted. Just the range is indicated in the label. The dependent variable, FEV_1, is mapped into the *z*-dimension and the reader looks at the graph from above. The values of FEV_1 are coded as heights above the age-height plane and as colors.

Starting at the two extremes, one can see that the largest FEV_1 values are observed in the young and tall. The smallest values are observed in the old and short.

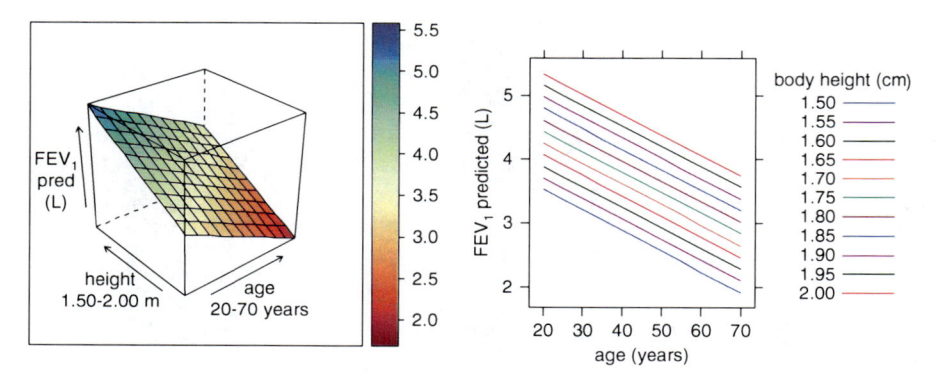

Fig. 1.11 Forced exhaled volume of air in 1 s (FEV₁) as a function of age and body weight, shown as surface (*left*) and as two-dimensional slices through the surface (*right*)

For a fixed age, one can see the increase of FEV$_1$ with height. For a fixed body height, one can see the decrease in FEV$_1$ with age. The values can be read with sufficient accuracy: a male person, 20 years of age and 2 m tall, exhales about 5.5 L in 1 s. That value will decrease to about 3.5–4 L at the age of 70.

If the particular values are of interest, one can slice the surface into two dimensions, plotting the parallel lines of the wireframe. The decrease of FEV$_1$ with age for different body heights, slicing the surface in the height dimension, is shown in Fig. 1.11b.

1.5 Examples of Applications

This section shows inspirational graphs for particular types of data and data analysis for dose/concentration-response, biomarker, and safety data.

1.5.1 Integration of Dose–Response and PK/PD

Consider a clinical study where patients were administered different doses of a drug, the drug concentration was measured in the blood stream, and the clinical response of interest was measured. Each patient contributes one record (the data are simulated). An Emax-type model was fit to the data to characterize the concentration-response (pharmacokinetic/pharmacodynamic or PK/PD) relationship.

Of interest is the PK/PD relationship but also the range of drug concentrations obtained with the different doses: what is the range or variability of concentrations and how much do the concentrations overlay for the different doses?

Figure 1.12 shows the information. The PK/PD relationship is shown in the top graph, the E_{max} model fit with the associated 90% confidence interval is overlaid and the concentrations that achieve 20, 50, and 80% of the maximally achievable effect,

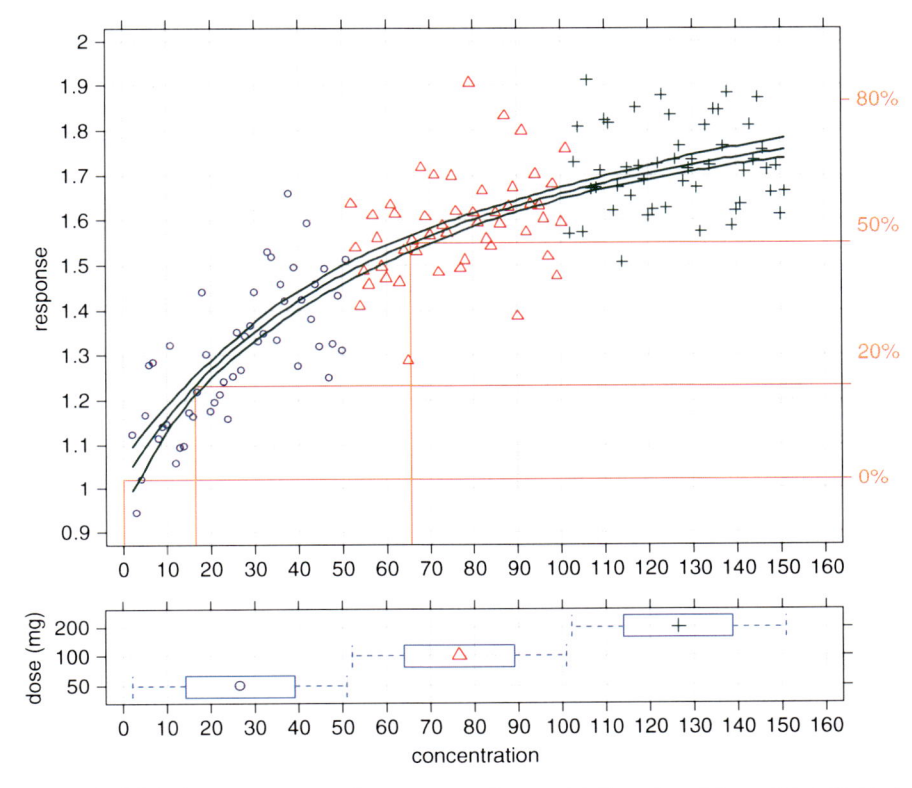

1 Concepts and Principles of Clinical Data Graphics   17

Fig. 1.12 Clinical response versus drug concentration, model fit, and confidence intervals (*top*). Dose versus drug concentration range at trough at steady state (*bottom*). Dose groups are indicated by *symbols* and *colors*

E_{max}, are indicated by lines that enable reading off the numerical values for the concentrations and the associated effects. Underneath, the relations between doses and concentrations are shown as box-and-whisker plots.

Note that the *x*-axes of the two graphs are aligned to allow cross-referencing between the two graphs. Note also that the median concentrations per dose group in the lower graph are indicated as colored symbols. The same colored symbols are used in the top graph to indicate the dose group for each patient. That association eliminates the need for a legend in this figure: it is obvious from the graph.

1.5.2 A Six-Dimensional Scatterplot of Biomarker Data

Figure 1.13 shows how quite complex information can be condensed into a graph that is easily understood. In this particular example, the data are displayed in a form

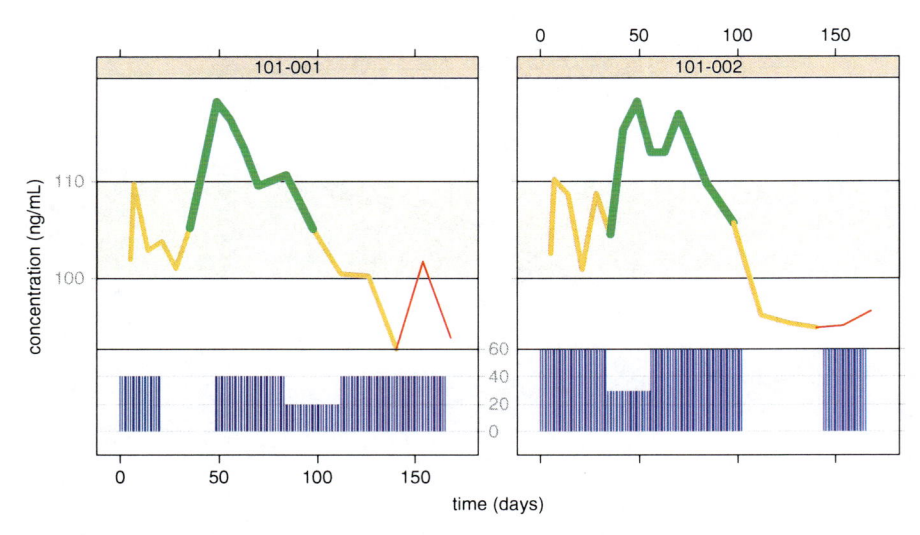

Fig. 1.13 A biomarker over time. Clinical response is coded by *color* and *line width*. Normal ranges are indicated as *horizontal lines*. Dosing history is indicated *below the profile*

that made the construction of a statistical model obsolete since the clients were able to derive the results from the display of the raw data.

The data originate from a single patient. The patient is identified by the number in the panel strip above the graph. The observation period lasted for about 180 days. The drug doses are indicated by the blue vertical lines at the bottom. Several dose adjustments become apparent.

Blood samples were taken and the concentration of a marker in the blood was measured at various time points. The normal range is indicated by black horizontal lines.

The patient's well-being was measured at each visit: it was assessed whether the disease was progressing, stable, or improving. The disease status is associated with an intuitive color scheme: red (warning color) indicates progressive disease, yellow indicates stable disease, and green (positive association) indicates an improving patient status. The graph can do without a legend because the color scheme is intuitive.

For gray scale prints, the clinical response is also mapped into the line widths: the thicker the line, the more positive the response.

One could argue that in total this scatterplot has six dimensions: the patient number, the time axis, the blood concentration of the marker, the normal range, the patient's disease status, and the drug dosing history.

Similar graphs can be developed for model fits, for example originating from population PK/PD modeling: the data can be shown in addition to the fits (population and individual fit), and if further information is available that changes over time, it might as well be incorporated by means of color or line widths. Further characteristics might also be marked in the graph: outliers can be circled and trough measurements can be shown as the letter "T".

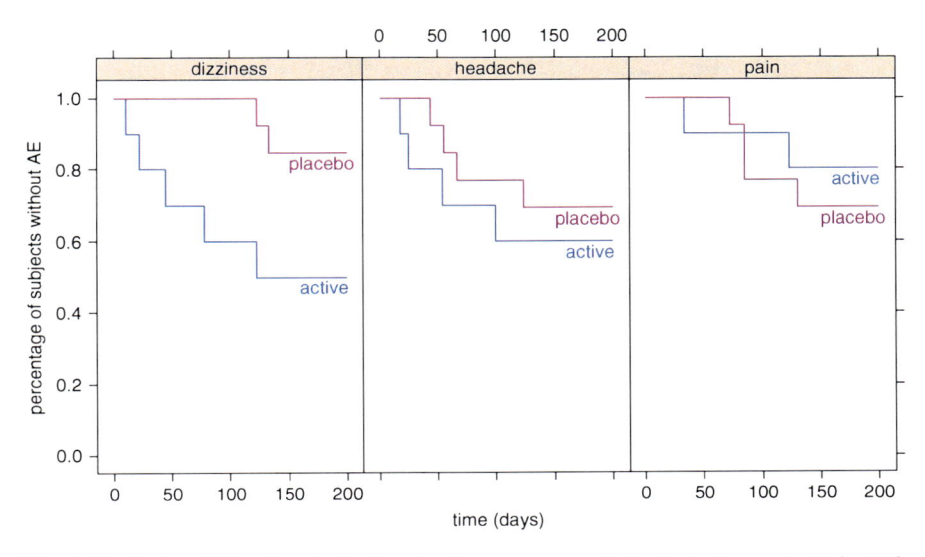

1 Concepts and Principles of Clinical Data Graphics

Fig. 1.14 Kaplan–Meier (survival) plots for the occurrence of the first adverse event in each patient

1.5.3 Safety Data: Kaplan–Meier Plots

Even if the frequency of the adverse event is similar between treatments, it might well be that they occur at different points in time. If it is of interest when safety problems occur, survival plots such as Kaplan–Meier curves can be used to show, for a given time point after start of treatment, how many of the subjects on treatment are still without adverse event. Figure 1.14 shows such a graph.

1.6 Discussion and Conclusion

Graphical tools are basically available since the computer became widespread. Even before computers were available for the creation of graphics, misleading graphics were discussed in the 1940s and 1950s (Haemer 1947, 1951). Graphical and exploratory data analysis concepts go back at least to the 1970s when Tukey (1977) published his landmark book. Cleveland and Tufte since the 1990s (Cleveland 1993, Tufte 2002, 2007) established general principles for visualization of quantitative information.

Tufte in fact made the distinction between graphics and text less rigid: he introduced tables in the form of semi-graphics layout with the famous "estimates of % survival rates" (Tufte 2007, pp. 174–176). A further step toward integrating concise graphics into floating text to replace stand-alone figures is taken by Tufte's "sparklines" or "intense, simple, word-sized graphics" (Tufte 2007, pp. 47–63) that he illustrates with medical record data. The R package YaleToolkit allows the creation of sparklines in R (Emerson and Green 2012).

Software to graph data has become widely available with popular office software, and statistical software provides graphics for several decades. Software like R is even available freely.

Many industries have discovered graphics as powerful tools to detect patterns easily. The pharmaceutical industry is still a fairly conservative industry despite mounting pressure to analyze data more thoroughly. In particular in clinical phases II and III, analysis plans are written and computer programs are finalized before the data become available. Once the database is locked, results are generated by running the programs and controlling the quality of the output. The primary goal is frequently to meet the strict timelines.

Creating useful graphs is sometimes only possible after having seen the data, such that more time is required to gain the necessary insights. This creates a dilemma for the analysts who are required to report the results in time. Recently, various initiatives were started to standardize graphs and define them up-front. In particular, automated or semi-automated reporting tools have become popular where a report template is defined and once the data become available, tables and graphs are created automatically.

Such rigorous planning procedures can pose a hurdle for data-driven analyses. The same holds for tables and listings, but it seems that a graph easily reveals its usefulness—or uselessness—whereas a table can frequently hide that it could have been done better. Tufte (2007) shows an impressive example of how to improve a table of cancer survival rates (Tufte, 2007, pp. 47–63). As a consequence, tables are produced in all sorts of varieties and combinations and thrown at medical reviewers, statisticians, safety boards, and other parties. The timelines are kept, even if a signal is missed.

This paper is aimed at demonstrating the usefulness of graphics to detect signals early in clinical drug development. Some steps were taken by the authorities to analyze safety data more rigorously, and graphics have proven to be invaluable tools for such analyses.

Thus, recent steps lead in the right direction but there is still much to be improved to learn more efficiently from the large data sets generated in clinical studies.

References

Becker RA, Cleveland WS, Shyu MJ (1996) The visual design and control of Trellis displays. J Comput Graph Stat 5(2):123–155

Bock J (2001) Graphical presentation of single patient results. In Millard and Krause

Cleveland WS (1993) Visualizing data. Hobart Press, Summitt, NJ

Cleveland WS. (1998) Trellis home page. http://cm.bell-labs.com/cm/ms/departments/sia/project/trellis/

Emerson JW, Green W (2012) Package "YaleToolkit." http://cran.r-project.org/web/packages/YaleToolkit/YaleToolkit.pdf. Accessed 11 Feb 2012

Haemer KW (1947) The perils of perspective. Am Stat 1(3):19

Haemer KW (1951) The pseudo third dimension. Am Stat 5(4):28

Heiberger RM, Holland B (2004) Statistical analysis and data display: an intermediate course with examples in S-PLUS, R, and SAS. Springer, New York, NY

Ihaka R (2003) Colour for presentation graphics. In Hornik K, Leisch F, Zeileis A (eds) Proceedings of the 3rd international workshop on distributed statistical computing. Vienna, Austria. http://www.ci.tuwien.ac.at/Conferences/DSC-2003/Proceedings/

Krause A (2009) Taking it to higher dimensions. Stat Comput Graphics Newslett 20(1):11, http://stat-computing.org/newsletter/

Lumley T (2006) Color coding and color blindness in statistical graphics. ASA Stat Comput Graphics Newslett 17(2):4–7, http://www.amstat-online.org/sections/graphics/newsletter/Volumes/v172.pdf

Millard S, Krause A (eds) (2001) Applied statistics in the pharmaceutical industry (with case studies using s-plus). Springer, New York, http://www.elmo.ch/doc/statistics-in-pharma/

Morris JF, Koski A, Johnson LC (1971) Spirometric standards for healthy nonsmoking adults. Am Rev Respir Dis 103:57–67

Norman DA (2002) The design of everyday things. Basic Books, New York, NY

R Development Core Team (2009) R: a language and environment for statistical computing. Vienna, Austria. http://www.R-project.org/

Robbins NB (2005) Creating more effective graphs. Wiley, Hoboken, NJ

Robbins NB (2009) Comments on "Taking it to higher dimensions". Stat Comput Graphics Newslett 20(2):24

SAS Institute Inc. (2009) SAS/GRAPH® 9.2: Graph Template Language User's Guide, 2nd edn. Cary, NC, 2009. http://support.sas.com/documentation/cdl/en/grstatug/62464/PDF/default/grstatug.pdf

S-PLUS (2005) S-PLUS® 7 User's Guide, Insightful Corporation, Seattle, WA

Tufte ER (2002) The visual display of quantitative information, 2nd edn. Graphics Press, Cheshire, CT, 2nd printing

Tufte ER (2007) Beautiful evidence. Graphics Press, Cheshire CT, 2nd printing

Tukey JW (1977) Exploratory data analysis. Addison-Wesley, Reading, MA

Unwin A, Theus M, Hofmann H (2006) Graphics of large data sets: visualizing a million. Springer, New York, NY

Wikipedia: The Free Encyclopedia. Wikimedia Foundation Inc. Updated 27 Feb 2012, 11:27 UTC. Encyclopedia on-line. Available from http://en.wikipedia.org/wiki/statistical-Graphics. Internet. Retrieved 9 Oct 2012

Zeileis A, Meyer D, Hornik K (2007) Residual-based shadings for visualizing (conditional) independence. J Comput Graph Stat 16(3):507–525

Zeileis A, Hornik K, Murrell P (2009) Escaping RGBland: selecting colors for statistical graphics. Comput Stat Data Anal 53:3259–3270, Presentation at Use-R 2009: http://www.agrocampus-ouest.fr/math/useR-2009/slides/Zeileis+Hornik+Murrell.pdf

Chapter 2
Graphics for Exploratory Analysis and Data Discovery in the Life Sciences

Michael O'Connell, Ian Cook, Difei Luo, and Josh Patel

Abstract Life Sciences data are complex. In the pharmaceutical industry, vast quantities of patient-level safety and efficacy data are collected, managed, analyzed, and reported in bringing new medicines to market. Clinical trial operations data, involving sites, investigators, milestones, and budgets are similarly complex, especially in larger companies and contract research organizations where many trials are managed simultaneously. Upstream of this, much biology and chemistry data are generated, screened, and modeled to characterize targets and identify new molecular entities. Downstream of research and development there are many large patient-level and provider-level databases analyzed by pharmaceutical companies, industry consortia and regulators to understand real world safety and efficacy of therapies, and by commercial organizations to optimize sales and marketing efforts. Similar large and complex insurance claims and medical records databases are analyzed by healthcare organizations to manage their business and provide efficient and effective healthcare.

This chapter provides a review of exploratory and reporting graphics in the life sciences industry. Graphics principles are outlined in context; and pointers to other chapters in the book are provided along the way.

M. O'Connell (✉)
Tibco Software Inc., 632 Broadway Floor 3A, New York, NY 10012, USA
e-mail: moconnel@tibco.com

D. Luo
Tibco Software Inc., Braywick Rd, Maidenhead, Berkshire SL6 1DA, UK

I. Cook • J. Patel
Tibco Software Inc., 200 W Franklin St Suite 250, Chapel Hill, NC 27516, USA

A. Krause and M. O'Connell (eds.), *A Picture is Worth a Thousand Tables*:
Graphics in Life Sciences, DOI 10.1007/978-1-4614-5329-1_2,
© Springer Science+Business Media New York 2012

2.1 Introduction

Graphics promote scientific insights, better business, and more effective cross-functional communications. In pharmaceutical clinical development, report-style graphics are used in scientific presentations and publications, clinical study reports and regulatory submissions; while graphical exploratory analysis is widely used for instream data review, medical monitoring, and safety assessment.

Exploratory and reporting graphics are quite different. Report graphics typically need to be self-contained and documented with source data and output file references. They stand as static summaries of an analysis and need to be readily reproducible in the future.

Exploratory graphics are typically interactive and not self-contained in a document. They draw upon context and environment, where the user explores points/regions of graphs with brushing and drill-down through a sequence of visualizations; to derive insights and understand root causes of scientific and business phenomena. In pharmaceutical clinical development, this enables viewing of population trends, subpopulation effects, and detailed exploration of exceptional individual subjects, sites, and investigators. In pharmaceutical sales and marketing, exploratory graphics enable analysis of products across markets, providers, and healthcare systems. Exploratory graphics enable management by exception, highlighting insights in to the unknown that drive rapid progression of science and business.

This chapter provides a review of exploratory and reporting graphics in the life sciences industry. Several use cases are included as case studies including (a) graphical analysis of a phase 2 clinical trial that was used by the CDISC ADaM (http://www.cdisc.org) subcommittee in a mock submission to the FDA (Food and Drug Administration), (b) analysis of a collection of clinical trials from an operational perspective, (c) analysis of pharmaceutical sales and marketing data, and (d) examples from life science companies specializing in data management, analysis, and reporting services. Graphics principles are outlined in context; and pointers to other chapters in the book are provided along the way. Given the emphasis on reporting graphics in other chapters, we focus on exploratory graphics in this chapter.

2.2 Pharmaceutical Research and Development: Context of Chapters in This Book

Pharmaceutical research and development is a lengthy and complex process, as outlined in Fig. 2.1. During drug discovery, tens of thousands of compounds are typically explored in context of a disease pathway and a set of potential targets. Complex chemistry and biology experiments and knowledge are brought to bear, and innovative analysis and visualization is used to uncover safety and efficacy biomarkers and effects Zucknick et al. (Chap. 8) present a variety of effective report graphics from the discovery process including the heatmap (Fig. 8.4) and principal component biplots (Fig. 8.7) that display biomarker-defined subpopulations identified via unsupervised learning methods. They also show the utility of explor-

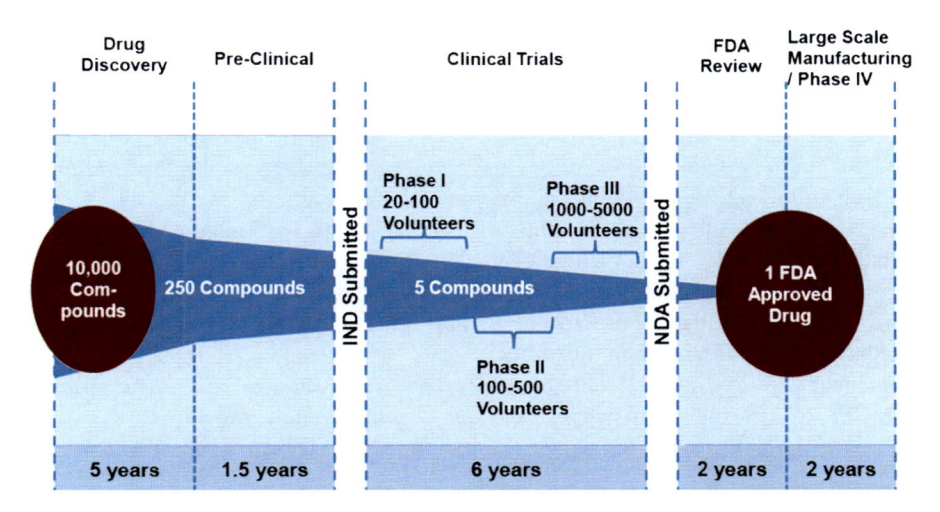

Fig. 2.1 Pharmaceutical Research and Development Funnel, showing four stages: Drug Discovery, Preclinical Research, Clinical Trials (Phases 1, 2, 3), and FDA Review (Source PhRMA)

atory graphics (their Figs. 8.17 and 8.18) in discovery research, with drill-down from the biomarker-defined subpopulations to survival charts of individual subjects from associated clinical data.

As 10,000 compounds in discovery funnel into approximately 200 compounds entering preclinical research, a large set of in vitro and in vivo experiments, models, and simulations are performed to understand drug metabolism and pharmacokinetics (DMPK). This work culminates in filing of an IND (investigational new drug) submission to regulatory agencies such as the FDA. The approval of an IND filing enables the drug to be studied in humans, and this DMPK work continues as the drug enters early-phase clinical trials. Roosen et al. (Chap. 6) present a set of innovative report graphics on early-phase clinical data highlighting interesting subject responses in context of experimental subpopulations, indicated through differently styled lines and shaded regions (Figs. 6.7 and 6.11). Krause (Chap. 1) illustrates the creative use of Trellis displays in report graphics (Cleveland 1993) with a time course presentation of early-phase subject-level biomarker data (Fig. 1.13) incorporating clinical response via color and dosing as a sub-panel along the bottom of the subject panels. Girgis and Mohanty (Chap. 13) give a modeling perspective on QT prolongation based on a three-way crossover study in 40 healthy subjects.

From the thousands of compounds explored in discovery only a handful make it into late-phase clinical trials. These compounds have shown efficacy and dose–response in early-phase trials and have passed numerous safety tests in small human populations and through biomarker assessments. Late-phase trials involve much larger populations and compounds are studied extensively through many years of research, comprising data from many dimensions, e.g., efficacy endpoint measurements; safety data including adverse events, labs, ECG and vital sign measures; and medical history and concomitant medications. Exploratory graphics play a crucial role in the analysis and communication of safety data during the late-phase trials.

Report graphics similarly play a key role in scientific presentations, publications, clinical study reports, and regulatory submissions.

Safety effects are typically studied in at least two buckets, sometimes referred to as targeted and designated safety effects. Targeted effects involve safety issues that are thought to be related to the drug class and/or the disease pathway and are often pre-specified for detailed analysis. The risk of these targeted events is typically tracked over the course of the studies (e.g., Xia et al. 2011). Designated effects are mandated for study by regulatory agencies. These events are often discussed as adverse events of special interest (Crowe et al. 2009, CIOMS, 2005). In particular, cardiac and liver effects are front of mind and covered by at least two FDA guidances: the E14 guidance for cardiac effects and QT prolongation; and the DILI guidance for drug induced liver injury (Food and Drug Administration, 2005, 2009). The SPERT working group suggests a 3-tier system for analyzing adverse events:

- Tier 1: Prespecified detailed analysis and hypothesis testing; these overlap with the AESI's and designated medical events
- Tier 2: Signal detection among common events
- Tier 3: Descriptive analysis of infrequent adverse events

Anziano (Chap. 18) presents a number of exploratory graphics for late-phase assessment of cardiac side effects; and Tornoe (Chap. 16) provides a focused analysis of QT prolongation according to the E14 guidance. Tornoe's report graphics communicate salient information very effectively; see, for example Fig. 16.4, showing QT v. RR with confidence intervals and bands. Merz (Chap. 7) provides a detailed review of liver lab analysis, including an innovative set of interactive exploratory graphics; e.g., Fig. 7.9 that uses symbol size and color to show time sequence detail of liver enzyme elevations in a modified eDISH plot (evaluation of drug-induced serious hepatotoxicity); and Figs. 7.17 and 7.18 showing relationships between elevated liver enzymes with concomitant medications and adverse events. Austin (Chap. 14) extends the cross-data-domain analysis with detailed patient-level report graphics including the excellent patient-profile (Fig. 14.9) covering multiple data domains (drug dose, AEs, Labs, Conmeds). Several other chapters cover graphical analysis of safety effects in late-phase trials: Amit and Lane (Chap. 17) provide a comprehensive review of report graphics across safety data domains; and Southworth (Chap. 10) provides an exploratory analysis combining graphs and tables with drill-down to the subject level.

Miskell et al. (Chap. 5), Looby (Chap. 11), Wang and Mohanty (Chap. 12) and Gilder (Chap. 9) provide reviews of report graphics for metabolic, pulmonology, neurology, and oncology therapy areas respectively. Gilder's chapter provides graphical analysis for the entire development lifecycle including adaptive designs (Figs. 9.3 and 9.4), interim analysis (Fig. 9.5), delta plots for final analysis (Figs. 9.6 and 9.7), forest plots from meta-analyses (Figs. 9.8 and 9.9), and innovative survival and event graphics (Figs. 9.11, 9.12, 9.13, 9.14, 9.15, 9.16, 9.17).

Snyder (Chap. 19) provides an overview of exploratory analysis for both clinical and trial operations data, especially important for late-stage clinical trials. From an operational perspective, enrolment, milestones, and budget are critical components for tracking progress of clinical trials. Snyder's Figs. 19.7, 19.8, 19.9, 19.10 provide informative views on enrolment, and Figs. 19.11 and 19.12 show data management milestones.

There are several chapters on post-marketing graphical analyses. Lane et al. (Chap. 15) describe report graphics for meta-analysis including various Forest plots (Figs. 15.1, 15.2, 15.3) and funnel plots (Figs. 15.4, 15.5, 15.6). Ryan (Chap. 21) provides an exploratory graphical analysis of insurance claims data that highlights safety effects through patient-profiles (e.g., Fig. 21.4), and treemaps of drug prevalence (e.g., Fig. 21.6). Weerahandi et al. (Chap. 20) provide an innovative exploratory analysis of sales and marketing data including optimization of field force effectiveness (Fig. 20.10) and interactive maps showing lift in prescriptions and profit (Fig. 20.12).

Our chapter attempts to complement the excellent set of graphical analyses and case studies outlined above, focusing on exploratory graphics for real-time insights that drive scientific understanding and business value across a range of use cases.

2.3 Graphical Analysis of Clinical Data

Large clinical trials involve collection, management, analysis, and reporting on vast quantities of patient-level safety and efficacy data from case report forms, interactive voice/web response systems, central and local laboratories, and imaging systems. Systems for coding/decoding adverse events (e.g., MedDRA) and concomitant medications (e.g., WHODD) are utilized; and adoption of CDISC standards, from CDASH to SDTM to ADaM (http://www.cdisc.org), enables consistent data elements for analysis. These days, most of the data flow electronically and fairly efficiently from study and protocol design, electronic data capture, data cleaning, and source data verification, into analysis and reporting systems.

There are several stages in the evolution of clinical data; and multiple personas and functional areas that are intimately engaged with the data. Graphical analysis plays a vital role throughout. Study managers play a somewhat pivotal role, interacting with data managers re. data cleaning, with clinicians re. scientific and medical monitoring, and with clinical research associates (CRAs) re. site monitoring.

Study managers examine the raw data, from eCRFs (electronic case report forms), central and local labs, to ensure they make clinical sense. For example they may review sequential lab measures for consistency, review protocol adherence, and treatment discontinuations. They may make operational adjustments and dispatch CRAs to sites for source data verification or site training; or instruct clinical data managers to initiate a query on source systems. They closely monitor protocol adherence and deviations by site and region; looking for missing data, protecting endpoints, and enabling subsequent studies to benefit from patterns observed.

Clinicians typically have a deeper level of medical insight and review raw data and downstream data, e.g., CDISC SDTM and ADaM-like constructs, as prepared by data managers and programmers. They also monitor and review nuances of protocols and statistical analysis plans (SAPs) with statisticians. For example, clinicians may review concomitant medication data for a drug that should have been prescribed to treat an observed adverse event (AE).

Data managers do their own data review and clean/manage the data from raw through SDTM and ADaM-like constructs in a fundamental manner, handling data

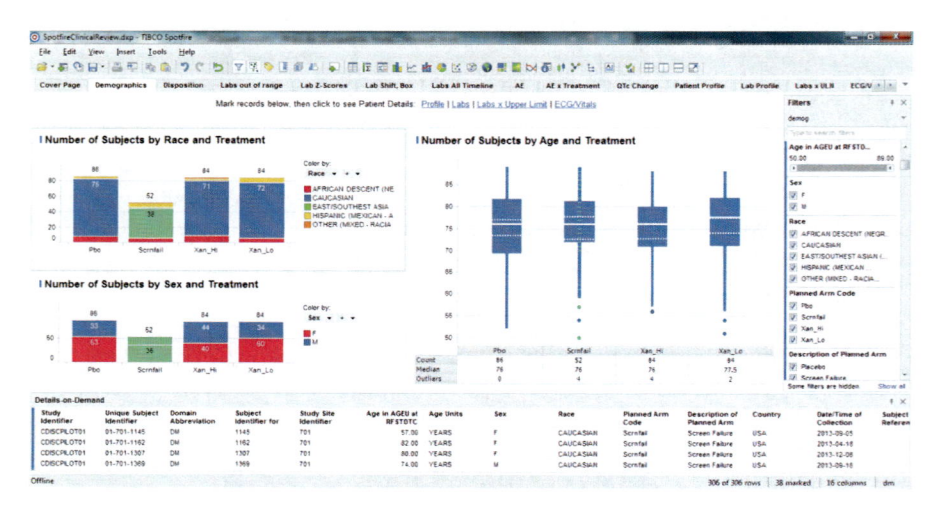

Fig. 2.2 Exploratory Analysis of Xanomeline study: Number of subjects in trial by Sex, Race, Age, and Treatment (unblinded). Caucasian subjects who failed screening are marked (*green*); enabling drill-down to other visualizations for this marked subpopulation. The filter panel on the right hand side enables rapid filtering to subpopulations and changes to graphic style and elements through drag and drop operations

queries, safety reconciliation, edit checks, and discrepancies analysis, e.g., ensuring there are matching AEs when a subject shows discontinuation for an AE. Statisticians work primarily with downstream data (SDTM and ADaM-like) and work closely with clinicians to explore, review, analyze, and report on the blinded and unblinded data. Programmers play an important role in data provisioning, ETL and analysis; transforming raw data into the derived SDTM and ADaM-like fields required for analysis, review, and reporting.

We illustrate some of the data review and graphical exploratory analyses described above using data from the CDISC SDTM/ADaM Pilot Project (http://www.cdisc.org/content1037), originally from a prospective, randomized, multi-center, double blind, placebo-controlled, parallel-group study of the Lilly drug Xanomeline. The trial was on a population of patients with probable mild to moderate Alzheimer's disease and used a three-arm, placebo-controlled design of 26 weeks duration. The objectives of the study were to evaluate the efficacy and safety of two doses of active drug as compared to placebo delivered transdermally.

Figure 2.2 shows the breakdown of subjects in the study by sex, age, race, and treatment (unblinded). Caucasian subjects that failed screening are marked in green (by mouse), enabling drill-down to other visualizations for this subpopulation. Similar views may be achieved on raw data, before unblinding. The *Filters* on the right enable rapid subsetting to subpopulations, and changes to graphic style through drag and drop operations. For example, one may uncheck sex, race or treatment components or modify colors by dragging a component of the filter panel onto a graph.

Figure 2.3 shows disposition by treatment and attrition of subjects over time. Note the large group of subjects in low and high dose who discontinued due to adverse events (red band in the stacked barchart).

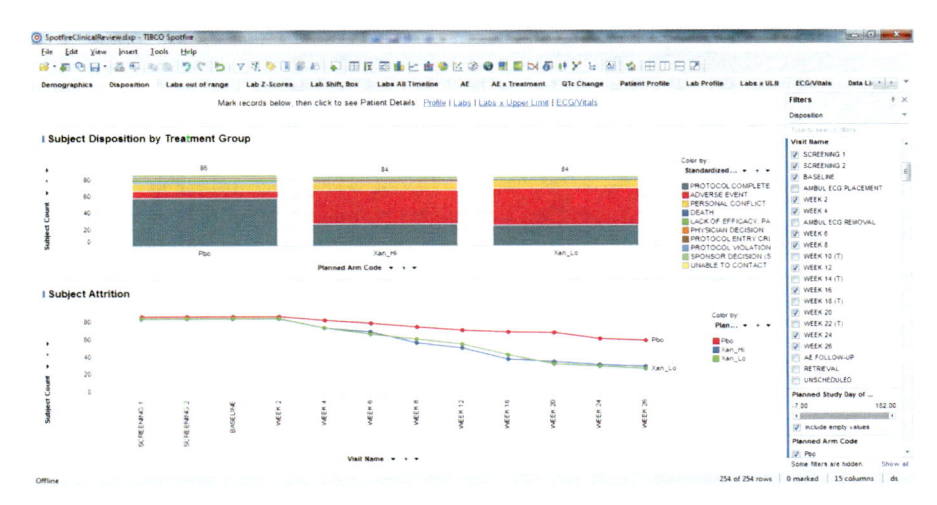

Fig. 2.3 Exploratory analysis of Xanomeline study: subject disposition and attrition

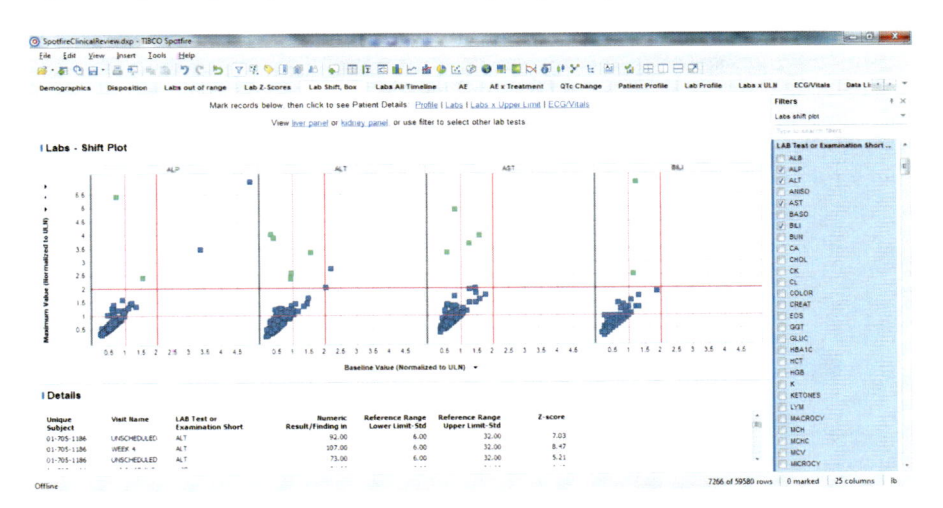

Fig. 2.4 Exploratory analysis of Xanomeline study: shift plot of liver lab values

Figures 2.4 and 2.5 show a drill-down sequence highlighting subjects whose liver labs were elevated during the trial. Figure 2.4 is a Trellised (Cleveland 1993) shift plot (Amit et al. 2008) for the four liver labs, ALP, ALT, AST, and BILIRUBIN from left to right. X-axis is the baseline value and Y-axis is maximum on-treatment value; all data have been normalized to upper limit of normal (/ULN). Subjects in the top left corner of each Trellis panel have been marked (via mouse) in green; these subjects had baseline values below 2× ULN (upper limit of normal) and maximum on-therapy values above 2× ULN. Once these subjects are marked, they form a subpopulation and the underlying subject list may then be passed across different graphics.

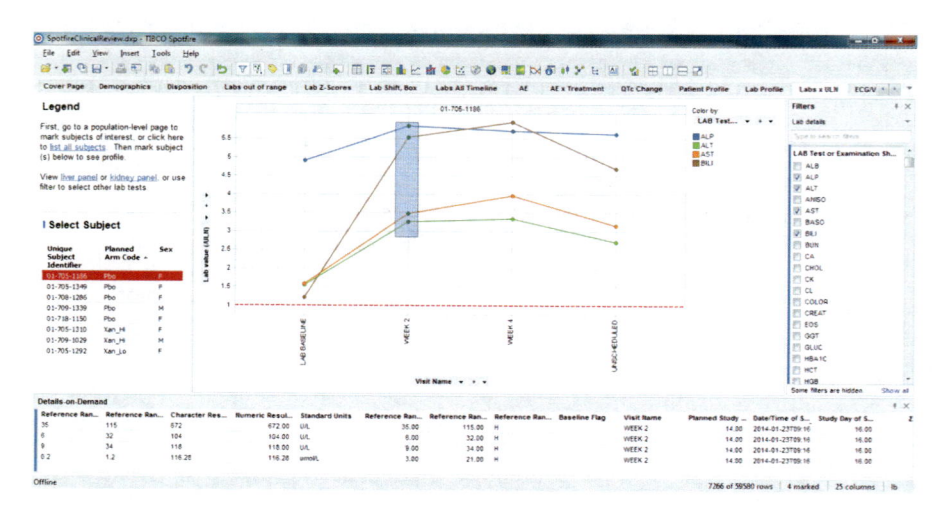

Fig. 2.5 Exploratory analysis of Xanomeline study: drill-down from shift plot of liver lab values (subjects marked green in Fig. 2.4) to normalized line plot of all four liver labs

Figure 2.5 shows a line plot of all four liver labs normalized to their upper limits of normal (/ULN) for the subject selected in the *Select Subject* list on the left, that has been populated from the marking on the shift plots. The normalized liver lab values for the week 2 visit are highlighted on the line plot and these values are listed in the *Details-on-Demand* table below. The *Filters* on the right show the active labs on the plot—these can be modified by checkbox operations. Note that this subject has elevated transaminase and bilirubin values above 3× ULN and probably should have been withdrawn from the study. The *Select Subject* list on the left shows that this is a placebo subject; however, this drill-down graphics sequence may be done on blinded data equally easily.

The drill-down in Figs. 2.4 and 2.5 produces time sequences of all (liver) lab elevations. This is clinically important as discussed by Merz (Chap. 7). The FDA guidance (Drug Administration 2005) specifies that bilirubin elevations above 2× upper limit of normal with 30 days of an elevation of AST/ALT above 3× upper limit of normal are of critical concern, based on Zimmerman's landmark work (Zimmerman 1978).

Figures 2.6, 2.7, and 2.8 show a drill-down sequence highlighting subjects with a treatment-related adverse event, pruritus. Figure 2.6 shows results of an adverse event treatment-emergence analysis using a bagging model, (Southworth & O'Connell 2009). The left-hand panel shows subject counts of adverse events sorted in order of count and colored by treatment. Three adverse events appear to be treatment-emergent, each related to skin irritation (right-hand panel). This is not surprising given that the treatment was applied transdermally. Subjects in the high dose with pruritus are marked on the left hand panel for drill-down to Fig. 2.7; and the Tooltip (black box above the marked group) describes this subgroup. Note that as in all displays, the *Filters* on the right can be used to filter events (e.g., to leave

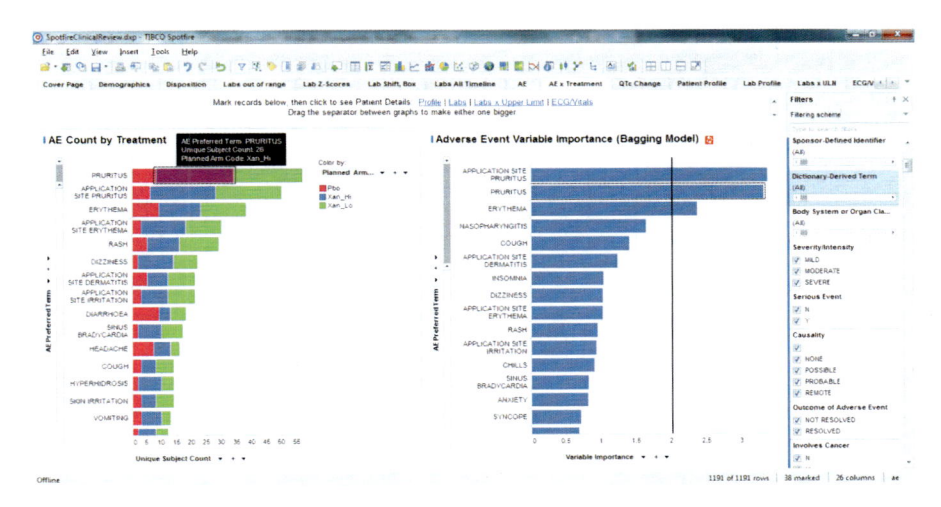

Fig. 2.6 Exploratory analysis of Xanomeline study: adverse event treatment-emergence analysis using bagging model (Southworth and O'Connell 2009)

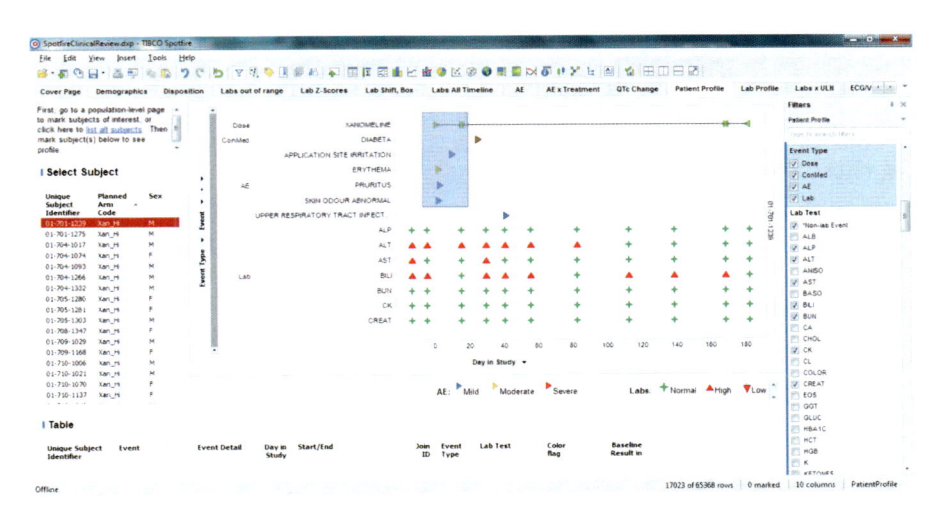

Fig. 2.7 Exploratory analysis of Xanomeline study: patient-profile is shown for subject highlighted at the top of the Select Subject list on *left*. The patient-profile includes dosing, concomitant medications, adverse events, and labs on a common *x*-axis timeline, days in study. Elevated labs are shown in *red symbols* and adverse events as *blue, yellow and red for mild, moderate, and severe* AEs

out mild events, or events with causality = none), or to color or Trellis the graphs with drag-drop operations.

Figure 2.7 shows a patient-profile for an individual subject highlighted at the top of the *Select Subject* list on the left. The patient-profile includes dosing, concomitant medications, adverse events, and labs on a common x-axis timeline, days in

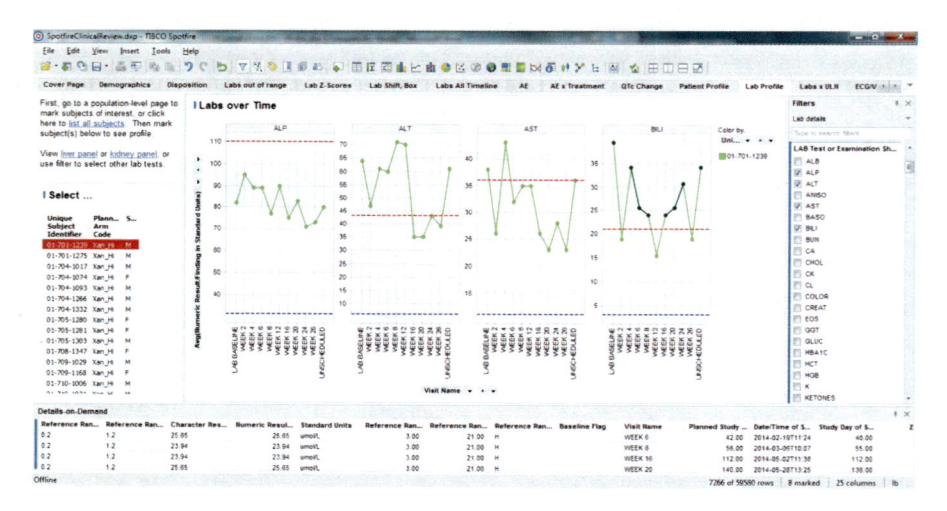

Fig. 2.8 Exploratory analysis of Xanomeline study: time-sequenced line plot for elevated labs identified in Fig. 2.7

study. Elevated labs are shown in red symbols and adverse events as blue, yellow, and red for mild, moderate, and severe AEs. Note the moderate erythema (yellow symbol) at the same time as the initial dose of Xanomeline.

The patient-profile in Fig. 2.7 can be readily modified using the *Filters* on the right, e.g., to add or subtract event types (adverse event, conmed, lab, dosing); or to add/subtract individual lab tests. Liver and kidney labs are included in Fig. 2.7, and these checks in the *Filter Panel* are readily modified.

The subject highlighted in Fig. 2.7 shows elevation of some of the (liver) labs in the patient-profile. A time-sequenced line plot of these labs by visit is shown in Fig. 2.8. The Bilirubin points on the graph are highlighted and values are shown in the table at the bottom of the display.

The drill-down through Figs. 2.6, 2.7, 2.8 is an interactive extension of the report version of the patient-profile described by Austin (Chap. 14). Austin does an amazing job to fit maximal information on multiple data domains on a single graph/page; but the page does have finite real-estate! In the exploratory analysis setting we are not constrained by the need to produce graphics for inclusion in a (paper) report. This allows the flexible drill-down and interactivity in the multidomain patient profiling.

The exploratory analysis above is done on unblinded CDISC ADaM data from the CDISC SDTM/ADaM Pilot Project (http://www.cdisc.org/content1037). Similar interactive graphical analyses may be done on blinded data—raw and SDTM-like—to address the needs of study managers, clinical research associates, data managers, clinicians, statisticians, and programmers as described above.

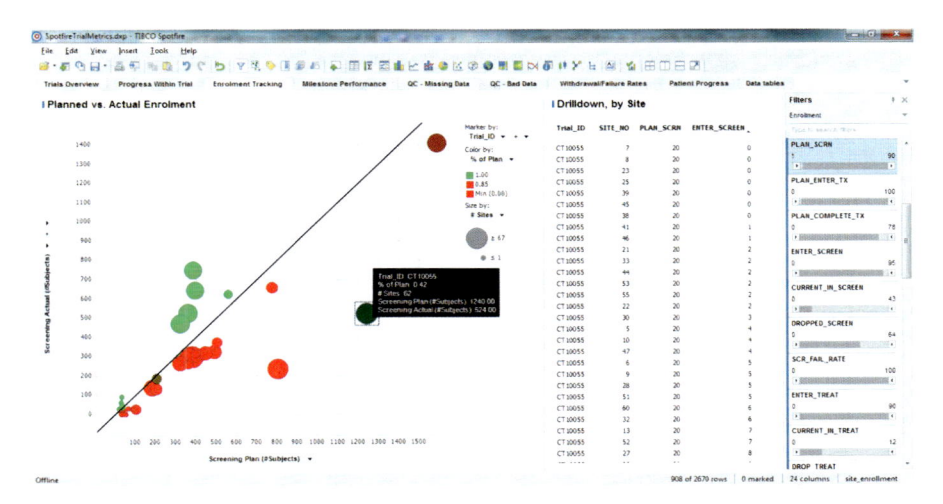

Fig. 2.9 Exploratory analysis of operations data: planned versus actual number of subjects entering screening. The graph symbols are colored according to % of plan and sized by the number of sites initiated. Trials below the line are below plan on screening. Site details for the trial marked are shown on the right, sorted to show sites that are most behind plan. Metadata on the marked trial are shown on the tooltip. Axes on the graph can be reconfigured to show other planned versus actual metrics by dragging appropriate filter devices from the right hand panel on to the axis labels

2.4 Graphical Analysis of Operations Data

The management of a large clinical development program typically includes graphical dashboarding of enrolment, milestones, and budget. Key Performance Indicator metrics (KPIs) for milestones can relate to specific stages of the trial, e.g., Screening to First Visit, First Patient In (FPI) to Last Patient In (LPI), Last Patient Last Visit (LPLV) to Study Close (SC); or to data cleaning cycle status, e.g., query opened to answered or source data verification coverage (SDV). Consumers of the dashboards include study managers, medical directors, data managers, and clinical research associates.

Figure 2.9 shows planned versus actual number of subjects successfully completing the screening visit for a collection of clinical trials in progress. The graph symbols are colored according to % of plan and sized by the number of sites initiated. The axes can be readily changed to any enrolment or milestone metric by dragging an appropriate filter device from the *Filter Panel* on to the axis labels. For example, the graph is readily configured to planned versus actual subjects who have entered treatment by dragging filter devices for these metrics on to the graph.

Figure 2.10 shows various key milestone metrics including Screening to First Visit, First Patient In (FPI) to Last Patient In (LPI), Last Patient Last Visit (LPLV) to Study Close (SC). The trial in focus may be selected from the panel of trials on the left of the visual or from another graph, e.g., from Fig. 2.9.

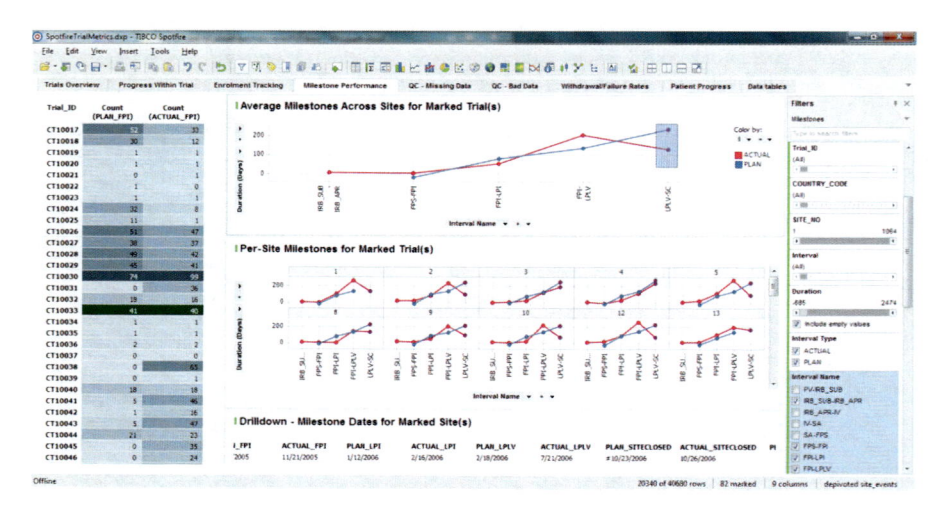

Fig. 2.10 Exploratory analysis of operations data: Planned (*blue*) versus actual (*red*) milestones (see text for milestone abbreviations) for selected trial (CT10033). The top graph shows milestone comparison across the trial; the Trellis graph below shows milestones for individual sites. The table at the bottom provides dates for planned and actual milestones. The trial may be selected from the panel on the left or from markings on other graphs, e.g., Fig. 2.9

In large trials, it is helpful to be able to predict enrolment, especially the time of total enrolment as determined by sample size required to achieve designated power in the primary endpoint analysis. Since functional area resources/capacity typically depend on enrolment (e.g., more statistics resources are required early and late in the life cycle), such enrolment prediction enables planning of resources required to manage a collection of trials. For pharmaceutical companies, this translates in to head count required in functional areas and contract research organization (CRO) contracts/contractors needed to manage the portfolio of trials in clincial development.

Figure 2.11 shows a 95% prediction interval for the end of a study with 1,000 subjects. The prediction was done at 287 days into the trial when enrolment was at approx. 200 subjects. The predictions are based on a model that assumes a Poisson process for site initiation and another Poisson process for recruitment within each site. The rate of the Poisson process for site initiation is allowed to vary by country, with a cap on planned maximum sites per country. The rate of the Poisson process for recruitment within sites is also allowed to vary by country, based on observed country recruitment. This is an extension of work done by Anisimov and Federov (2007) who assumed Poisson recruitment at individual sites with rate parameters drawn from a Gamma distribution. We have found that recruitment rates typically do not follow a Gamma distribution very well, hence our more empirical-based approach. In cases where there is little enrolment, there is likely some utility in the Gamma prior distribution. The enrolment and end-of-study predictions in Fig. 2.11 are based on 10,000 simulations from the appropriate nested Poisson processes.

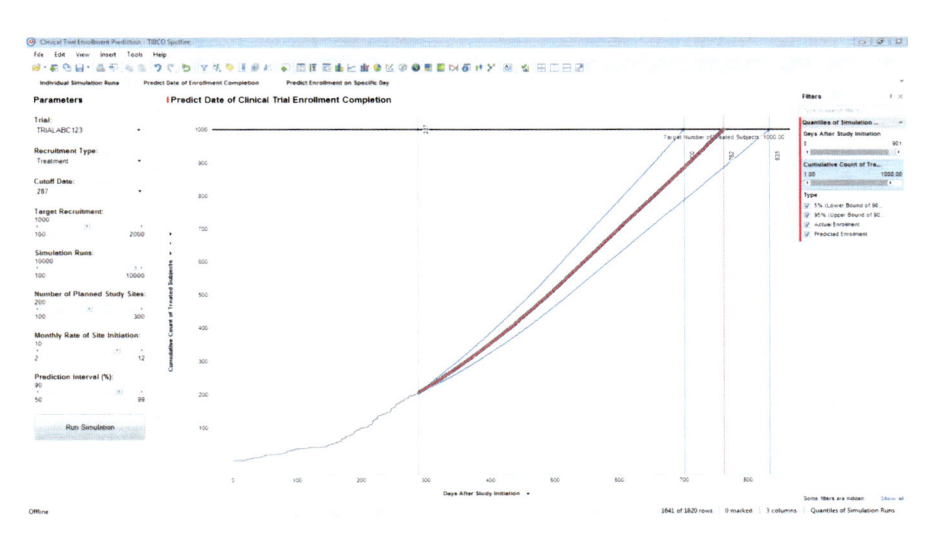

2 Graphics for Exploratory Analysis and Data Discovery in the Life Sciences   35

Fig. 2.11 Enrolment and end-of-study prediction. The prediction interval on end-of study was obtained from percentiles of 10,000 simulated enrolment curves from a relevant set of nested Poisson processes for site initiation and recruitment within sites

Snyder (Chap. 19) provides additional graphical analysis of operations data that show progress in data management and status of the clinical data lifecycle. This includes graphical presentations of query aging by query type (edit check, raised through review) and data verification/monitoring status.

2.5 Graphical Analysis of Sales and Marketing Data

Weerahandi et al. (Chap. 20) describe exploratory analysis of sales and marketing data including optimization of field force effectiveness. Brand managers, business analysts, and marketers need to monitor the performance of their products in various markets and geographic regions and predict their near term business. They like to track share of their products in different markets and assess the effects of marketing campaigns executed at local, regional, and national levels. Sales managers like to track performance of their sales reps in order to effectively manage their targets.

Figure 2.12 shows sales curves of pharmaceutical products by product and region with forecasts based on ARIMA (autoregressive integrated moving average) models and Holt-Winters exponential smoothing. The implementation allows automated selection of parameters based on AIC (Akaike's Information Criteria) and MLE (maximum likelihood estimation) respectively, as well as user-specified autoregressive and moving average parameter selection for ARIMA and smoothing parameter selection for Holt-Winters. The user experience allows for choice of product and region combinations, with resulting graphical displays in Trellis format by region.

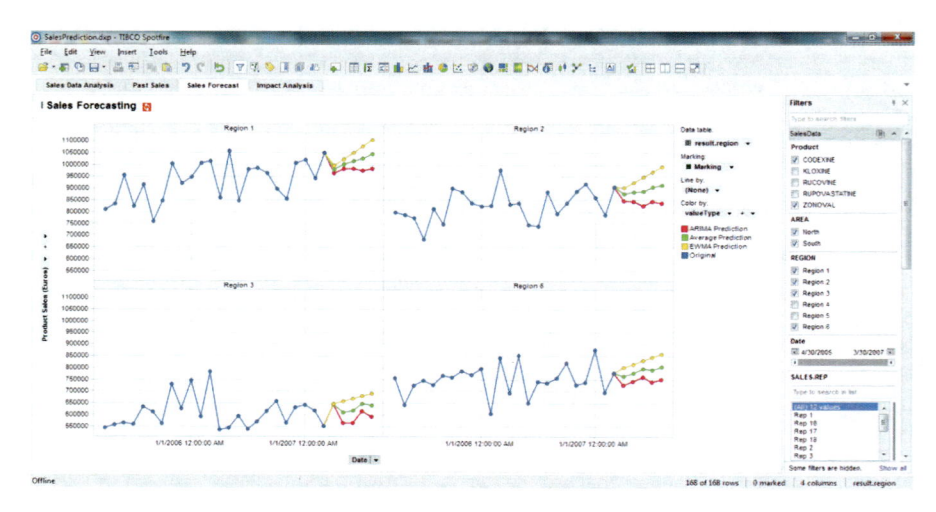

Fig. 2.12 Forecasting of pharmaceutical sales by-product and region; with drill-down to sales rep

The predictions in Fig. 2.11 incorporate seasonality, which is very important in life sciences due to increases in healthcare services in the winter months. The drill-down to sales reps and individual product sales by region enables managers to assess both product and sales rep performance at fine level of detail, e.g., sales rep by-product by account (e.g., hospital, clinic, etc.) within region. This helps to identify areas for improvement and cross-sell of products in accounts.

2.6 Graphical Analysis Services

This section provides examples from life science companies specializing in data management, analysis, and reporting services to the industry. In particular, we highlight a commercial clinical data graphical analysis application - Medidata's Insights™.

Medidata Insights is a cloud-based business analytics offering that enables life science organizations to measure performance on a range of key performance indicators alongside industry benchmarks. Insights leverages operational data from other Medidata products, such as Medidata Rave® and Medidata Grants Manager®, to capture and display performance metrics. Insights' data visualization and filtering capabilities allow users to drill down to data that are relevant to the decision at hand, across studies, phases, and therapeutic areas. By aggregating and anonymizing data collected from the thousands of clinical trials conducted using Medidata technologies, industry benchmarks are developed and displayed in context with the company performance. These benchmarks follow the same filtering rules as the company data, allowing users to conduct apples-to-apples comparisons across metrics.

Fig. 2.13 Medidata Insights application, showing source data verification coverage by therapy area and phase, with cross-sectional snapshot in the upper panel and longitudinal trending in the lower panel. The *red lines* indicate company metrics, with industry benchmarks shown in lines of other colors

Figure 2.13 shows some visual analyses of source data verification coverage for a subscriber pharmaceutical company to the Insights service. The company's performance for select therapy areas is shown as red line on the bottom graph with industry benchmark as the blue line.

Medidata, contract research organizations (CROs), and other pharmaceutical data providers are actively expanding their electronic/online offerings to create efficiencies in clinical research, and to shorten time to market for new medicines while providing transparency in operations and clinical information.

2.7 Software

All graphics in this chapter were created using Spotfire®, TIBCO Software, version 4. Spotfire workbooks, once saved in the library on the Spotfire server, may be opened and interactively utilized in a web browser; this is shown in Fig. 2.13. All other graphs are shown as screen captures from the Spotfire Professional client.

Spotfire incorporates TERR™ (Tibco Enterprise Runtime for R), R and S-PLUS™ as statistical engines and enables registration and management of R, S-PLUS™, SAS™, and Matlab™ scripts in the Spotfire library on the Spotfire server. The safety signal detection analysis shown in Fig. 2.6 uses a bagging algorithm based on the Forests S-PLUS library. The enrolment prediction analysis shown in Fig. 2.10 is based on Monte Carlo simulations using TERR. For large trials, algorithms are readily parallelizable and run on multiple CPUs. The time series forecasts shown in Fig. 2.11 use the S-PLUS Finmetrics™ library.

2.8 Conclusion

Exploratory graphics are an invaluable tool for efficient and accurate analysis and interpretation of life sciences data. Exploratory graphics have a rich history from Tukey's landmark book, Exploratory Data Analysis (Tukey 1977), through Tufte (Tufte 1983) and Cleveland (Cleveland 1993); and now in to modern exploratory data analysis environments such as Spotfire.

Intelligent use of exploratory graphics leads to better science and more effective communication of information. In this chapter, we have highlighted exploratory analysis of pharmaceutical clinical safety, operations and sales, and marketing data. A typical pharmaceutical company spends approximately one-third of their expense budget on research and development, one-third on sales and marketing, and one-third on manufacturing. In recent times, with public focus on drug safety, much attention is being given to safety planning, evaluation, and reporting. Exploratory and report graphics are playing an important role in setting standards for safety analysis (Crowe et al. 2009). In general, effective use of concise and compelling graphical analysis in life sciences ultimately results in more efficient use of patient data, cost savings, faster approval of drugs, and improved patient care.

Acknowledgments Michael O'Connell thanks all of the contributors to this book, most of whom I have collaborated with over the past 10 years. I also thank all of my colleagues at TIBCO, especially those in the Spotfire product family for the fantastic collaborations and contributions to Life Sciences graphics and data discovery. In addition to my coauthors, I particularly acknowledge Andrew Berridge, Robert Collins, Peter McKinnis, David Mosenkis, Christof Gaenzler, Gerard Conway, Mark Demesmaeker, Kerstin Pietzko, Don Sullivan, Matt McGowan, Colin Blackmore, and others in the TIBCO Spotfire community.

References

Amit O, Heiberger RM, Lane PW (2008) Graphical approaches to the analysis of safety data from clinical trials. *Pharmaceutical Statistics* 7:20–35.
Anisimov V, Fedorov V (2007) Modelling, prediction and adaptive adjustment of recruitment in multicentre trials. Stat Medicine 26:4958–4975
Cleveland W (1993) Visualizing data. Hobart Press, Summit, NJ
Crowe BJ, Xia HA, Berlin JA, Watson DJ, Shi H, Lin SL et al (2009) Recommendations for safety planning, data collection, evaluation and reporting during drug, biologic and vaccine development: a report of the safety planning, evaluation, and reporting team. Clinical Trials 6(5):430–440
Southworth H, O'Connell M (2009) Data mining and statistically guided clinical review of adverse event data in clinical trials. J Biopharm Stat 19:803–817
Tufte E (1983) The visual display of quantitative information. Graphics Press, Cheshire, CT
Tukey JW (1977) Exploratory data analysis. Addison-Wesley, Boston, MA
US Food and Drug Administration (2005) Guidance for industry, E14 Clinical Evaluation of QT/QTc Interval Prolongation and Proarrhythmic Potential for Non-Antiarrhythmic Drugs

US Food and Drug Administration (2009) Guidance for Industry, Drug-Induced Liver Injury: Premarketing Clinical Evaluation

Xia A, Crowe B, Schriver R, Oster M. and Hall D (2011) Planning and core analysis for periodic aggregate safety data reviews. Clin Trials online, January 2011

Zimmerman HJ (1978) Drug-induced liver disease. In: Hepatotoxicity, the adverse effects of drugs and other chemicals on the liver, 1st edn. Appleton-Century-Crofts, NY, pp 351–3

Chapter 3
Grables: Visual Displays That Combine the Best Attributes of Graphs and Tables

Thomas E. Bradstreet

Abstract A grable combines the emergent features of a graph with the precise quantities of a table into a single display. Its purpose is to accommodate a wider variety of visual tasks and a possibly wider audience, than either a graph or a table can address alone. The best principles of visual perception from both graph and table design and construction should be considered when designing and constructing grables. We present some proposed visual and cognitive strengths and weaknesses of graphs and tables, the visual tasks that each is best suited for, and some specific guidelines for their design and construction. We use these guidelines and principles of perception to design and construct a variety of grables. We also provide some general guidelines for software selection.

3.1 Introduction

A grable combines the emergent features of a graph with the precise quantities of a table into a single display. Its purpose is to accommodate a wider variety of visual tasks and a possibly wider audience, than either a graph or a table can address alone (Hink et al. 1996, 1998). We present two introductory examples.

3.1.1 Example 1: Oral Contraceptive Interaction Study: Pharmacokinetic Data

A well-known grable often used for exploratory data analysis is the stem-and-leaf display (Tukey 1977). This display not only provides a histogram of the sample

T.E. Bradstreet (✉)
Experimental Medicine Statistics, Merck Research Labs (UG 1D-44),
PO Box 1000, 351 North Sumneytown Pike, North Wales, PA 19454, USA
e-mail: thomas_bradstreet@merck.com

Fig. 3.1 Individual AUC ratios. Individual ratios of AUCs comparing Drug D administered with an oral contraceptive (OC+D) versus the oral contraceptive alone (OC) in 22 female subjects. The oral contraceptive is comprised of two components, ethinyl estradiol (EE) and norethindrone (NET). Ratios shown are (EE+D)/EE and (NET+D)/NET

(EE+D)/EE		(NET+D)/NET
0	**1.5**	
33	**1.4**	
880	**1.3**	
821	**1.2**	0226
743221	**1.1**	2
863	**1.0**	01235699
9741	**0.9**	1256
	0.8	01679

distribution, but it also documents the data values. In Fig. 3.1, the back-to-back stem-and-leaf plot displays the ratios of areas under the plasma-concentration-versus-time-curves (AUCs) for 22 female subjects who completed a 2-treatment, 2-period, complete crossover oral contraceptive interaction trial. The objective of the trial was to determine if the concomitant administration of Drug D with an oral contraceptive (OC+D) perturbs the usual pharmacokinetic profile of the oral contraceptive alone (OC). The oral contraceptive is comprised of two components, ethinyl estradiol (EE) and norethindrone (NET).

In Fig. 3.1, the ratios correspond to each component of the oral contraceptive, (EE+D)/EE and (NET+D)/NET. The stem (vertical rectangular box) contains the *bolded black stem* values **0.8** to **1.5**, which are the observed ratios accurate to one decimal place. The leaves are shown for EE and NET hanging to the left and to the right of the *stem values*, respectively, and they provide additional accuracy to two decimal places. For example, the smallest ratio is **0.80**, (NET+D)/NET, and the largest ratio is **1.50**, (EE+D)/EE. At each *stem value*, the leaves are sorted from smallest to largest, starting from the stem.

Figure 3.1 immediately provides information about the sample distributions of the ratios for EE and NET. Both distributions appear to be truncated on the left as does NET on the right, but EE is somewhat skewed right. All of the ratios for NET are contained in the narrower and lower range, **0.80** to **1.26**, versus **0.9**1 to **1.50**, for EE. The sample median for NET is (**1.01** + **1.02**)/2 = **1.01**5, smaller than the corresponding value, (**1.13** + **1.14**)/2 = **1.135** for EE. Both distributions have their modal value on the stem, **1.0** for NET, which is lower than **1.1** for EE. However, the modal values for NET are **1.09** and **1.22**, for EE they are **1.12**, **1.38**, and **1.43**. Neither distribution contains an outlier. A detailed analysis of the original data can be found in Bradstreet and Panebianco (2004). A further generalization of the stem-and-leaf display to multi-way tables can be found in Schenker et al. (2007).

3.1.2 Example 2: Iontophoresis Induced Pain: Pharmacodynamic Data

Sixteen subjects completed an 8-treatment, 8-period, complete crossover trial investigating whether or not a proposed iontophoresis induced pain model is valid.

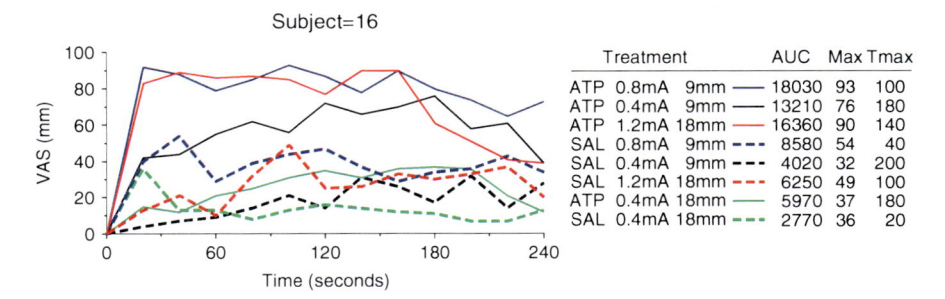

Fig. 3.2 VAS Pain Scores for Subject 16. The pain responses over time for ATP treatments are shown with *solid lines* (—), and saline treatments are shown with *dashed lines* (-----). Matching pairs of ATP and saline treatments are shown with the same color. Tabled along side each line are the treatment label and the corresponding summary statistics. The treatment labels are ordered from top to bottom according to the last VAS score reported at 240 s for each treatment

Treatment			AUC	Max	Tmax
ATP	0.8mA	9mm	18030	93	100
ATP	0.4mA	9mm	13210	76	180
ATP	1.2mA	18mm	16360	90	140
SAL	0.8mA	9mm	8580	54	40
SAL	0.4mA	9mm	4020	32	200
SAL	1.2mA	18mm	6250	49	100
ATP	0.4mA	18mm	5970	37	180
SAL	0.4mA	18mm	2770	36	20

Each iontophoresis treatment comprised a combination of either ATP solution or saline solution (SAL), paired with one of 3 electrical currents (0.4, 0.8, 1.2 mA), using either a 9 or 18 mm iontophoresis chamber. Eight of 12 possible factorial combinations were evaluated. Given the same current and chamber size, ATP should cause more pain than saline.

Subjects scored pain on a 100 mm visual analogue scale (VAS). The VAS is a horizontal line 10 cm in length labeled with a 0 at the left end and the number 100 at the right end. Every 20 s for 4 min in each treatment period, the subjects scored their pain level by marking a vertical line on the VAS. Among the objectives of the study were to identify electrical current level and iontophoresis chamber size combinations that induce sufficient pain; to identify a range in time over the 240 s iontophoresis period where the results look most promising; and to identify within-subject summary measures computed over the individual time points that are the most sensitive and clinically meaningful.

Figure 3.2 presents the results for Subject 16. The graph portion of the grable displays the subject's VAS response for each of the eight factorial treatments. The pain responses over time for ATP treatments are shown with solid lines (—), and saline treatments are shown with dashed lines (-----). Matching pairs of ATP and saline treatments are shown with the same color. For example, the 1.2 mA current and 18 mm chamber size combination is shown in red. Tabled along side each VAS line is the treatment label and the corresponding summary statistics: AUC (area under the curve), Max (maximum VAS), and Tmax (time in seconds that Max was observed). The treatment labels are ordered from top to bottom according to the last VAS score reported at 240 s for each treatment.

For Subject 16 we see that in general, each ATP treatment induces more pain than its corresponding saline treatment. The difference between ATP and saline is notable in three treatments: ATP, 0.8 mA, 9 mm (solid blue); ATP, 1.2 mA, 18 mm (solid red); and ATP, 0.4 mA, 9 mm (solid black). A similarly designed grable can be used to display information summarized across the 16 subjects.

In Sect. 3.2 we present some proposed visual and cognitive strengths and weaknesses of graphs and tables, including comparisons between the two groups of displays regarding the visual tasks that each is best suited for when presenting information. Specific guidelines for their design and construction follow. We provide additional references containing additional principles of visual perception and construction of graphs and tables. In Sect. 3.3 we present some ideas on designing and constructing grables. In Sect. 3.4 we present four more examples of grables. Some of the examples are more graph than table, some are as much table as graph, and one shows a grable that is more like a text table than a graph. In Sect. 3.5 we provide some guidance on selecting software for constructing and displaying grables. In Sect. 3.6 we close with a discussion.

3.2 Graphs Versus Tables

Before focusing on grables, it is important to become familiar with proposed visual and cognitive strengths and weaknesses of graphs and tables as these should be considered when designing and constructing grables. The literature comparing graphs to tables spans a wide range of disciplines including statistics, computer science, management information systems, industrial engineering, business, management science, information science, psychology, education, and political science. Many competing theories exist as to which display formats are better than others. These theories include, but are not limited to, analytic models (e.g., Tufte 1983, 2001; Kosslyn 1989, 1994), compatibility models (e.g., Vessey 1991, 1994), and cognitive process models (e.g., Cleveland and McGill 1984, 1986, 1987; Cleveland 1985, 1994; Meyer et al. 1997; Meyer 2000).

Several visual task experiments were conducted under a wide range of experimental designs, conditions, limitations, and restrictions. Many evaluated the simplest graphs and tables under relatively simple conditions. A large portion of the results are either inconsistent or inconclusive which is not surprising given the wide range of experimental designs, endpoints, and statistical analyses performed. There are also several surveys and meta-analyses comparing graphs to tables. For example, see Carter (1947), Powers et al. (1984), Lalomia and Coovert (1987), Coll (1992), Hwang (1995), Harvey and Bolger (1996), Meyer et al. (1997, 1999), Meyer (2000), and Porat et al. (2009).

Professional opinions vary as to exactly when and why a graph is better to use than a table (Ehrenberg 1978; Gelman et al. 2002; Scott 2003; Kastellec and Leoni 2007) or a table better than a graph (Ehrenberg 1978). Sometimes neither display is deemed as appropriate and a description using only text is best as for small data sets (Carswell and Ramzy 1997), or in larger data sets when only limited results are of interest (Ehrenberg 1978). This debate is far from over (e.g., Gelman 2011). Researchers continue to conduct experiments (e.g., Porat et al. 2009). Useful observations and guidelines are emerging. We highlight some of these.

Visual tasks are divided into two categories. They are:

1. *Spatial tasks*: require making associations between values or perceiving relations in the data
2. *Symbolic tasks*: involve the extraction of individual data values

Cognitive fit exists when both the visual representation of the data and the visual task are both spatial or both symbolic. For lower level visual tasks, cognitive fit produces increased speed and accuracy in problem solving, decision making, and information retrieval (Vessey 1991; Meyer 2000).

Graphs efficiently present spatially related information identifying associations, trends, relationships, deviations, minima, maxima, and orders of magnitude in the data, facilitating a mostly qualitative view at a glance without addressing the individual elements separately or analytically. Tables efficiently present symbolic information quantitatively representing individual data values and facilitating tasks such as locating, reading, extracting individual data values, and performing exact computations such as differences and ratios on the selected values (Ehrenberg 1977c, 1978; Lalomia and Coovert 1987; Vessey 1991; Coll 1992; Harvey and Bolger 1996; Meyer 2000; Kastellec and Leoni 2007; Porat et al. 2009). Tables can also simultaneously display multiple variants of the data such as the original data, transformed data, means, proportions, differences, ratios, and percentages, in a compact area, but this must be done with care (Bradstreet et al. 2008).

The visual effectiveness of both tables and graphs can be improved by sorting the data according to purpose. For look-up and documentation capabilities in tables, sorting by patient number, alphabetically, or by one or more demographics can be productive. For understanding what the data have to say, sorting the data by the magnitude of a desired effect or trend can be insightful (Friendly and Kwan 2003; Bradstreet and Palcza 2012).

When time pressure on the viewer is low, the effectiveness of graphs and tables depends upon the type and complexity of the visual task. But with increasing time pressure, graphs generally are favored (Hwang 1995), and graphs can improve visual task performance with increased complexity.

The observers' prior and accumulating knowledge and experience with a particular graph or table format can favor that format over others (Powers et al. 1984; Meyer 2000). Rightly or wrongly, the most familiar form of data presentation is often perceived as the easiest to comprehend, even among pairs of competing table designs, or similarly among pairs of competing graph designs. Indeed we observed such familiarity bias when introducing box-and-whisker plots, schematic plots, dot charts, and several other visual displays to collaborators as alternatives to those which were their standard at the time such as pie charts and segmented bar charts. The new displays were initially met with resistance. But once the users understood the advantages of the new graphs, with use, the new graphs became familiar friends. Then the new graphs were requested routinely, even in some cases where not appropriate. People will also choose one display format over another if it requires the least effort to perform the visual task (Porat et al. 2009) regardless of its ability to correctly communicate information.

The relative efficiency of competing displays depends on one or more variables. Meyer et al. (1997) identified and summarized seven categories of variables for considering the relative efficiency of competing displays. They are:

1. Type of display (graph vs. table)
2. Variations within display type (e.g., line graph versus bar graph)
3. Conditions of presentation (e.g., visual angle, room illumination, display-background contrast, time pressure)
4. Complexity of displayed data (number of points in the display, configuration of the points in the display, regularity or order in the data displayed)
5. Information sought by the user (e.g., evaluation of trend versus extraction of specific numerical values)
6. Characteristics of the user population (e.g., users' experience with competing displays)
7. Criterion for choosing a display (e.g., speed of extraction, accuracy of information obtained, quality of decisions, understanding of complex relations between variables, aesthetic appeal, users' subjective preferences)

3.2.1 Guidelines for Graphs

Effective graphs exhibit combinations of the following qualities (Chambers et al. 1983; Cleveland 1985, 1994; Tukey 1990, 1993; Wainer 1997; Bradstreet et al. 2008). Effective graphs:

1. Serve a defined purpose: exploration, understanding, or communication
2. Show the data
3. Tell the truth
4. Encourage comparison of different pieces of data
5. Reveal a large amount of quantitative information in a small area
6. Reveal the data at several levels of detail; effectiveness increases with the complexity of the data
7. Are only as complex as required by the task that they are designed to perform; they avoid pomposity
8. Provide impact: communication with clarity, precision, and efficiency
9. Are a visual metaphor for the data
10. Are closely integrated with statistical and verbal descriptions of the data

When designing and constructing graphs, quantitative and categorical information is encoded by symbols, geometry, and color. Graphical perception is the visual decoding of this encoded information. Ten graphical-perception tasks can be ranked from best to worst on how accurately we perform those tasks in decoding quantitative information from graphs (Cleveland 1985, 1994). They are:

1. Position along a common scale
2. Position along identical, nonaligned scales

3. Length
4. Angle
5. Slope
6. Area
7. Volume
8. Color hue
9. Color saturation
10. Density of information

In addition, we must be able to detect all intended graphical elements (as is not the case with either coincident points or superimposed curves), and we must be able to judge distance accurately. These 10 elementary graphical-perception tasks along with detection of graphical elements and judging distance accurately, should be considered when designing and constructing a graph. Data should be encoded in the graph so that visual decoding involves tasks as high as possible in the ordering of the graphical-perception tasks (Cleveland 1985, 1994).

A large share of the ink in a graph should present data-information, with the ink changing as the data change. Data ink is the non-erasable core of the graph, the non-redundant ink arranged in response to variation in the numbers represented. The data ink ratio is the ratio of the data ink to the total ink used to print the graph. For a clear and efficient graph, the data ink ratio should be maximized by erasing both non-data ink and redundant data ink, within reason. This includes eliminating chartjunk. Chartjunk is the unnecessary, often default, but as often intended, graphical decorations found in conventional graphical design and software which clouds and stagnates the flow of important quantitative messages from the graph, and does not tell the viewer anything new. A particularly prolific form is moiré vibration, the undisciplined and distracting appearance of vibration and movement due to cross hatchings and visually distracting patterns injected into graphical elements. Graphs should be information rich in that the amount of data is large relative to the area that the graph covers with high data density. Many graphs are comparative, often constructed from a series of small multiples, i.e., many shrunken plots per page that show shifts in variable relationships as the index variable changes (Tufte 1983, 1990, 2001).

Other guidelines for graph construction and principles of visual perception will be pointed out as required for each of the examples later in the chapter.

3.2.2 Guidelines for Tables

Each table should have a specific purpose (Ehrenberg 1975). We posit that generally there are 3 reasons to construct a table of data. They are:

1. To communicate key findings
2. To organize summaries of statistical analyses to facilitate interpretation
3. To document and store detailed information such as the original data

The amount, type, arrangement, and degree of accuracy of data displayed in tables vary according to purpose. For example, a table that presents key results generally should be constructed from only three of four columns and rows, contain a dozen or fewer highly rounded data values, and the information in the rows and columns should be arranged comparatively, ordered either by addressing a hierarchy of questions of interest or by effects observed in the data. These types of tables should follow Ehrenberg's (1977a, c) strong criteria for a good table in that patterns and exceptions in the data should be obvious at a glance or, at least meet his weak criterion that the patterns and exceptions should be obvious at a glance once the viewer has been informed as to what they are, perhaps with a caption. Conversely, a documentation table would contain most or all of the raw data, perhaps accompanied by some descriptive summary statistics, with only selected rounding, if any, presenting exact data values. A documentation table would be organized by combinations of clerical aspects of the data such as patient numbers and time, making for an easy look-up and extraction of one or more individual datum.

Some of the many guidelines for constructing tables are listed below (Ehrenberg 1975, 1977a, b, c, 1982). The guidelines should be considered as is appropriate for the table attributes of the grable.

1. Place or order the data to compliment the graph part of the grable (Bradstreet et al. 2008)
2. Make it easy to compare relevant numbers. Put numbers that have to be compared close to together. Arrange row order so that if mental arithmetic needs to be performed vertically, it is easy to do so. Consider ordering columns and rows based upon prior knowledge about the table content. Or, rows and columns can be ordered qualitatively by the magnitude of some aspect of the data such as means
3. Numbers are easier to read down a column than across a row, especially for a large quantity of numbers
4. Align the data values vertically according to decimal points or other features common to the data that are meaningful
5. Unless exact values are needed for documentation, generally round numbers to two effective digits. Round to a variable number of digits when necessary
6. The parallel concept to chartjunk (Tufte 1983, 2001) is tablejunk (Bradstreet et al. 2008). Labels should be clear, brief, and have meaning independent of the text. There is no need to rule off every column (or row) with a separate line. Too many or incorrectly placed vertical grid lines can interrupt eye movements. Irregular spacing of rows and columns can be particularly distracting. Too much space between rows or columns can force the eye to move too much making patterns more difficult to see and remember
7. Horizontal and vertical lines, and also gaps of white space, should be used sparingly, to parse major divisions in a table. Occasional regular gaps can help guide the eye and emphasize patterns. Single spacing with occasional gaps is an easy rule to adopt

8. *Bold* and *light* typeface can help distinguish between data falling into two categories. They can also be used to visually separate column and row headings from the data (Wright 1973)
9. A brief written summary should be given for every table to bring out the main qualitative features

Other guidelines for table construction and principles of visual perception will be pointed out as required for the examples later in the chapter.

3.2.3 *More Guidelines and Examples: Recommended Reading*

The guidelines for construction of graphs and tables and principles of visual perception highlighted above and illustrated in the examples, are not meant to be exhaustive. But instead we hope to provide readers with some initial display tools, and stimulate readers to learn more. Additional information on the proper and improper design and construction of graphs and tables, and principles of visual perception, can be found in Ehrenberg (1975, 1982), Tufte (1983, 1990, 1997, 2001, 2006), Schmid (1983), Cleveland (1985, 1993, 1994), Kosslyn (1994), Henry (1995), Wainer (1997, 2005, 2009), Harris (1999), Gelman et al. (2002), Few (2004, 2006, 2009), Wilkinson (2005), Robbins (2005), Chen et al. (2008), Freeman et al. (2008), and Wong (2010). These references are rich with principles, guidelines, and examples, and they present a diversity of authors' opinions and areas of interest.

3.3 Grables

The challenge at hand is to design and construct data displays that in meaningful ways best deliver the messages that the data contain, while simultaneously addressing the viewers' needs for a clear understanding. In some cases the viewers' needs may suggest constructing either a graph or a table, both, or a combination of both. For example, in the assessment of average bioequivalence, the viewer needs to know how the values of the geometric mean ratio and confidence interval from the statistical analysis relate relative to the values of the regulatory limits for establishing average bioequivalence. Further, there is an interest in individual subjects' responses, especially those subjects with extreme data who demonstrate a large subject-by-formulation interaction. Another example is graphing individual subject safety data, and simultaneously tabling and graphing the corresponding group summary statistics. Again, even if the average results look favorable, there is an interest in identifying those individual patients whose data are extreme suggesting a potential safety issue.

Tullis (1981) found that combinations of graphs and tables produce faster but an equally accurate level of understanding as tables constructed in either a narrative or a structured format. Lucas (1981) found that subjects receiving both graphical and tabular output had a higher level of understanding than subjects receiving only graphs, and the subjects found the combined information more useful to them. Powers et al. (1984) found slower but more accurate performance by subjects when given both a graph and a table as compared to either graphs or tables alone. Also, presenting both a graph and a table provides the viewer and the presenter with the option of focusing on the format that they are most familiar with (Powers et al. 1984).

A strategy that takes the simultaneous presentation of the information in a graph and table a step further is to construct a grable. A grable combines the emergent features of a graph with the precise quantities of a table into a single display (Hink et al. 1996, 1998). A grable accommodates a wider variety of visual tasks and a possibly wider audience, than either a graph or a table can address alone. Hink et al. (1998) showed when considering both accuracy and time simultaneously, that grables and tables were favored over conventional graphs alone. Subsequently, Calcaterra and Bennett (2003) showed improved performance in subjects when specific data values were added to configural displays (displays that map multiple individual variables into a single graphical format). Tufte's (2006) sparklines are a successful implementation of the grable strategy.

Given the combination of ink from both a graph and a table, a grable must be designed and constructed with even greater care so as not to clutter up the display and hinder the clarity and accessibility of important information contained in the data. The best principles of visual perception from both graph and table design and construction should be considered.

3.4 More Grables

The following examples continue to present a variety of grables. Some are more graph than table, some are as much table as graph, and one is more like a text table than a graph. For each example, the grable characteristics are discussed followed by principles of good (and bad) graph and table design and construction. The examples work cumulatively in that characteristics and principles pointed out in an earlier example may not be highlighted again in a subsequent example, but they may be implicit in their use in the subsequent example. Electronic versions of the data used in Examples 1 (Bradstreet and Liss 1995), 4 (Bradstreet 1994), and 6 (Bradstreet and Short 2001) can be found at a website continuously maintained by Short (2006).

3.4.1 Example 3: Evaluating Dosing Regimens: Reflux in GERD Patients

This example provides some foundations for the others which follow. It demonstrates a transition from a table to several grables in a step-by-step fashion, pointing

3 Grables: Visual Displays That Combine the Best Attributes of Graphs and Tables   51

Table 3.1 Anti-Rankit mean percent reflux time

Placebo	40 mg h.s.	20 mg b.i.d.	40 mg b.i.d.
11.3%	7.4%	5.9%	2.5%

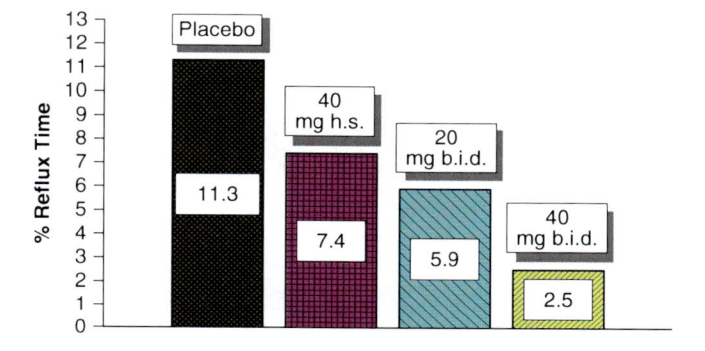

Fig. 3.3 Mean percent reflux time. First attempt at constructing a grable

out some favorable, and some not so favorable, principles of graph construction and visual perception. It also emphasizes that constructing effective grables (or graphs or tables) can be an iterative process with the final grable constructed dependent upon a combination of visual tasks, visual perception, the structure of the data, human preferences, and software capabilities. But once the serious work of communicating effectively is completed, the final grable can become a standard display for similar studies to follow.

Twelve gastroesophageal reflux disease (GERD) patients completed a 4-treatment, 4-period, complete crossover trial to evaluate 20 mg b.i.d. (twice daily), 40 mg b.i.d., and 40 mg h.s. (at bed time) doses of a drug targeted at the reduction of GERD symptoms as compared to placebo, and as compared to each other. The percent reflux time was measured for each of the three doses and placebo when each patient was in the upright position. For more information on the design and statistical analysis of data from higher order crossover studies see Ratkowsky et al. (1993), Jones and Kenward (2003), Brown and Prescott (2006), and Bradstreet et al. (2010). Table 3.1 documents the results of the study.

A first attempt at constructing a grable might look like the data labeled bar chart in Fig. 3.3.

Some major design, construction, and visual perception maladies are worth noting. In general, use of bar charts is tricky when differences are of interest since viewers tend to visually place shorter bars on top of taller bars and estimate proportional differences rather then additive differences. As scaled, the visual slope does not equal the algebraic slope among the placebo and the b.i.d. dosing regimens and this is not communicated. If the b.i.d. dose response evaluation is of key importance, then it can be argued that the 40 mg h.s. results should be visually detached from the b.i.d. results. There is a plethora of non-data ink, redundant data ink, and chartjunk

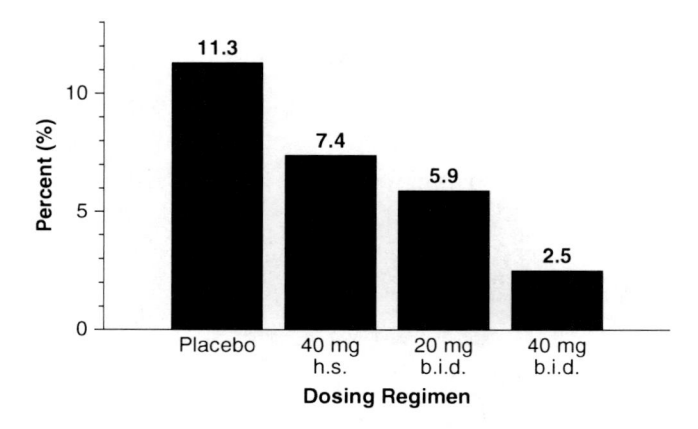

Fig. 3.4 Mean percent reflux time. Second (improved) attempt at constructing a grable

wrapped around just four bivariate data points—only eight data values. Examples of wasted ink include the shadow boxed dosing regimen labels, the moiré vibration in the cross hatching patterns inset into each bar, and there are too many tick marks and tick mark labels, with misguided emphasis on the tick marks. The data density in Fig. 3.3 is unacceptably low. Colors could be selected to visually group and compare the dosing regimens in a more meaningful fashion, and for many scientific publications there are other colors which might be viewed as more appropriate.

Figure 3.4 improves upon Fig. 3.3. The shadow boxes were removed, the dosing regimen labels were moved to the x-axis, and the corresponding x-axis label was added. The cross hatchings were removed and replaced, for now, with a uniform color. The y-axis is marked by regularly spaced major and minor tick marks, only the major tick marks have labels, and the tick marks are no longer visually distracting. The y-axis label has been shortened to just the key information. The data labels were moved from the middle of the bars, to the top of the bars, nearer to their value on the y-axis.

Figure 3.5 incorporates further improvements. The data values were removed from just above the bars and used as tick mark labels for tick marks at irregular intervals corresponding only to the most important data values. Bar colors were changed to visually link the two b.i.d. dosing regimens for comparison to each other, while setting apart both the 40 mg h.s. dosing regimen and the placebo for comparisons among themselves and to the b.i.d. dosing regimens. The tick mark labels were color coded to match the corresponding bar.

From here, there are several directions in which we might proceed including a dot chart, but we illustrate a dot chart in Example 5. The two strategies we chose both concentrate on removing most of the redundant data ink and non-data ink in the bars. The first strategy in Fig. 3.6 emphasizes the vertical distances between the dosing regimens as ordered in this case, by the response, mean percent reflux time. Colors again emphasize the four dosing regimens, and the corresponding vertical differences as visualized by the bolded, two sided, tick marks. This grable might be considered as more a text table than a graphic—a tablic.

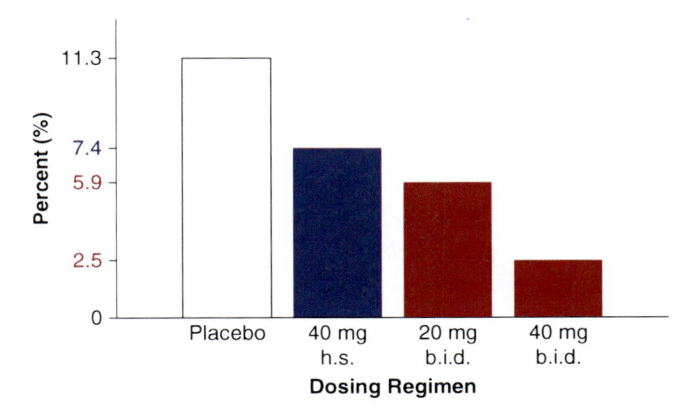

Fig. 3.5 Mean percent reflux time. Third (improved) attempt at constructing a grable

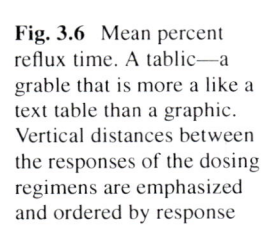

Fig. 3.6 Mean percent reflux time. A tablic—a grable that is more a like a text table than a graphic. Vertical distances between the responses of the dosing regimens are emphasized and ordered by response

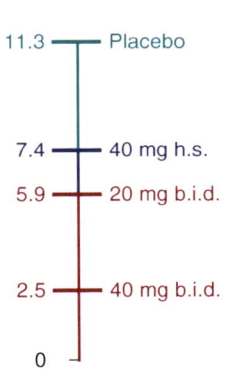

The second strategy shown in Fig. 3.7 emphasizes the b.i.d. dose response aspect of the study while providing a pairwise comparison between the 40 mg h.s. and 40 mg b.i.d. dosing regimens. To demonstrate the second strategy, we choose an arithmetic scaling on the x-axis for the placebo and two b.i.d. doses so that the distance between adjacent doses is 20 mg. We also insert a full axis break between the 40 mg h.s. dose and the other doses to indicate that the 40 mg h.s. dose is not part of the dose response analysis, but is still part of pairwise comparisons.

The slope portrayed in Fig. 3.7 is still too steep. Most software packages automatically default to, or give the user an easy choice among either landscape, portrait, or square orientations, which do not except by coincidence, ensure that the graph portion of the grable is proportionally correct. To make our point, we arbitrarily chose the square orientation taken over the entire y-axis and over the entire x-axis including the 40 mg h.s. dose, since it is a common default and it is intermediate between landscape and portrait.

When possible given the physical dimensions of the hardcopy page or computer screen, and when reasonable given the story that the data are telling, the physical slope on the graph portion of the grable should match the algebraic slope given by

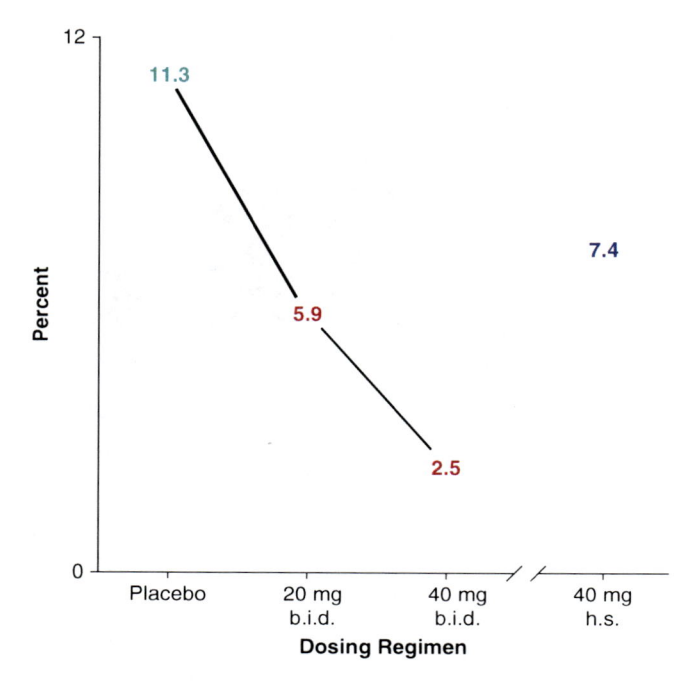

Fig. 3.7 Mean percent reflux time. A grable emphasizing the dose response between the placebo, 20 mg b.i.d., and 40 mg b.i.d. doses, and visualizing pairwise comparisons with the 40 mg h.s. dose. The slope of the dose response is too steep

the data (Bradstreet et al. 2006, 2008). Consider on each axis the ratio of the distance traveled in units (e.g., %) to the physical distance traveled (e.g., inches). The general idea is that the ratio must be the same on both the x- and y-axes. The implementation of this seemingly simple idea is highly dependent upon the specific details of a given grable.

In our example, suppose the placebo, 20 mg b.i.d., and 40 mg b.i.d. doses were spaced along 10 in. of the x-axis, from the placebo tick mark to the 40 mg b.i.d. tick mark. Then for the correct physical slope, the physical length of the y-axis from the 0 tick mark to the 12 percentage point tick mark, is the solution for d in the equality, $(40-0)/10 = (12-0)/d$. Solving for d gives $d = 3$ in. Similarly, if the 40 mg difference between the placebo and 40 mg b.i.d dose spanned 5 in. on the x-axis, then solving for d in $(40-0)/5 = (12-0)/d$, the 12 percentage points from zero to 12 should span 1.5 in. on the y-axis.

In some cases, the proportionally correct grable will not be as scientifically revealing as other combinations of dimensions with different aspect ratios (Cleveland 1985, 1994) or when banking to 45° (Cleveland 1994). It may not be physically possible to construct a proportionally correct grable given available space. Or the software being used either is not be able to do this or it may be extremely difficult to get the software to perform accordingly. In these cases, additional information should be provided as to what a grable with proportionally correct visual slope

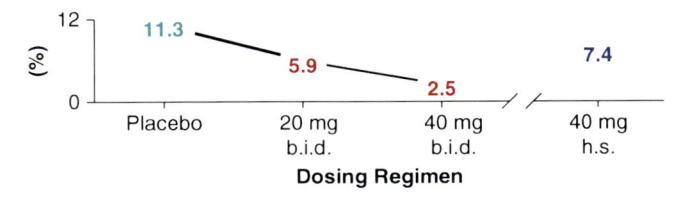

Fig. 3.8 Mean percent reflux time. A proportionally correct grable visualizing the correct dose response between the placebo, 20 mg b.i.d., and 40 mg b.i.d. doses

would look like relative to the one shown (Bradstreet et al. 2006, 2008). This information can take several forms. One is to provide an additional miniature grable which shows the correct slope, either nearby or possibly inset into the original grable. Another visual indicator sources from geometry and non-digital clock faces. Proximal to, or inset into the original grable, provide a visual representation of a pair of rays originating from a common point like the hands on a clock. One ray represents the physical slope, the other represents the algebraic slope, with each ray labeled accordingly. Or, instead, but preferably in addition to visual cues, provide a written notification as how to adjust the slope in your visual mind. A grable with proportionally correct slope is shown in Fig. 3.8. Figure 3.8 leaves a much different impression and interpretation of dose response than Fig. 3.7.

If in Fig. 3.7 we had originally chosen the portrait orientation instead of the square orientation, then the change in the physical slope from Fig. 3.7 to Fig. 3.8 would have been even more dramatic.

Whatever scaling is chosen for statistical analysis (e.g., arithmetic, ordinal, logarithmic), the observed mean results for the placebo, 20 mg b.i.d., and 40 mg b.i.d. doses should be connected with line segments only if the statistical analysis estimates or describes dose response directly incorporating the observed sample means. An example is partitioning the sums of squares due to treatments in the corresponding ANOVA into single degree-of-freedom contrasts for linear and quadratic curvature using orthogonal polynomial coefficients. However, for linear and polynomial regression, the best fitting function is obtained by least squares minimization of the vertical distances from the individual data points. The estimated function may or may not pass through one or more sample means. In this case, plot the estimated function, possibly with a confidence band, and the individual data points (Bradstreet et al. 2006, 2008).

Quite often the most important information to display is the relative difference between some or all of the treatments. In this situation, it is important to construct a grable where mental calculations are either minimized or eliminated (Bradstreet et al. 2006, 2008). In the current example, the primary interest is in the responses of the 3 active treatment regimens relative to the placebo, and secondarily relative to each other in either a pairwise or dose response fashion. A common strategy is to create 2 grables, one showing the observed data and one showing the differences from placebo. These would be arranged either spatially side-by-side (preferred), or shown temporally in time one after the other (less desirable). However, a single grable which effectively displays both the observed data, and the differences from

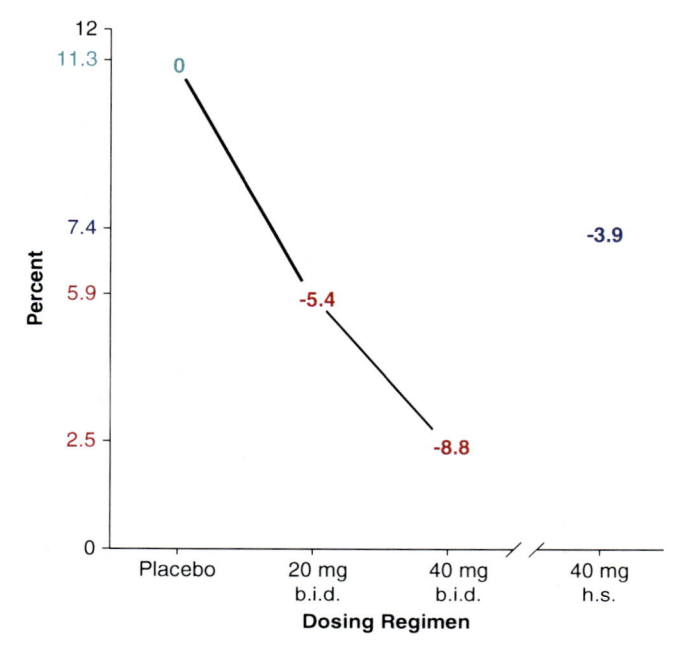

Fig. 3.9 Mean percent reflux time vs. placebo. A *line plot* which displays both the observed data and the difference from placebo

placebo, can be the best approach. Figures 3.9 and 3.10, illustrate strategies for showing both the observed levels of response and the differences from placebo.

Depending on the target audience and the amount of previous use, various hybrids of Fig. 3.9 can be reasonable. For example, it may be effective to plot the observed values in the plotting area and the corresponding differences from placebo on the *y*-axis, labeling the tick marks with the differences. A bit more advanced hybrid of Fig. 3.7, but possibly confusing to a naïve audience, would be to remove the tick mark and the tick mark label at 11.3, and replace the zero difference, 0, with 11.3, the observed value for placebo. Although not technically correct given the title and the scale, this gives the impression of starting at the 11.3 value and sliding downward to the right by the stated differences. The selection of the best suited version requires the careful consideration of technical accuracy versus the combination of an informative figure caption and the familiarity of the audience with the different versions.

3.4.2 Example 4: Evaluating Bioequivalence: Pharmacokinetic Data

Twenty-six healthy male subjects completed a 2-treatment, 2-period crossover bioequivalence trial to determine if the pharmacokinetic characteristics of one 40 mg

Fig. 3.10 Mean percent reflux time vs. placebo. A tablic which displays both the observed data and the differences from placebo

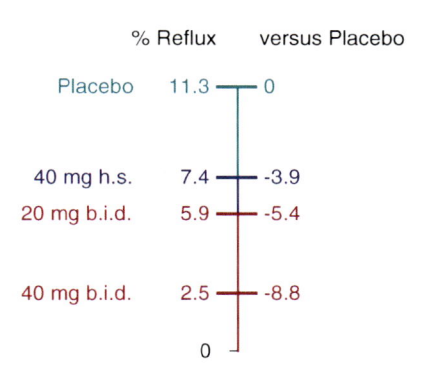

```
                              % Reflux        versus Placebo

              Placebo          11.3 ─────────── 0


              40 mg h.s.        7.4 ─────────── -3.9
              20 mg b.i.d.      5.9 ─────────── -5.4


              40 mg b.i.d.      2.5 ─────────── -8.8

                                0
```

capsule of a drug made by Company A are the same as the concurrent administration of two 20 mg capsules of the same drug made by Company B. The pharmacokinetic variable, area under the plasma-concentration-versus-time-curve (AUC), was calculated (ng×h/mL) for each subject for each formulation from drug levels (ng/mL) assayed from plasma samples taken over time. For more information on the design, conduct, statistical analysis, and the display of results from a 2-treatment, 2-period crossover bioequivalence studies, see Bradstreet and Dobbins (1996), Pikounis et al. (2001), Food and Drug Administration (2001, 2003), Jones and Kenward (2003), Bradstreet and Panebianco (2004), and European Medicines Agency (2009).

In Fig. 3.11 the open circles (**O**) represent the ratio (Company A, 1×40 mg/ Company B, 2×20 mg) of AUCs for each subject. The solid dot (•) indicates the estimated geometric mean ratio and the vertical bar with horizontal endpoints (I) represents the corresponding 90% confidence interval. On the *y*-axis the Food and Drug Administration's regulatory limits for average bioequivalence of (**0.80, 1.25**) are labeled as is the ratio of **1.00**.

Visually, it is immediately clear that average bioequivalence was not concluded since the upper confidence limit of the 90% confidence interval (I) lies above the upper bioequivalence limit (– – –).

The exact numerical results are also of interest, particularly so in cases like this indicating a notable degree of subject-by-formulation interaction, which is further magnified by two extreme AUC ratios, and with the upper confidence limit close to the upper bioequivalence limit. Therefore, the geometric mean symbol (•) is labeled with its value, 1.12, as are the limits of the 90% confidence interval (0.98, 1.27). The 2 up arrows (↑↑) signal that there are 2 subjects with AUC ratios lying above the upper end of the *y*-axis with values of 2.32 and 2.70. These arrow indicators for outliers were first suggested to us by John W. Tukey (personal communication). Importantly, the arrow indicators allow a detailed view of the behavior of the majority of the data. Graphing the data to scale including the two outliers would condense much of the data into a series of blue ink blobs that would not provide much useful information.

Other principles for graph and table design, and visual perception, were used in constructing Fig. 3.11. They include:

Fig. 3.11 Average bioequivalence analysis. The *open circles* (**O**) represent the ratio (1×40 mg/2×20 mg) of AUCs for each subject. The *solid dot* (•) indicates the estimated geometric mean ratio and the vertical bar with horizontal endpoints (I) represents the corresponding 90% confidence interval. On the *y*-axis, the Food and Drug Administration's regulatory limits for average bioequivalence of (**0.80**, **1.25**) are labeled as is the ratio of **1.00**

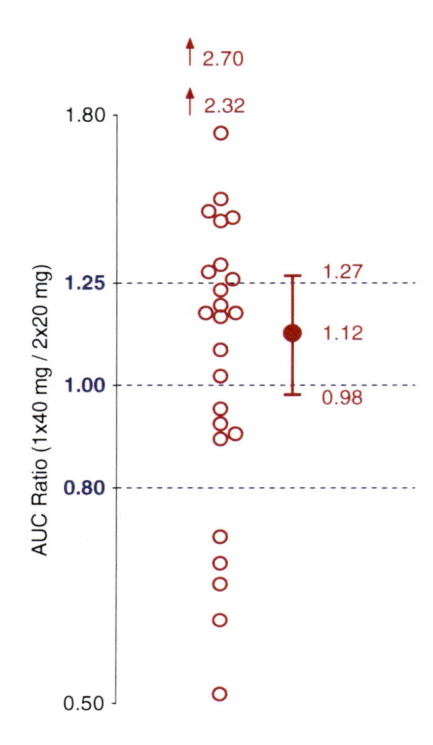

1. Spending data ink wisely; minimal non-data ink and redundant data ink
2. Plotting data on the log scale to align correctly with the statistical analyses
3. Labeling tick marks, summary statistics, and individual ratios with antilog values which are more easily accessible to a wider audience
4. Using an open plotting symbol (**O**) to lessen confusion due to overplotting
5. Jittering plotting symbols horizontally to lessen confusion due to overplotting
6. Clearly indicating in the caption that error bars represent a 90% confidence interval and not another interval measure such as standard deviation or standard error
7. Assigning thicker lines to more important graphing elements (e.g., 90% CI) and thinner lines to less important ones (e.g., *y*-axis)
8. Using reference lines to indicate important values across the entire graph
9. Constructing the reference lines with texture and width so as not to distract from the data, and placing the lines behind the data
10. Placing and labeling only those tick marks, critical to understanding the data and making a decision on bioequivalence
11. Choosing distinct color combinations either for emphasis (red, blue) or without emphasis (black), that are not problematic for some viewers (e.g., red, green), and not relying solely on color to transmit information
12. Heavily, but intelligently, rounding exact data values. Note that in this example, there is no need to display the 90% confidence limits to 3 decimal places as neither is close enough to the regulatory limits for average bioequivalence for rounding to matter in the decision
13. Using relatively simple sans serif fonts

3.4.3 Example 5: First in Man Evaluation: Clinical Lab Safety Data

Two panels of 6 male subjects enrolled in an alternating panel, fixed-rising-dose, safety study. Each subject received placebo and three of six possible doses of a drug. Panel A (**O**) received 0.2 mg, 1 mg, and 5 mg, and Panel B (**Δ**) received 0.5 mg, 2 mg, and 10 mg. For more information on the design and analysis of alternating panel fixed-rising-dose studies, see Rodda et al. (1988), Bolognese (1991), and Jin and Sun (2008).

Figure 3.12 combines a dot chart (Cleveland 1985, 1993, 1994) of individual subject values with a table of the corresponding summary statistics. Each line in the dot chart portion of the grable displays each subject's percent change from baseline in basophils at 24 h (**O Δ**) with the mean value (**X**) for that group of subjects. Open circles (**O**) represent subjects in Panel A, and in Panel B subjects are represented by open triangles (**Δ**). Tabled to-the-right on the same line are the corresponding number of subjects, mean, standard deviation, minimum value, and maximum value.

Other principles of graph and table design, and visual perception, were used in constructing Fig. 3.12. They include:

1. The dot chart takes advantage of the higher level, more accurate, visual decoding of information positioned along a common scale
2. Spending data ink wisely; minimal non-data ink and minimal redundant data ink
3. Using open and clearly distinct plotting symbols (**O Δ X**) to lessen confusion due to overplotting
4. Jittering plotting symbols vertically to lessen confusion due to overplotting
5. Using prominent graphing elements to represent the data values (**O Δ**) and summary statistics (**X**) while downplaying less important non-data structure such as the x- and y-axes
6. Positioning a reference line (y-axis) to indicate an important value (zero) that applies across the entire graph, but placing it in the background with texture, width, and color chosen so as not to interfere with the data
7. Encoding categorical information (Panel A and Panel B) with combinations of symbols and colors (**O Δ**), not relying solely on color to transmit information
8. Choosing distinct color combinations either to emphasize (blue, cyan, red) or deemphasize (black, gray) components of a grable, that are not problematic for color challenged viewers (e.g., red, green)
9. Placing and labeling only the necessary tick marks
10. Ordering rows monotonically, from bottom to top, by dose
11. Placing the data values according to the graph part of the grable
12. Heavily, but intelligently, rounding exact data values
13. Decimal aligning data values in columns
14. Using white space, not vertical grid lines, to separate columns of data values
15. Removing unnecessary leading digits in data values
16. Providing a brief, insightful, verbal summary of the grable in the caption
17. Using relatively simple **sans serif fonts**

Dose		N	Mean	SD	Min	Max
10 mg		3	0.47	.06	.4	.5
5 mg		6	-0.23	.31	-.6	.2
2 mg		6	0.28	.31	-.3	.6
1 mg		6	0.63	.65	-.4	1.5
0.5 mg		6	0.02	.35	-.5	.4
0.2 mg		6	-0.22	.26	-.6	.0
Pbo (B)		5	0.06	.26	-.3	.4
Pbo (A)		6	0.13	.37	-.4	.5

Change from Baseline (%)

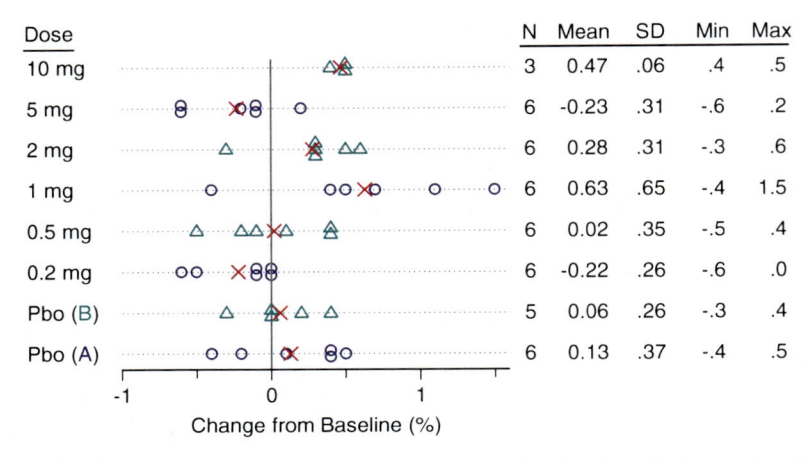

Fig. 3.12 Clinical laboratory data: Basophils. Each line of the *dot chart* displays each subject's change from baseline (%) in basophils at 24 h (**O Δ**) with the mean value (**X**) for that group of subjects. Panel A is represented by *open circles* (**O**), Panel B by *open triangles* (**Δ**). Tabled on the same line are the corresponding summary statistics: the number of subjects (*N*), the mean, the standard deviation, the minimum value, and the maximum value

In a similar fashion, Fig. 3.13 shows lengths of PQ intervals (ms) at 3 time points post dose (baseline, 2 h, 24 h) incorporating the comparative small multiples strategy.

3.4.4 Example 6: Evaluating Dose Proportionality: Pharmacokinetic Data

A total of 12 healthy male and 12 healthy female subjects completed a 4-treatment, 4-period crossover dose proportionality trial to determine if the pharmacokinetic characteristics of four oral doses (2.5, 5, 10 and 15 mg) of a drug are dose proportional. The pharmacokinetic variable, area under the plasma-concentration-versus-time-curve (AUC), was calculated (ng×h/mL) for each subject for each dose of drug calculated from drug levels (ng/mL) assayed from plasma samples taken over time. For more information on the design, conduct, and statistical analysis of dose proportionality studies, see Haynes and Weiss (1989), Yuh et al. (1990), Gough et al. (1995), Smith (1997), Smith et al. (2000), and Sethuraman et al. (2007).

There are at least 3 general strategies for visualizing and assessing dose proportionality at the individual subject level (Bradstreet et al. 1999, 2008). In the first, arithmetic AUC (*y*-axis) is plotted versus arithmetic dose (*x*-axis), with the AUC values connected by line segments. Dose proportionality is indicated for a subject if the line segments form a straight line with a positive slope which also passes through the origin (0, 0). In the second strategy, log-transformed AUC (*y*-axis) is plotted versus log transformed dose (*x*-axis), again connecting the log AUC values with line

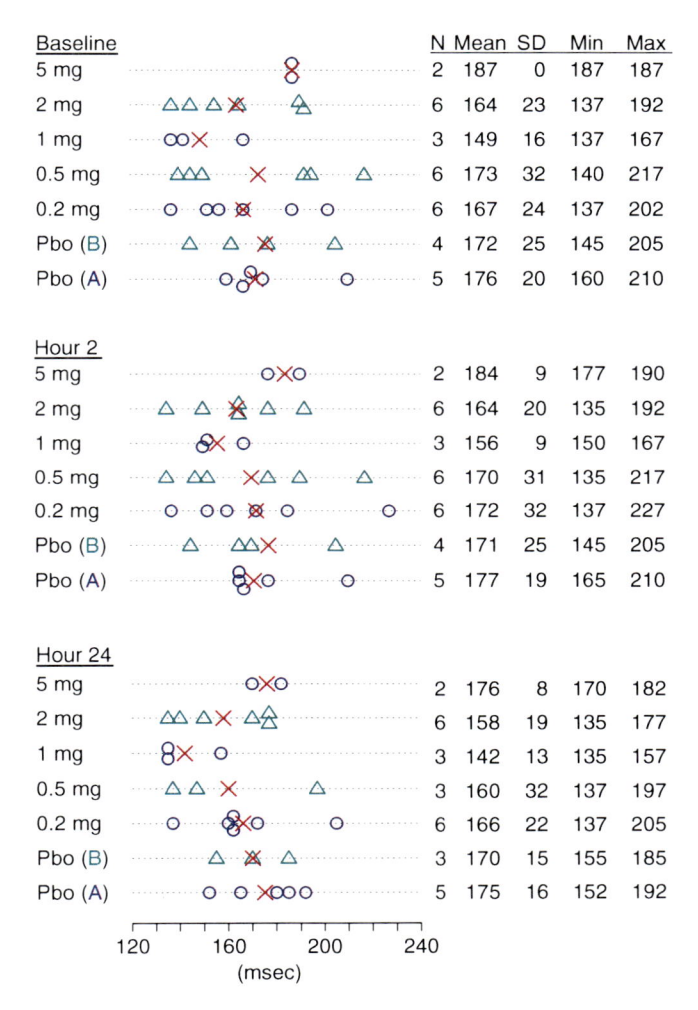

Baseline		N	Mean	SD	Min	Max
5 mg		2	187	0	187	187
2 mg		6	164	23	137	192
1 mg		3	149	16	137	167
0.5 mg		6	173	32	140	217
0.2 mg		6	167	24	137	202
Pbo (B)		4	172	25	145	205
Pbo (A)		5	176	20	160	210

Hour 2		N	Mean	SD	Min	Max
5 mg		2	184	9	177	190
2 mg		6	164	20	135	192
1 mg		3	156	9	150	167
0.5 mg		6	170	31	135	217
0.2 mg		6	172	32	137	227
Pbo (B)		4	171	25	145	205
Pbo (A)		5	177	19	165	210

Hour 24		N	Mean	SD	Min	Max
5 mg		2	176	8	170	182
2 mg		6	158	19	135	177
1 mg		3	142	13	135	157
0.5 mg		3	160	32	137	197
0.2 mg		6	166	22	137	205
Pbo (B)		3	170	15	155	185
Pbo (A)		5	175	16	152	192

(msec)

Fig. 3.13 Clinical laboratory data: PQ intervals. Data are arranged in small multiples sorted by time and then by dose within time. Each line of the dot chart displays each subject's individual value (**O Δ**) with the mean value (**X**) for that group of subjects. Panel A is represented by *open circles* (**O**), Panel B by *open triangles* (**Δ**). Tabled on the same line are the corresponding summary statistics: the number of subjects (*N*), the mean, the standard deviation, the minimum value, and the maximum value

segments. Dose proportionality is indicated for a subject if the line segments form a straight line with slope equal to 1. The intercept is not of immediate interest but it provides useful subject specific information. In the third strategy, each arithmetic AUC value is divided by the corresponding arithmetic dose which standardizes the AUC values to 1 mg, or the AUCs can be standardized to a particular dose such as 10 mg. Then the AUC/dose values (*y*-axis) are plotted versus the arithmetic dose values (*x*-axis), connecting the AUC/dose values with line segments. Dose proportionality

Table 3.2 AUC standardized to 1 mg of drug—males ($n = 12$)

Subject	2.5 mg	5 mg	10 mg	15 mg
1	9.8	6.5	8.6	10.5
2	10.7	6.4	7.7	10.5
3	8.3	8.5	7.2	7.5
4	3.3	4.4	5.2	4.9
5	4.4	6.6	6.2	8.4
6	2.0	3.5	4.1	4.1
7	8.7	9.2	9.3	11.9
8	3.4	4.4	3.4	4.2
9	7.2	7.2	8.1	7.9
10	6.2	8.6	9.3	10.9
11	7.8	7.5	10.8	13.3
12	4.9	6.0	6.6	7.8

is indicated for a subject if the line segments form a straight line with slope equal to 0. Again, the intercept is not of immediate interest but it provides useful subject specific information.

We use the third strategy, AUC/dose versus dose, to demonstrate the construction of a grable from the corresponding table of individual subject data. The dose adjusted to 1 mg AUC data for the males are documented in Table 3.2. The original AUC/dose values were rounded to one decimal place to be more easily read.

Table 3.2 is sorted by subject number and dose, facilitating the documentation and look-up of individual values. The grable in Fig. 3.14 not only documents the individual values, but importantly provides an initial assessment of dose proportionality for each subject, and compares responses among the subjects. For clarity, the data were sorted vertically, from bottom to top, by the AUC/dose values for the 2.5 mg dose (Friendly and Kwan 2003; Tufte 2006; Bradstreet and Palcza 2012). Since there are only 10 subjects, this arrangement should not increase look-up speed, especially when considering the additional information provided on dose proportionality.

The data values in Fig. 3.14 were rounded to the first decimal place to retain enough accuracy for documentation and look-up. But the trailing decimals to some degree, inhibit readability and they slow down even simple mental arithmetic. To address this, exact values could be plotted but labeled instead with AUC/dose values which are rounded excluding the decimal. However, this would generate line segments with non-zero slope visually connecting the same rounded data label, an awkward position to be in. Alternatively, the rounded data labels could be plotted, but this is too much rounding for the accuracy desired given the range of the data is from 2.0 to 13.3. A possible solution is to standardize the data values to another dose, say to the 10 mg dose. Figure 3.15 displays this arrangement. The desired accuracy is achieved, readability is increased, and mental arithmetic is simplified. We remind ourselves that if either differences or ratios among the doses were of primary interest, these could be plotted avoiding the mental arithmetic.

Fig. 3.14 AUC standardized to 1 mg of drug. Males, sorted vertically from bottom to top, by value of AUC/dose for 2.5 mg dose

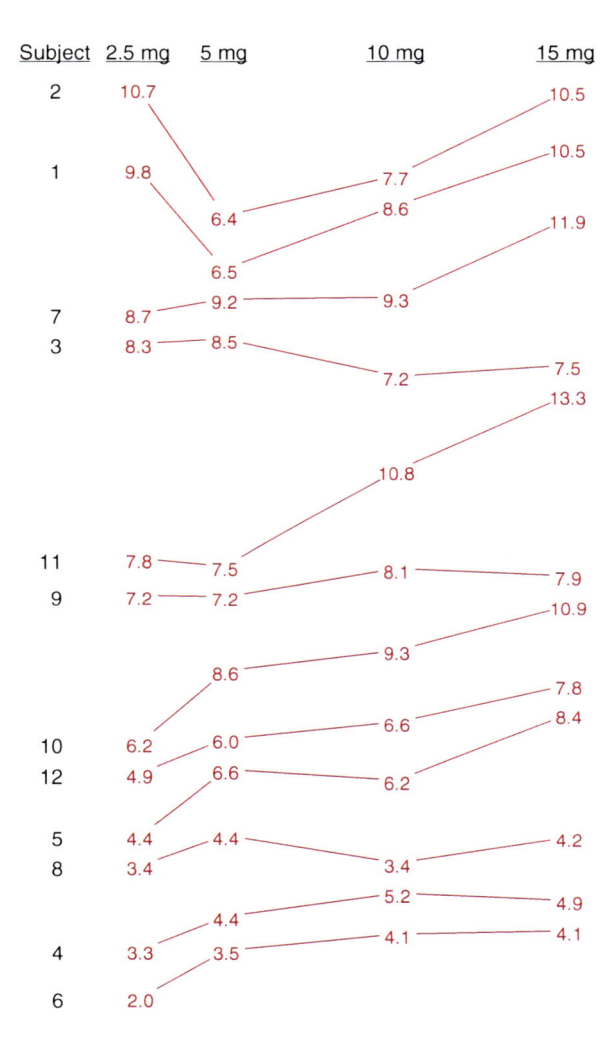

Subject	2.5 mg	5 mg	10 mg	15 mg
2	10.7			10.5
1	9.8		7.7	10.5
		6.4	8.6	11.9
		6.5		
7	8.7	9.2	9.3	
3	8.3	8.5	7.2	7.5
				13.3
			10.8	
11	7.8	7.5	8.1	7.9
9	7.2	7.2		10.9
		8.6	9.3	7.8
			6.6	8.4
10	6.2	6.0		
12	4.9	6.6	6.2	
5	4.4	4.4		4.2
8	3.4		3.4	
			5.2	4.9
		4.4	4.1	4.1
4	3.3	3.5		
6	2.0			

3.5 Software

To construct a well-designed grable, or as equally important a well-designed graph or table, requires software with the prerequisite capabilities, which some software packages may not possess. In addition, the default settings of many software packages are not conducive to producing effective visual displays immediately. However, an initial investment of time will pay off for the visual task at hand as well as for subsequent runs of the same or similar displays.

It is not our intention to condemn or promote particular software packages, but instead to provide a list of qualities to consider when selecting software. These considerations should be framed within your particular needs and local computing environment. Some desired software characteristics include (Bradstreet et al. 2008):

Fig. 3.15 AUC standardized to 10 mg of drug. Males, sorted vertically from bottom to top, by value of AUC/dose for 2.5 mg dose

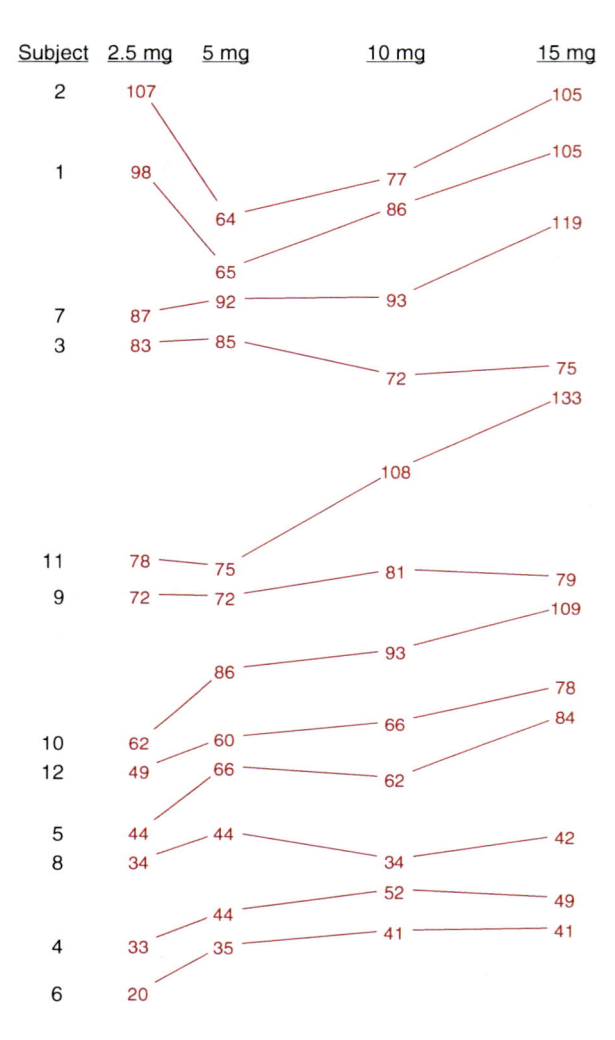

Subject	2.5 mg	5 mg	10 mg	15 mg
2	107			105
1	98		77	105
		64	86	119
		65		
7	87	92	93	
3	83	85		
			72	75
				133
			108	
11	78	75	81	79
9	72	72		109
			93	
		86		78
			66	84
10	62	60		
12	49	66	62	
5	44	44		42
8	34		34	
			52	49
		44	41	41
4	33	35		
6	20			

1. Capable and flexible enough to construct grables correctly
2. Relatively easy to learn and program
3. Modest complexity to run
4. A GUI (*G*raphics *U*ser *I*nterface) may be helpful for users with lesser programming skills, provided it allows for virtually the same capabilities and flexibility as constructing code from first programming principles
5. Highly portable, both electronically and physically
6. Amenable to automation to support production as well as one-off environments
7. Must integrate well with other graphics, statistical, and word processing software
8. Satisfies data analysis as well as presentation and publication requirements

Traditionally, no one software package will meet all of your needs. Consider choosing one that meets most of your needs while sacrificing on lower priorities, or shop for a complimentary set that meets all of your needs. It is also useful to organize a local group of software users who are similarly dedicated to implementing the principles of visual perception in the design and construction of effective grables.

3.6 Discussion

When presenting patient data, many situations require showing spatial relationships and also displaying, highlighting, or extracting individual data values. Spatial relationships like trends, associations, and other visual patterns typically are best displayed with a graph. Displaying, highlighting, or extracting one or more data values typically is best accomplished with a table. Because of personal familiarity, or a path of least effort, presenters and viewers may arbitrarily favor one display format over the other.

This dual display dilemma can often be solved with a grable. A grable combines the emergent features of a graph with the precise quantities of a table into a single display. Its purpose is to simultaneously accommodate a wider variety of visual tasks and a possibly wider audience, than either a graph or a table can address alone.

Proposed visual and cognitive strengths and weaknesses of graphs and tables should be considered when designing grables, as should proposed guidelines for their construction. Designing and constructing a grable can be more challenging than for either a graph or a table alone. The best practices selected from each visual format must be complimentary when used in combination, which is not guaranteed.

We provided examples of grables highlighting principles of design, construction, and perception. Although rather simple, these grables and the guidelines for graph and table construction provide initial guidance on how to get started. Additional guidance and examples can be found in the recommended readings.

Careful consideration should be given to software selection. It can be productive and rewarding to collaborate with users who have a similar desire to efficiently produce high-quality grables.

Grables are not automatic visual panaceas for perception. Like well-constructed graphs and tables, they require careful thought in design and construction. Several iterations may be required before the final design is achieved. Once completed, the final display or variations of it, can be used for future clinical studies.

Acknowledgments The author thanks Christine Stocklin for her indispensible help in creating the grables using an S-PLUS GUI.

References

Bolognese JA (1991) Statistical issues for the initial human safety study. In: 1991 Proceedings of the American Statistical Association, Biopharmaceutical Section, American Statistical Association, Alexandria, VA: 274–283

Bradstreet TE (1994) Favorite data sets from early phases of drug research—part 3. In: 1994 Proceedings of the American Statistical Association, Statistical Education Section, American Statistical Association, Alexandria, VA: 247–252

Bradstreet TE, Dobbins TW (1996) When are two drug formulations interchangeable? Teach Stat 18:45–48

Bradstreet TE, Liss CL (1995) Favorite data sets from early (and late) phases of drug research—part 4. In: 1995 Proceedings of the American Statistical Association, Statistical Education Section, American Statistical Association, Alexandria, VA: 335–340

Bradstreet TE, Palcza JS (2012) Digging into data with graphics. Teach Stat 34:68–74

Bradstreet TE, Panebianco DL (2004) An oral contraceptive drug interaction study. J Stat Educ 12: www.amstat.org/publications/jse/v12n1/datasets.bradstreet.html. Accessed 24 September 2011

Bradstreet TE, Short TH (2001) Favorite data sets from early phases of drug research–part 5. In: 2001 Proceedings of the American Statistical Association, Statistical Education Section, American Statistical Association, Alexandria, VA: [CD-ROM]

Bradstreet TE, Goldberg M, Porras A (1999) Interdisciplinary issues in the design, analysis, and interpretation of dose proportionality studies. Handout, Biopharmaceutical Section, Joint Statistical Meetings, Baltimore, MD, August 8–12, 1999

Bradstreet TE, Nessly M, Short TS (2006) Effective displays of data need more attention in statistics education. Handout, Statistical Education Section, Joint Statistical Meetings, Seattle, WA, August 6–10, 2006, http://biostat.mc.vanderbilt.edu/wiki/pub/Main/StatGraphCourse/TEB.pdf. Accessed 24 September 2011

Bradstreet TE, Nessly M, Short TS (2008) Effective displays of data for communication, decision making, and ACMs. Course Notebook, Merck Research Laboratories

Bradstreet TE, Panebianco DL, Maganti L, Maes A (2010) Selecting covariance structures in 3, 4, and 6 period pK and pD crossover trials. Handout, Biopharmaceutical Section, Joint Statistical Meetings, Vancouver, CA, July 31–August 5, 2011

Brown H, Prescott R (2006) Applied mixed models in medicine, 2nd edn. Wiley, Chichester

Calcaterra JA, Bennett KB (2003) The placement of digital values in configural displays. Displays 24:85–96

Carswell CM, Ramzy C (1997) Graphing small data sets: should we bother? Behav Inform Technol 16:61–71

Carter LF (1947) An experiment on the design of tables and graphs used for presenting numerical data. J Appl Psychol 31:640–650

Chambers JM, Cleveland WS, Kliner B, Tukey PA (1983) Graphical methods for data analysis. Duxbury Press, Boston, MA

Chen C, Härdle W, Unwin A (eds) (2008) Handbook of data visualization. Springer, Berlin

Cleveland WS (1985) The elements of graphing data. Wadsworth, Monterey, CA

Cleveland WS (1993) Visualizing data. Hobart Press, Summit, NJ

Cleveland WS (1994) The elements of graphing data, revised edition. Hobart Press, Summit, NJ

Cleveland WS, McGill R (1984) Graphical perception: theory, experimentation and application to the development of graphical methods. J Am Stat Assoc 79:531–554

Cleveland WS, McGill R (1986) An experiment in graphical perception. Int J Man–Machine Stud 25:491–500

Cleveland WS, McGill R (1987) Graphical perception: the visual decoding of quantitative information on graphical displays of data. J R Stat Soc A 150:192–229

Coll JH (1992) An experimental study of the efficacy of tables versus bar graphs with respect to type of task. Inform Manag 23:45–51

Ehrenberg ASC (1975) Data reduction—analyzing & interpreting statistical data. Wiley, London

Ehrenberg ASC (1977a) Some rules of data presentation. Statistical Reporter 305–310

Ehrenberg ASC (1977b) Three exercises in data presentation. Bias 4:53–65

Ehrenberg ASC (1977c) Rudiments of numeracy. J R Stat Soc Ser A 140:277–297

Ehrenberg ASC (1978) Graphs or tables? The Statistician 27:87–96

Ehrenberg ASC (1982) A primer in data reduction—an introductory statistics textbook. Wiley, Chichester

European Medicines Agency (2009) Guidelines for the investigation of bioequivalence. http://www.ema.europa.eu/docs/en_GB/document_library/Scientific_guideline/2009/09/WC500003011.pdf. Accessed 2 September 2011

Few S (2004) Show me the numbers—designing tables and graphs to enlighten. Analytics Press, Oakland, CA

Few S (2006) Information dashboard design—the effective visual communication of data. O'Reilly Media, Sebastopol, CA

Few S (2009) Now you see it—simple visualization techniques for quantitative analysis. Analytics Press, Oakland, CA

Freeman JV, Walters SJ, Campbell MJ (2008) How to display data. Blackwell Publishing, Malden, MA

Friendly M, Kwan E (2003) Effect ordering for data displays. Comput Stat Data Anal 43:509–539

Food and Drug and Drug Administration (2001) Guidance for industry—statistical approaches to establishing bioequivalence. http://www.fda.gov/downloads/Drugs/Guidance Compliance RegulatoryInformation/Guidances/ucm070244.pdf. Accessed 2 September 2011

Food and Drug and Drug Administration (2003) Guidance for industry—bioavailability and bioequivalence studies for orally administered products—general considerations. http://www.fda.gov/downloads/Drugs/GuidanceComplianceRegulatoryInformation/Guidances/ucm070124.pdf. Accessed 2 September 2011

Gelman A (2011) Why tables are better than graphs. J Comput Graph Stat 20:3–7

Gelman A, Pasarcia C, Dodhia X (2002) Let's practice what we preach: turning tables into graphs. American Statistician 56:121–130

Gough K, Hutchinson M, Keene O, Byrom B, Ellis S, Lacey L, McKellar J (1995) Assessment of dose proportionality: report from the Statisticians in the Pharmaceutical Industry/Pharmacokinetics UK Joint Working Party. Drug Inform J 29:1039–1048

Harris RL (1999) Information graphics—a comprehensive illustrated reference. Oxford University Press, New York

Harvey N, Bolger F (1996) Graphs versus tables: effects of data presentation format on judgemental forecasting. Int J Forecast 12:119–137

Haynes JD, Weiss AI (1989) Modeling pharmacokinetic dose-proportionality data. In: 1989 Proceedings of the American Statistical Association, Biopharmaceutical Section, American Statistical Association, Alexandria, VA: 85–89

Henry GT (1995) Graphing data—techniques for display and analysis. Sage Publications, Thousand Oaks, CA

Hink JK, Wogalter MS, Eustace JK (1996) Display of quantitative information: Are grables better than plain graphs or tables? Proceedings of the Human Factors and Ergonomics Society 40th annual meeting 1155–1159

Hink JK, Eustace JK, Wogalter MS (1998) Do grables enable the extraction of quantitative information better than pure graphs or tables? Int J Ind Ergon 22:439–447

Hwang MI (1995) The effectiveness of graphic and tabular presentation under time pressure and task complexity. Inform Resour Manag J 8:25–31

Jin B, Sun P (2008) Linear models for the analysis of alternating panel rising dose designs. BARDS Technical Report Series, #132. Merck Research Labs

Jones B, Kenward MG (2003) Design and analysis of cross-over trials, 2nd edn. Chapman & Hall, Boca Raton

Kastellec JP, Leoni EL (2007) Using graphs instead of tables in political science. Perspect Polit 5:755–771

Kosslyn SM (1989) Understanding charts and graphs. Appl Cognit Psychol 3:186–226

Kosslyn SM (1994) Elements of graph design. WH Freeman, New York

Lalomia MJ, Coovert MD (1987) A comparison of tabular and graphical displays in four problem-solving domains. ACM SIGCHI Bull 19:49–54

Lucas HC (1981) An experimental investigation of the use of computer-based graphics in decision making. Manag Sci 27:757–768

Meyer J (2000) Performance with tables and graphs: effects of training and a Visual Search Model. Ergonomics 43:1840–1865

Meyer J, Shinar D, Leiser D (1997) Multiple factors that determine performance with tables and graphs. Hum factors 39:268–286

Meyer J, Shamo MK, Gopher D (1999) Information structure and the relative efficiency of tables and graphs. Hum Factors 41:570–587

Pikounis B, Bradstreet TE, Millard SP (2001) Graphical insight and data analysis for the 2,2,2 crossover design. In: Millard SP, Krause A (eds) Applied statistics in the pharmaceutical industry. Springer, New York

Porat T, Oron-Gilad T, Meyer J (2009) Task-dependent processing of tables and graphs. Behav Inform Technol 28:293–307

Powers M, Lashley C, Sanchez P, Shneiderman B (1984) An experimental comparison of tabular and graphic data presentation. Int J Man–Machine Stud 20:545–566

Ratkowsky DA, Evans MA, Alldredge JR (1993) Cross-over experiments—design, analysis, and application. Marcel Dekker, New York

Robbins NB (2005) Creating more effective graphs. Wiley, New York

Rodda BE, Tsianco MC, Bolognese JA, Kersten MK (1988) Clinical development. In: Peace KE (ed) Biopharmaceutical statistics in drug development. Marcel Dekker, New York

Schenker N, Monti KL, Cobb GW, Fesco RS, Chmiel JS (2007) Combining features of a frequency table and a stem-and-leaf plot to summarize the American Statistical association's strategic activities. American Statistician 61:245–247

Schmid CF (1983) Statistical graphics—design principles and practices. Wiley, New York

Scott DW (2003) The case for statistical graphics. AMSTAT News 315:20–22

Sethuraman V, Leonov S, Squassante L, Mitchell T, Hale M (2007) Sample size calculation for the power model for dose proportionality studies. Pharmaceut Stat 6:35–41

Short TH (2006) http://www.jcu.edu/math/faculty/TShort/Bradstreet/index.html. Accessed 2 September 2011

Smith R (1997) A statistical criterion for dose proportionality. Handout, ENAR Biometric Society Meetings, Memphis, TN, March 24–26, 1997

Smith BP, Vandenhende FR, DeSante KA, Nagy AF, Welch PA, Callaghan JT, Forgue ST (2000) Confidence interval criteria for assessment of dose proportionality. Pharmaceut Res 17:1278–1283

Tufte ER (1983) The visual display of quantitative information. Graphics Press, Cheshire CT

Tufte ER (1990) Envisioning information. Graphics Press, Cheshire CT

Tufte ER (1997) Visual explanations. Graphics Press, Cheshire CT

Tufte ER (2001) The visual display of quantitative information, 2nd edn. Graphics Press, Cheshire CT

Tufte ER (2006) Beautiful evidence. Graphics Press, Cheshire CT

Tukey JW (1977) Exploratory data analysis. Addison-Wesley, Reading, MA

Tukey JW (1990) Data-based graphics: visual display in the decades to come. Stat Sci 5:327–339

Tukey JW (1993) Graphical comparisons of several linked aspects: alternatives and suggested principles (with discussions and rejoinder). J Comput Graph Stat 2:1–49

Tullis TS (1981) An evaluation of alphanumeric, graphic, and color information displays. Hum Factors 23:541–550

Vessey I (1991) Cognitive fit: a theory-based analysis of the graphs versus tables literature. Decision Sci 22:219–240

Vessey I (1994) The effect of information presentation on decision making: a cost-benefit analysis. Information Manag 27:103–119

Wainer H (1997) Visual revelations—graphical tales of fate and deception from Napoleon Bonaparte to Ross Perot. Copernicus/Springer-Verlag, New York

Wainer H (2005) Graphic discovery—a trout in the milk and other visual adventures. Princeton University Press, Princeton

Wainer H (2009) Picturing the uncertain world—how to understand, communicate, and control uncertainty through graphic display. Princeton University Press, Princeton

Wilkinson L (2005) The grammar of graphics, 2nd edn. Springer, New York
Wong DM (2010) The Wall Street Journal guide to information graphics—the do's and don'ts of presenting data, facts, and figures. W. W. Norton & Company, New York
Wright P (1973) Research in brief: understanding tabular displays. Visible Lang 7:351–359
Yuh L, Eller G, Ruberg SJ (1990) A stepwise approach for analyzing dose proportionality studies. In: 1990 Proceedings of the American Statistical Association, Biopharmaceutical Section, American Statistical Association, Alexandria, VA: 47–50

Chapter 4
The Use of Figures in Epidemiological Publications: A Survey of Current Practice and Consequent Recommendations

Elisabeth Wreford Andersen and Stuart J. Pocock

Abstract The aims of this survey are to document the current use of figures in epidemiological publications and to make proposals for future practice. To do this, the authors identified all 181 analytical epidemiology articles from 10 major medical journals in the period June to August 2008. For each article the number and type of figures were ascertained and each figure was studied for style and contents. The mean number of figures per article was 0.98. Eighty-four articles (46%) had no figures and most others had just one figure. The most common types of figures were plots of estimates, Kaplan–Meier plots, flow diagrams, smooth or model based curves and distributional plots. These 5 groups of plots accounted for 89% of the figures in the survey. For each of these 5 types of figures, examples of good practice were chosen and commented on. From this overview of current practice some general suggestions regarding the use of figures were given. Well-constructed figures greatly add value to the presentation of the study results. However, many authors choose not to include figures and there is room for improvement in the content and presentation of figures that are included.

4.1 Introduction

Graphical data display is a valuable tool for presenting the results of an epidemiologic study. In general, figures have a major visual impact, and if employed properly they can catch the attention of the reader in illustrating and supporting the main results.

E.W. Andersen (✉)
London School of Hygiene and Tropical Medicine, London, UK

Technical University of Denmark, Department of Informatics, Data Analysis,
DK-2800 Kgs. Lyngby, Denmark
e-mail: ewan@imm.dtu.dk

S.J. Pocock
London School of Hygiene and Tropical Medicine, London, UK
e-mail: Stuart.Pocock@lshtm.ac.uk; ewa@ssi.dk

In principle, modern computing power makes the construction of figures straightforward, though not necessarily in a form suitable for journal publication. There is much guidance in the general use of graphics both as part of the dynamic process of data analysis and in formally presenting results (Cleveland 1994; Robbins 2004; Tufte 2001; Wilkinson 2005). Much of the focus in medical and epidemiology journals, however, is on specific types of figures, e.g., Kaplan–Meier plots (Pocock et al. 2002) or figures in meta-analyses (Bax et al. 2009). The STROBE initiative published guidelines with advice as to how to report observational studies in epidemiology (Vandenbroucke et al. 2007; von Elm et al. 2007) but these do not mention the use of figures in any detail.

A previous survey has explored the use of figures in clinical trials in major medical journals (Pocock et al. 2007, 2008). The current manuscript extends this work in considering the use of figures in observational studies in epidemiology. We reviewed epidemiologic studies published in 10 general medical and epidemiology journals from June to August 2008, with the goal of highlighting the types of figures in use, and making recommendations for improvements in practice.

4.2 Materials and Methods

The focus of this survey was analytical epidemiology, that is epidemiology relating health outcomes to exposures in individuals. The authors identified all 181 articles that could be termed analytical epidemiology, published in June to August 2008 in 10 journals. Five specialised epidemiological journals were chosen: American Journal of Epidemiology, Annals of Epidemiology, Epidemiology, International Journal of Epidemiology and Journal of Clinical Epidemiology as well as five major medical journals: Annals of Internal Medicine, British Medical Journal (BMJ), Journal of the American Medical Association (JAMA), Lancet and New England Journal of Medicine.

For each article we noted the number and types of figures used, concentrating on figures presenting data. Photographs and diagrams lacking data were not included in the survey nor were figures from meta-analyses or randomised trials. The first author went through the chosen volumes and identified all the articles that could be classified as analytical epidemiology and if any doubts arose over the classification of an article the remaining author was consulted. Key points for each study were noted such as type of study design (mainly cohort, case–control or cross-sectional) and number and types of figures.

Each figure was then considered carefully by the current authors to assess its content and appropriateness of appearance. To aide these considerations a list of desirable features was drawn up, partly based on the recommendations for the use of figures in clinical trials (Pocock et al. 2007, 2008). This list was concerned with general aspects of figures (e.g., including measures of uncertainty) but also with specific types of figures often included in epidemiological articles (e.g., Kaplan–

Meier plots). Having gone through the sample of articles the classification of figures was simplified and common issues regarding each type of figure were identified using the list of desirable features leading to the recommendations at the end of this paper. Some figures were chosen as examples of good practice to illustrate the points the authors wish to make about the use of figures.

In this survey we have not dealt with genetic epidemiology. This field has its own set of figures, which would also merit investigation.

4.3 Results

4.3.1 Overall Survey Findings

Table 4.1 shows the main findings from our survey. In this 3-month period there were 181 analytical epidemiology papers identified, of which 62% appeared in the epidemiology journals and 38% in the general medical journals. The American Journal of Epidemiology had the largest number of such articles (53). Overall there were 97 articles (54%) with at least one figure. The mean number of figures per article was 0.98. Most articles had zero or one figure. As regards study design, cohort studies were most common (54%) and 97% of articles described cohort, cross-sectional, case–control or nested case–control studies. Whether or not figures were used did not appear to vary by study type. Twenty-five of the articles describe studies conducted in a clinical setting, such as hospitalised patients followed for recurrence of the disease of interest. The use of figures was more common in studies of this type: figures were used in 72% of these studies compared to 51% of the articles describing research done in a non-clinical setting ($P = 0.047$).

The most common types of figures were plots of estimates (39 articles) of outcome/risk factor association, which includes forest plots; Kaplan–Meier plots (23 articles) showing failure time outcomes; flow diagrams (20 articles) describing the flow of study subjects through the study; smooth or model based curves (20 articles) displaying the results of a statistical model fitted to the data; and distributional plots (14 articles) describing the study population. These five groups of plots accounted for 89% of the figures in the survey and they will be the ones we will focus on.

In addition, 6 articles had figures showing population incidence or mortality data, mainly as age-standardised rates. Four articles had individual data. This is not so common in analytical epidemiology as the studies often include a large number of subjects and plots showing individual data become very dense. Plots of repeated measures over time were only found in three articles, whereas this type of figure was one of the most common for clinical trials (10) where it is common to measure the treatment effect at certain fixed time points. This seldom occurs in analytical epidemiology.

Table 4.1 Characteristics of the Survey in 181 Articles

Journal	Number of articles (%)	
Epidemiological		
American Journal of Epidemiology	53 (29)	
Annals of Epidemiology	22 (12)	
Epidemiology	18 (10)	
International Journal of Epidemiology	17 (9)	
Journal of Clinical Epidemiology	3 (2)	
General medical		
British Medical Journal	23 (13)	
Journal of the American Medical Association	22 (12)	
Lancet	10 (6)	
Annals of Internal Medicine	9 (5)	
New England Journal of Medicine	4 (2)	
Number of figures in each article	Number of articles (%)	
None	84 (46)	
One	50 (28)	
Two	24 (13)	
Three	14 (8)	
Four	8 (4)	
Five	1 (1)	
Type of study design	Number of articles (%)	At least one figure
Cohort	105 (58)	61 (58)
Cross-sectional	43 (24)	19 (44)
Case–control	27 (15)	12 (44)
Case-crossover	3 (12)	3 (100)
Case–cohort	2 (1)	1 (50)
Twin study	1 (1)	1 (100)
Type of figure[a]	Number of articles (%)	
Plot of estimates	39 (30)	
Kaplan–Meier plot	23 (18)	
Flow diagram	20 (15)	
Smooth or model based curve	20 (15)	
Distributional plot	14 (11)	
Incidence/mortality	6 (5)	
Individual data	4 (3)	
Repeated measures	3 (2)	
3D plot	1 (1)	

[a]Each type of figure is only counted once per article

Table 4.1 indicates that most figures published in epidemiological articles show associations between outcomes and risk factors. Apart from flow diagrams, purely descriptive figures are not that common. The style and content of the 5 most commonly used types of figures (from our survey) are now discussed with examples.

4.3.2 Plot of Estimates

The most common type of figure was a plot of point estimates of outcome/risk factor associations. These are typically from analyses of binary or time-to-event type outcomes showing odds ratios, relative risk or hazard-ratio estimates. Though they do not always do so, it is desirable that such figures include confidence intervals to display the statistical uncertainty in each point estimate (Frikke-Schmidt et al. 2008; Snape et al. 2008) and hence to avoid overinterpretation of the data. In some articles columns have been used to show the estimates, but since it is really each point estimate that needs to be shown a single symbol is preferred.

A good example of a plot of estimates with confidence intervals is seen Fig. 4.1 in (Frikke-Schmidt et al. 2008) which is a plot of hazard-ratios for ischemic heart disease as a function of high-density lipoprotein. The figure is clear, the axes are well chosen with good labeling and a caption that explains exactly what has been plotted. Another useful feature is that the number of events and total number in each group have been included in the figure making it easier for the reader to draw conclusions from the results: this was generally lacking in other articles. In general the combination of tabular and graphical presentation can add much-needed context to figures and is recommended. In this figure the hazard ratios are plotted on a log-scale allowing details to be seen more clearly and giving symmetric confidence intervals.

Forest plots are included in this type of figure (8 of the 39 plots of estimates). What makes a forest plot slightly different from other figures containing point estimates with confidence intervals is that the forest plot usually shows estimates across different sub-groups or different studies, and sometimes also an overall estimated effect of the exposure of interest. An example of a forest plot is seen here in Fig. 4.1 (Zhang et al. 2008). This figure contains results from a pooled analysis of hair-dye use and non-Hodgkin lymphoma. It is helpful that the size of the square plotted for each estimate is proportional to the number of cases. Also it helps that the odds-ratios are plotted on a log-scale, so that the distance from 0.5 to 1 is the same as from 1 to 2 and each confidence interval is symmetric about the point estimate. It would, however, have been desirable if the numbers of cases and controls that led to each estimate had also been included in the figure. In general, the combination of graphical and tabular data can make a figure much more informative. The overall estimate has been plotted using a different symbol making it easy to distinguish from the individual estimates. It is not clear from the figure or the caption, however, how the overall estimate was calculated. In general it is recommended that the figure and caption stand on their own, providing context and support for the estimates displayed.

4.3.3 Kaplan–Meier Plot

Kaplan–Meier plots are used to show time-to-event data by groups of interest (e.g., with or without the exposure being studied). The event of interest can be death, but

All non-Hodgkin lymphoma

Study	OR (95% CI)
Total	
EpiLymph	1.2 (1.0, 1.6)
NCI	1.2 (0.9, 1.6)
UCSF	1.0 (0.8, 1.3)
Yale	1.2 (0.9, 1.5)
Pooled	1.2 (1.0, 1.3)
Use starting before 1980	
EpiLymph	1.5 (1.1, 1.9)
NCI	1.2 (0.9, 1.7)
UCSF	1.0 (0.7, 1.3)
Yale	1.3 (1.0, 1.7)
Pooled	1.3 (1.1, 1.5)
Use starting in 1980 or later	
EpiLymph	1.2 (0.9, 1.5)
NCI	1.2 (0.8, 1.7)
UCSF	1.0 (0.7, 1.5)
Yale	1.0 (0.7, 1.2)
Pooled	1.1 (0.9, 1.3)

Odds ratio

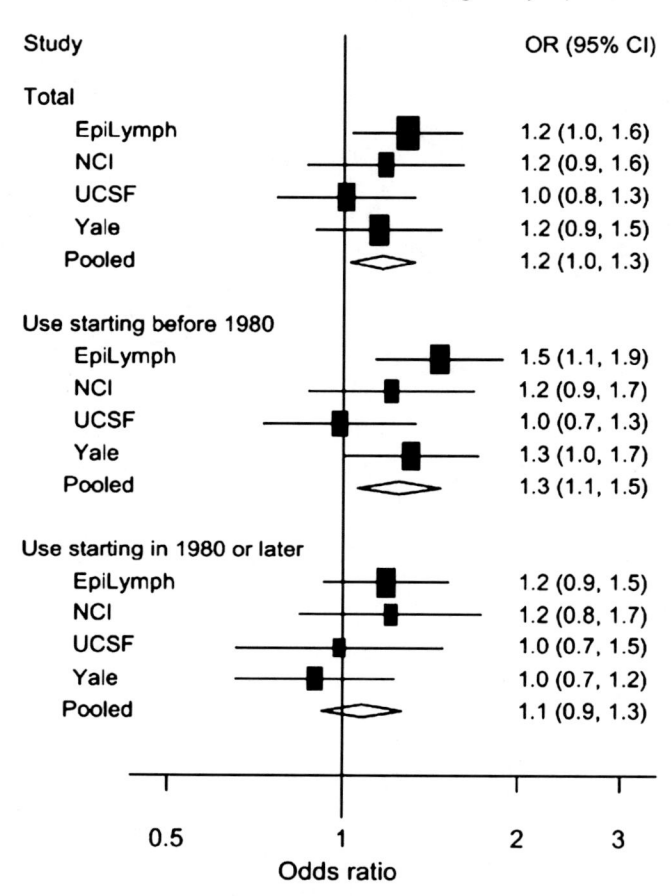

Fig. 4.1 Odds ratios (ORs) for non-Hodgkin lymphoma (NHL) among women, by study center, in a pooled analysis of hair-dye use and NHL, 1988–2003. *Boxes* show results from individual studies; *diamonds* indicate pooled data. Bars, 95% confidence interval (CI) (Zhang et al. 2008)

often it is disease incidence and sometimes a positive event such as recovery. Kaplan–Meier type plots can show either the probability of being event-free over time (the curves will go down) or the cumulative incidence (and the curves will go up). The plots of cumulative incidence are sometimes referred to as Nelson–Aalen plots.

A good example of a Kaplan–Meier type plot is found in the figure in (Wood et al. 2008) where the cumulative incidence of mortality has been plotted for injection drug users and non-injection drug users. The two groups are well identified and the axes are clearly labeled with the same axes used for the plots of all-cause mortality and non-accidental mortality allowing visual comparisons to be made (here only the all-cause mortality is shown). Under the horizontal axis the numbers at risk in the two groups are shown at appropriate intervals, another useful combination of graphical and tabular display. The uncertainty of the plots is shown by including confidence intervals at regular intervals (this is rarely done, but as above, acknowledging uncertainty is highly recommended), though it would have been clearer had the two groups' confidence intervals been slightly staggered to avoid confusing overlap. Also an overall significance test comparing the two groups is included on the figure. If neither the confidence intervals nor a significance test was included in the Kaplan–Meier plot, it would be easy to over interpret any apparent differences in outcome between the exposure groups.

Sometimes the Kaplan–Meier plot is arguably extended for too long when there are very few subjects still at risk and the estimates become very uncertain (e.g., in Limaye et al. 2008). The Kaplan–Meier plot in (Williams et al. 2008) shows survival and the vertical axis is cut off at 0.5, which could make the differences between the groups seem deceptively larger than they really are. It may be better to plot the cumulative incidence (i.e., plot going up) in this type of situation when only a small part of the scale from 1 to 0 is used.

4.3.4 Flow Diagram

Flow diagrams are used to display the flow of the study subjects through a study. It is an essential part of study reporting for clinical trials (Begg et al. 1996) and can also be of value in many epidemiological studies (Vandenbroucke et al. 2007). In the article (Ix et al. 2008) a flow diagram is used to describe a case–cohort study of incident diabetes mellitus in older persons. The study uses a case–cohort design and the flow chart is effective in making the design clearer for the reader. The sub-cohort is clearly identified and the numbers of exclusions and numbers of incident cases of diabetes during follow-up are documented. In general a flow diagram can be very helpful in understanding the structure of an epidemiological study in both, describing the study subjects (e.g., responders and non-responders to questionnaires in a cross-sectional study), and in explaining more clearly any nuances of study design.

4.3.5 Smooth or Model Based Curve

A potentially useful way of summarising the results of a study is to show a fitted curve from the statistical model. This can be an eye-catching illustration, but there can be problems for the reader. There is often no information about the model's goodness of fit and such figures seldom include any numbers, so it can be difficult to know how much to trust the results. Again, inclusion of parsimonious numerical information perhaps arranged in a table, can help matters here. It can also happen that the plotted curve covers a spread of the exposure variable's distribution where

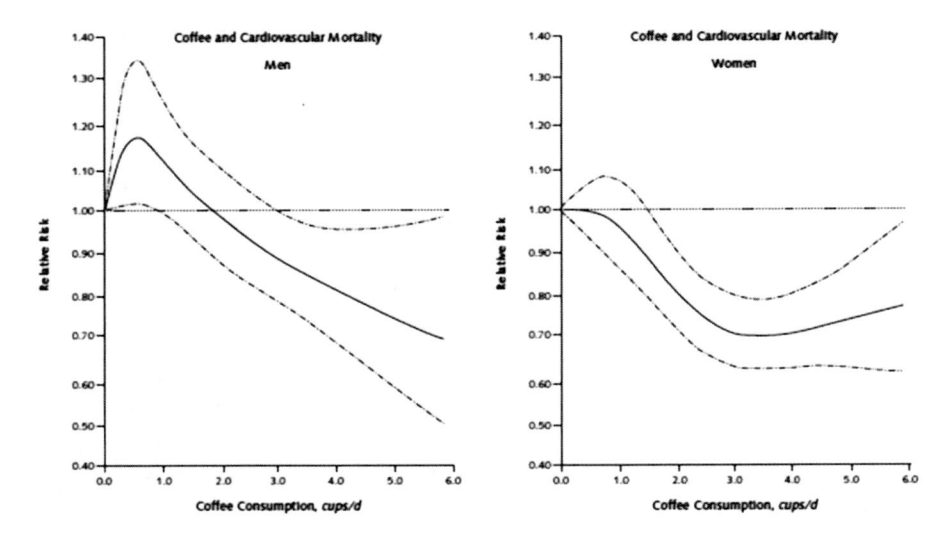

Fig. 4.2 Non-linear relationship between coffee consumption and cardiovascular mortality (Lopez-Garcia et al. 2008)

there are few (or even no) data points, thus encouraging unjustified extrapolation. As noted previously, acknowledging the uncertainty in the data, either through the use of confidence regions or by displaying the raw data itself in addition to a smooth curve is critical. Despite these caveats, plotting smooth curves can be a creative way to explore variation in the conclusions of an analysis under different conditions. For instance, a figure showing a combination of a fitted curve from a statistical model with simple point estimates and confidence intervals (e.g., treating an exposure as continuous but also showing results from a categorical analysis) could add much value to both types of analyses.

Figure 4.2 (Lopez-Garcia et al. 2008) shows relative mortality risk by amount of coffee consumption compared to persons drinking no coffee. In this figure 95% confidence interval curves have been included to convey the extent of uncertainty in the estimates. From other results in the article it is seen there are very few individuals drinking more than four cups of coffee per day, so the figure's extension out to six cups per day seems unwarranted. It is hard to know what to believe from such figures (e.g., how strong is the evidence of non-linearity in this case?), but as a supplement to the tables where the data are described and analysed in more detail (as in the tables in Lopez-Garcia et al. 2008) they can be useful illustrations.

4.3.6 Distributional Plot

In this survey the authors found few purely descriptive figures (e.g., histograms, bar charts or box plots). This is probably due to the fact that in an article for a journal

Fig. 4.3 Distribution of interpregnancy intervals in the study population ($n = 533$), Collaborative Perinatal Project, United States, 1959–1965 (Mikolajczyk et al. 2008)

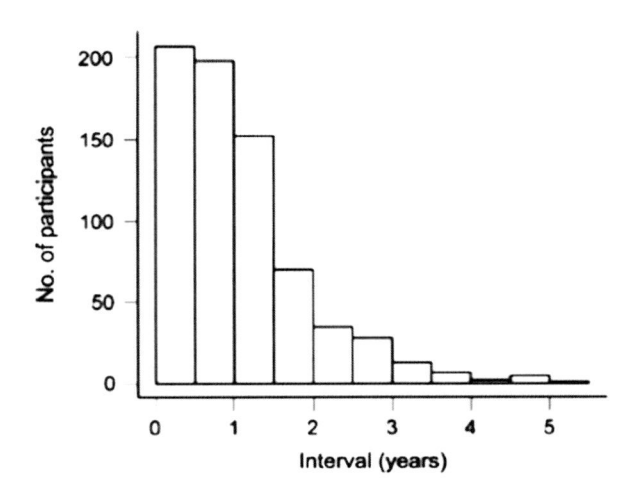

the number of tables and figures allowed can be limited, so often a figure showing the main results will be preferred. Nevertheless, some articles included descriptive plots to good effect.

Figure 4.3 (Mikolajczyk et al. 2008) shows the distribution of interpregnancy intervals in the study population. This figure conveys the point clearly that the interpregnancy intervals were usually relatively short and the distribution is skewed to the right. The number of women in the study is usefully included in the caption.

4.4 Discussion

A good figure can be very effective in the presentation of results and is more likely to be noticed and remembered by the reader than a dense table full of results. As such it is therefore critical that the impressions conveyed in the graphical format are accurate and properly acknowledge the limits of the data in question. Analytical epidemiology articles can study several exposure/outcome associations, so it is up to the authors of an article to choose wisely which results should be presented as figures. In a study with many results choosing some key findings to present in a figure can help focus the reader's attention on the main points the authors wish to make.

In this survey we were surprised to find that 46% articles did not include a figure even though it would often have been useful to do so. Given the speed of modern computing and availability of software, choosing not to present a graphical display may undersell a result, where for relatively little cost, a greater emphasis on the main finding could be obtained. For instance, a well-written article (Villamor et al. 2008) about the risk of oral clefts contains no figures. The results presented in the article (their tables 2–4) are odds ratios of cleft palate or cleft lip during second pregnancy by change in mother's BMI since first pregnancy or by months since first pregnancy. We think the results would have been more readily appreciated by the reader if the odds ratios (with 95% confidence intervals) from the primary analyses were presented in a figure similar in style to the plot of estimates in (Frikke-Schmidt et al. 2008).

As in a prior survey of figures in the reporting of clinical trials (Pocock et al. 2007), the images examined here were of many different types and styles, but the vast majority could be classified into a small number of groups: These included plots of estimates, Kaplan–Meier plots, flow diagrams, smooth or model based curves and distributional plots. How best to present a figure is of course partly a matter of personal taste, but having done this survey we would like to make some general and some more specific points for the construction of figures in the future. Some of these recommendations seem like plain common sense, but it can still be useful to state them clearly.

4.5 Recommendations

Much has been written about presenting figures in general (Cleveland 1994; Robbins 2004; Tufte 2001; Wilkinson 2005) and such advice is also valid for analytical epidemiology. There are, however, a few points that could be emphasised in this particular setting:

- It is important that each figure can largely stand alone. That is, key information needed to understand a figure should be included either on the figure itself, in the caption or as a footnote
- Each exposure group should be clearly labeled within the figure itself, so groups are easily distinguishable. It is also important to choose any colours and symbols carefully as many readers will be seeing the article in black and white
- Figures should include appropriate measures of uncertainty such as confidence intervals or standard errors. It may also be useful to include appropriate P-values on the figure to help the reader understand the extent to which any association could plausibly be due to chance

4.5.1 Plots of Estimates

- Plots of estimates are best presented as points with 95% confidence intervals rather than as bar charts/columns
- The scale used for the plot should be chosen to give enough detail. Consider using a log-scale especially for hazard-ratios, odds-ratios and risk-ratios, so that confidence intervals are symmetric around the point estimate
- Any plot of estimates should also include some tabulations to give a better understanding of the data, e.g., number of events and number of subjects in each group. Far too many plots fail to provide such simple information

4.5.2 Kaplan–Meier Plots

- Kaplan–Meier plots should include the numbers at risk at regular time points under the horizontal axis
- When possible plots should include confidence limits at regular time points to give an impression of the uncertainty of the curves
- The plots should not extend too far over time where there are few subjects left at risk and estimates become unreliable
- If incidence rates are not high, a plot going up of the cumulative incidence is preferable as this allows differences between exposure groups to be seen more clearly

4.5.3 Flow Diagram

- A flow diagram should include the number of subjects and their flow through the study
- A flow diagram should help communicate the nature of the study design, e.g., number of cases and controls in a case–control study
- It is important to include the numbers excluded from the analysis in each group and the reason for exclusion (e.g., ineligible prevalent cases in a study of incidence). If a questionnaire is used also include the number of non-responders in each group

4.5.4 Smooth or Model Based Curves

- The caption or footnote should include key information about which model has been used to create the figure, and also about the goodness of fit of this model
- Include confidence intervals to show the variability of the estimated curves
- It is useful for the reader to know how the exposure data behind the curve are distributed, this could be indicated under the horizontal axis
- If possible combine a figure of a fitted curve from a statistical model with simple point estimates and confidence intervals (e.g., treating an exposure as continuous but also showing results from a categorical analysis) as this would add much value to both types of analyses

4.5.5 Distributional Plots

Much is well known about how to present such descriptive plots (Robbins 2004; Tufte 2001). However, there are a couple of specific points arising from our survey.

- Bar charts are sometimes used to show summary statistics, e.g., means or percentages, but it may be better to have such background data in a table
- When using box plots it should be stated in the caption or footnote what the end of the whiskers represent. Having too many individual points outside the whiskers can be distracting

4.6 Conclusion

We would like to encourage a wider use of insightfully informative figures in articles on analytical epidemiology. We hope that our survey of current practice and consequent recommendations prove useful for the construction of such figures in future articles.

References

Bax L, Ikeda N, Fukui N et al (2009) More than numbers: the power of graphs in meta-analysis. Am J Epidemiol 169:249–255
Begg C, Cho M, Eastwood S et al (1996) Improving the quality of reporting of randomized controlled trials. The CONSORT statement. JAMA 276:637–639
Cleveland WS (1994) The elements of graphing data. Hobart Press, Summit, NJ
Frikke-Schmidt R, Nordestgaard BG, Stene MCA et al (2008) Association of loss-of-function mutations in the ABCA1 gene with high-density lipoprotein cholesterol levels and risk of ischemic heart disease. JAMA 299:2524–2532
Ix JH, Wassel CL, Kanaya AM et al (2008) Fetuin-A and incident diabetes mellitus in older persons. JAMA 300:182–188
Limaye A, Kirby KA, Rubenfeld GD et al (2008) Cytomegalovirus reactivation in critically ill immunocompetent patients. JAMA 300:413–422
Lopez-Garcia E, Van Dam RM, Li TY et al (2008) The relationship of coffee consumption with mortality. Ann Intern Med 148:904–914
Mikolajczyk RT, Zhang J, Ford J et al (2008) Effects of interpregnancy interval on blood pressure in consecutive pregnancies. Am J Epidemiol 168:422–426
Pocock SJ, Clayton TC, Altman DG (2002) Survival plots of time-to-event outcomes in clinical trials: good practice and pitfalls. Lancet 359:1686–1689
Pocock SJ, Travison TG, Wruck LM (2007) Figures in clinical trial reports: current practice and scope for improvement. Trials 8:36
Pocock SJ, Travison TG, Wruck LM (2008) How to interpret figures in reports of clinical trials. BMJ 336:1166–1169
Robbins NB (2004) Creating more effective graphs. Wiley-Interscience, Hoboken, NJ
Snape MD, Kelly DF, Lewis S et al (2008) Seroprotection against serogroup C meningococcal disease in adolescents in the United Kingdom: observational study. BMJ 336:1487–1491
Tufte ER (2001) The visual display of quantitative information. Graphics Press, Cheshire, CT
Vandenbroucke JP, von Elm E, Altman DG et al (2007) Strengthening the Reporting of Observational Studies in Epidemiology (STROBE): explanation and elaboration. Epidemiology 18:805–835
Villamor E, Sparen P, Cnattingius S (2008) Risk of oral clefts in relation to prepregnancy weight change and interpregnancy interval. Am J Epidemiol 167:1305–1311

von Elm E, Altman DG, Egger M et al (2007) The Strengthening the Reporting of Observational Studies in Epidemiology (STROBE) statement: guidelines for reporting observational studies. Lancet 370:1453–1457

Wilkinson L (2005) The grammar of graphics. Springer, New York

Williams P, Van Dyke R, Eagle M et al (2008) Association of site-specific and participant-specific factors with retention of children in a long-term pediatric HIV cohort study. Am J Epidemiol 167:1375–1386

Wood E, Hogg RS, Lima VD et al (2008) Highly active antiretroviral therapy and survival in HIV-infected injection drug users. JAMA 300:550–554

Zhang Y, Sanjose SD, Bracci PM et al (2008) Personal use of hair dye and the risk of certain subtypes of non-hodgkin lymphoma. Am J Epidemiol 167:1321–1331

Part II
Preclinical and Early Clinical Development

Chapter 5
Use of Graphics for Studies with Small Sample Sizes: A Simulated Case Study of an Early-Phase Asset for Treatment of Type 2 Diabetes Mellitus

Denise Shortino, Ann Walker, and Andrew Miskell

Abstract Graphics play an important role in understanding clinical data and the metabolic area of drug research is no exception. Challenges of early-phase clinical research include unpowered studies with small sample sizes (approximately 8–12 subjects per group), relatively short study durations, and the use of surrogate clinical endpoints or biomarkers to investigate pharmacological effects. Careful examination of individual patient data along with summary data is vital to interpreting the data. Graphical displays are the most effective means of accomplishing this goal enabling one to quickly assess effects between and within both subjects and treatments.

5.1 Introduction

Discovery of more effective treatments for Type 2 Diabetes Mellitus (T2DM) is a major focus of research in the metabolic disease area. Although currently there are a number of treatment options, there is still a need for drugs with better risk/benefit profiles. T2DM is a progressive metabolic disorder that is a growing healthcare issue worldwide and it is estimated that globally there are approximately 246 million people (about 6% of the adult population) with diabetes. By the year 2025, this number is expected to expand to about 380 million (Diabetes Atlas 2006). People with diabetes have an increased risk of suffering serious complications, including heart attack, stroke, kidney failure, blindness, and lower limb disorders which can lead to foot amputations.

The American Diabetes Association (ADA) defines diabetes as:

- HbA1c greater than or equal to 6.5% or
- Fasting plasma glucose greater than or equal to 126 mg/dL (7.0 mmol/L) or

D. Shortino (✉) • A. Walker • A. Miskell
GlaxoSmithKline, Clinical Statistics, 5 Moore Drive, Research Triangle Park,
Durham, NC 27709, USA
e-mail: Denise.D.Shortino@gsk.com

- Random (non-fasting) blood glucose level—diabetes is suspected if higher than 200 mg/dL and accompanied by the classic symptoms of hyperglycemia or
- Oral glucose test—diabetes is diagnosed if glucose level is higher than 200 mg/dL after 2 h

The goal of treatment is to improve long-term glycemic control. The current recommendations by the American Diabetes Association (ADA) of the target glycemic goals for diabetes control are HbA1c <7% or fasting/pre-prandial glucose of 80–120 mg/dL (American Diabetes Association, 2011). For regulatory filings, glycemic control is based on changes in hemoglobin A1c (HbA1c). This is a surrogate endpoint to reducing long-term microvascular complications.

A challenge faced in early phase clinical development is identifying assets that do not warrant continued investment based on the lack of potential to provide a better safety and efficacy profile than currently marketed products. Glucose is a key endpoint used in early-phase clinical development to assess potential efficacy. Not only do changes in glucose predict changes in HbA1c, the registerable endpoint, but these effects tend to occur in glucose before HbA1c (Nathan et al. 2008). In early phase studies, different aspects of glucose response are investigated: fasting, post-prandial, and the weighted mean over a 24 h period. A reduction in mean glucose of approximately 20 mg/dL predicts a reduction of 0.7% HbA1c at 12 weeks (Nathan et al. 2008).

Clinical pharmacology studies are relatively short-duration studies where the primary goal is to assess the safety and tolerability of the investigational drug, characterize the pharmacokinetic profile, and estimate pharmacodynamic effects. In addition to glucose, the assessment of metabolic hormones, such as insulin, gastric inhibitory polypeptide (GIP), glucagon-like peptide (GLP-1), glucagon, and peptide YY (PYY) plays an important role in evaluating the mechanism of action identified in preclinical models.

The examples included in this chapter are based on simulated data to represent a 14-day repeat-dose parallel group study with three active treatment arms (15 mg, 30 mg, and 45 mg) and a placebo arm, where 24 h profiles of metabolic hormones were collected on Days -1 and 14. The graphs are generated from SAS™, S-PLUS™, and Spotfire™. In our department, statistical analysis software such as SAS™ and S-PLUS™ are typically used for production of displays included in regulatory submissions because the software can be executed in an internal reporting system which meets the regulatory requirements of a closed-validated system. Other software, such as Spotfire, is sometimes used for exploratory purposes due to its interactive capabilities.

5.2 Individual Glucose and Insulin Concentration Profiles Versus Time

The relationship between glucose and insulin is important to understand. The purpose of glucose is to provide cells with energy. The uptake of glucose by the cells is

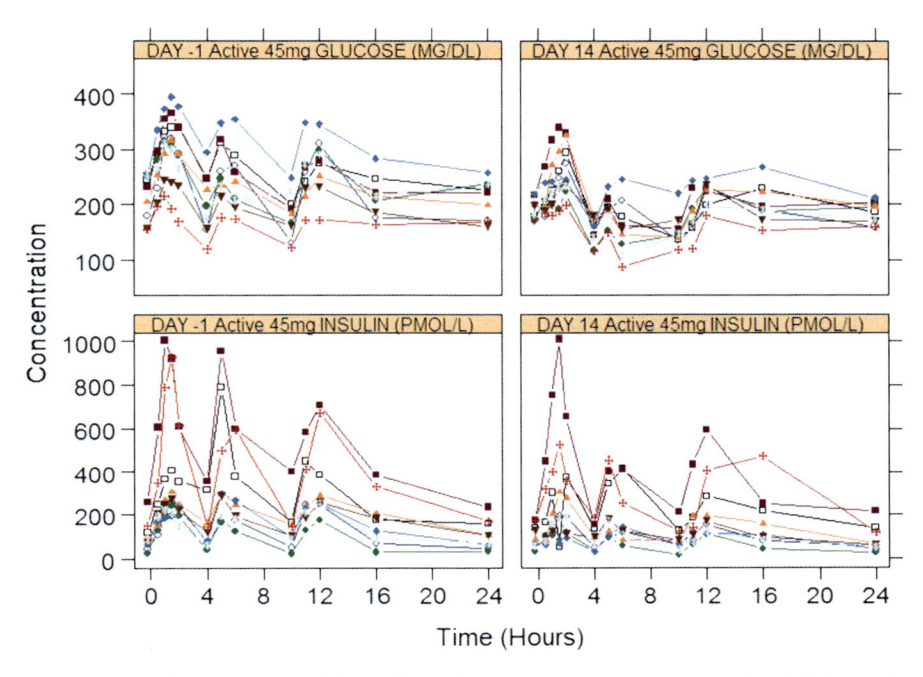

5 Use of Graphics for Studies with Small Sample Sizes...

Fig. 5.1 *Individual glucose (mg/dL) and insulin (mmol/L) concentrations vs. time (h) for active 45 mg dose*. The *top-left* plot is the individual glucose profiles on Day -1 before 45 mg dose of active treatment was administered. The three peaks in glucose correspond to the meals provided at hours: 0, 4, and 10. *Top-right* is the corresponding individual glucose profiles after 14 days of treatment. *Bottom left* is the individual insulin profiles on Day -1 (*baseline*) and *bottom right* is the individual insulin profiles after 14 days of treatment

regulated by insulin, a hormone produced by the pancreas. Diabetic patients have elevated glucose levels compared to non-diabetics for one of two reasons (1) insulin resistance (insulin is being released but the cells are not responding to it) or, (2) not enough insulin is being produced (Gerich 1988).

Relative changes in glucose and insulin can help elucidate the mechanism of action of a drug. Figure 5.1 shows individual concentration profiles of plasma glucose and insulin. These individual profiles allow one to examine the following:

- Relationship at the subject-level between glucose and insulin at baseline and whether this relationship is altered after 14 days of treatment
- Overall response within a treatment group after 14 days of treatment
- Effect of 14 days of treatment on glucose and insulin separately
- Variability between individuals
- Implausible values to be queried

Comparing the plots of the individual glucose profiles before and after 14 days of treatment, the 45 mg dose of the active treatment appears to have some clinical activity as the glucose is lower for most subjects. One can also see less variability between

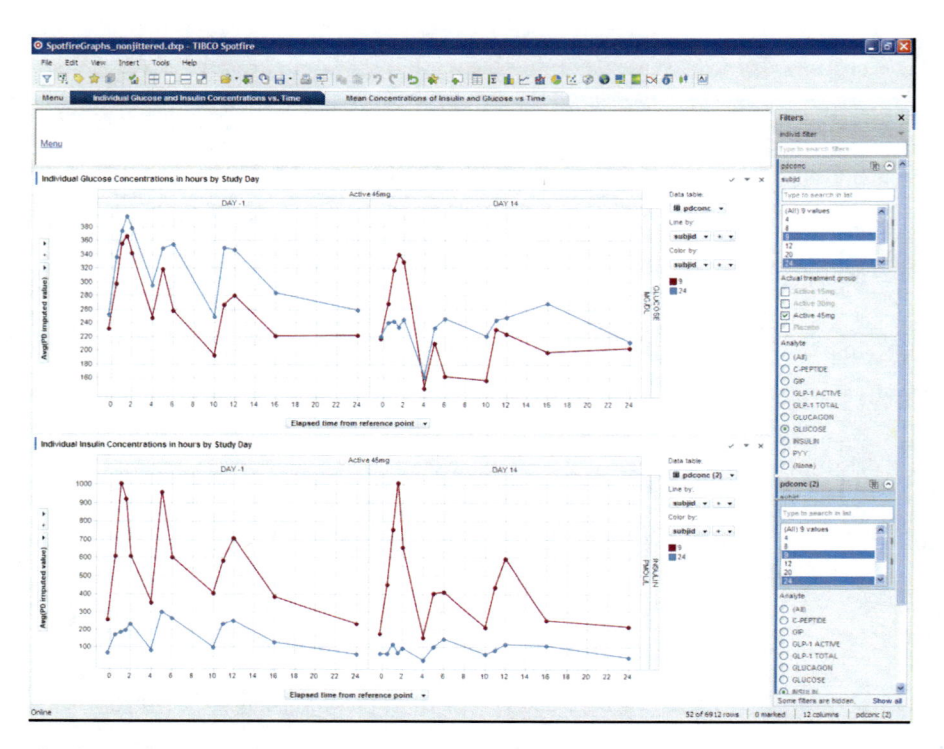

Fig. 5.2 Similar graph to Fig. 5.1 produced using Spotfire

subjects on Day 14 compared to Day -1. The post-prandial peak glucose is blunted on Day 14 relative to Day -1. The insulin profiles are consistent with the glucose profiles with peaks relative to the meals and large between subject variability.

Interactive software is useful in exploring the data in more detail. Figure 5.2 displays the same information as Fig. 5.1 but sub-setting it for subjects 9 and 24. It was created using Spotfire instead of S-PLUS™. The filter panel on the right side of the display enables the user to filter on a subject number, change the treatment group or the specific pharmacodynamic parameter displayed, thus being more interactive. The filter panel can be customized to include any other variables within the dataset to subset the data.

5.3 Mean (+SE) Glucose and Insulin Concentration Versus Time

Figure 5.3, Showing the mean glucose and insulin profiles over a 24 h time period, enables one to examine:

- Whether treatment groups are balanced at baseline (Day -1)
- Whether there is a treatment effect on glucose or insulin at any dose level relative to baseline and/or placebo

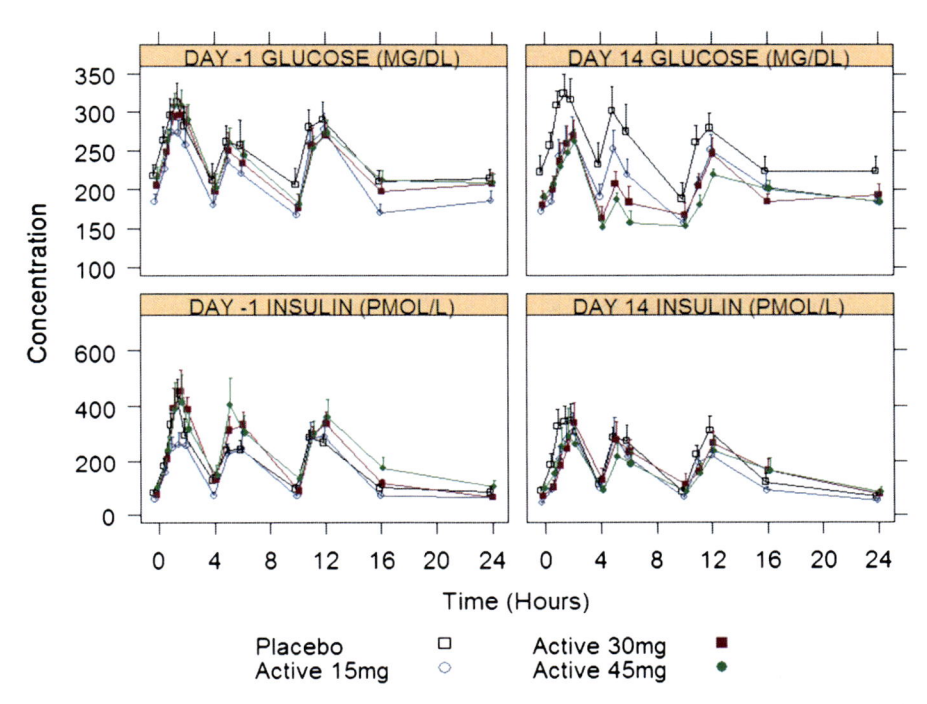

5 Use of Graphics for Studies with Small Sample Sizes...

Fig. 5.3 *Mean (+SE) glucose (mg/dL) and insulin (mmol/L) concentrations vs. time (h).* The top-left plot is the mean glucose profile for each treatment group at Day -1. The three peaks correspond to the post-prandial response after meals. The *top-right* plot is the corresponding mean glucose profile after 14 days of treatment. *Bottom left* is the mean insulin profile on Day -1 (baseline) and *bottom right* is the mean insulin profile after 14 days of treatment

- Within each treatment group, whether the relationship between glucose and insulin changes with 14 days of treatment
- The dose–response relationship

The mean glucose and insulin profiles at Day-1 are relatively similar between treatment groups. After 14 days of treatment there is a dose-response relationship seen in the mean glucose profiles that is less apparent in the insulin profiles.

5.4 Individual Glucose Concentration Profiles Versus Time

Figure 5.4, showing the individual glucose concentration profiles at Day -1 and Day 14, enables one to examine within a treatment group the:

- Magnitude of response for each individual after 14 days of treatment (within a panel)
- Consistency of response across individuals (e.g., is there a response during each post-prandial period)

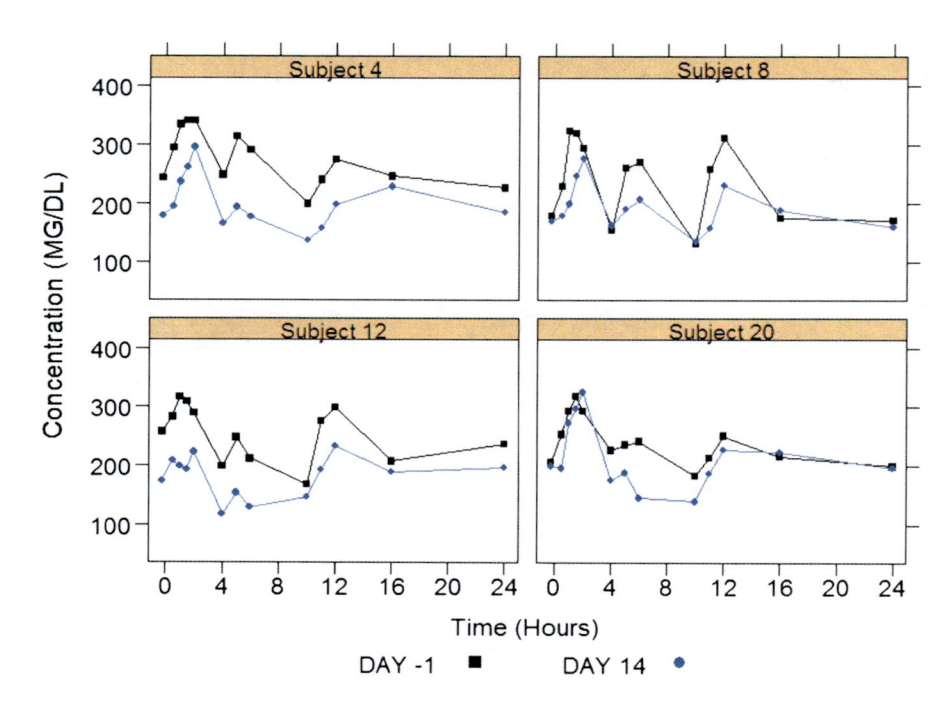

Fig. 5.4 *Individual glucose concentrations (mg/dL) vs. time (h).* Each quadrant represents a different subject with the *black line* representing Day -1 glucose concentrations and the *blue line* representing Day 14 glucose concentrations

- Responders and non-responders
- Implausible values to be queried

This graph demonstrates the magnitude of individual responses to treatment and the consistency of response across subjects. The 4 subjects all received the 45 mg dose of active treatment. Subjects 4, 8, and 12 had a more consistent post-prandial response than Subject 20. In addition, a reduction in fasting glucose (0 h) is apparent in Subjects 4 and 12.

5.5 Mean (+SE) Glucose Concentrations Versus Time

Figure 5.5, showing the mean glucose concentrations over a 24 h time period, enables one to examine the:

- Magnitude of mean response for each treatment group after 14 days of treatment (within a panel)
- Differences in mean response across treatments
- Variability

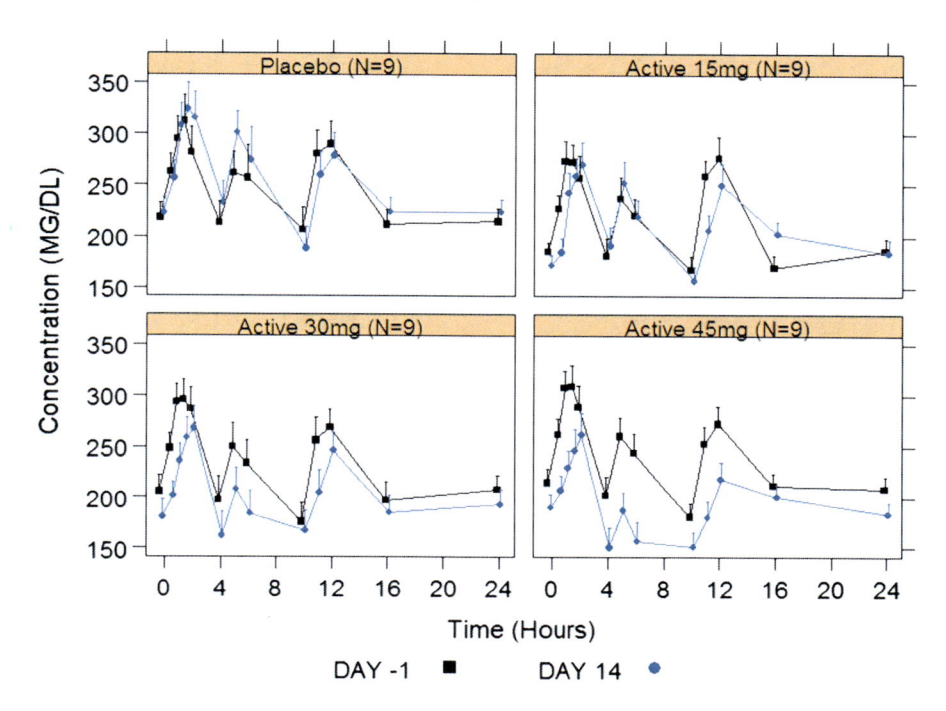

5 Use of Graphics for Studies with Small Sample Sizes...

Fig. 5.5 *Mean (+SE) glucose concentrations (mg/dL) vs. time (h)*. Each quadrant is a separate treatment group with the *black line* representing the Day -1 mean glucose profile and the *blue line* representing the Day 14 mean glucose profile

This graph is useful in comparing the drug effect within and between treatment groups. Figure 5.5 shows a greater magnitude of response in the highest dose group (45 mg). The 30 mg dose group is also demonstrating reductions in mean glucose relative to Day -1.

5.6 Comparative Plot of Individual Weighted Mean Glucose

Since changes in mean glucose predict changes in HbA1c (the registerable endpoint for T2DM), it is a key endpoint analyzed in short-duration studies (Nathan et al. 2008). Figure 5.6, a plot of individual values, allows one to examine the:

- Magnitude of individual subject response within and across treatment groups after 14 days of treatment
- The relationship between magnitude of response and baseline
- Whether there are responders and non-responders
- Variability

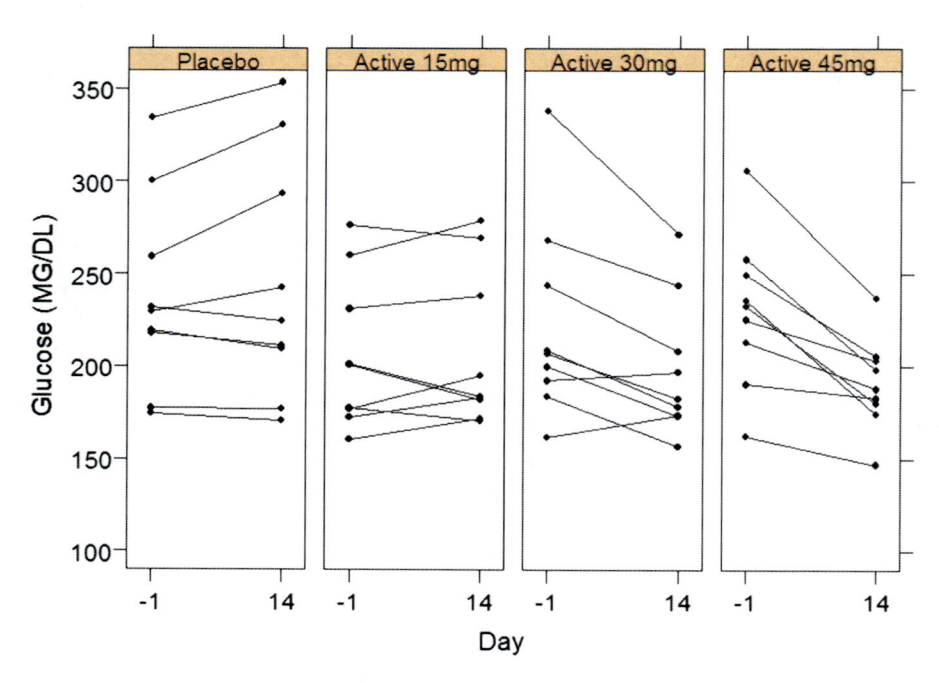

Fig. 5.6 *Comparative plot of individual weighted mean glucose (mg/dL).* Within each treatment group, individual mean glucose values for Day -1 and Day 14 are plotted for each subject with a line connecting the values

This graph demonstrates both a greater magnitude of response and a greater number of subjects with a reduction in glucose at the 2 highest doses of the active treatment compared to placebo and the low dose. All subjects in the highest dose group experienced a reduction in mean glucose, whereas, in all other treatment groups at least 20% of subjects had an increase.

5.7 Box-Plot of Weighted Mean Glucose with Individual Values Overlaid

Box-plots graphically provide descriptive statistics of numerical data (minimum, lower quartile, median, upper quartile, and maximum). These graphs are non-parametric in nature, as they show differences in the population without making any assumptions of the underlying distribution.

Figure 5.7 shows the summary information depicted in a box-plot with the individual values overlaid, which enables one to examine the:

- Distribution of weighted mean glucose at baseline and after 14 days of treatment
- Magnitude of median response for each treatment group after 14 days of treatment (within a panel)

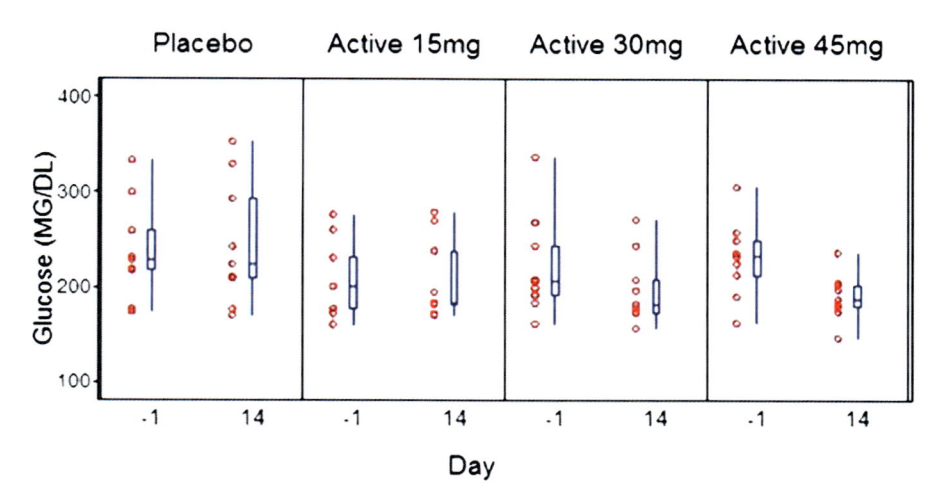

Fig. 5.7 *Box-plot of weighted mean glucose (mg/dL) with individual values overlaid.* Within each treatment group, individual weighted mean glucose values are plotted adjacent to the box-plot of those same values

Although this plot does not enable one to see individual changes, it does provide information on the distribution of individual values at baseline and 14 days after treatment and demonstrates the shift in central tendency from baseline to Day 14 within each of the treatment groups. As with the comparative plot above, the reduction in glucose in the two highest active dose groups is apparent.

5.8 Multi-Panel Plot of Change from Baseline Weighted Means for Selected PD Parameters Versus Glucose

Figure 5.8, a multi-panel plot of change from baseline weighted means for selected pharmacodynamic parameters versus glucose, enables one to examine:

- Whether there is a relationship between change in glucose and the change in the selected incretin hormones
- Whether the relationship differs across treatments

Vertical and horizontal reference lines at zero and −20 mg/dL (glucose only) allow one to focus on meaningful changes indicative of response.

Significant glucose reductions of at least 20 mg/dL, all in the two highest dose groups, tend to be associated with decreases in insulin indicating that the drug is not acting as an insulin secretagogue and may be taking pressure off of β-cells. There also appears to be a relationship between glucose reductions and increases in PYY which warrants further investigation.

There is no apparent relationship between changes in glucose and changes in GIP, Total GLP-1 or glucagon.

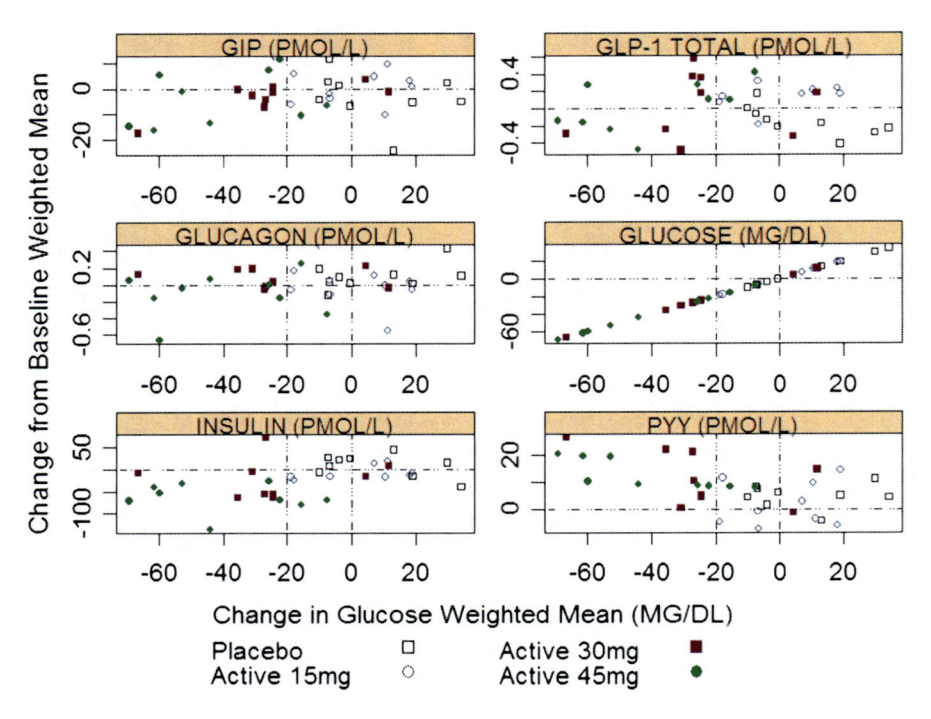

Fig. 5.8 *Multi-panel plot of change from baseline weighted means for selected PD parameters vs. glucose (mg/dL).* Scatter plot of individual change from baseline of selected incretins vs. glucose. Glucose responders are located to the left of the vertical reference line at −20 mg/dL

5.9 Forest Plot of Least Squares Means of Change from Baseline for Weighted Mean Glucose

Figure 5.9 provides a visual of the statistical analysis results (ANCOVA) by displaying the point estimate and 95% CI of the least squares means within each treatment group. It allows comparison of mean change from baseline across treatments. Reference lines are provided at 0 and at a target effect size of −20 mg/dL. Confidence intervals that include zero indicate that the change from baseline was not statistically significant. Significant reductions were observed in both the 30 and 45 mg Active dose groups with mean reductions from baseline greater than or equal to the target of 20 mg/dL. These dose levels warrant further study.

5.10 Conclusion

The graphs from this chapter display the variety of graphical techniques employed in the assessment of early–phase clinical assets. These small studies provide an initial indication of the developability of a new compound and the data can be used

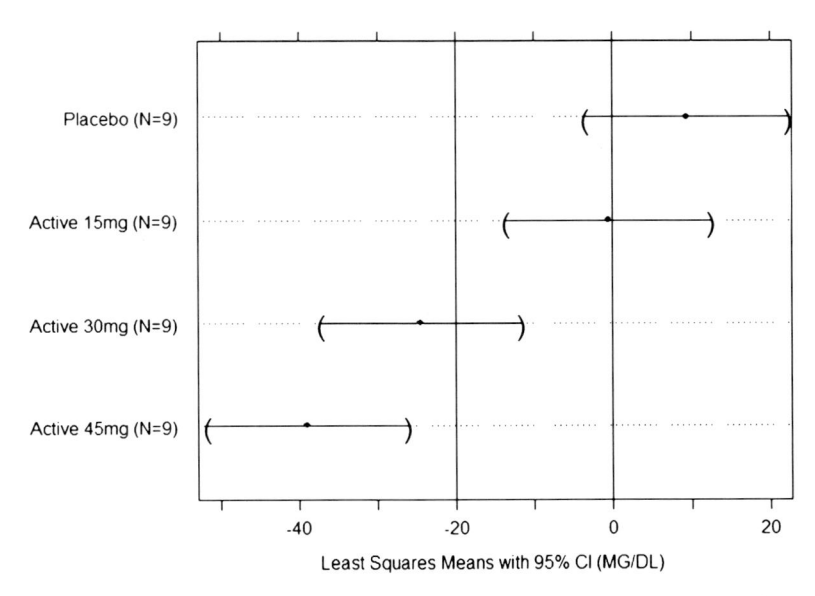

Fig. 5.9 *Forest plot of least squares means from ANCOVA of change from baseline weighted mean glucose (mg/dL)*

in the design of subsequent studies. Graphs are essential in evaluating the mean response, consistency of response, the influence of individual subjects, and the relationship between endpoints. The application of graphics is invaluable in the evaluation of the mechanism of action of a new drug, highlighting relationships that would otherwise be difficult to discern.

References

American Diabetes Association: Standards of Medical Care in Diabetes (2011) Diabetes Care 34(Suppl 1)
Diabetes Atlas (2006) International Diabetes Federation, 3rd Edition
Gerich J (1988) Role of insulin resistance in the pathogenesis of type 2 (non-insulin-dependent) diabetes mellitus. Baillières Clin Endocrinol Metab 2(2):307–326
Nathan D, Kuenen J, Borg R et al (2008) Translating the A1C assay into estimated average glucose value. Diabetes Care 31(8):1473–1478
SAS is a registered trademark for SAS Institute Inc. Statistical software, SAS-PC version 9.1.3
S-PLUS (version 7.0) and Spotfire (version 3.2) are statistical software from TIBCO

Chapter 6
Exploring Pharmacokinetic and Pharmacodynamic Data

Charles Roosen, Richard Pugh, and Andrew Nicholls

Abstract The recent trend towards Bayesian and adaptive study designs has led to a growth in the field of pharmacokintetics and pharmacodynamics (PK/PD). The mathematical models used for PK/PD analysis can be extremely computationally intensive and particularly sensitive to messy data and anomalous values. The techniques of paneling and creating polygon summaries help to bring clarity to potentially messy graphics. An understanding of the expected shape of the data, combined with the right choice of graphic can help identify unusual patterns of data.

6.1 Introduction

Pharmacokinetics and pharmacodynamics (PK/PD) describe the manner in which a compound enters and exits the body and the effect the compound has on the body using mathematical–statistical models. The often complex mathematical models used in PK/PD analysis provide quantitative descriptions of compounds, which are critical for dose finding and safety assessments during drug development. With fewer drugs making it to market and pipelines seemingly drying up, the focus has turned towards Bayesian and adaptive study designs and towards PK/PD analysis as a means of accelerating the early phases of clinical drug development.

C. Roosen, PhD
Mango Solutions AG,
Aeschenvorstadt 36, CH-4051, Basel, Switzerland
e-mail: Charlie.roosen@zurich.com

R. Pugh • A. Nicholls(✉)
Mango Solutions Ltd,
2 Methuen Park, Chippenham SN14 0GB, UK
e-mail: rpugh@mango-solutions.com

Specifically a typical pharmacokinetic (PK) analysis involves modeling the concentration of an administered drug over time using structural models (often compartmental models) and exploring covariate relationships using statistical techniques. Pharmacodynamic (PD) analysis attempts to describe the relationship between the concentration of a compound and some measurable clinical response that represents an improvement in patient well-being. Like PK data, PD data is often temporal with the pharmacodynamic effect measured at discrete time points.

The primary aim of the PK/PD exploratory data analysis (EDA) process is to establish the quality of the data and to confirm that the intended structural model for the system is appropriate. Here we describe some general principles for graphing longitudinal PK/PD data and present some custom graphs applied to PK/PD data. The intent is not only to present specific types of plots but also to demonstrate general principles of creativity for efficient EDA. The techniques are of particular interest for data involving multiple measurements made for each individual over time.

6.2 Datasets and Graphics

Graphics have been drawn using a combination of simulated datasets that have been created using the MSToolkit package (http://r-forge.r-project.org/projects/mstoolkit/) and a popular simulated dataset distributed with the Xpose package (Jonsson and Karlsson 1999). The MSToolkit package is a suite of R functions developed specifically in order to simulate trial data and manage the storage of the simulated output. The MSToolkit package has been selected due to the ease in which PK/PD data including covariate interactions can be generated. The Xpose package data has been selected due to a commonly occurring challenge with PK/PD data. It consists of measurements over time for numerous individuals (subjects), which can make trends difficult to spot due to the large amount of information available on the same data range.

All plots are created with either the lattice package (Deepayan Sarkar 2009) or with the ggplot2 package (Hadley Wickham 2009) in R (R Development Core Team 2008).

6.3 Repeated Measures Data

Clinical trials are typically planned such that for each subject the primary endpoint(s) of interest (and perhaps a number of other key endpoints) is collected at baseline and then again at key points throughout the study. A pharmacodynamic analysis seeks to understand the effect that dose/concentration has on the endpoint over time. A PK analysis is similarly concerned with repeated measures over time; however, the aim of the PK analysis is to understand the change in concentration over time. In either case, such repeated measures data are not independent; there is

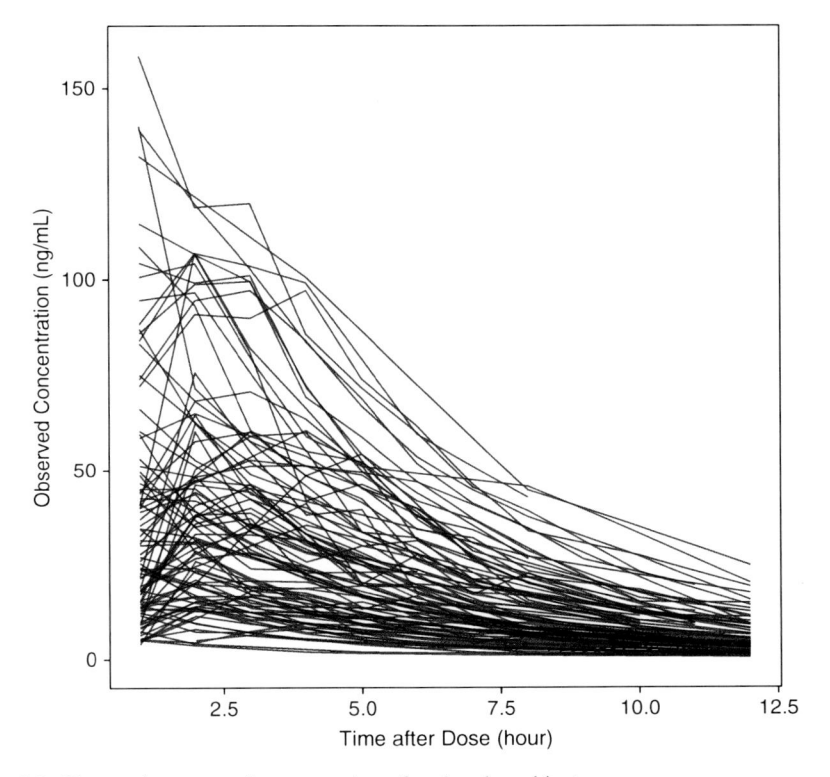

6 Exploring Pharmacokinetic and Pharmacodynamic Data

Fig. 6.1 Observed concentration versus time after dose by subject

correlation between the measurements of each individual, and the statistical techniques used must reflect this.

We focus first on PK data and start by plotting individual subject concentration-time profiles together in a single plot. Figure 6.1 shows the change over time in observed concentration for a number of simulated subjects. Usually this "spaghetti" plot serves only to confirm that our data is in range as any points not in range would skew the axes. For a large number of subjects the spaghetti plot can be difficult to read due to the number of intersecting lines on the plot and even if semi-transparent lines are used it is difficult to see how one subject compares with the rest of the population. As a solution to this we partition our plot into panels where each panel contains a single subject's time profile. In Fig. 6.2 time profiles for the first four subjects in our dataset have been plotted in a 2 ×2 grid. This is sufficient to ascertain the shape of the profile but the free scaling of the axes provides little indication of scale or location. It is therefore difficult to compare subject profiles.

In Fig. 6.2, each subject profile is drawn a free scale which is automatically generated using data ranges applicable only to that particular subject. As a by-product of this, 2 profiles appearing to be almost identical at first glance, for example the

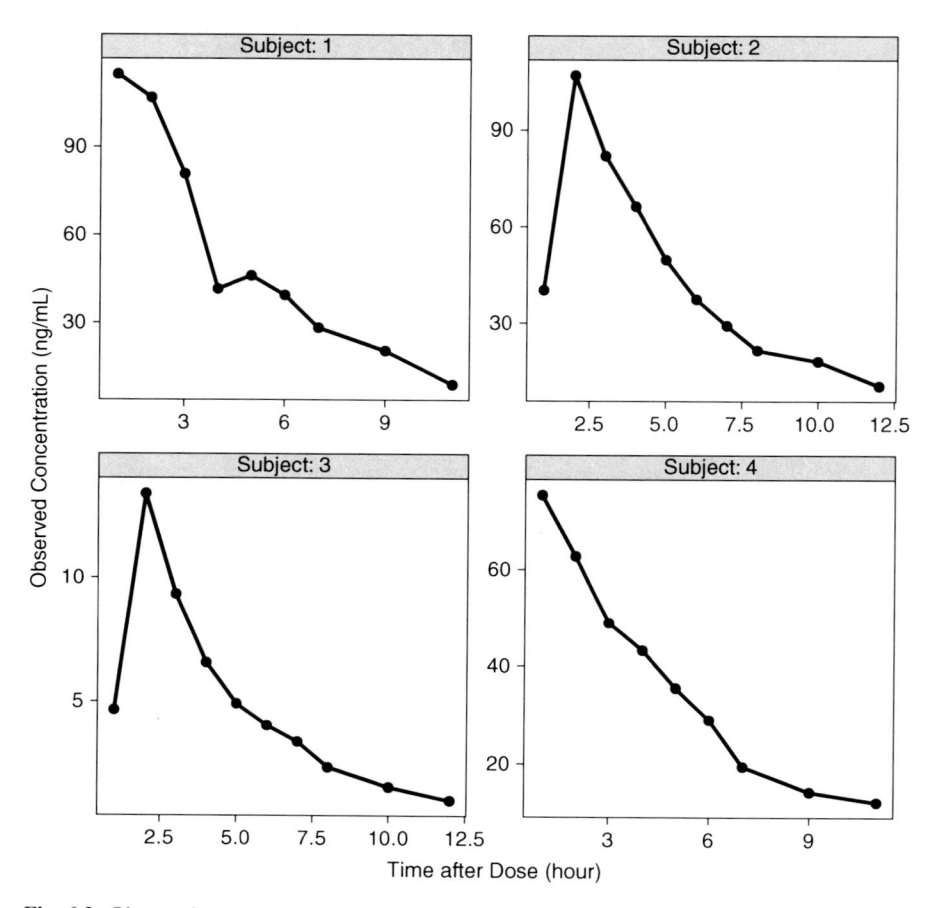

Fig. 6.2 Observed concentration versus time after dose by subject

profiles for subjects 2 and 3, may actually be very different when closer attention is paid to the axes. In fact the maximum concentration for subject 2 is approximately 8 times higher than the maximum concentration for subject 3. By default, most software packages fix the scaling such that the data ranges for the full population are used when constructing the panels. This example highlights the danger of overriding this default. It should be noted, however, that for non-homogeneous data, for example for data in which the subjects took different doses, fixing the scales can hide data characteristics such as the shape of the concentration-time curve for subjects taking a low dose.

In Fig. 6.2 only the first 4 subjects from our simulated pharmacokinetic dataset have been selected. It is natural to plot these four profiles in a 2 ×2 grid in order to fill the available space. For an EDA for which the primary intention is to establish the quality of the data by comparing an individual subject profile to the rest of the data, this approach usually suffices. However, when the primary aim is to make

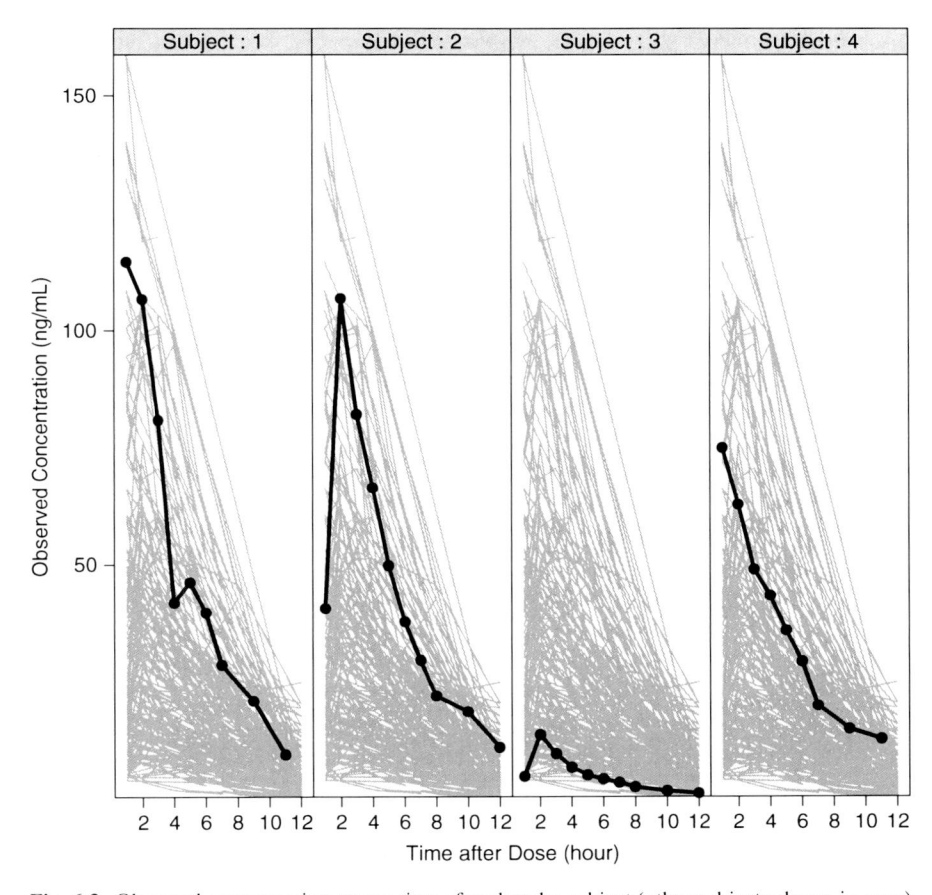

6 Exploring Pharmacokinetic and Pharmacodynamic Data 103

Fig. 6.3 Observed concentration versus time after dose by subject (other subjects shown in *gray*)

comparisons between panels the layout is far more important. For example, in Fig. 6.2 the primary variable of interest is concentration, which is shown on the y-axis. It is much easier to compare concentrations between two subjects when the y-axes are aligned. That is to say when the two plots are located side by side as opposed to vertically stacked. In order to make comparisons between panels, the panels should be stacked in the direction that best facilitates the comparison of interest.

For an exploratory analysis it is useful to compare individual profiles to the remaining population so that we can ascertain the relative location of the profile and identify outlying profiles more quickly. One technique that will enable such a comparison is to plot the overall data as a background "shadow." An illustration of this technique is given in Fig. 6.3. As in Fig. 6.2, the first 4 subjects from our simulated data have been plotted in separate panels. In each panel the subject of interest has been drawn in a thick, black line, with the remaining subjects from the simulated dataset plotted in gray. For each subject we see instantly how their profile compares to the rest of the population. Note also that plotting the four panels side by side also now facilitates comparison between these four subjects.

These techniques form the foundation for an exploratory repeated measures analysis. Later we see how they can be further extended in order to investigate dose and other covariates.

6.4 Data Quality

Many data quality issues are simple to spot, particularly incorrectly formatted values and data values that are significantly out of range. However, incorrect values that are still within the overall data range (such as values that have been transpose/swapped) are more difficult to identify. In this case, we must rely on the structural information we anticipate within the data in order to spot outlying values.

If an incorrectly coded value is still within the anticipated data range, a basic univariate analysis is not able to identify the issue. In this case, some errors may be impossible to spot without quality control (QC) steps against the data source. We can, however, use basic graphical methods to spot observations that do not seem to follow similar trends. These observations may be erroneous or may be highly informative.

In this section we concentrate on 2 key variables: the observed concentration variable to be modeled and "time after dose," taken as the primary independent variable.

In the field of PK, we typically understand the ways in which the direction of trend of the concentration curve should change, based on the dosing mechanism employed. For example, after a drug is orally administered the concentration typically increases sharply to a maximum concentration, then smoothly decreases as the drug is metabolized and eliminated.

As such, we can often spot erroneous data in a dependent versus independent relationship by simply analyzing the sign of the gradient at each step within a graph of observed concentration versus time after dose. The gradient is the slope of the line connecting two observations, so a switch in sign indicates a switch in the direction.

Figure 6.4 graphs observed concentration versus time after dose, split by the number of changes in the sign of the gradient during the time period studied. In this plot, an upward facing green triangle is used for a value that is larger than the previous value and downward facing red triangle for a value that is less than the previous value. A change in the direction occurs at locations where the plotting symbol changes.

As you can see here, the majority of the data has at most one change in the sign of the gradient. Selecting subjects where there are more shifts in gradient sign often exposes data groups which may require further analysis. Figure 6.5 reproduces the above graphic for selected subjects with a high number of gradient sign changes.

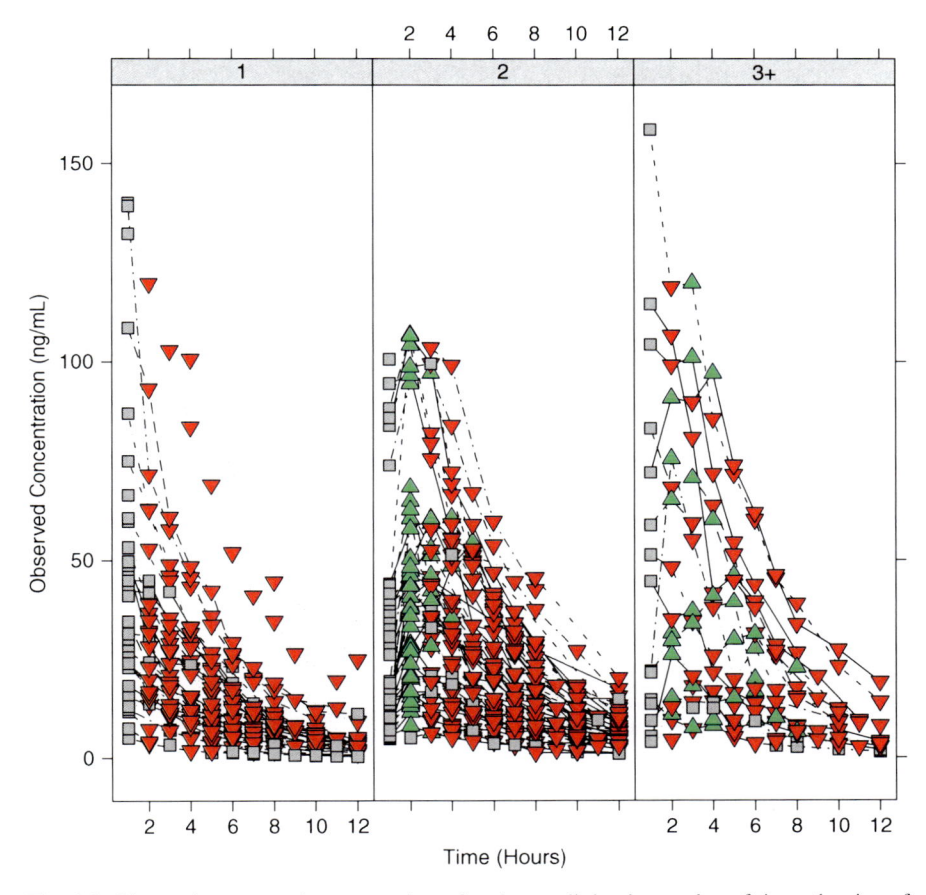

6 Exploring Pharmacokinetic and Pharmacodynamic Data

Fig. 6.4 Observed concentration versus time after dose, split by the number of times the sign of the gradient changes direction

We expect that these short-term increases are due to measurement or reporting inaccuracies.

Another way to look for potentially erroneous data that could be contained within the acceptable range of the data is to scale the data against a model-agnostic smoother (such as a loess smoother). In order to perform this scaling, the following steps are used:

- Log the concentration data
- Fit a smooth line to the logged data
- Calculate differences from this smooth line
- Calculate the mean "difference from smooth line" for each subject, and subtract this from the differences (effectively "centering" each subject's data)

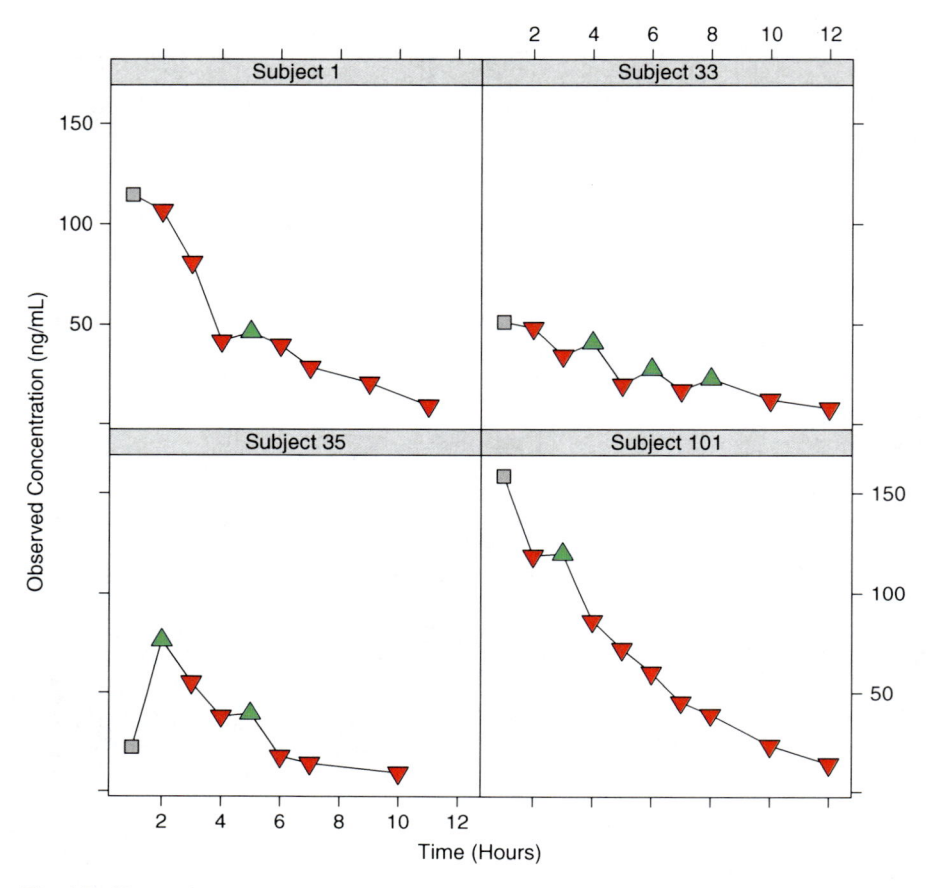

Fig. 6.5 Observed concentration versus time after dose, split by the number of times the sign of the gradients changes direction (including only selected subjects)

- For each time point, subtract the mean of the data and divide by the standard deviation
- Plot the calculated values vs. time and look for outlying values

Figure 6.6 presents the scaled data generated versus "time after dose." In this graphic, the subject identifiers are shown in dark blue, with subject groups linked by light gray lines. The change of color here focuses the reading on the subject values themselves (darker color) as opposed to the "trend" of each subject (lighter color). Horizontal grid lines have been added to allow ease of reading.

From this analysis, we can identify subjects that may warrant further analysis based on large absolute differences from 0. For example, the following subjects have at least 1 value that is more than 3 standard deviations from the mean time point center in the above plot: 10, 25, 36, 39, and 141. Creating a plot of observed

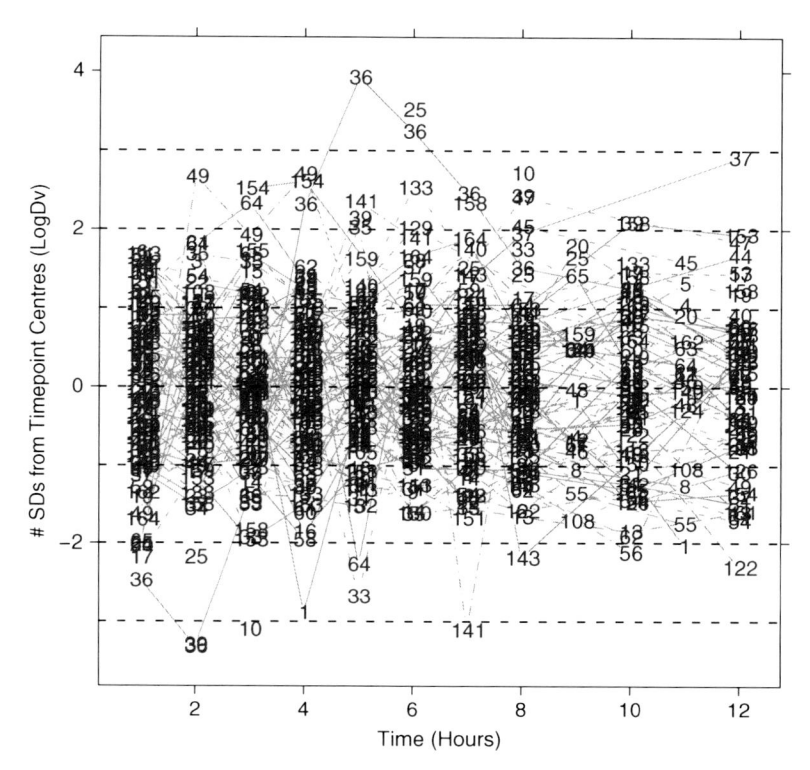

6 Exploring Pharmacokinetic and Pharmacodynamic Data

Fig. 6.6 Scaled standard deviations from time point centers versus time after dose, with subject identified

concentration versus time after dose for these subjects produces the graph shown in Fig. 6.7.

In this instance, we want to illustrate how these subjects compare to the general trend of the data. If we were to include all subjects in this graph, the visualization might look too "busy," and the focal message could be confused. As such, in order to represent the "general trend" of the data, we have instead included polygon "shadows" of color representing the 0, 25, 75, and 100 percentiles of the total data set on each plot; these are represented as light and dark blue polygons (0th and 100th percentiles drawn in light color, overlapped with dark polygons representing the 25th to 75th percentiles). In addition, the "center" of the data is included as a (black) median line. To contrast against this background, the individual subject data has been graphed as bright red points linked by a line.

From this analysis, subjects 10 and 36 certainly warrant further investigation. In particular, we should look at the observation at time point 3 for subject 10 (which appears to be comparatively low) and the reasons that the concentrations peak comparatively late for subject 36 when compared with the general trend of the data (since most concentrations peak at 2 h).

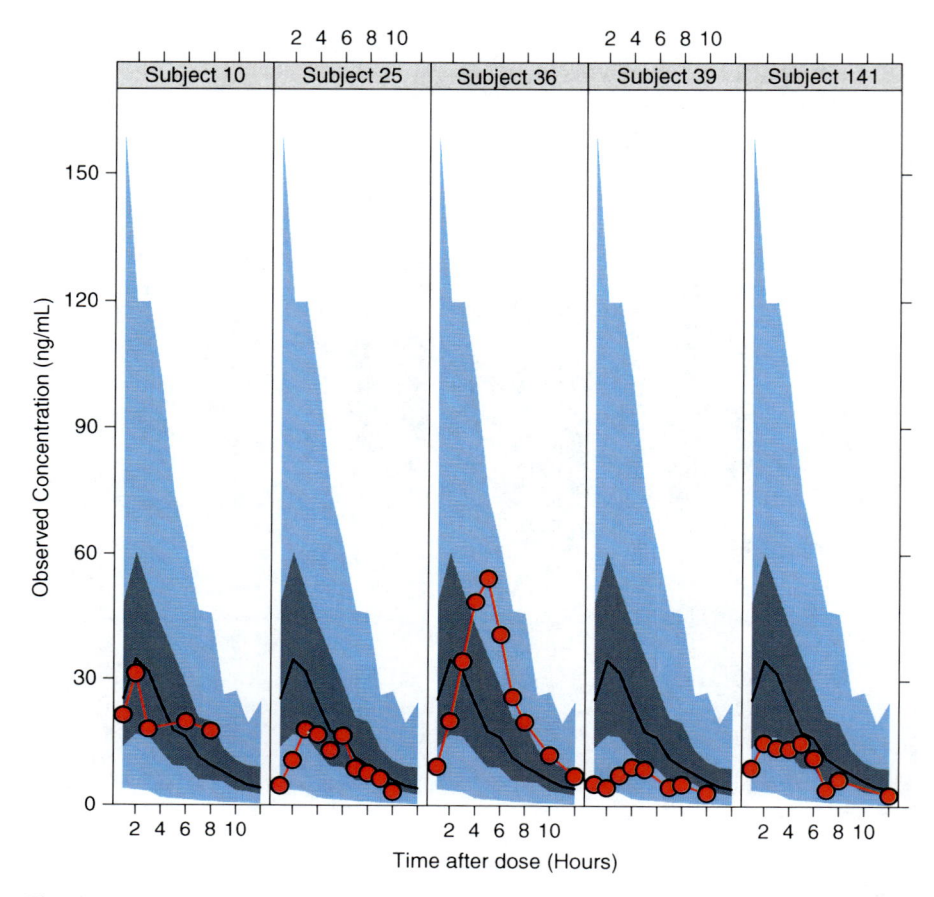

Fig. 6.7 Observed concentration vs. time after dose including only those subjects flagged in the above analysis, with *blue shaded areas* representing the range and interquartile range for the overall data and the overall median drawn in *black*

6.5 Baseline Effects

The techniques discussed thus far are applicable to both PK and PD repeated measures data.

Baseline is almost always a highly important covariate when modeling a PD endpoint. If the baseline distribution of the dependent variable differs between dose groups, it can often be seen to undermine the findings of a study even when the baseline effects have been accounted for in our model. This is particularly true of smaller early phase studies for which the more extreme baseline values have a greater influence on population summaries.

For the EDA, the effect of baseline can blur the effects of other covariates and unless we are interested investigating interactions with baseline we should look

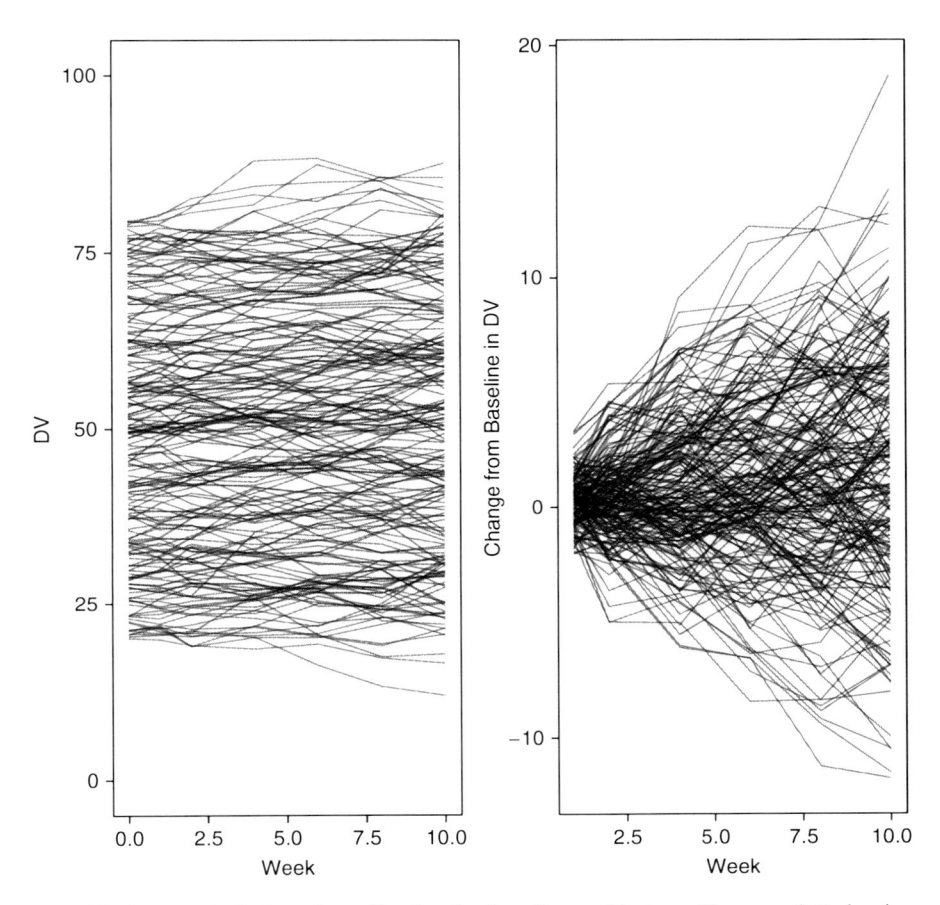

Fig. 6.8 Improved clarity after adjusting for baseline, subject profiles are plotted using semi-transparent lines

to try and remove the baseline effect in order to focus on the covariate of interest. For longitudinal data, this usually means plotting the change from baseline over time such that all profiles begin from the same place. In Fig. 6.8 we first plot our dependent variable over time and then the change from baseline in the dependent variable over time. In the left-hand plot the change in the response variable appears random over time. In the right-hand plot however, it is apparent that there is increased variability over time, as would be expected for a controlled trial in which subjects must meet certain baseline criteria in order to be randomized to the study.

When plotting change from baseline we highlight the challenge of choosing an appropriate y-axis range. It is often easy to neglect this choice since almost all software packages automatically generate a y-axis based on the observed range of our data, occasionally adding a small amount of "white space" to this range in an attempt to aesthetically enhance the default plots. It is not uncommon for the observed bounds of our data to closely coincide with this default range and so this default range is usually sufficient for our needs. When we plot change from baseline data however, the possible range is often much larger than the observed range. Therefore if we adopt the default range, then a small change in the profile of our data might appear to be much larger than it actually is, potentially leading us down the wrong path. Often the range we choose lies somewhere between a change of potential clinical importance and a plausible range.

6.6 Investigating Covariates

Once we are satisfied with the quality of our data, the next step is to investigate dose and any other covariates that we are considering for inclusion in our model. Any covariate that we choose to investigate (including dose) falls into one of the two categories, continuous and discrete. We look at each in turn, beginning with discrete covariates.

When plotting by dose or any other discrete covariate we have a choice to make. We can either use grouping techniques to distinguish between the levels within a single plot or partition the plot into panels. The advantage of grouping within a single plot is that all profiles are on the same axis, and thus it is theoretically easier to compare the profiles. For a spaghetti plot of individual subject profiles, color or line type can be used to differentiate between the groups. However, with a large number of subjects or when the data is highly variable it can be difficult to extract any differences between groups using this method. For this reason it is common to plot average data profiles for each group, for example median values. When plotting using a graphics device that allows semi-transparent coloring we may also decide to plot "ribbons" or error regions to give an idea of the range of the data. However the use of error bars or ribbons can crowd the plot, particularly when the covariate has many levels. Even with just three levels a plot can look crowded if the differences between the groups are small. For that reason we are usually limited to plotting a single summary statistic for each subgroup over time. Figure 6.9 shows the change over time in a simulated response variable by country. In this particular dataset there are 9 levels of the country variable. A simple plot of the median values for each country highlights a sudden rise in response between weeks 8 and 10 for China in contrast to the majority of countries for whom the response tends to decrease between weeks 8 and 10.

In the previous example plotting median values helped highlight an interesting feature of our data. However without any indication of variability we cannot be sure whether this is truly interesting or simply random fluctuation. Figure 6.10 plots the change from baseline over time in the dependent variable by dose group. The blue regions within each panel represent the range and interquartile range for the dose group with the median represented by a black line. In addition, the range of response values across all doses is shown in gray. The position of the blue area in relation to the shadow helps to identify a dose effect with increased response in the higher dose group.

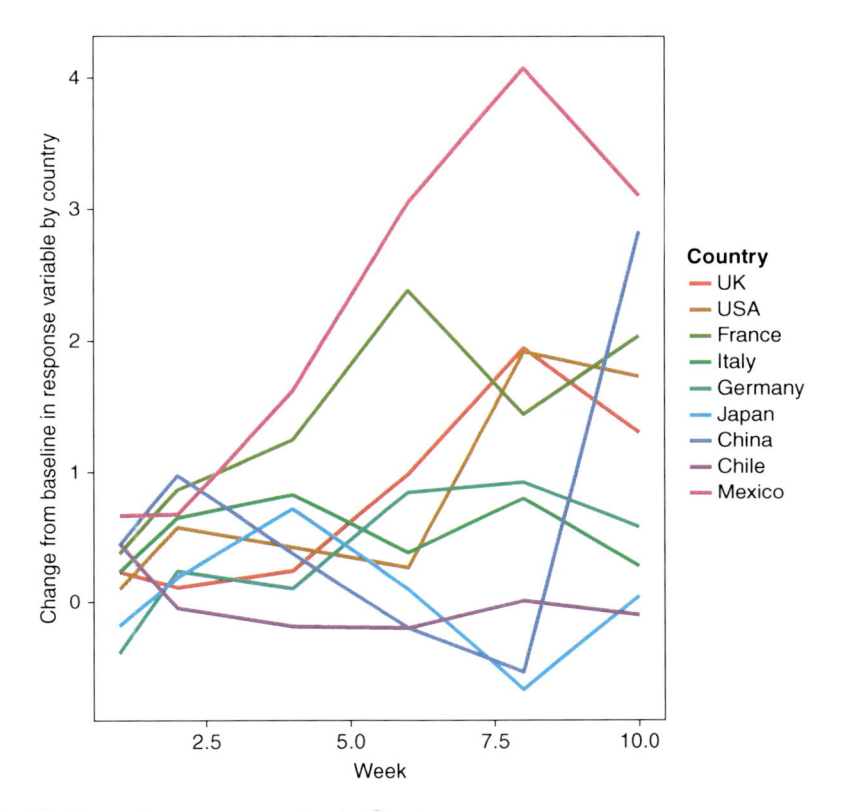

6 Exploring Pharmacokinetic and Pharmacodynamic Data   111

Fig. 6.9 Change in response over time by Country

The approach is one that is easily extended into higher dimensions. In Fig. 6.11 we look at two discrete covariates, the sex of the subjects and whether or not they smoke. Again, the 0th, 25th, 50th, 75th, and 100th percentiles are displayed for each covariate group and as with the last graphic, the 0–100 range is in light blue, 25–75 in darker blue, and a black line shows the median. Figure 6.11 shows higher concentrations for female subjects, and a later peak in concentrations for male subjects. The graph also suggests a slight increase in concentration values for smokers. The consistency across the plots plot would also seem to indicate a lack of any interaction between sex and smoking habit. Note that to facilitate comparison of the observed concentration values, the plots are aligned from left to right as opposed to a 2 ×2 grid layout.

When the covariate of interest is continuous we no longer have the option of paneling available to us since to panel by a variable implies that it is discrete. We can of course split the continuous variable into discrete categories of interest and apply the same techniques already discussed for categorical variables. In a modeling scenario we generally wish to avoid partitioning our data in this way as we create artificial boundaries between data points. However for an EDA this actually may help us to spot an effect, and if we do intend to categorize a continuous variable, then we can use paneling as a way of selecting the breakpoints. Of course it is vital

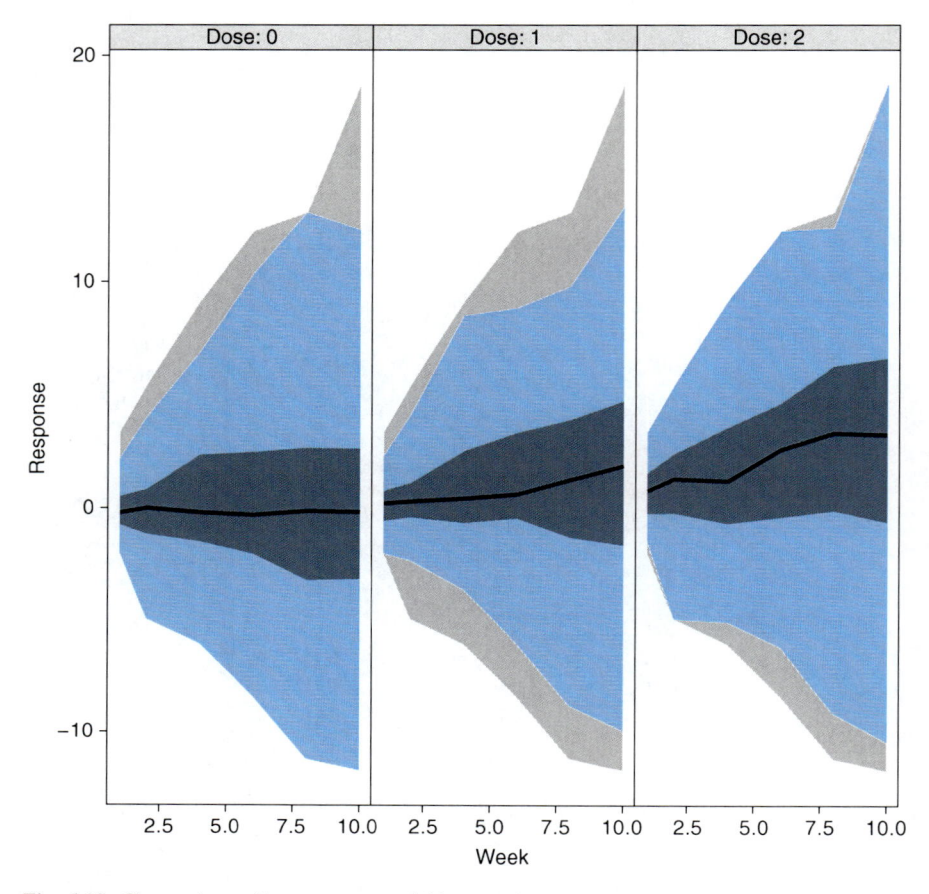

Fig. 6.10 Change in median response variable over time by dose group with *blue shaded areas* representing the range and interquartile range for each dose group and the overall data range included as a *gray shadow* on the plot

that the natural order is retained when a discrete variable has been formed from a continuous one.

If we decide not to cut our continuous covariate up into a discrete variable, then we face a dimensionality challenge; we are effectively already treating both the response variable and time as continuous variables and the within-subject correlation adds further dimensionality. When faced with this challenge, it can be tempting to move to a three-dimensional coordinate system. Given that we only have 2 dimensions available to us, this approach should be adopted with caution. In particular the angle of rotation has a strong effect on our interpretation of the graphic and interesting features of the data can easily be hidden. An alternative approach is to treat time as a discrete variable. This is possible for clinical trial data since the response information is collected at scheduled (discrete) times. We can then investigate the response variable against our covariate of interest via a sequence of scatter plots. If

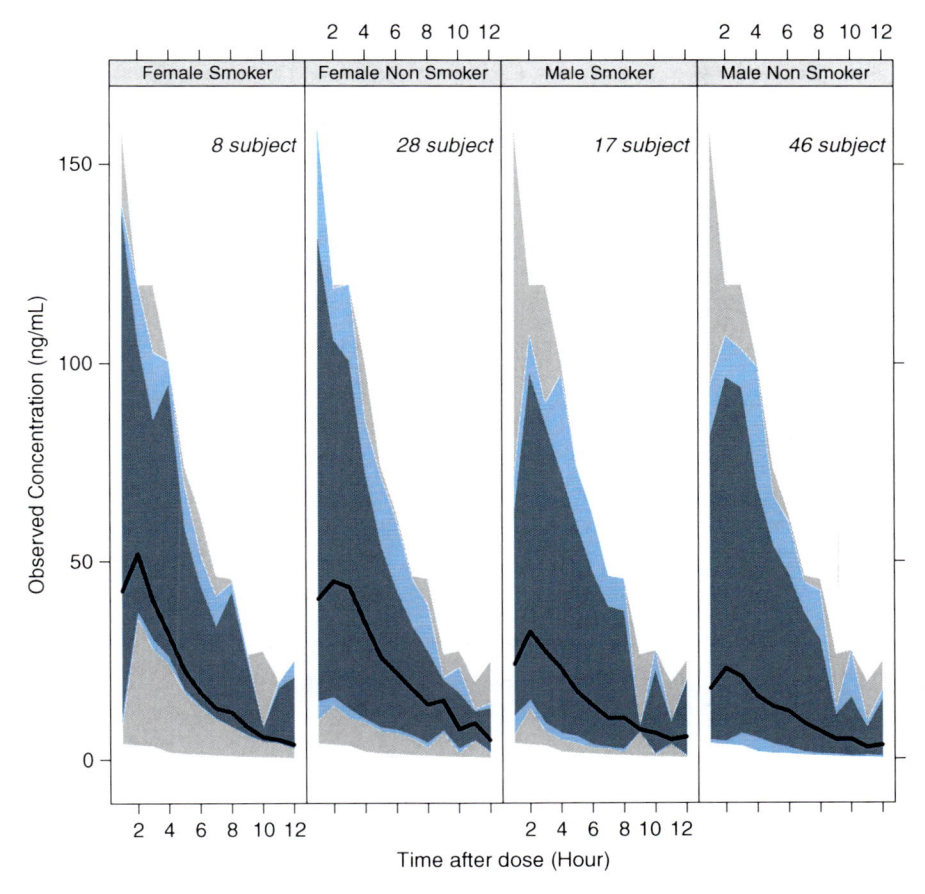

Fig. 6.11 Observed concentration vs. time after dose plots split by sex/smoking Indicator, with *blue shaded areas* representing the range and interquartile range for the subgroup and the overall data range included as a *grey shadow* on the plot

a relationship exists between the two variables, we would expect to see a pattern develop as we move from left to right across the panels. As with all ordinal categorical variables it is vital that when we panel by time we maintain the order and arrange the panels from left to right.

In Fig. 6.12 we have paneled by time in order to investigate how the relationship between Y and X changes over time. Varying the plotting symbol and color by dose, we see how the dose groups begin to separate over time.

This method is not without its drawbacks. The length of time between the discrete data collection points is usually wider towards the end of a subject's participation in a study. Care must therefore be taken when paneling by the discrete time points to preserve the distance between time points. In Fig. 6.12 therefore the first time point has been removed so that the difference between each time point is equal. Alternatively, spacing could be used to indicate larger time intervals.

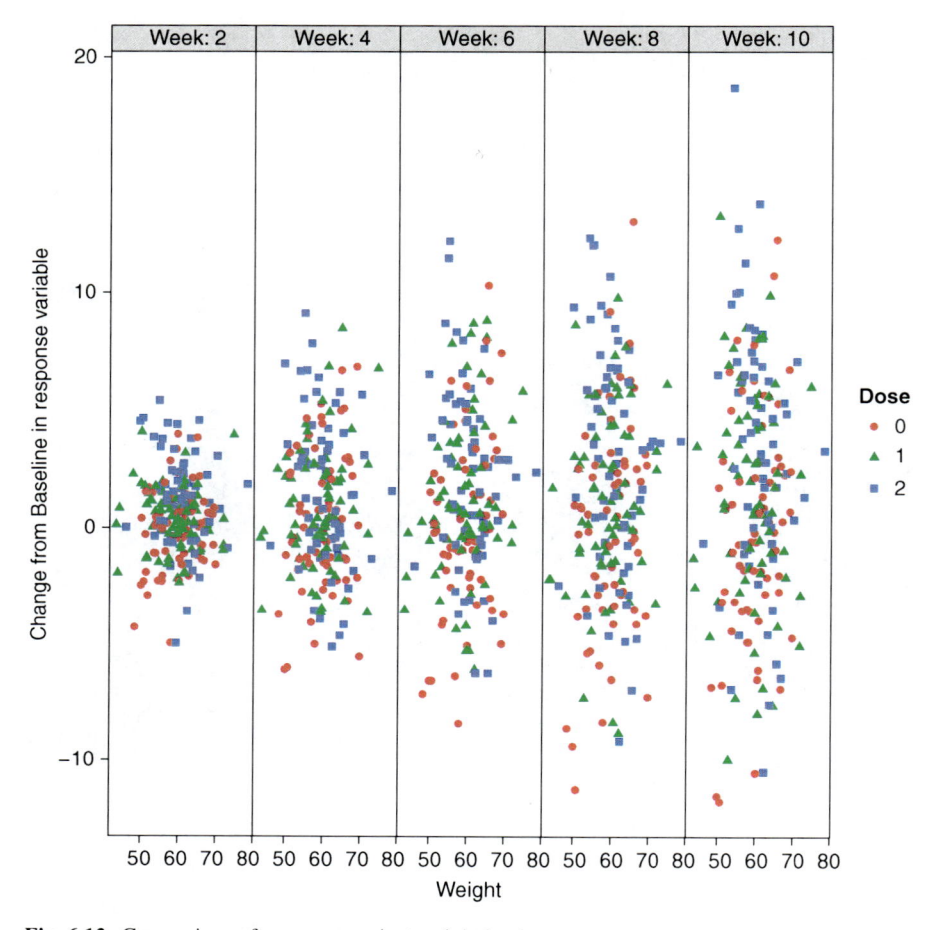

Fig. 6.12 Comparison of response against weight by dose

6.7 Conclusion

For the PK/PD modeler, the EDA is the first step in the model building process. Exploring the data graphically presents many challenges due to the within-subject correlation that results from repeated measures data. An understanding of this structure can help us to break the data down and spot troublesome with-range anomalies. The aim of the EDA is to attain a better understanding of the data prior to commencing a model building process. It is therefore important to focus the exploratory work on identifying patterns within the data that may influence this process. Plot change from baseline data when applicable and consider the comparison of subgroups with the population as a whole in addition to with each other.

References

Cleveland WS, Grosse E, Shyu WM (1992) Local regression models. In: Chambers JM, Hastie TJ (eds) Statistical models in S, Chap. 8. Wadsworth and Brooks/Cole, Pacific Grove, CA

Jonsson EN, Karlsson MO (1999) Xpose–an S-PLUS based population pharmacokinetic/pharmacodynamic model building aid for NONMEM. Comput Meth Programs Biomed 58(1):51–64. (Available from http://xpose.sourceforge.net/)

R Development Core Team (2008) R: A language and environment for statistical computing. R Foundation for Statistical Computing, Vienna, Austria

Sarkar D (2009) Lattice: multivariate data visualization with R. Springer, New York

Wickham H (2009) ggplot2: Elegant graphics for data analysis. Springer, New York

Chapter 7
(Interactive) Graphics for Biomarker Assessment

Michael Merz

Abstract Biomarkers have become cornerstones of modern drug development, improving assessment of both efficacy and safety of new molecules. Use of graphical data exploration can help to maximize information output from biomarker data and improve efficiency of the data analysis process. The chapter describes a systematic workflow for interactive graphical data exploration focusing on liver safety biomarkers as an example. Using interactive software, the workflow allows research and development teams to jointly explore their data, generate and test hypotheses, and identify the most suitable graphs for study reports or regulatory reporting.

7.1 Introduction

7.1.1 Biomarker Definitions

Decisions in drug development and clinical medicine alike are increasingly based on results of biomarker measurements, providing the opportunity to obtain relevant safety or efficacy readouts long before rock-solid clinical endpoints have been reached.

The term "biomarker" comprises a variety of different measurements such as genetic markers, used to either predict therapeutic response or risk for specific side effects, clinical chemistry variables such as blood cholesterol or blood sugar, or cardiovascular parameters such as blood pressure, heart rate, or specific elements of ECG readings.

M. Merz (✉)
Novartis Institutes for BioMedical Research, Translational Sciences,
Basel, Switzerland
e-mail: michael.merz@novartis.com

Examples of drugs that gained market access based on biomarker data include the breast cancer drug Herceptin™, which was approved in 1998 along with a genetic biomarker predicting if a patient would be likely to benefit from treatment (Bouchie 2004), or anti-diabetic drugs which usually are approved depending on their effect on blood glucose measurements.

Biomarker categories as defined by the FDA in 2005 (US Food and Drug Administration 2005) are as follows.

- Class I known valid biomarker
  A biomarker that is measured in an analytical test system with well-established performance characteristics and for which there is widespread agreement in the medical or scientific community about the physiologic, toxicologic, pharmacologic, or clinical significance of the results.

  Known valid biomarkers are for example certain pharmacogenetic markers predicting response to treatment or risk of side effects. A comprehensive list of such markers in the context of approved drug labels is available on the FDA's website:
  http://www.fda.gov/Drugs/ScienceResearch/ResearchAreas/Pharmacogenetics/ucm083378.htm.

- Class II probable valid biomarker
  A biomarker that is measured in an analytical test system with well-established performance characteristics and for which there is a scientific framework or body of evidence that appears to elucidate the physiologic, toxicologic, pharmacologic, or clinical significance of the test results.

  A probable valid biomarker may not have reached the status of a known valid marker because, for example, of any one of the following reasons:

  - The data elucidating its significance may have been generated within a single company and may not be available for public scientific scrutiny
  - The data …, although highly suggestive, may not be conclusive
  - Independent verification of the results may not have occurred

- Class III exploratory biomarker
  A biomarker that does not match criteria I or II.

Most of the biomarkers used for clinical diagnosis or safety testing in drug development are somewhere in between Class I and II, mostly not having been formally qualified, but based on extensive knowledge and evidence to justify their routine use.

This chapter illustrates the graphical analysis of biomarker data within a systematic workflow, which should be applicable to a wide range of different biomarker types. In order to demonstrate the added value of a defined workflow, systematic liver safety biomarker assessment is used as an example.

Regulatory decisions to reject approval of new drugs due to concerns about liver side effects are increasingly based on careful analysis of liver safety biomarker data rather than clinically manifest serious liver damage. Examples include drugs such as the beta blocker dilevalol or the anti-diabetic vildagliptin, which were both not approved in the United States based on liver test abnormalities, or the anti-diabetic troglitazone, which was withdrawn from the market due to several cases of serious liver injury, which might have been predicted based on liver safety biomarker data from clinical trials.

Liver safety biomarkers fall into two different categories, *enzyme* and *function* tests. Liver enzymes are catalytic proteins located in liver cells, which, upon injury to the organ, are released into the blood stream. Increased plasma activities of liver enzymes thus mostly indicate injury to the liver.

Liver function tests, primarily bilirubin, are dependent, as the name implies, on proper organ function. One of the functions of the liver is elimination of bilirubin, a degradation product of hemoglobin, into the bile system. Decrease of hepatic function, e.g., due to drug-induced liver injury, if sufficiently severe, ultimately leads to increased bilirubin levels in blood.

7.2 Why Use Graphics for Biomarker Data?

A wealth of evidence suggests that the human brain's capability to process visual information significantly exceeds processing of text information (Goolkasian 1996). This holds true as well when comparing visual to tabular presentation of numerical data, such as biomarker data. To illustrate this, Table 7.1 provides a tabular overview on simulated blood pressure measurements in clinical studies across two drug development programs. The data show average proportional change from baseline for systolic blood pressure, compared between active treatment and control groups.

Figure 7.1 shows the same data as treemap, with size of the rectangles corresponding to the number of patients per treatment group; color coding is by average proportional change from baseline for systolic blood pressure. Hierarchy headers are by compound (letters) and study (number code).

The fact that active treatment across studies in program II seems to be associated with an increase in systolic blood pressure can be captured much easier and literally in the twinkling of an eye, as compared to the tabular representation: rectangles representing program 2 studies are clearly shifted more toward red for active treatment groups as compared to controls. It is also immediately obvious from the graphical display of the data which of the studies in program II show most pronounced differences between treatment groups, i.e., nos. 4, 103, 107, and 2105.

Some characteristics of biomarker data make them particularly amenable to graphical exploration:

- In most cases in clinical practice and pharmaceutical research, maximum information can be obtained by analyzing and assessing biomarkers not individually, as single markers, but rather as *marker panels*, since individual markers may describe differential aspects of a given clinical event. Thus, for example, during drug development, the standard panel of markers used for liver safety assessment

Table 7.1 Proportional change from baseline for systolic blood pressure by pooled treatment groups across two different drug development programs

Program	Study	Active treatment	Control
I	3	1.01	1.00
	4	1.00	1.00
	7	1.01	0.99
	102	1.00	0.97
	107	1.03	0.99
	108	1.01	1.01
	1204	1.01	1.02
	2102	1.01	1.00
	2103	1.01	0.97
	2106	0.99	1.00
	2302	1.00	1.00
II	2	1.04	0.99
	3	1.03	0.98
	4	1.15	1.03
	7	1.05	0.99
	102	1.06	0.99
	103	1.17	1.04
	107	1.17	1.01
	2102	1.02	1.04
	2103	1.02	1.03
	2105	1.17	0.98

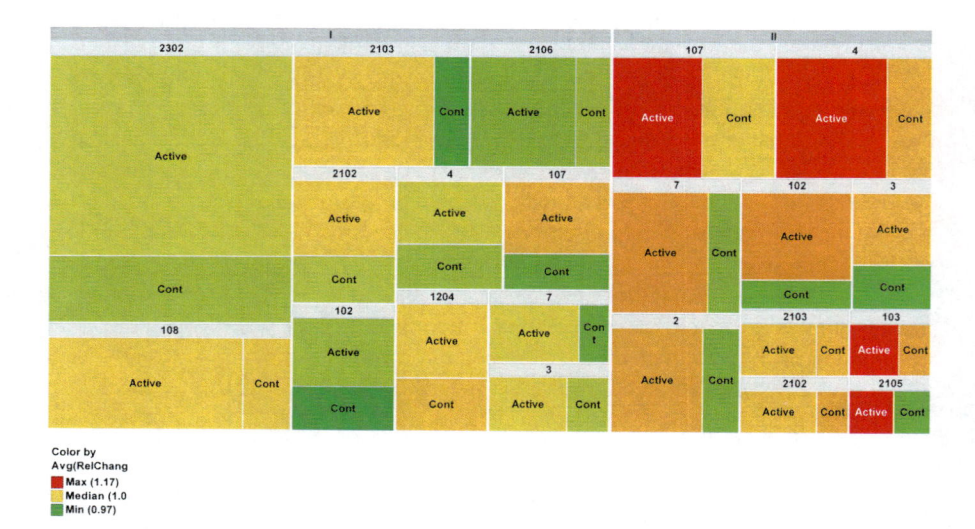

Fig. 7.1 Treemap displaying proportional change from baseline for systolic blood pressure, size of rectangles proportional to sample size, color corresponding to average increase in blood pressure per pooled treatment group

comprises alanine aminotransferase (ALT), aspartate aminotransferase (AST), gamma glutamyl transferase (GGT), alkaline phosphatase (ALKPHS), and total bilirubin (TBIL) (in Europe, the terms SGPT and SGOT are more generally used for ALT and AST, respectively)

Elevation of ALT alone is a very sensitive, yet not highly specific signal for liver injury. The additional elevation of bilirubin increases specificity of the ALT signal. The combined elevation of ALT exceeding 3 times the upper limit of normal, and bilirubin exceeding two times the upper limit of normal is widely known as "Hy's law," a term coined by FDA expert Dr Robert Temple in the 1990s, and named after Dr Temple's late colleague Dr Hyman Zimmermann, who had described this predictive liver safety signal originally.

- In addition to being closely correlated with other biomarkers, many safety biomarkers are *dependent on demographic covariates* such as age, gender, body mass index (BMI), etc.
- Important as well for proper assessment of biomarker results is thorough *assessment of time courses of individual and combined markers*. A significant elevation of ALT followed by an elevation of bilirubin may indicate severe liver injury. The reverse sequence however, sometimes observed for example in young male patients with a harmless genetic disorder of bilirubin excretion ("Gilbert's syndrome"), may not indicate any relevant injury at all

Thus, proper analysis of biomarker data is *multivariate by nature* and has to *take into account time dependency of observations*. Although biomarker assessment via summary tables and individual data listings may be feasible as well, use of graphics, ideally in the setting of a defined, systematic workflow, is by far more efficient and reliable.

The examples presented in this chapter have been elaborated using TIBCO Spotfire™, a software tool facilitating interactive, graphical data exploration, allowing project teams to *jointly* assess their data as a group, instead of individual analysis of static standard plots.

7.3 Prerequisites

7.3.1 Normalization

Adequately assessing biomarker data often needs comparison between different continuous variables, across different studies, different laboratories, etc. To facilitate that, normalization of continuous data is helpful. For variables being reported along with normal ranges, an efficient algorithm allowing to easily derive where in relation to the normal range the un-corrected value had been, is given in Table 7.2:

Figure 7.2 displays normalized vs. raw values for a marker with LLN = 10 and ULN = 50:

Using the above algorithm, a normalized value of e.g.,

Table 7.2 Algorithm to normalize continuous biomarker data in relation to the normal range; *LLN* lower limit of normal; *ULN* upper limit of normal

Raw value (RV)	Normalized value (NV)
RV < LLN	NV = 100 (RV/LLN−1)
LLN < RV < ULN	NV = (RV-LLN)/(ULN-LLN) * 100
RV > ULN	NV = RV/ULN * 100

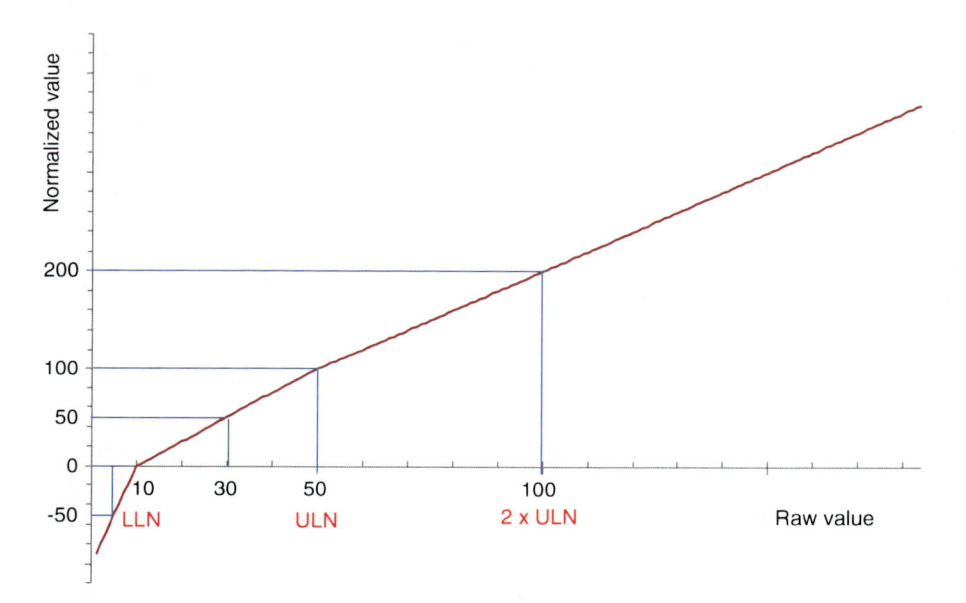

Fig. 7.2 Normalization of continuous biomarker data, relative to lower and upper limit of normal

- 100 corresponds to a raw value at ULN
- 200 corresponds to a raw value at 2×ULN
- 50 corresponds to a raw value in the middle between LLN and ULN
- −50 corresponds to a raw value in the middle between 0 and LLN

7.3.2 Units

Biomarker data across different laboratories or studies may be reported in different units. Before pooling data, units need to be converted to standard or preferred units, if available, to facilitate comparison across studies or different study centers and laboratories. However, normalization relative to the given normal ranges, as shown above, eliminates the need for conversion across different measurement units.

7.3.3 Data Types

Although the vast majority of biomarker data are continuous variables such as enzyme activities, analyte concentrations in serum, etc., adequate interpretation of biomarker results often requires simultaneous assessment of other data such as concomitant medication, adverse events, medical history, etc. Thus, graphical exploration of biomarker data should also include plots providing a synoptic view on biomarker, concomitant medication, and adverse event data per patient.

Analyzing raw biomarker values only may be insufficient to see the complete picture. Frequently, including derived variables such as absolute and relative changes from baseline, maximum values on treatment, flags for exceeding predefined threshold values, etc. are required to adequately interpret biomarker results. Thus, before starting biomarker data exploration, a set of derived variables required needs to be defined and calculated.

7.3.4 Data Structure

Typically, datasets for biomarker analysis have study identifier, subject identifier, visit numbers and visit names, parameter names, parameter results, lower and upper limits of normal ranges, units, and relevant covariates such as age, gender, BMI, ethnicity displayed by column. For most of the graphics used for biomarker exploration, this structure is sufficient. However, in order to address specific questions such as shape of bivariate distributions, shifts from baseline by visit, etc., transposing the dataset by parameter names or by visits may be necessary. In order to allow an efficient workflow, it is helpful to define individual steps of the workflow and required data structures upfront and make sure analysis datasets are available in all formats required.

7.4 Questions to be Addressed

Key questions to address when analyzing biomarker data comprise:
- Are distributions of biomarker values different across treatment groups?
- Is incidence of out-of-range values different across treatment groups?
- What is the number of patients exceeding certain threshold values across treatment groups?
- Are shifts from baseline different between treatment groups?
- Is there any evidence for a dose–response-relationship for biomarker effects?

- How are changes across different biomarkers correlated, and how do those correlations differ between treatment groups?
- What do time profiles of individual biomarkers or biomarker panels, such as liver enzymes, look like?
- Are biomarker changes observed during treatment transient or progressing while a patient is on treatment?
- What do time profiles look like after stop of treatment?
- How does intake of certain concomitant medication or occurrence and/or resolution of certain adverse events relate to time profiles of biomarkers of interest?
- How do biomarker results compare across studies within a clinical program, or across different development programs focusing on different drugs that have the same molecular target?

To systematically address these questions, a set of standard graph templates can be used and customized as required. The following examples focus, as mentioned, on assessment of liver safety biomarkers, in particular the enzymes ALT, AST, GGT, and ALKPHS, and, as a marker of hepatic excretory function, TBIL.

7.5 Graph Templates and Systematic Workflow

7.5.1 Distributions Across Treatments

Frequency distribution of liver tests across different treatment groups can be displayed as Trellis histograms, with biomarker names across columns and treatment groups across rows. Tools for interactive graphical data exploration ideally allow for easy setup of any kind of conditional ("Trellis") plot, providing opportunities for display of additional dimensions such as gender (color coded in the example below), as well as distribution characteristics (average and standard deviations in the example).

Figure 7.3 shows histograms of liver enzymes ALKPHS, ALT, and AST. To account for the underlying log-normal distributions of liver enzyme data, x-values are presented on a logarithmic scale. This excludes representation of negative values, which, based on the normalization algorithm used may occur, but without having any clinical relevance: of primary interest are values beyond the upper limit, rather than below the lower limit of normal. The plot indicates rightward shifts particularly for ALT and AST, plus presence of more extreme values at the right tail of the distribution for both active treatment groups. Mean ALT values, represented by broken green lines, are higher during active treatment as compared to placebo. Color coding by gender does not suggest a relevant difference of effects between male and female patients.

Thus, this initial step in a systematic workflow, using simple histograms, provides already meaningful information about a potential treatment induced liver effect.

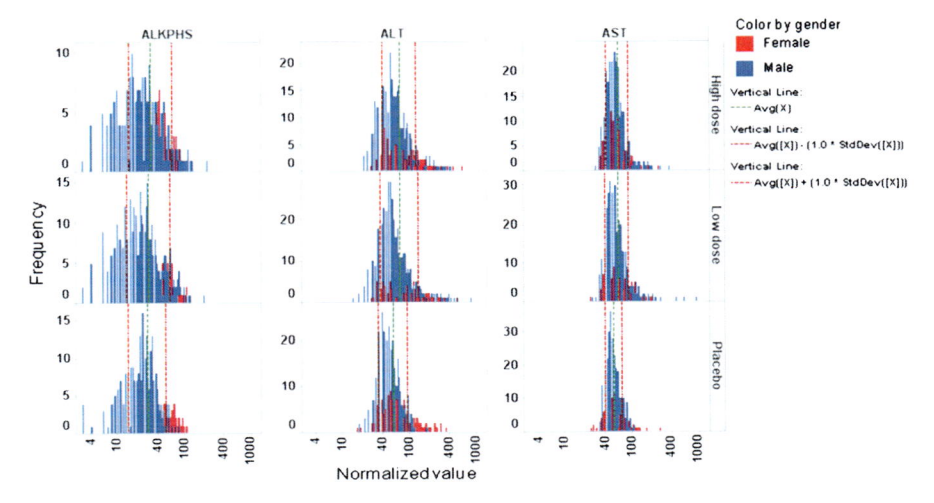

Fig. 7.3 Distributions of normalized biomarker results, parameters by column, treatment groups by row, color coding by gender

7.5.2 Incidence of Out-of-Range Values

A key piece of information when analyzing safety biomarker data usually is how many patients showed results outside the normal range. For liver safety analysis, of particular importance are values exceeding the upper limit of normal (ULN). As severity of liver injury may correspond to levels of enzyme activities and bilirubin concentrations, a standard process in safety assessment is to count values exceeding ULN, $3 \times$ ULN, $5 \times$ ULN, etc.

Graphically, this is displayed easily using bar charts, conditioned by liver test name and treatment group, as shown in Fig. 7.4.

On the x-axis, value classes are displayed ("0": inside normal range, "1," exceeding $1 \times$ ULN, etc.), the y-axis holds percent of subjects per value class.

The plot shows higher incidences of values > ULN for ALT, AST, and ALKPHS in both active treatment groups as compared to placebo. For ALT, incidence of values exceeding $3 \times$ ULN is higher in both active treatment groups as well.

Another way to show the same information, but making treatment differences more obvious is demonstrated in Fig. 7.5, using color coding for treatment groups and placing treatment groups next to each other.

7.5.3 Shifts from Baseline

Biomarker results always have to be viewed in the context of their respective baselines to allow adequate assessment of treatment or disease effects. This can be done

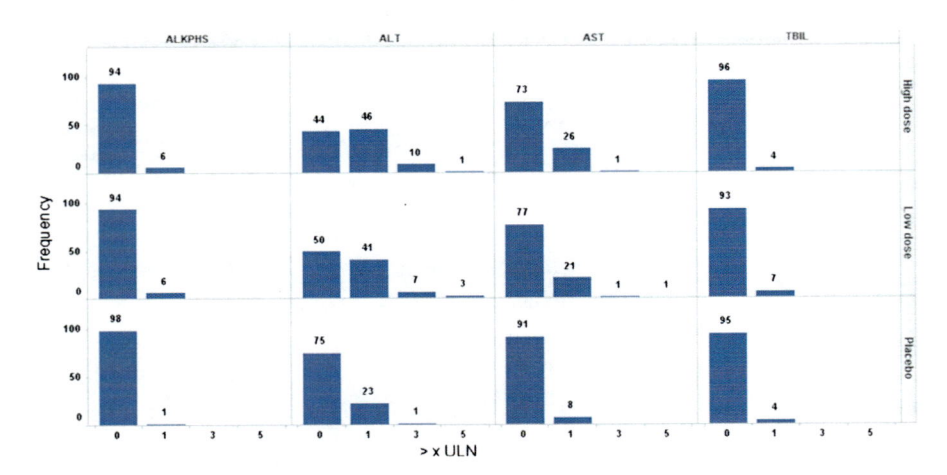

Fig. 7.4 Number of patients within and above the normal range, parameters by column, treatment groups by row

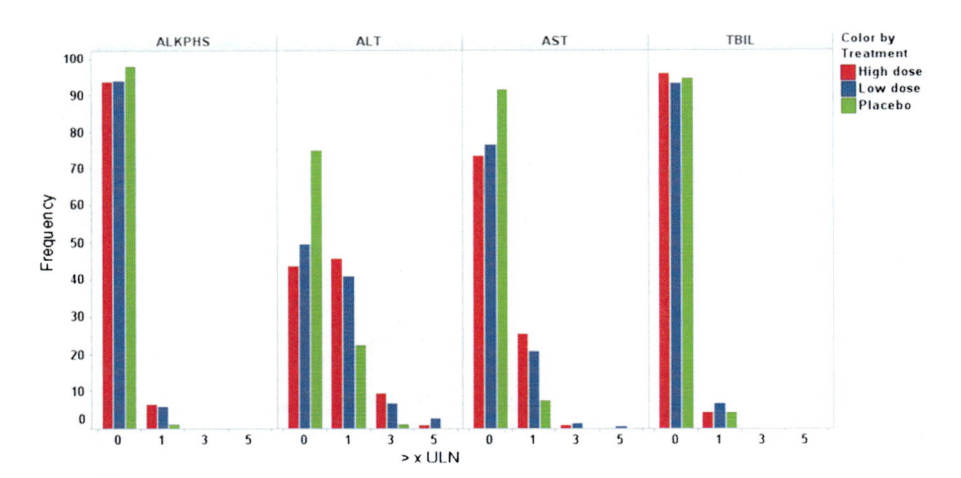

Fig. 7.5 Frequency of patients within and above the normal range, parameters by column, color coding by treatment

either by analyzing absolute and relative changes from baseline, or by using scatter plots with baselines on the *x*-axis and e.g., maximum post-baseline values on the *y*-axis. When plotting only *maximum* post-baseline values on the *y*-axis, however, careful consideration needs to be given to the number of post-baseline measurements per patient, particularly when no control groups are available for comparison across treatments: the larger the number of post-baseline observations per patient, the more biased the plot will be toward values increasing from baseline.

Figure 7.6 shows a respective example with four post-baseline observations per patient. Again, this is a Trellis plot with treatment groups across rows and biomarker

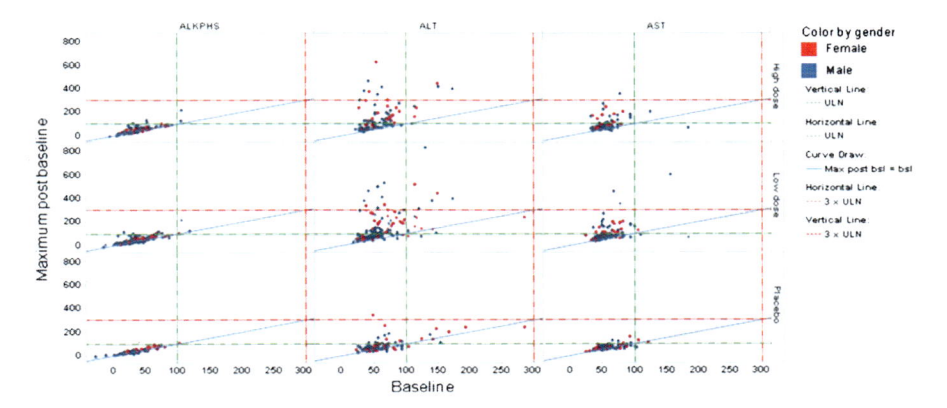

7 (Interactive) Graphics for Biomarker Assessment

Fig. 7.6 Shifts from baseline, parameters by column, treatment groups by row, color coding by gender

names across columns. Color coding is by gender. The blue diagonal line in each panel represents the line of identity, i.e., each value on the line corresponds to maximum post-baseline equaling baseline, each point above the line is an increase, points below the line a decrease from baseline, respectively.

In addition, the plot allows to assess the number of patients exceeding certain threshold values, represented by the green (=ULN) and red (3×ULN) horizontal and vertical broken lines in each panel.

In this example, there is a clear trend for higher shifts from baseline, i.e., elevations, for ALT and AST in both active treatment groups. However, even the placebo group displays some elevations from baseline at least for ALT. Although this is a phenomenon not uncommon in clinical studies and may be explained by effects of diet, physical exercise, concomitant disease or comedication, the effect may at least partially also be due to a bias introduced by the number of post-baseline measurements. To avoid that, an alternative would be to plot *all* post-baseline values per patient, instead of selecting the maximum values only. This, however, may substantially hamper legibility of the data.

The overall effect in the active treatment groups is not dramatic; however, some individual values exceed 5×ULN at post-baseline which would warrant close monitoring.

At initial inspection, there is no obvious effect of gender on shifts from baseline, nor is there a strong baseline dependence of high post baseline values for ALT and AST.

7.5.4 Dose–Response-Relationship

In order to assess dose effects more quantitatively than feasible via scatter plots, box plots may be used for absolute or relative changes from baseline and compared across treatment groups. Figure 7.7 shows maximum absolute changes from baseline

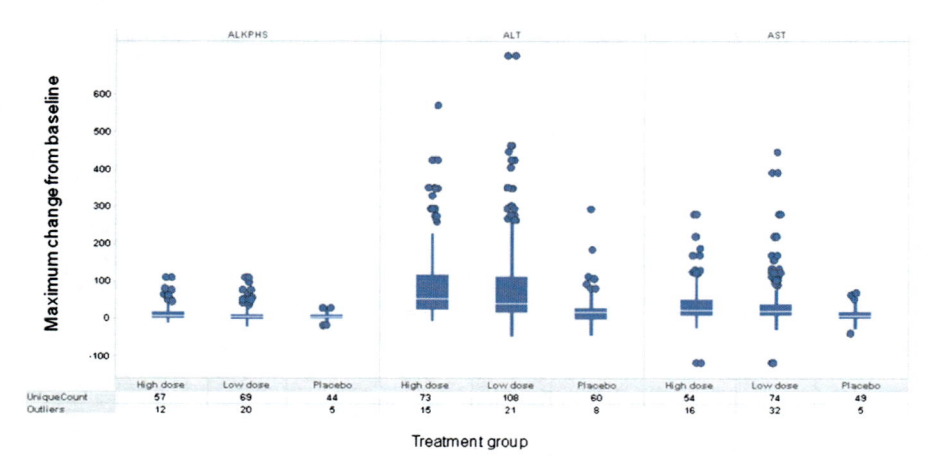

Fig. 7.7 Maximum absolute changes from baseline across treatment groups, parameters by panel, treatment groups by column per panel

per patient for liver enzymes across treatment groups. Plots per treatment group are defined by median (white line), lower and upper quartiles (box), lower and upper adjacent values (whiskers), and outliers (individual data points). Outliers are jittered on the x-axis to improve visibility.

The plot suggests differences for maximum elevations from baseline of ALT and AST as compared between both active treatment groups and placebo treatment.

For ALKPHS, only a potential trend toward higher elevations with active treatment as compared to placebo can be observed.

7.5.5 Correlations

Often it is helpful to explore correlations either between biomarkers that are or could be linked from a physiological perspective, or between biomarkers and potential covariates of interest such as age, BMI, gender, ethnicity, etc.

For liver safety assessment, a key graph used by the FDA to detect safety signals of concern is the so-called "eDISH" (evaluation of Drug-Induced Serious Hepatotoxicity (Gelperin et al. 2008)) plot, a log-log display of peak TBIL vs. ALT, both in multiples of ULN, with horizontal and vertical lines indicating Hy's law thresholds, i.e., ALT = 3 × ULN and TBIL = 2 × ULN. Combination of elevated ALT > 3 × ULN and TBIL > 2 × ULN, as mentioned in the introduction, is a pretty specific prognostic indicator for the risk to develop severe drug-induced liver injury. The "eDISH" plot allows to easily spot cases potentially matching Hy's law criteria, all located in the upper right quadrant of the graph. Data points in the lower right quadrant, i.e., exceeding 3 × ULN for ALT, but being below 2 × ULN for TBIL, suggest an increased risk for liver injury as well, if incidence is differing between

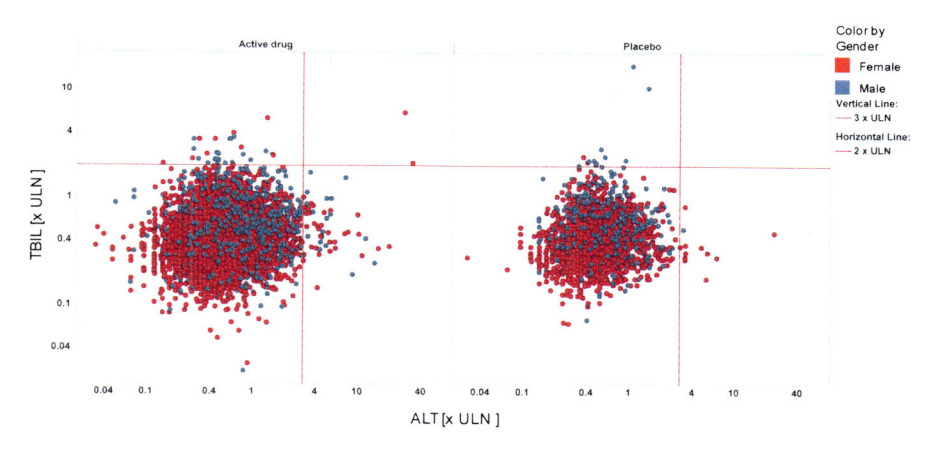

Fig. 7.8 eDISH plot, TBIL [×ULN] vs. ALT [×ULN] on a log/log scale, treatment by panel, pooled active vs. placebo control, color coding by gender

active treatment and control groups, however, not to the same extent and with less specificity as compared to Hy's law. These cases in the lower right quadrant are sometimes referred to as "Temple's corollary" cases, named after Dr. Robert Temple, a fellow expert of the late Dr. Hyman Zimmerman at the FDA.

Figure 7.8 shows an example of an eDISH plot, comparing pooled active study drug against placebo data. Color coding is by gender, horizontal, and vertical lines indicate Hy's law thresholds.

Patients with active treatment show a higher incidence of values in the lower right quadrant and two potential Hy's law cases in the upper right quadrant, thus suggesting a potential risk for severe drug-induced liver injury associated with this drug, without apparent association to gender. As stated in the FDA's guidance on Drug-Induced Liver Injury, "… Finding one Hy's Law case in the clinical trial database is worrisome; finding two is considered highly predictive that the drug has the potential to cause severe DILI when given to a larger population" (US Food and Drug Administration 2009).

A limitation with the standard eDISH plot is its lack of displaying the sequence of maximum observed values for ALT and bilirubin, i.e., which of both was first, as well as length of time intervals between maximum observed values. However, from a clinical perspective, these data are highly relevant, since only bilirubin elevations *parallel to or following* ALT elevations may indicate loss of hepatic function due to liver injury. Moreover, a long time interval, exceeding 2–4 weeks, between both peaks may also speak against a causal correlation. Thus, it would be valuable to have this information included in the graphical display, as well. Figure 7.9 shows a proposed modification to the eDISH plot, using color coding for sequences of ALT and bilirubin peaks, and size coding for the time interval between both peaks. In order to make the most relevant data points easily visible, the more concerning sequence of bilirubin parallel to or following ALT peak is coded in red, the time interval is coded as 1/interval to make shorter time intervals being displayed as

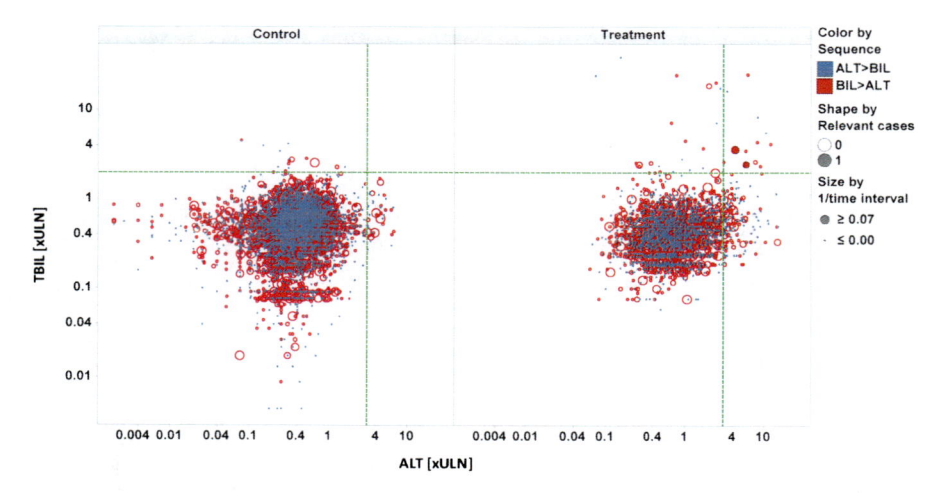

Fig. 7.9 Modified eDISH plot, color by sequence of peak values, size by one/time interval between peaks, shape by clinical relevance

larger data points. Filled circles refer to data located in the Hy's law quadrant, with ALT peak simultaneous to or followed by bilirubin peak, and time interval between both being less than 4 weeks. Thus, the points to watch out for primarily are large, filled, red circles.

In the above example, 9 of 12 data points in the Hy's law quadrant for patients in active treatment groups show the sequence of interest, i.e., bilirubin following or simultaneously elevated with ALT peak, but only two of those have a time interval of less than 4 weeks between both peaks. Thus, using this modified version of the eDISH plot, 10 out of 12 potential Hy's law cases can be immediately identified as being likely less relevant.

Other correlations of interest when exploring liver safety profiles of new drugs are those between different liver enzymes, i.e., ALT/AST, ALT/GGT, and ALT/ALKPHS. Whereas in the healthy liver, ALT and AST are closely correlated, ALT and the 2 other enzymes usually are not. However, in some cases of drug-induced liver injury, elevations of ALPKPHS and/or GGT may correlate with increased ALT activities, providing some hints about the underlying pathology, i.e., cholestasis or mixed type cholestatic/hepatocellular injury. Isolated elevations of GGT activities without associated ALT or ALKPHS changes may sometimes indicate enzyme induction rather than cell injury, as observed, e.g., in cases of chronic alcohol abuse.

Figures 7.10, 7.11, 7.12, and 7.13 display examples of bivariate distributions for maximum post-baseline values of AST vs. ALT, ALKPHS vs. ALT, GGT vs. ALT, and GGT vs. ALKPHS. Color coding is by treatment group, horizontal and vertical lines indicate ULN (green line) and 3×ULN (red line), respectively. The blue diagonal line indicates identity, e.g., in Fig. 7.10 maximum ALT equaling maximum AST.

Figure 7.10 shows the typical correlation between ALT and AST, ALT generally displaying higher activities than AST. In cases of e.g., alcohol-induced liver injury

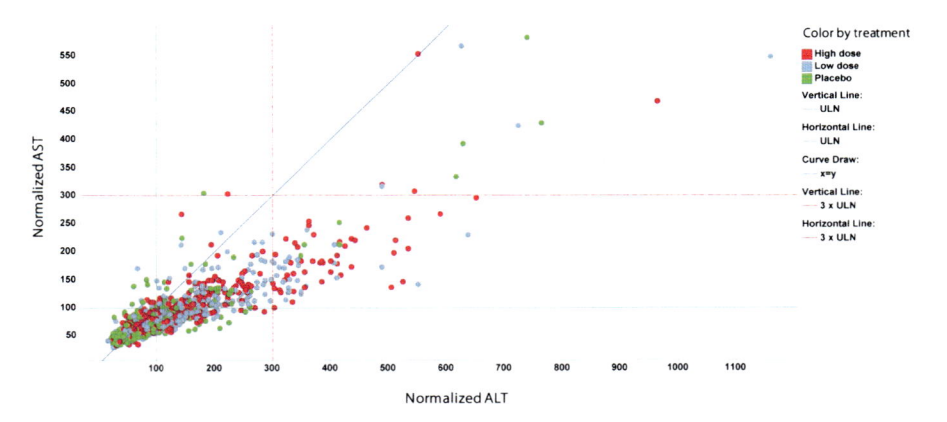

Fig. 7.10 Post-baseline AST vs. ALT, color coding by treatment group, ULN and 3×ULN indicated as *green* and *red lines*, respectively, line of identity indicated as *blue* diagonal

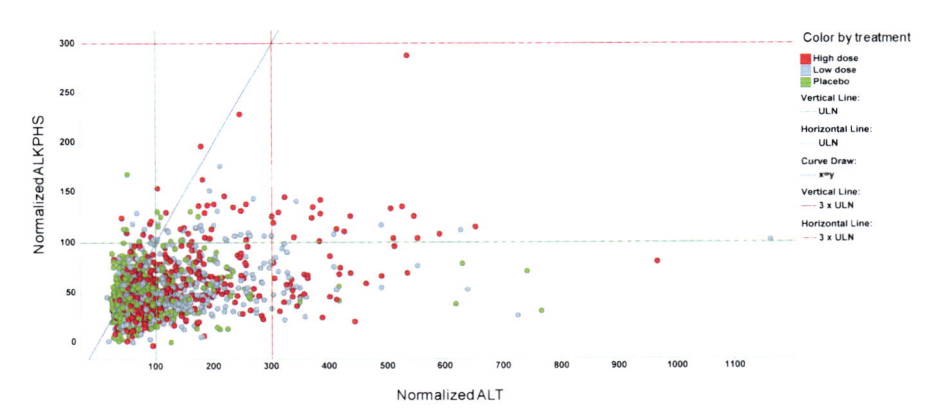

Fig. 7.11 Post-baseline ALKPHS vs. ALT, color coding by treatment group, ULN and 3×ULN indicated as *green* and *red lines*, respectively, line of identity indicated as *blue* diagonal

or muscular injury, this correlation may be reversed, i.e., AST>ALT, supporting differential diagnosis of the underlying pathology.

Of note, elevations of both ALT and AST are visible not only in both active treatment groups (red and blue data points), but also for some placebo patients (green points). Maximum ALT values are exceeding 10×ULN, thus being clinically significant.

Figure 7.11 displays some ALKPHS elevations without clear association with ALT activities, Fig. 7.12 demonstrates significant GGT elevations not correlated with ALT elevations, and Fig. 7.13 points to a potential, albeit rather weak correlation between ALKPHS and GGT.

This data overall may point toward mild to moderate cholestatic or mixed type cholestatic/hepatocellular liver injury, suggested by ALKPHS and GGT elevations.

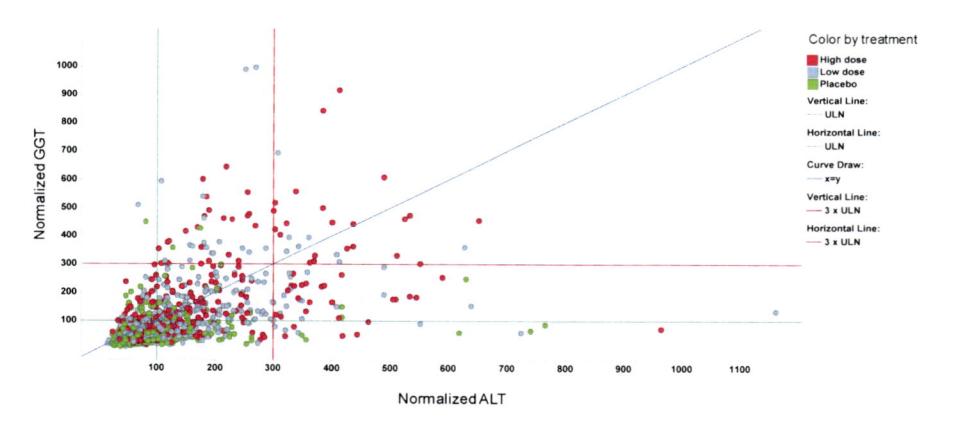

Fig. 7.12 Post-baseline GGT vs. ALT, color coding by treatment group, ULN and 3×ULN indicated as *green* and *red lines*, respectively, line of identity indicated as *blue* diagonal

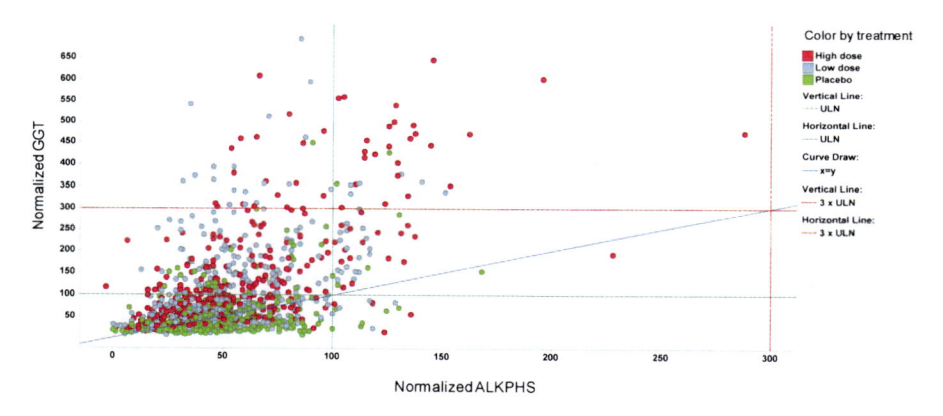

Fig. 7.13 Post-baseline GGT vs. ALKPHS, color coding by treatment group, ULN and 3×ULN indicated as *green* and *red lines*, respectively, line of identity indicated as *blue* diagonal

However, some involvement of enzyme induction, indicated by GGT elevations being substantially higher than ALKPHS elevations, cannot be ruled out.

7.5.6 Time Profiles

Changes of biomarkers over time can provide crucial information on both underlying pathology and causal relationship to drug treatment. Line plots of either individual markers or marker panels are most useful to assess biomarker time profiles, particularly if combined with elements indicating start and/or end of drug treatment.

Figure 7.14 provides an overview on ALT profiles over time of 49 patients in a clinical study who showed ALT elevations while on treatment, one panel per patient.

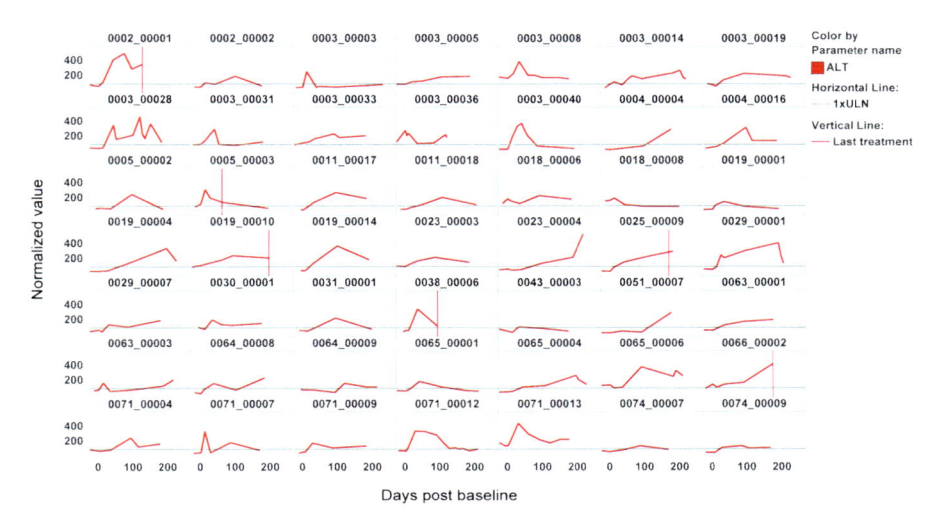

Fig. 7.14 Time profiles of ALT, panel by patients, treatment end indicated by *vertical red line*, ULN indicated by *horizontal green line*

Treatment end is indicated by a vertical red line, the horizontal green line represents ULN. As displayed in the plot, most of the patients showed short-lived, transient peaks of ALT, with serum activities decreasing despite continued treatment. Only few patients had to be taken off treatment due to continuous or worsening elevations of ALT.

Overall, for the drug under investigation, this could be rather reassuring information, suggesting that most patients may adapt to a drug-related effect on the liver, if any.

Panels of standard liver tests ALT, AST, ALKPHS, and TBIL are displayed jointly by patient in Fig. 7.15, presenting a subset of the patients shown in Fig. 7.14. Color coding is by liver test, horizontal lines represent ULN (green line) and $3 \times$ ULN (red line), respectively. This plot allows assessing time-wise association of different biomarker effects by patient. Of interest for evaluation of liver safety for example is to see if bilirubin or ALKPHS elevations follow or precede ALT peaks. Elevated bilirubin levels following or paralleling ALT elevations may indicate hepatic dysfunction, increased ALKPHS activities post-ALT elevations may point to cholestasis secondary to hepatocellular injury.

Only patient 0004_00016 shows discrete elevation of ALKPHS in parallel with peak ALT and AST, pointing toward a cholestatic component of liver injury. There are no apparent elevations of bilirubin parallel or subsequent to ALT elevations in any of the patients, confirming the rather benign nature of liver enzyme changes observed in the study.

A crucial question in liver safety assessment of clinical trials is how many patients show elevated ALT $> 3 \times$ ULN and TBIL $> 2 \times$ ULN either at the same time or with TBIL elevation following ALT peaks within a few days up to 1 month, as mentioned above. Whereas the standard "eDISH" plot displays peak elevations, or

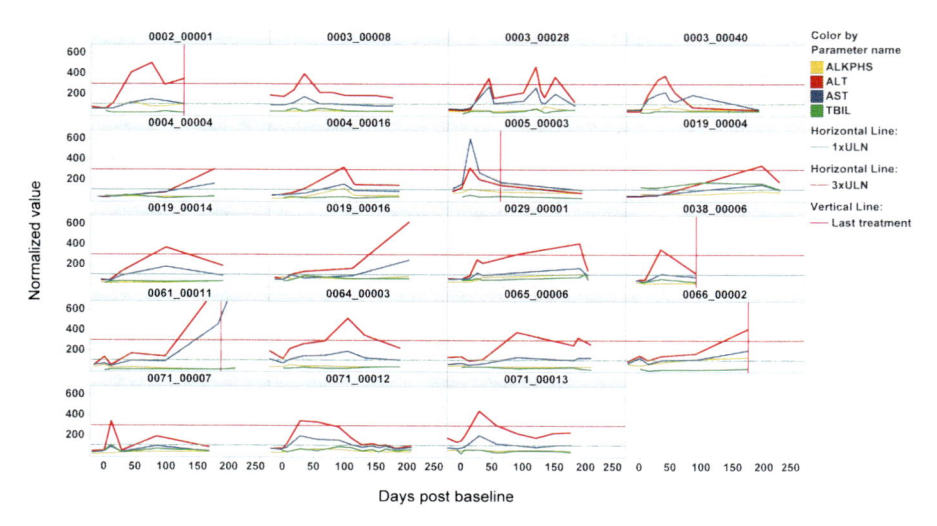

Fig. 7.15 Time profiles of ALT, AST, ALKPHS, and TBIL, panel by patients,, treatment end indicated by *vertical red line*, ULN and 3×ULN indicated by *horizontal green* and *red line*, respectively. Color coding by liver test

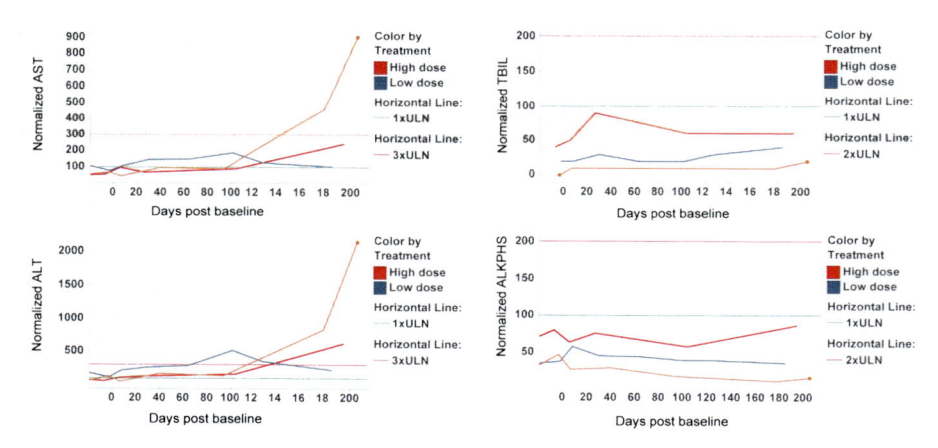

Fig. 7.16 Time profiles of AST, ALT, ALKPHS, and TBIL, panels by lab test, ULN and 3×ULN (2×ULN for TBIL and ALKPHS, according to clinical practice) indicated by *horizontal green* and *red line*, respectively. Color coding by treatment group. Selected patient marked in *orange*

rather maximum observed values of both parameters irrespective of the time interval between those peaks, and thus may come up with false positive Hy's law cases, individual line plots by parameter, using an interactive software, allow to mark, e.g., all patients exceeding 3×ULN for ALT and see the same patients' profile for bilirubin highlighted at the same time.

Figure 7.16 shows line plots for AST, ALT, TBIL, and ALKPHS, panels by liver test and lines by patient, across three patients. Color coding is by treatment, horizontal lines represent ULN (green line) and 3×ULN (red line), respectively. The

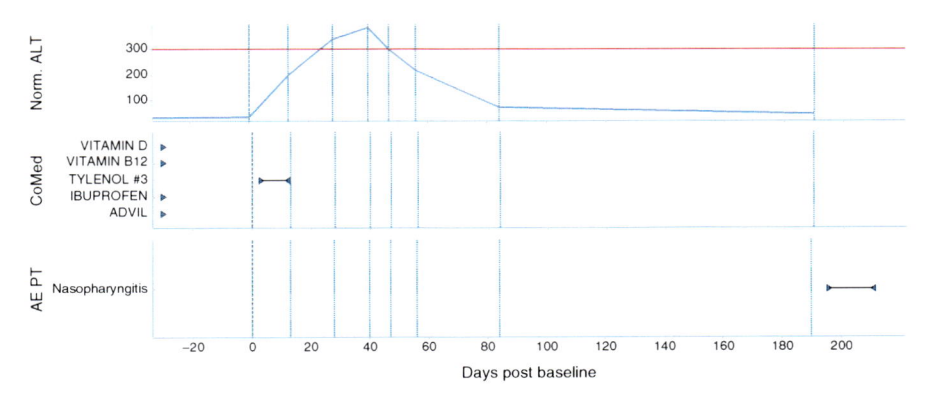

Fig. 7.17 Synoptic line plot of ALT, concomitant medication, and adverse events; start and end date of adverse events and concomitant medications indicated by *blue triangles*, patient no.1

orange line with markers at both ends is a patient having been marked in one line plot and being highlighted automatically in the other plots. Prerequisite for this kind of display is to have parameters stored by column in the dataset, so selecting one record picks up all four parameter values.

The patient marked in this example has significantly elevated ALT and AST values, but no associated bilirubin elevations, thus is not a Hy's law case.

7.5.7 Association with Concomitant Medication and Adverse Events

A particularly helpful graph to analyze association of biomarker changes with adverse events and concomitant medication is a synoptic presentation of line plots for all three items along a shared time axis. With an interactive analysis tool, this can be done using synchronized x-axes.

Figures 7.17 and 7.18 display ALT time profiles of two individual patients from different clinical trials on top of the plot, concomitant medication, and adverse events beneath. The horizontal red line in the top plot represents $3 \times$ ULN for ALT. In the two lower plots, start and end times of concomitant medication intake and adverse events are displayed as blue triangles, the black line between associated triangles indicates ongoing concomitant medication or adverse event, respectively.

As can be seen from Fig. 7.17, this patient had taken Tylenol, i.e., acetaminophen, an analgesic drug well known to cause liver injury (Lee 2008), for several days prior to the peak in ALT values. Thus, it is very likely that the ALT elevation was due rather to acetaminophen than to the study drug in this clinical trial.

Figures 7.18 shows a different patient, who had taken an acetaminophen containing medication, Nyquil, before the first ALT peak, but no suspicious comedication

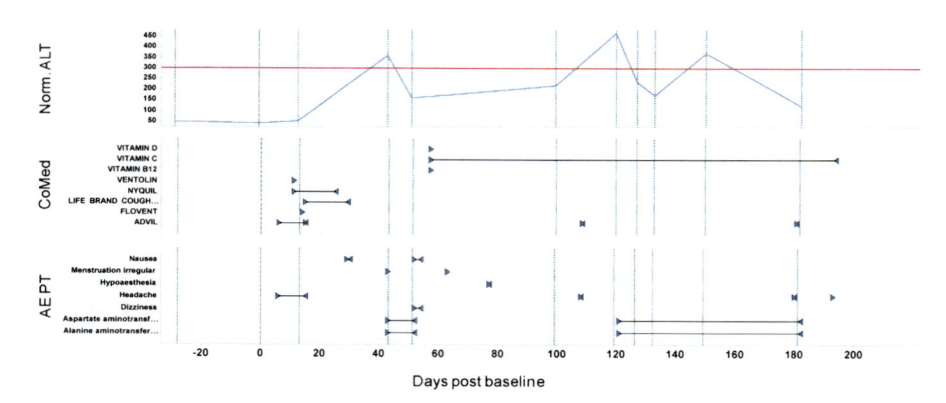

Fig. 7.18 Synoptic line plot of ALT, concomitant medication, and adverse events; start and end date of adverse events and concomitant medications indicated by *blue triangles*, patient no.2

around the time of the second ALT peak. However, the patient reported several adverse events of headache during the trial, one particularly preceding the second ALT peak. Upon questioning the patient specifically if she might have taken acetaminophen to treat her headache but forgotten to report this as concomitant medication, the patient confirmed that this indeed had been the case.

Thus, a synoptic display of ALT profiles, concomitant medication and adverse events may sometimes help substantially to identify causes for clinically relevant changes in liver safety biomarkers.

7.5.8 Biomarker Profiles Across Programs and Compounds

Going beyond analysis of individual trials, graphics can also help a lot to understand potential safety signals originating from biomarker responses across an entire development program, consisting of a range of clinical studies, or a portfolio of projects, e.g., related to a specific compound class.

As in the introductory example, Fig. 7.19 uses a treemap to display differences in average ALT values across pooled active vs. control treated patients in a range of clinical studies across different compounds. Size of the rectangles corresponds to number of patients per treatment group; color coding is by average post-baseline ALT values per treatment group, green indicating smaller values, and red indicating larger values. Hierarchy headers are by compound (letters) and study (number code), respectively.

Fig. 7.19 Treemap of average maximum post-baseline ALT values across different compounds and studies, size of rectangles proportional to sample size, color corresponding to average maximum ALT per pooled treatment group

The graph clearly indicates substantially higher post-baseline ALT values for compound H in the active treatment group vs. control. Similarly, though less pronounced, compounds D, K, and Q show a potential liver safety signal, based on average ALT values under active treatment.

7.6 Summary

Biomarker data usually are complex, multivariate by nature, and typically include multiple measurements over time. Particularly in the context of clinical drug safety assessment, association of biomarker effects with clinical adverse events, concomitant diseases, and concomitant medication needs to be accounted for. Graphics can help significantly to understand cause-effect relationships, mechanisms of toxicity, and may support both risk assessment and management of drug side effects.

A systematic workflow, including predefined graphical templates as outlined for liver safety assessment in this section, helps to ensure completeness of evaluations, supports hypothesis generation and testing, and facilitates identification of the most suitable graphics for publishing and regulatory reporting.

The use of interactive graphics software instead of focusing on static graphs will enable a project team to jointly assess biomarker data, help to thoroughly query the data from different perspectives, and foster team ownership of both analysis and conclusions.

References

Bouchie A (2004) Cancer trials get set for biomarkers. Nat Biotechnol 22(1):6–7
Gelperin K, Guo T, Senior J (2008) A simple graphic tool for assessing serious liver injury cases in a clinical trial—eDISH. AASLD-FDA-NIH-PhRMA Hepatotoxicity Special Interest Group Meeting. http://www.fda.gov/Drugs/ScienceResearch/ResearchAreas/ucm076901.html
Goolkasian P (1996) Picture-word differences in a sentence verification task. Mem Cognit 24(5):584–594
Lee WM (2008) Etiologies of acute liver failure. Semin Liver Dis 28(2):142–152
US Food and Drug Administration, March (2005) Guidance for Industry, Pharmacogenomic Data Submissions. pp 17–18
US Food and Drug Administration, July (2009) Guidance for industry, drug-induced liver injury: premarketing clinical evaluation

Chapter 8
Graphical Displays for Biomarker Data

Manuela Zucknick, Thomas Hielscher, Martin Sill, and Axel Benner

Abstract Analysis of high-dimensional biomarker data is in general of exploratory nature and aims to discover or dissect subgroups of patients sharing a specific pattern of biomarker measurements. One major challenge is to extract the relevant markers from the extremely large pool of measured markers. Specific techniques such as grouping and ordering and dimension reduction allow to aggregate huge amounts of data into single meaningful graphics. These graphics can guide the direction of exploration during the analysis. We present graphical tools for unsupervised and supervised objectives based on gene expression data of multiple myeloma patients which are part of the MAQC-II project.

Keywords High-dimensional • Biomarker • Genomic

8.1 Introduction

The National Institutes of Health (NIH) Biomarkers Definitions Working Group has defined a biological marker or biomarker as "a characteristic that is objectively measured and evaluated as an indicator of normal biological processes, pathogenic processes, or pharmacological responses to therapeutic intervention" (Biomarkers Definitions Working Group 2001). Thus, as a measure of biological function, a biomarker can help to unravel biological mechanisms and pathways or it may help to predict the future course of a disease. High-throughput biomarker data sets consist of an extremely large number of potential markers measured in relatively few samples or patients.

M. Zucknick (✉) • T. Hielscher • M. Sill • A. Benner
Division of Biostatistics, German Cancer Research Center,
Im Neuenheimer Feld 280, 69120 Heidelberg, Germany
e-mail: m.zucknick@dkfz-heidelberg.de

The general approach to identify true biomarkers from a large list of candidates is to select a list or combination of the most promising candidates with respect to some specific objective such as tumor subtype classification, risk stratification, or treatment response, assuming that the majority of measurements is uninformative noise. In a second step, these candidates are further evaluated or validated with a low-throughput technique on a larger set of patients, and potentially used in a prospective setting of a clinical trial. In that, the analysis of high-throughput biomarker data currently starts in an exploratory manner in most cases but can provide useful tools of clinical relevance.

One example is the so-called MammaPrint® test that was developed based on high-dimensional gene expression data measurements. It allows the identification of breast cancer patients that will likely benefit—or not benefit—from certain types of chemotherapy and thus assists in therapy guidance. Alternative molecular data sources for potential biomarkers are chromosomal aberrations or genetic signatures that help in combination with established clinicopathological risk factors to describe risk groups of patients more precisely. Based on this risk classification, patients are then treated with therapies that are more or less aggressive.

The visualization techniques we present here are mostly generic in the sense that they can be used irrespectively of the underlying statistical method. In a few cases graphics are specific to a certain method or class of methods. This will be pointed out. We will briefly describe the method used without discuss its benefits and disadvantages at length.

There are various types of high-throughput biomarker data, which in summary are called omics data, derived in various fields of molecular biology such as genetics and genomics, transcriptomics, epigenomics, and metabolomics. Among the most common ones are gene expression data measured on microarrays. A microarray allows to measure the expression levels of large numbers of genes simultaneously. Genes can be represented by multiple transcripts on the microarray. On the Affymetrix platform, these transcripts are called probe sets. Here we will use the generic term "feature" to refer to probe sets, transcripts, genes and biomarkers, if not otherwise stated.

Throughout the chapter we use 2 data sets of gene expression measurements from a total of 554 multiple myeloma patients (Barlogie et al. 2006). These data sets are part of the MicroArray Quality Control Project (MAQC-II) (Shi et al. 2010) and publicly available on the Gene Expression Omnibus (GEO) database (http://www.ncbi.nlm.nih.gov/geo) as series record GSE24080 from where we obtained the pre-processed and normalized data. Expression of transcripts (probe sets) has been measured with Affymetrix Human Genome U133 Plus 2.0 arrays for both sets of patients. Note that after pre-processing and normalization the expression data are on the $\log_2$ scale. The first set comprises 340 patients who have been enrolled on the Total Therapy II (TT2) trial between 2000 and 2004. This set will be used as training set. The second set consists of patients ($n=214$) enrolled on the subsequent study, Total Therapy III (TT3), between 2004 and 2006 and is used as validation set. For all patients clinicopathological and cytogenetic data as well as follow-up data have been collected. For example, we will look at the number of focal lesions measured with magnetic resonance imaging (MRI). (Walker et al. 2007) showed that patients with more than 7 focal lesions found in MRI constitute a high risk group.

In most situations it is reasonable to reduce the number of genomic features before performing any analysis as the majority of features is expected to show such a small amount of variation so that relevant effects cannot be reliably detected. We apply unspecific filtering selecting the 10,000 most variable features.

8.2 Unsupervised Methods

Unsupervised methods are used to find hidden structure in the data. Since unsupervised learning is independent of any associated external features or classes like clinicopathological parameters or disease prognosis, visualization is an important tool for the detection of structure.

Seriation methods (Arabie and Hubert 1996) are used to order features and/or samples based on their similarities. Similarly, clustering methods find subgroups based on the similarity between features and/or between samples. These methods take the data as they are and regroup them to make subgroups visible. Many clustering methods have been proposed in the literature. Here we consider two types of clustering methods that are most commonly used in practice, hierarchical clustering and partitioning methods. Hierarchical clustering results can be represented by dendrograms (Fig. 8.1), and partitioning methods can be visualized by plotting the distances of individual cluster members to the cluster centroids.

In addition to this approach of visualization by ordering and grouping similar features together, another possibility for making structures visible is to reduce dimensionality by removing noise. Dimension reduction methods like principal components analysis and multidimensional scaling map the data to a lower dimensional space, maximizing the represented variance with the aim of removing noise and making structures visible. It can also be helpful to visualize clustering results in a low-dimensional scatterplot which represents the data in terms of the first axes of the space, onto which the data were mapped by dimension reduction methods, and to color the features according to cluster membership (Fig. 8.7C).

8.2.1 Heatmaps and Clustering

The heatmap has been the most effective way of displaying high-dimensional array data ever since it was first proposed as being particularly suitable for gene expression array data in the seminal article by Eisen et al. (1998). A heatmap shows a data matrix organized in samples (columns) by features (rows), where individual data values are color-coded according to a color key as shown in Fig. 8.1. Typically, the data need to be sorted to make patterns visible, which is mostly done separately for

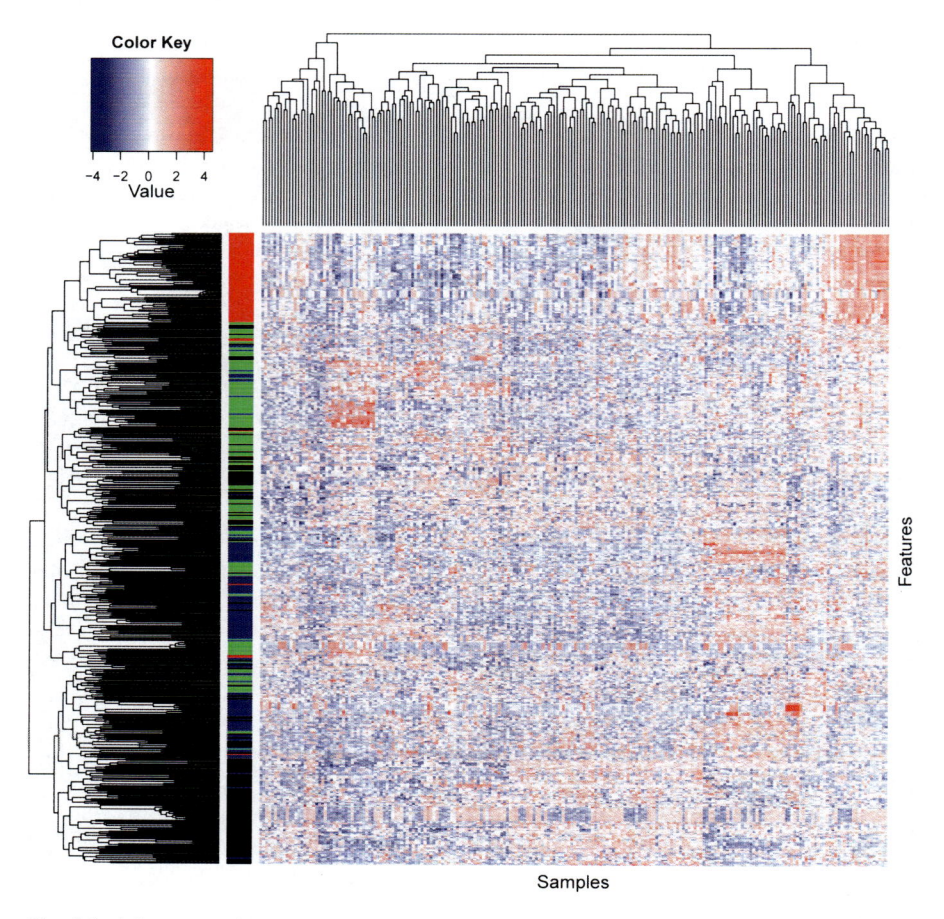

Fig. 8.1 A heatmap with hierarchical complete linkage clustering of samples and features using the Euclidean distance. Dendrograms indicate the hierarchical clustering results, while feature subgroups identified with k-means clustering ($K=4$) are highlighted by the colored stripe on the left of the heatmap

the rows and columns of the data matrix. The sorting is often achieved by hierarchical clustering methods, in which case the clustering results can be displayed alongside the heatmap as row and column dendrograms as in Fig. 8.1.

We demonstrate the use of unsupervised methods with the multiple myeloma test data set with 214 samples, where we reduced the number of features by unspecific filtering to the top 500 features with the largest variance. The data are centered and scaled so that all features have 0 mean and unit variance in order to avoid giving larger weight to features with a large variance across samples.

8.2.1.1 Hierarchical Clustering

For displaying the data in the form of a heatmap in Fig. 8.1, hierarchical clustering has been applied, first for ordering the samples to ensure that similar samples are arranged close together, but also for similar genes to be located close to each other. In particular, we chose complete linkage hierarchical clustering with a Euclidean distance metric. Complete linkage clustering is an agglomerative hierarchical clustering method, where the distance between two clusters is determined by the maximum distance between any two data points in the clusters. This method typically results in compact clusters with roughly equal diameters. Figure 8.1 is an example for a heatmap produced with the function `heatmap.2` from the R package `gplots` and Fig. 8.4 illustrates the use of function `heatmap_plus` from the R package `Heatplus`.

Both, sample-wise clustering and gene-wise clustering, are done independently, hence such heatmaps are not necessarily the best method for finding combined subsets of genes and samples, i.e., subsets of genes that show a specific expression pattern in only a subset of samples. For finding such combined subsets, biclustering methods are appropriate. They are introduced below.

8.2.1.2 Partitioning Methods

Partitioning methods map a distance matrix into a pre-specified number of clusters. As an example of partitioning methods we use k-means clustering (R function `kmeans`), which works by repeatedly assigning all samples to one of K clusters based on which cluster centroid it is closest to. The algorithm is started by choosing K data points randomly as the initial cluster centroids. After the first round of cluster assignments, the new cluster centroids are computed as the means of the newly found clusters. Then the algorithm is run iteratively until convergence, i.e., until the cluster assignments do not change anymore.

In Fig. 8.1 we have summarized the results of k-means clustering of the features with $K=4$ by the colors of a stripe next to the heatmap with hierarchical complete linkage clustering. Note that the k-means clustering results do not agree perfectly with hierarchical clustering results, even though in both clusterings a Euclidean distance metric was used. Figure 8.2 shows a silhouette plot of the k-means clusters (function `silhouette` from the R package `cluster`). The silhouette value of a data point is defined as the scaled difference between the average dissimilarity of a point to all points in its own cluster to the smallest average dissimilarity to the points of a different cluster; hence, large silhouette values indicate good separation (Rousseeuw 1987). Since the silhouette values are generally small and even negative for some data points assigned to clusters 1, 3, and 4, the clusters are not very compact and not well separated. The most compact cluster is cluster 2 that is equivalent to a distinct gene cluster in the hierarchical clustering dendrogram in Fig. 8.1. The color coding of the k-means clusters is the same throughout this section, i.e., cluster 1 = black, cluster 2 = red, cluster 3 = green, cluster 4 = blue.

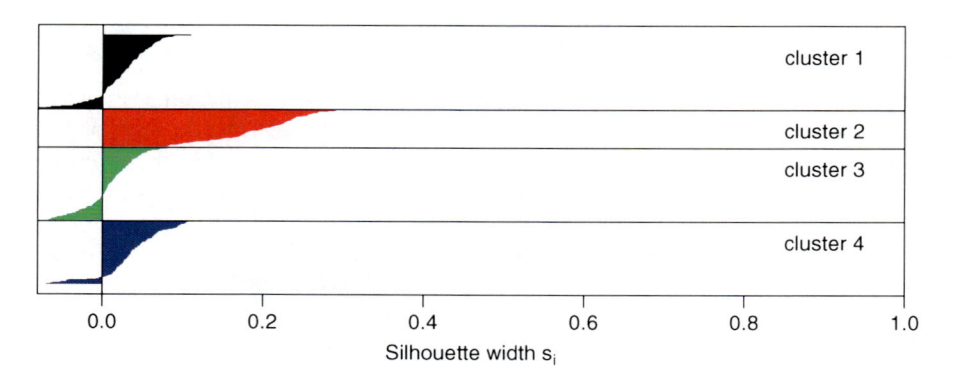

Fig. 8.2 A silhouette chart illustrating scaled average dissimilarities of genes within a k-means cluster relative to the nearest neighbouring cluster

While the silhouette plot can help to evaluate the compactness of individual clusters, it is not helpful for assessing how many clusters would be most useful. In our k-means example we have assigned all features into $K=4$ clusters, when a separation into 3 or 5 clusters might actually be more appropriate. Consensus clustering can be a useful tool in this context. Consensus clustering methods assess the stability of a particular cluster assignment by resampling methods (Monti et al. 2003). Summary statistics such as the areas under the curve (AUC) for cumulative density functions of consensus matrices and the corresponding AUC differences by cluster number (ΔK) can be displayed in "ellbow-type" plots to identify the optimal cluster number. However, in addition to the summary statistics, visualization of the consensus matrices themselves, for example by heatmaps, is always recommended (Fig. 8.3). The empirical density functions and corresponding ΔK-plot (Fig. 8.3A and B) indicate that at least 4 clusters are identified. The heatmap of the consensus matrix for $K=4$ shows that 4 clusters result indeed in a good separation between features, but there is some indication that a further separation into 5 or even 6 clusters might also be useful. Consensus clustering was performed using the R package `clusterCons`.

Graphics constructed with unsupervised learning methods can be overlaid with clinical data (Fig. 8.4). This enables to identify potential correlations between clinical factors and expression patterns identified in an unsupervised manner (exploratory analysis) or to visualize correlations that have already been established by supervised methods. Clinical data are shown for focal lesions (>7 lesions versus ≤7 lesions), cytogenetic aberrations (yes versus no), and $\log_{10}$ beta-2-microglobulin levels. The plot does not highlight any obvious correlations between sample cluster membership and clinical features, but a χ^2-test on the independence between the number of focal lesions (≤7 versus >7) and the 8 sample clusters rejects the null hypothesis at the 0.05 level ($p=0.04$), which is mostly due to the cluster located at the right (colored in dark orange) having more samples with >7 focal lesions than with ≤7 lesions. Note that this sample cluster seems to have consistently high gene

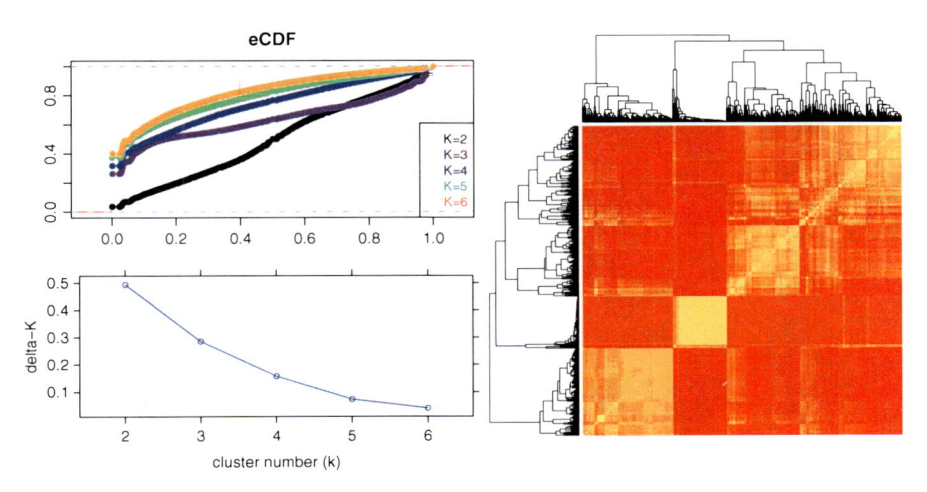

Fig. 8.3 Consensus clustering results for k-means clustering. The empirical cumulative density functions (eCDF) are displayed for k-means clustering with $k = 2,\ldots,6$ clusters (A, *top left*) with corresponding ΔK-values (B, *bottom left*). The consensus matrix for $K = 4$ clusters is displayed as a heatmap (C, *right*)

expression values in those features identified earlier as the compact K-means feature cluster 2 that is color-coded in red in the associated stripe plot in Fig. 8.1.

8.2.1.3 Biclustering

In case of traditional one-way clustering methods such as hierarchical clustering and k-means clustering, the similarity between samples is measured by means of distance metrics, e.g., the Euclidean distance. Usually, these distance metrics between 2 samples are a weighted sum of the pairwise differences between all features corresponding to these samples.

A typical problem arising in clustering high-dimensional data is that many features are probably irrelevant or redundant. These irrelevant features can confuse traditional one-way clustering approaches and therefore hide clusters that may be present in a more relevant lower dimensional space. In other words, a cluster of samples present in a microarray data set may share a common gene expression pattern only for a subset of relevant features. This cluster might not be detected by one-way clustering methods that define similarity between samples according to all features in the data set. Similarly, if the aim is to cluster the genes, a group of genes may only be coregulated within a subset of the samples. To find such clusters, other concepts that find similar samples in the relevant lower dimensional spaces are required, for example the concept of biclustering.

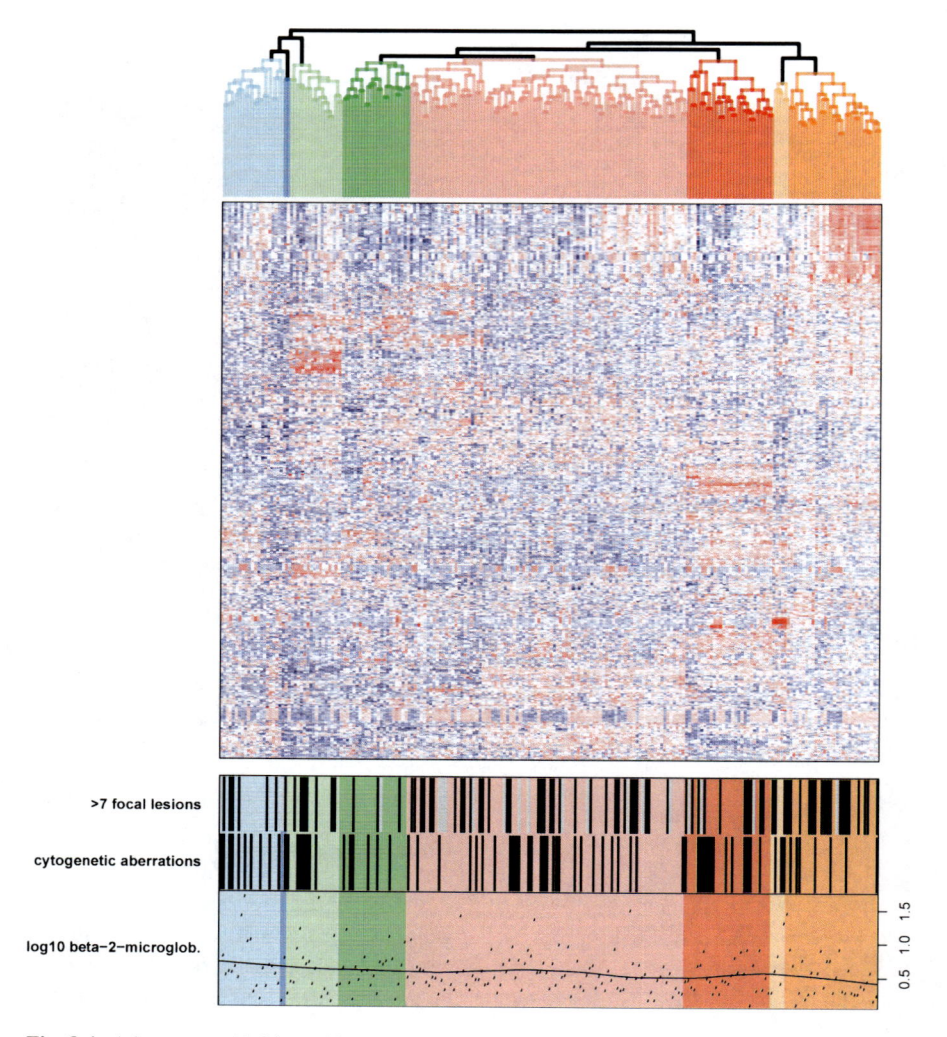

Fig. 8.4 A heatmap with hierarchical complete linkage clustering of samples and genes using the Euclidean distance metric (ordering identical to Fig. 8.1). Clinical data are shown for focal lesions (>7 lesions versus ≤7 lesions), cytogenetic aberrations (yes versus no) and $\log_{10}$ beta-2-microglobulin levels. The largest sample clusters are indicated by background colors in the plot representing the clinical data as well as in colored dendrogram arms

Biclustering is the simultaneous clustering of the rows and the columns of a data matrix, where the aim is to find submatrices whose entries show similar data patterns. Many biclustering algorithms allow the resulting biclusters to overlap, i.e., the same rows and columns can be assigned to different biclusters. Since genes are known to be involved in different biological pathways, the concept of overlapping biclusters is an additional reason why biclustering has gained popularity in the analysis of gene expression data. Comprehensive reviews about the concept of

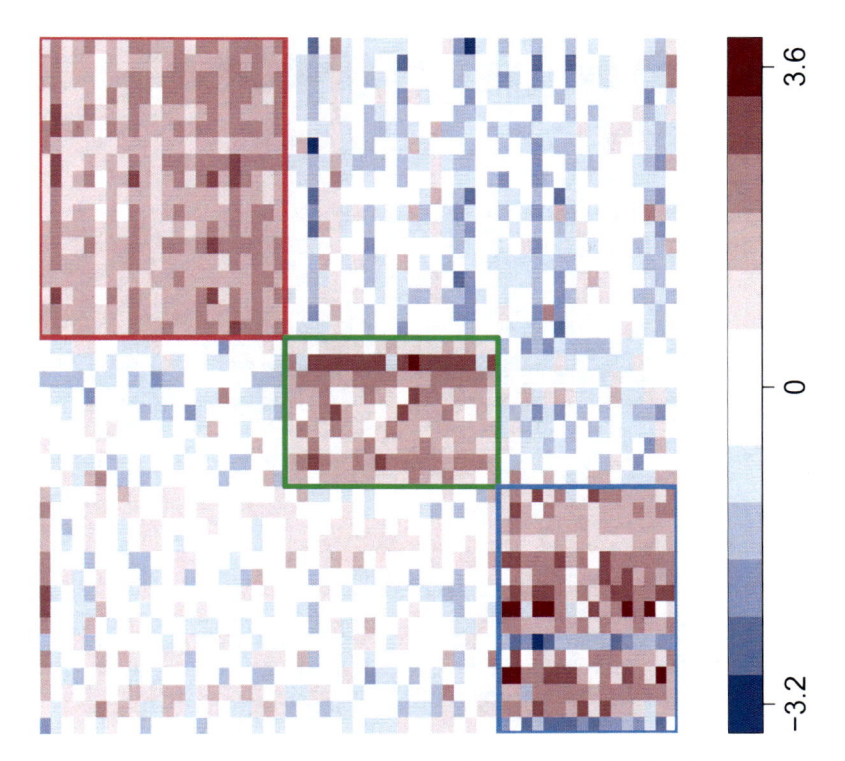

Fig. 8.5 A heatmap plot that shows three biclusters that have been identified in the example data set. The heatmap shows only those genes and samples that have been selected in at least one bicluster. The *colored rectangles* indicate the genes and samples that correspond to the three biclusters

biclustering and the different biclustering approaches were published by Madeira and Oliveira (2004) and Mechelen et al. (2004). The heatmap in Fig. 8.5 shows 3 biclusters that were identified by applying the recently proposed S4VD-algorithm (Sill et al. 2011) (R package s4vd).

Furthermore, Fig. 8.6 displays a parallel coordinates plot of one of the detected biclusters, more specifically the bicluster highlighted by a blue rectangle in Fig. 8.5. Parallel coordinates are a common visualization method to show multidimensional data points in 2 dimensions. Typically, each dimension in the data is visualized by an axis. The axes are organized as uniformly spaced vertical lines. Each data point is then visualized as a polyline connecting the axes. Figure 8.6 shows that if a bicluster is visualized by parallel coordinates the axes can either represent samples or features.

8.2.2 Dimension Reduction Methods

Dimension reduction methods attempt to map the data to a lower-dimensional space. One common method is principal components analysis (PCA, e.g., Jolliffe 2002).

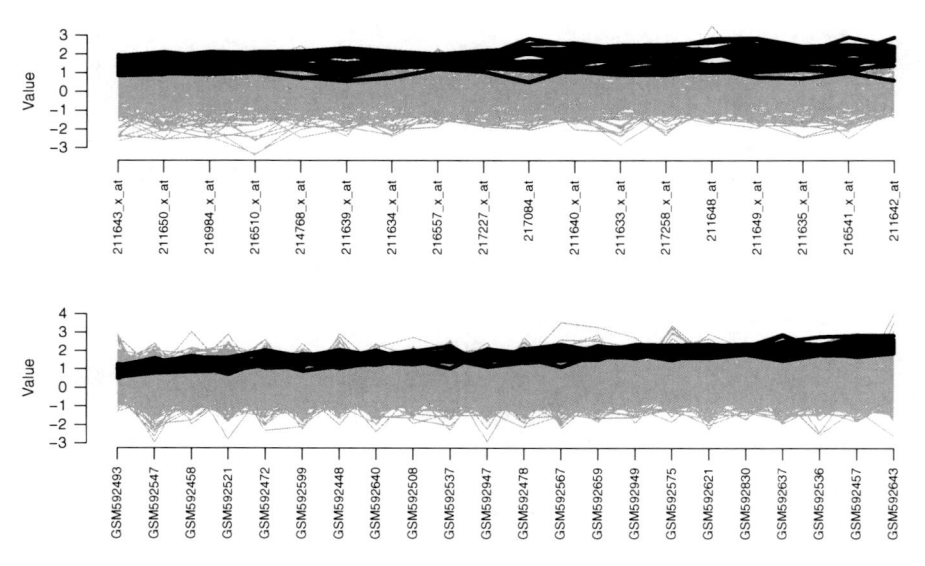

Fig. 8.6 A parallel coordinates plot of a single bicluster. In the upper parallel coordinates plot each *black line* corresponds to a sample that has been assigned to the bicluster while the *gray lines* correspond to the remaining samples in the data set. The lines display the gene expression values of the samples along the *x*-axis. Each tickmark of the *x*-axis represents a gene that has been assigned to the bicluster. In the lower parallel coordinates plot the lines correspond to genes and the tickmarks of the *x*-axis to the samples that have been assigned to the bicluster

PCA is based on singular value decomposition of the data matrix X with columns and rows representing samples and features, or equivalently the eigenvalue decomposition of the covariance matrix X^TX. Since each resulting orthogonal dimension v_i (= principal component) corresponds to an eigenvalue e_i and the sum of the eigenvalues give the total variance of the data set, each principal component (PC) v_i explains $e_i \times 100\%$ of the total variance.

In addition to PCA that uses a linear transformation of the data to the lower-dimensional space based on the Euclidean metric, there are many other dimension reduction methods. They can be linear or nonlinear and use many different metrics as well as non-metric approaches. Examples include kernel PCA, metric and non-metric multidimensional scaling, independent component analysis, self-organizing maps, and many more. In general, the results of all these methods can be visualized in the same manner by plotting samples and features in the reduced dimensional space using biplots, feature loadings plots, and sample score plots.

Figure 8.7 summarizes the results of dimension reduction by principal components analysis (R function `prcomp`). A very useful plot in this respect is the biplot, in which principal component contributions with regards to both observations and features are shown together. The biplot can be 2- or 3-dimensional, where any 2 (or 3) principal components (PCs) of interest can be used as the axes. Since most of the variation in the data is explained by the first few PCs, it usually makes sense to restrict plotting to these. Figure 8.7A shows the biplot for the first 2 components.

8 Graphical Displays for Biomarker Data

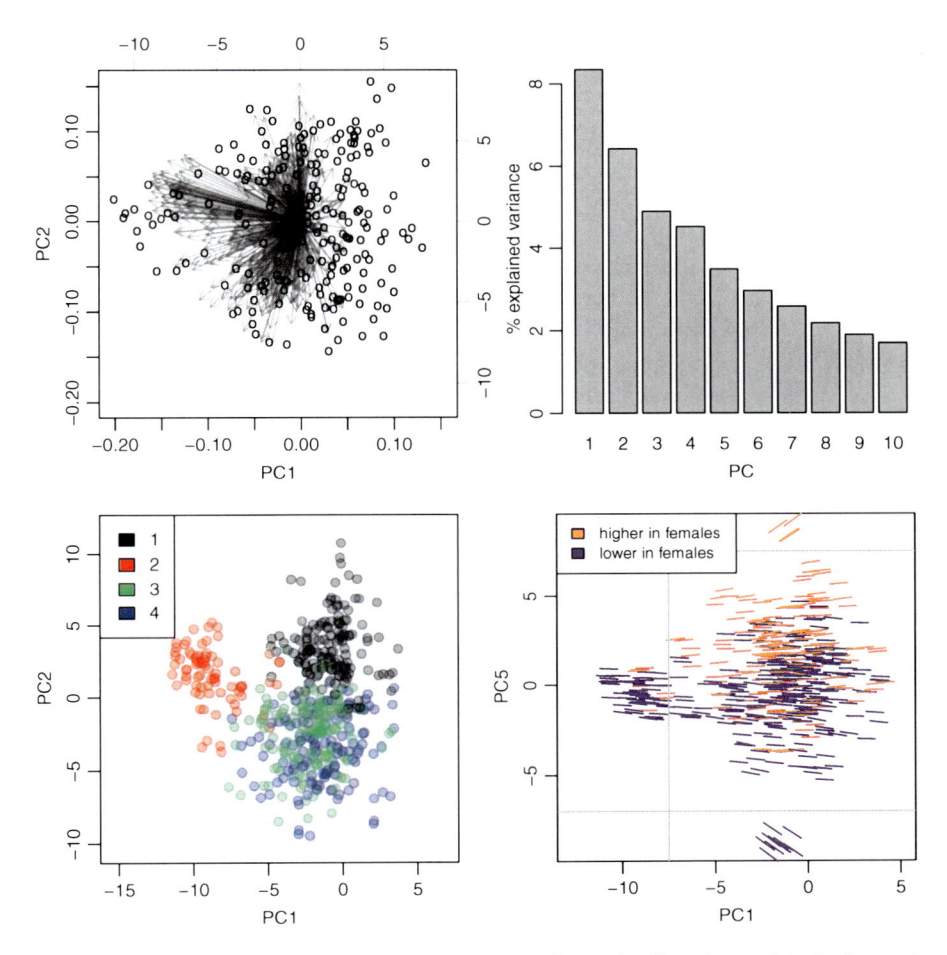

Fig. 8.7 A biplot showing the sample scores (*dots*) and feature loadings (*arrows*) in the first and second principal components (A, *top left*) and associated proportions of the total variation explained by the first ten principal components (B, *top right*). The gene loadings are again shown in gene plots (*bottom row*), where they are either colored reflecting the gene classes identified with k-means clustering ($K=4$) in Sect. 8.2.1.2 (C, *left*), or showing PC1 vs PC5 with additional patient data (D, *right*). In plot D, each gene is depicted by a line connecting the actual position of the gene in the PCA plot with the relative mean gene expression in males as compared to females, i.e., genes with higher mean expression in females vs. males are represented by a *line with a positive slope* (*in orange*), while genes with lower mean expression in females are shown as *falling lines* (*in purple*)

While the scores computed for samples are depicted as data points, the loading values associated with features are shown as arrows. Sample data points that are distant from the coordinates' origin have larger variations across feature values with respect to the principal components used as axes. When computing the distances of sample points from the axes origin, features with large loadings in the principal components of interest have larger weight than features that contribute little to these PCs. Since the coordinates of a tip of a feature arrow from the origin are proportional to the loading values of this feature in the corresponding principal components, the biplot identifies samples that are "different" from the majority of samples and at the same time illustrates nicely where these differences occur, i.e., for which features the samples show different values.

The biplot in Fig. 8.7A shows a subgroup of samples with large negative scores in the PC1 direction. At the same time, there is a subgroup of features pointing in the same direction, i.e., having large negative PC1 loadings values. The so-called gene plot in Fig. 8.7C shows the feature loadings again but plotted as colored dots, where the colors are chosen according to the k-means clusters identified in Sect. 8.2.1.2. This shows that the gene subgroup that we have identified visually by looking at the PCA biplot overlaps largely with one of the clusters identified by k-means clustering (cluster 2 colored in red).

The barplot of the proportions of variation explained by the first 10 individual principal components (Fig. 8.7B) shows that in this data set there seem to be many different sources of variation that cannot easily be explained by a few main principal components. In fact, the first 2 PCs together only explain 15% of the total variation in the data, while the first 10 PCs explain 39%. Therefore, it makes sense to look at plots visualizing more principal components than just the first 2. Figure 8.8, showing the pairwise gene loadings plots for the first 5 PCs (function pca from the R package pcaMethods), highlights an interesting pattern in the 5th component, created by 2 very small gene subsets far apart from the bulk of the genes. The gene plot of PC1 versus PC5 is also shown in Fig. 8.7D. As an additional feature, we demonstrate how to relate the gene expression data as summarized by the principal components with patient data, in particular with gender.

The 3 subgroups of genes, which we have identified as being "different" from the bulk of genes by principal components analysis (separating lines in Fig. 8.7D) are listed in Tables A.1 (subgroups in PC5) and A.2 (subgroup in PC1) in the appendix. If a subset of identified features is small, as is the case with the PC5 subsets ($n=6$ and $n=11$), it is easy to get an overview of all features by simply looking at the results in tabular form. However, if the number of interesting features gets larger, looking for information in a large table becomes cumbersome and confusing (e.g., Table A.2 for the PC1 subset), and graphical methods are necessary to visualize and summarize the results.

One example is the word cloud in Fig. 8.9 (R package wordcloud) that summarizes the words occurring in the gene titles in Table A.2. The word cloud visualizes the fact that the PC1 gene subgroup contains many immunoglobulin-related genes since the word immunoglobulin stands out most because it occurs most frequently in the text. This reflects biological knowledge that immunoglobulin levels are known to vary between multiple myeloma subtypes. Word clouds have been developed in the field of information visualization rather than statistical graphics (for an interesting debate on the difference between information visualization and statistical graphics and the use - and potential misuse - of infographics in statistical data analysis see Gelman and Unwin 2012). A word cloud is occasionally used as

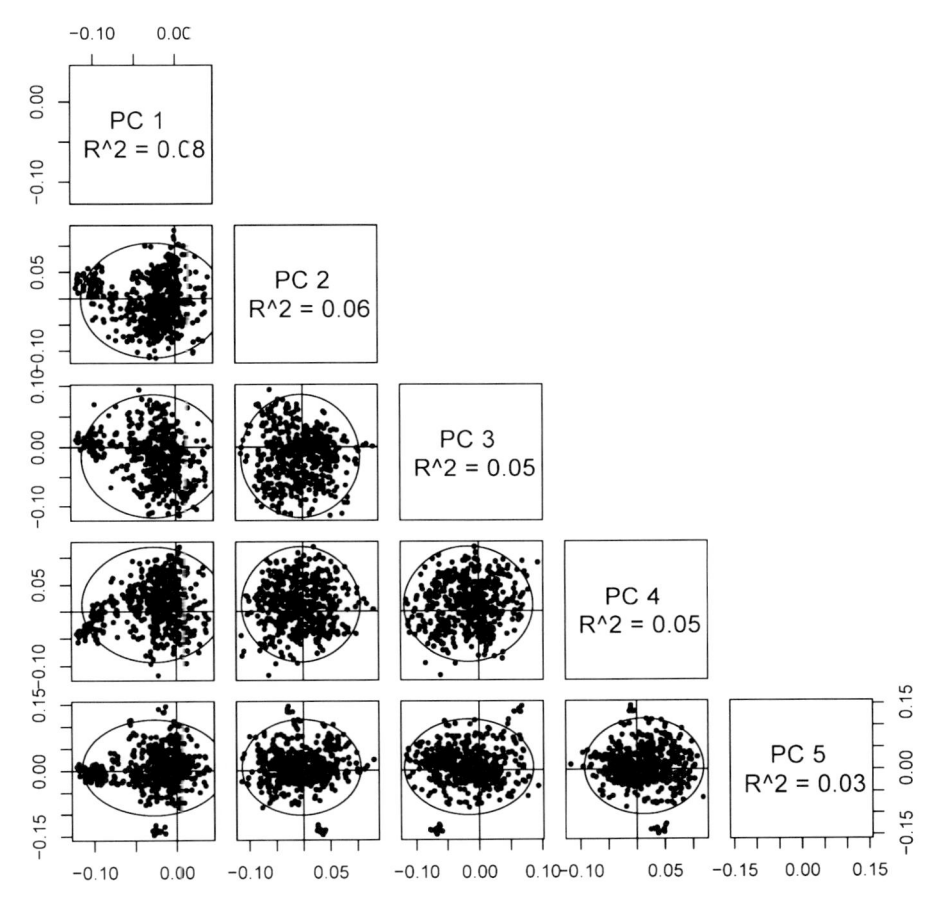

Fig. 8.8 Pairwise feature loadings plots of the first 5 principal components with 95% confidence ellipses based on the Hotelling T^2 statistic

a graphical tool in text mining, visualizing the contents of a frequency table through the use of font sizes and font colors to reflect word frequencies. A more biologically directed analysis of gene sets can be performed with gene set enrichment analysis (GSEA) and global tests. Such analyses are briefly introduced in Sect. 8.3.3.

Of course, since we have identified these subsets in a purely unsupervised manner, we do not know whether they are associated in any way with the disease under investigation, i.e., whether they correlate with disease subtypes or clinical progress of the disease.

We can use gene loadings plots to visualize any existing correlation of gene expression with patient data such as gender (or alternatively with more useful clinical information, e.g., patient survival times) by replacing the dots in Fig. 8.8 by colored icons (Fig. 8.7D). This approach of combining expression data in gene plots with corresponding phenotypical data via icons has been suggested by Pittelkow and

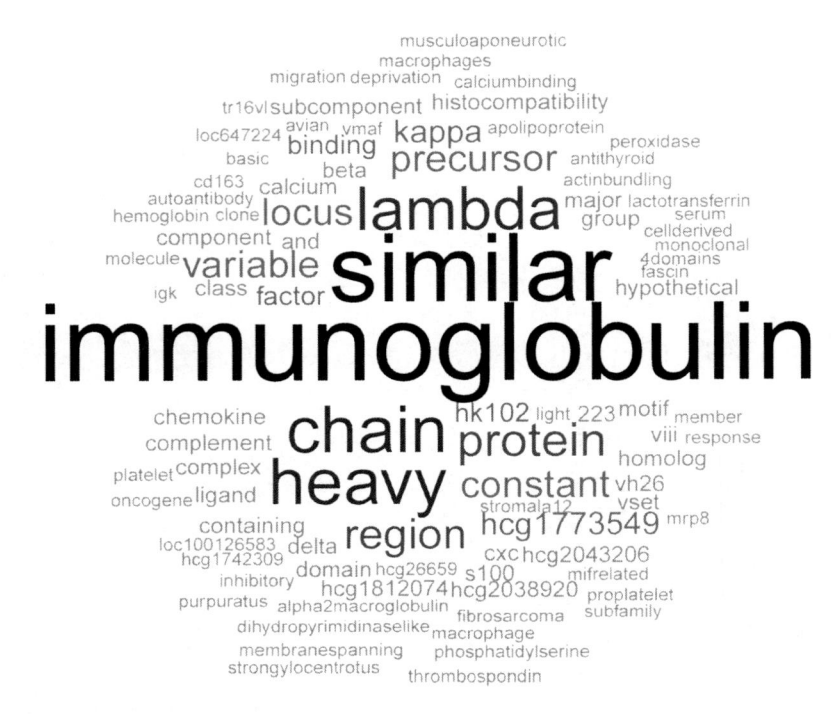

Fig. 8.9 Word cloud depicting the words contained in gene titles of gene subset 1 as identified by the first principal component. Font size and color choice reflect the word frequencies

Wilson [2003]. Figure 8.7D shows that one of the small gene subsets identified with respect to the 5th PC consistently shows higher gene expression in male patients (subset 2, PC5 < −7.0), while in the other subset higher gene expression is observed in female patients (subset 3, PC5 > 7.5). This is not surprising, as Table A.1 reveals, because all features in subset 2 lie on the Y chromosome and all features in subset 3 on the X chromosome.

8.2.3 Summary

By unsupervised analysis and associated graphical methods of the multiple myeloma test set with 214 samples and the 500 genes with largest variances we have identified several potentially interesting subgroups of genes and samples that behave differently from the bulk of features and samples. We have aimed at characterizing the identified sample subsets by linking them with clinical features of these patients (in a purely descriptive manner). We can gain more insight into communalities among the identified genes by feature set testing as illustrated in Sect. 8.3.3.

While it is an advantage of unsupervised analysis that one is free of prespecified questions and assumptions about the data, one disadvantage is the inherent danger of "over-mining" the data and over-interpretating the results. One has to keep in mind that the entire unsupervised analysis is exploratory and hypothesis-generating only. Validation of the results must always follow, for example, in terms of hypothesis testing in a controlled experiment.

8.3 Supervised Methods

The purpose of supervised methods is to identify features with differences between a priori known conditions (feature selection) or to model and develop rules to distinguish conditions (class prediction). Such conditions are typically externally defined phenotypes such as clinicopathological parameters, disease entities, or prognoses.

We illustrate 3 aspects of supervised methods in the following section. First, we are in interested in any single feature that is differentially expressed between 2 conditions or classes of samples. This is realized in an univariable feature selection. Instead of looking at single features, one can also compare the expression of a set of features between conditions (Sect. 8.3.3). This might be more adequate in situations where one expects many small rather than few large effects or if one is interested in the regulation of functionally defined feature sets such as genetic pathways.

Classification or class prediction aims at identifying a multivariable decision rule that allows a reliable discrimination of classes and predict new patients. For the sake of biological interpretation, we are interested in classification rules that consist only of a limited number of features and thus combine class prediction with feature selection. In our example we group patients according to the number of focal lesions measured by MRI.

8.3.1 Univariable Feature Screening

From the large variety of methods to identify differentially expressed features between 2 classes, we utilize the empirical Bayes approach by Smyth (2004) to fit a separate linear model for each feature. Testing for differential expression between 2 classes without accounting for additional factors yields a simple t-test problem. The empirical Bayes approach borrows information across genes to improve the stability of estimates of the gene-wise variability that is used to calculate the so-called moderated t-statistic test. Analyses are carried out with the `Bioconductor` package `limma` (Smyth 2005). We perform two-sided tests in order to select all features that are differentially expressed in patients with more than 7 focal lesions versus 7 or fewer focal lesions irrespective of the direction of regulation.

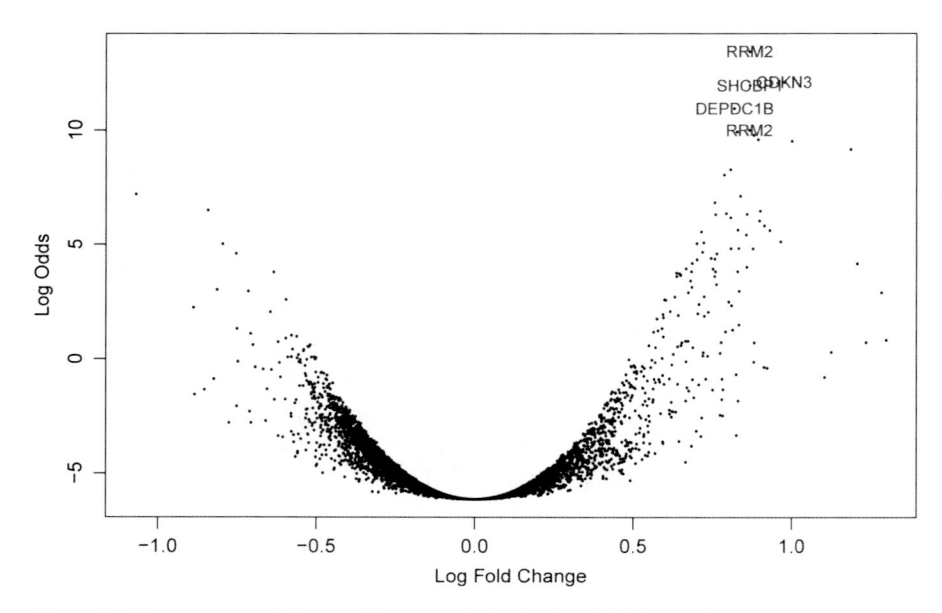

Fig. 8.10 Volcano plot of all tested features displaying log-fold change between groups versus log-odds of differential expression. The top 5 features are labeled with their gene symbol

One hundred and seventy features are identified after adjusting for multiple testing in order to control the false discovery rate using the Benjamini–Hochberg (1995) correction at the significance level of 0.05. Figure 8.10 shows a volcano plot of all single tests. This volcano plot displays the $\log_2$-fold change versus the log odds of the model, the latter being the log-posterior odds of differential expression. Instead of the log odds, any measure of statistical significance such as a test statistics or *p*-value can be used to create a volcano plot. Potentially interesting features are located at the upper left and right. We see that almost all (absolute) log fold changes are less than, and that more up- than down-regulation occurs in patients with many focal lesions. The 5 most regulated features have been highlighted and labeled with their gene symbol, all of them showing up-regulation. Two features belong to the same gene.

Simple boxplots (Fig. 8.11) are helpful to illustrate the effects of selected features in more detail. As already seen in the volcano plot, no extreme differences in terms of $\log_2$ ratio are observed between classes. Instead, significance seems to be driven by the large number of samples with fairly balanced groups.

8.3.2 Classification and Class Prediction

Classification or class prediction aims at finding a decision rule, in our case based on gene expression measurements, that allows to discrimate groups and to assign

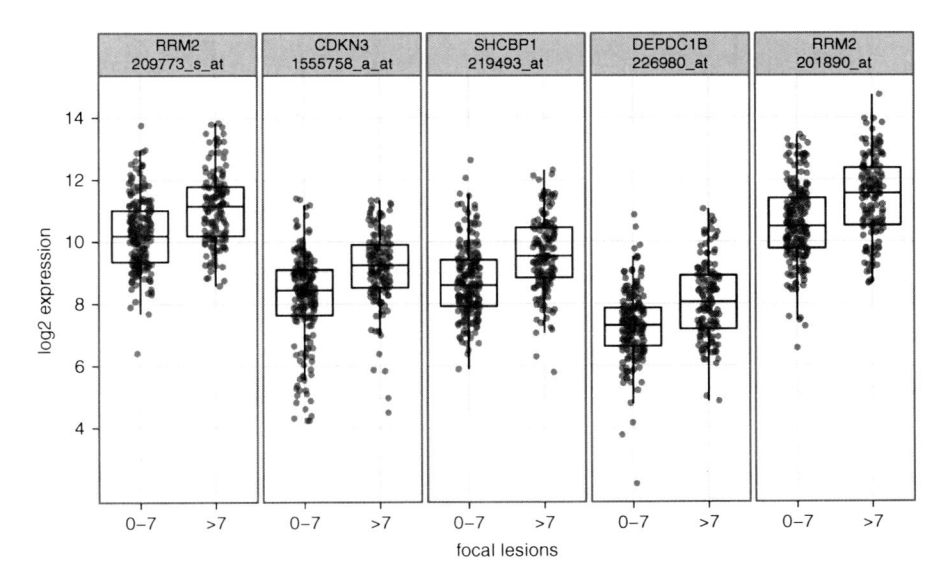

Fig. 8.11 Boxplots of log₂ expression for 5 most differentially expressed features. Two features are annotated to the same gene

new samples into the correct group. As for the univariable feature screening there are several methods available using very different approaches. We use the regularized logistic regression model with L_1-penalty, also known as lasso estimator (Tibshirani 1996), from the class of regularized regression models. Its solution is a multivariable logistic model with fewer features than samples, since the L_1-type penalty causes the majority of the model coefficients to be estimated as zero. We use the R package `glmnet` (Friedman et al. 2010) for calculations.

Most of the classification techniques that perform feature selection involve some sort of hyperparameter tuning. This hyperparameter controls the amount of shrinkage or dimension reduction. Typically, parameter tuning is based on optimizing a loss function and resampling methods such as cross-validation, permutation, or bootstrapping. For classification, possible loss functions include the misclassification rate, Brier score and—as in our case—the model deviance.

The lower panel of Fig. 8.12 displays the tenfold-cross-validated model deviance depending on the penalty parameter λ on a log scale. The larger λ the stronger the imposed regularization that results in fewer features being selected into the model as indicated on the upper x-axis. The solid vertical line marks the minimal model deviance that corresponds to a model including 27 features.

Plotting the regression coefficient path as in the upper panel of Fig. 8.12 illustrates the model building process depending on the amount of regularization. This type of visualization is specific for regularized regression models. Each horizontal line represents the coefficient profile path of a single feature. From left to right we see how the coefficients of all initially considered features are shrunk towards 0 as

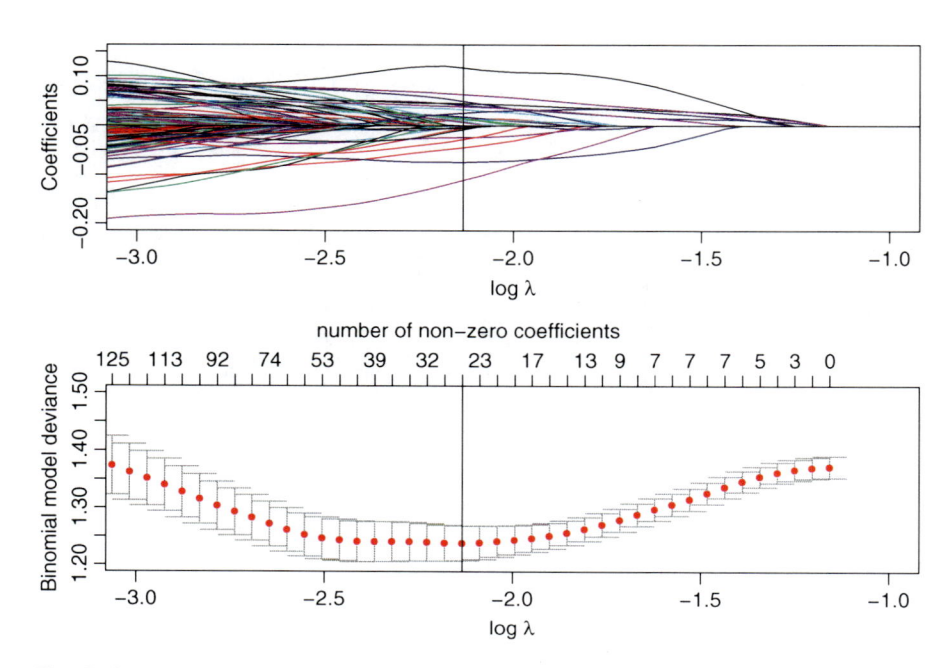

Fig. 8.12 *Upper panel*: lasso coefficient path depending on the penalty log λ. *Lower panel*: Cross-validated model deviance depending on the penalty log λ. The vertical line indicates the best model. The upper *x*-axis gives the number of nonzero coefficients included in the model

regularization increases. Since most of the coefficients are eventually estimated as 0, feature selection is implicitly performed. The vertical line indicates the selected model based on the cross-validated model deviance.

The accuracy of the classification is evaluated on the independent validation data set. Classification of new samples is not very convincing, which is not surprising since no clear separation was observed on the training data. A detailed picture of the classification is provided in the dot plots of predicted class probabilities (Fig. 8.13). The dot color indicates the classification as a patient with more than 7 (green) and up to 7 (red) focal lesions. A reasonable classifier would clearly separate both classes and consequently each pair of dots for each sample. The fact that most of the class probabilities are around 0.5 emphasizes the poor performance of the developed classifier. Plotting the ROC curve would be an alternative way of illustrating the classifier performance on the validation set.

In the case where no external validation data set is available, resampling methods need to be utilized to get a proper error estimation from the training data set. We use the R package `peperr` (Porzelius et al. 2009) to carry out these typically demanding computations. The .632+bootstrap estimator is calculated based on subsampling (Binder and Schumacher 2008) using only 50 bootstrap samples for illustration. The right panel of Fig. 8.14 displays the out-of-bag errors (dots) as well as the—too optimistic—apparent and the .632+bootstrap estimator of the misclassification rate

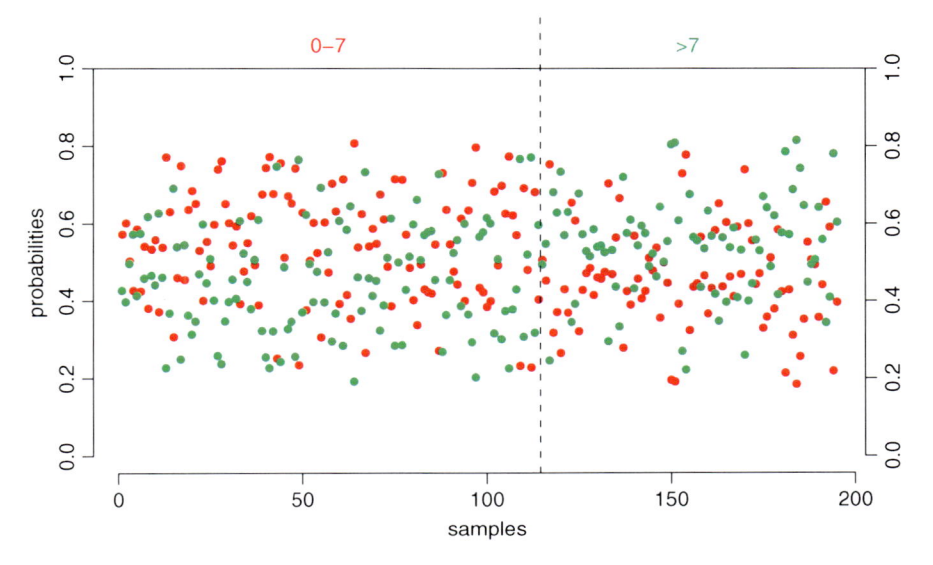

Fig. 8.13 Estimated class probabilities for each sample of the validation data set grouped by true class. Color coding indicates class assignment. A useful classifier would clearly separate each pair of *red/green dots* for each sample

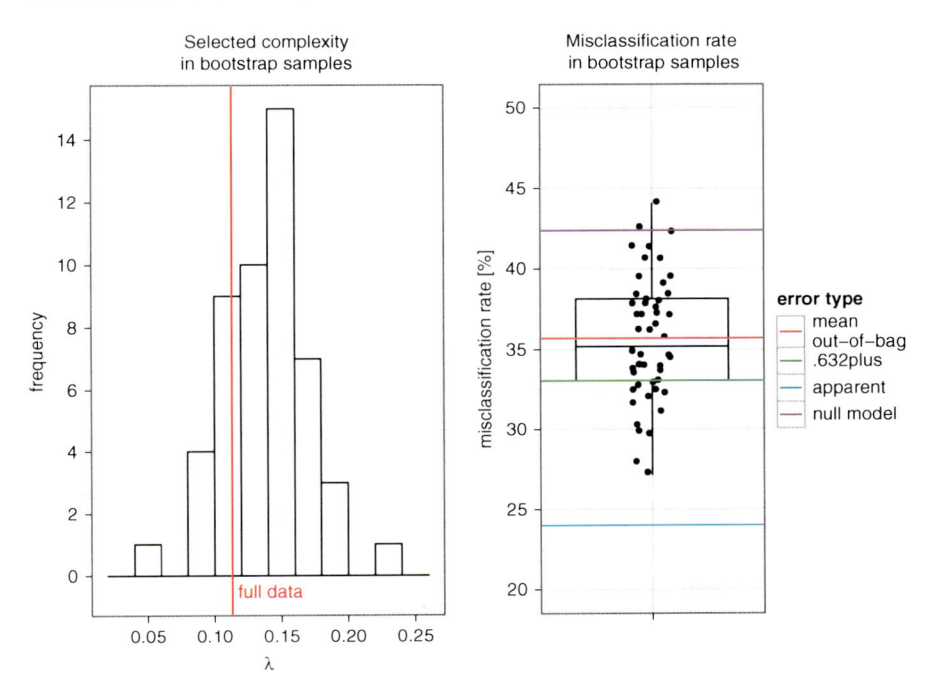

Fig. 8.14 *Left panel*: Selected penalty parameter λ. Penalty for the full model is indicated in *red*. *Right panel*: Boxplot of misclassification rates for out-of-bag bootstrap samples on training data. *Horizontal lines* indicate the average out-of-bag error, the apparent error when using the entire data for fitting and error estimation, the .632+ estimator as weighted average of both, and the error from the null model without any predictors

on the training data. The histogram of selected λ values is shown in the left panel. It allows the inspection of bootstrap fits and detection of possible issues due to the high-dimensional data structure. Depending on the choice of method and type of response, different complexity parameters and measures of prediction accuracy can be displayed.

8.3.3 Feature Set Testing

Once the regulated individual features have been identified by screening, one can analyze the biological function of the corresponding genes in order to deduce the main biological pathways. Alternatively, the opposite approach can be taken and biologically defined feature sets are tested as a whole for differences between conditions. This might be particularly helpful if regulation occurs due to many small effects. Another advantage is that it does not depend on an arbitrary threshold. On the other hand, it assumes well-curated feature sets that are possibly not available for less commonly used microarray platforms.

Again, this objective can be assessed by several methods. All these methods can be essentially divided into 2 branches investigating slightly different hypotheses as described by Goeman and Bühlmann [2007]. Methods testing the so-called self-contained null hypothesis try to determine if there is any association between the expression profile of the gene set and the condition of interest. Goeman's globaltest (Goeman et al. 2004) is one example for this class of tests. Alternatively, methods testing the so-called competitive null hypothesis assess the significance of the gene set in comparison with a randomly chosen subset of genes. GSEA is an example for this approach. One of the most popular methods for GSEA was introduced by Subramanian et al. [2005]. We present globaltest and GSEA as representatives of each approach. Figure 8.15 shows a covariates plot that summarizes the globaltest results for the cell cycle pathway (pathway 04110 in the KEGG data base, http://www.genome.jp/kegg/pathway.html) by the gene-wise z-scores that represent the contribution of each gene to the global test statistic. Strongly regulated genes have a larger impact that is reflected in a larger z-score.

The output of a GSEA typically consists of a table of gene sets and their enrichment statistics. For better interpretation of GSEA test results an enrichment map can be used (Merico et al. 2010), a weighted similarity network where the nodes represent gene sets and the edges denote the Jaccard similarity coefficient between 2 gene sets. Nodes are automatically arranged so that highly similar gene sets are placed close together. The node size represents the number of genes in the gene set. The enrichment score or enrichment p-value is mapped to the node color as a color gradient. In a two-class experiment design, node color ranges from red (high enrichment in one class) to white (no enrichment) to blue (high enrichment in the second class).

Figure 8.16 shows an enrichment map for the GSEA analysis for KEGG pathways with respect to the number of focal lesions using the R package HTSanalyzeR

04110 – Cell cycle

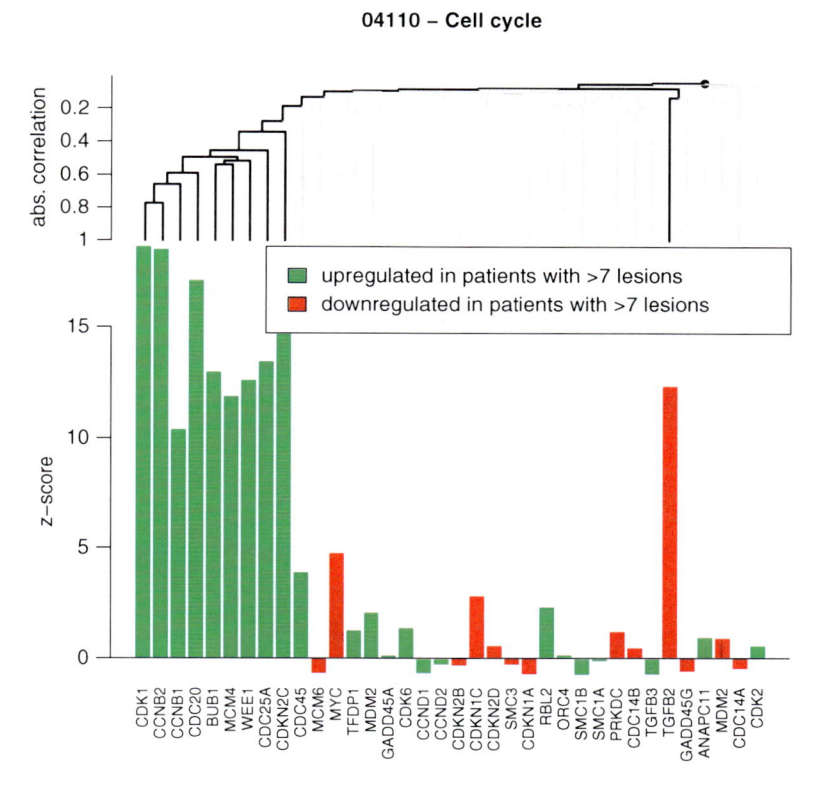

Fig. 8.15 Global test applied to gene sets from the KEGG database. The plot shows the contributions of genes to the result of the global test for the cell cycle pathway by the z-scores of the gene-wise test statistics. The coloring of the dendrogram is based on the multiple testing procedure of Meinshausen [2008] testing all subsets of genes induced by the clustering. Significant subsets are colored *black*, non-significant ones remain *gray*

(Wang et al. 2011). The enrichment map highlights biological relations between the pathways, e.g., the displayed pathways that are involved in the immune system are located together and a little separate from the other pathways (graft-versus-host disease, systemic lupus erythematosus, antigen processing and presentation).

8.4 Dynamic Graphics

The previous sections presented different graphics for the visualization of the results of supervised and unsupervised analyses of high-dimensional gene expression data and clinical data. Interactive graphics are another way to perform an explorative analysis of clinical data together with associated high-dimensional molecular data.

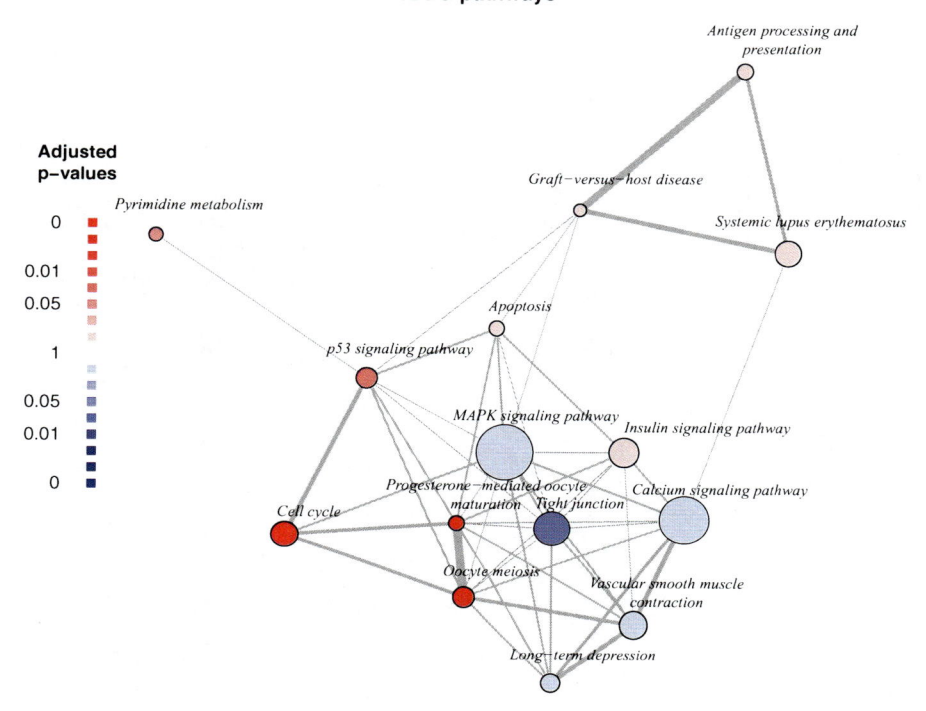

Fig. 8.16 Enrichment map of GSEA performed on KEGG pathways with respect to the number of focal lesions. Node colors are scaled according to the test *p*-values (the darker the more significant). Nodes are colored by the sign of the enrichment scores where positive scores are shown in red indicating that higher expression correlates with larger numbers of focal lesions, and negative scores are shown in *blue*. The size of nodes is proportional to the size of gene sets, while the width of edges is proportional to the overlap between gene sets, calculated using the Jaccard coefficient

An example for an interactive software that allows for an integrated explorative analysis of such data sets is SEURAT (Gribov et al. 2010).

SEURAT is a software tool that provides interactive visualization capability for the integrated analysis of high-dimensional molecular data. Gene expression data can be analyzed together with associated clinical data, array CGH (comparative genomic hybridization), SNP (single nucleotide polymorphism) array data and available gene annotations. To organize the different data types the software offers a comprehensive data manager.

The key idea of SEURAT is to overcome the typical limitations of graphics in the visualization of integrated high-dimensional data sets by linking different types of graphics. This means that all graphics generated are linked to one another, so that any transient selection of samples in one graphic will result in highlighting of the corresponding samples in all other associated graphics. Moreover, samples can be grouped and labeled using colors that will also be shown in associated graphics.

8 Graphical Displays for Biomarker Data

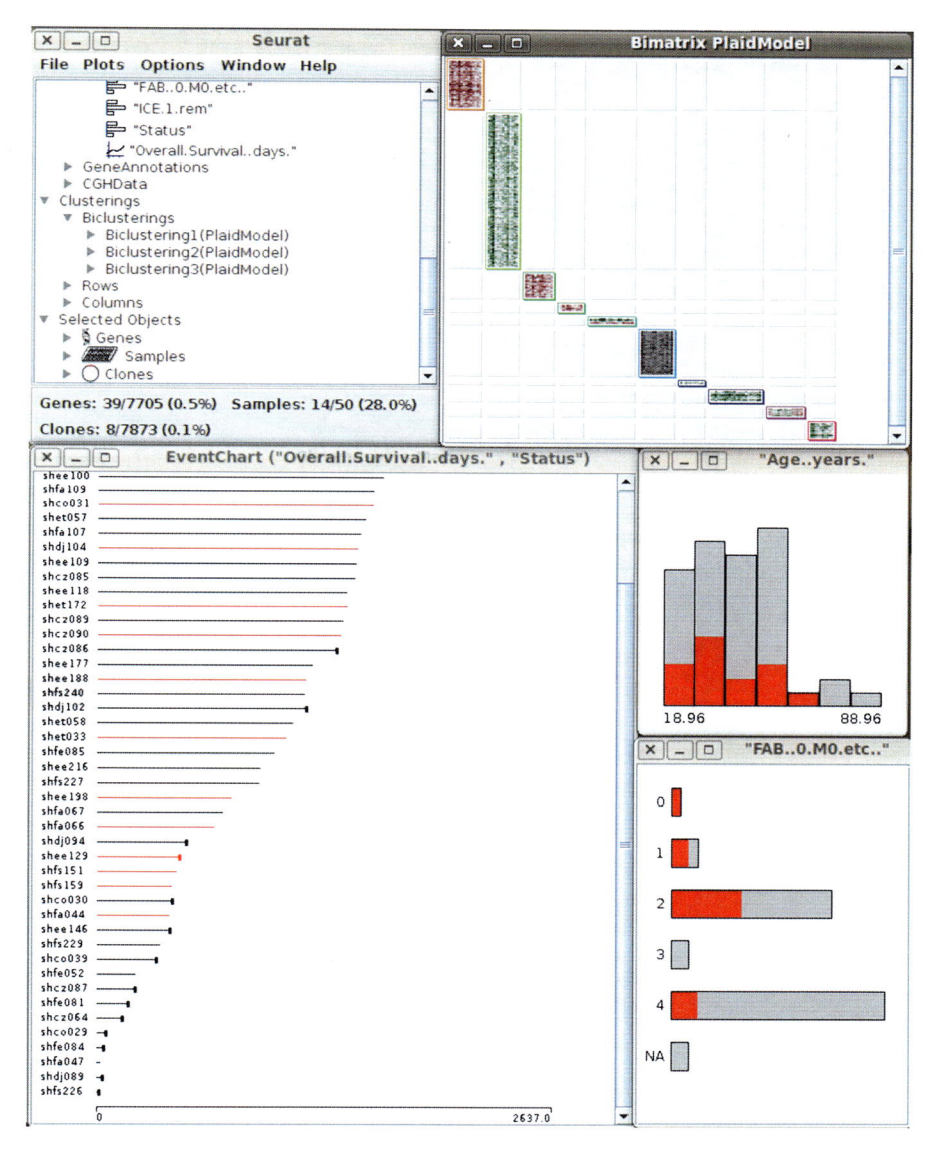

Fig. 8.17 Screenshot of an interactive analysis using SEURAT. The *upper left window* is the main window that organizes the different data sets and clustering results. The *upper right window* shows a biclustering result of a cluster analysis of a gene expression data set using the Plaid Model algorithm (Turner et al. 2005). One of the biclusters has been selected, indicated by *gray shading*. The *lower left window* shows an event chart that displays the overall survival times of the patients from which the gene expression data has been collected. The charts of the patients that have been assigned to the bicluster are highlighted in *red*. The *lower right windows* show a histogram and a bar chart plot that display additional clinical data, e.g., the age in years and the classification into disease classes. The *red highlighted parts* correspond, to the patients assigned to the selected bicluster in the *upper right plot*

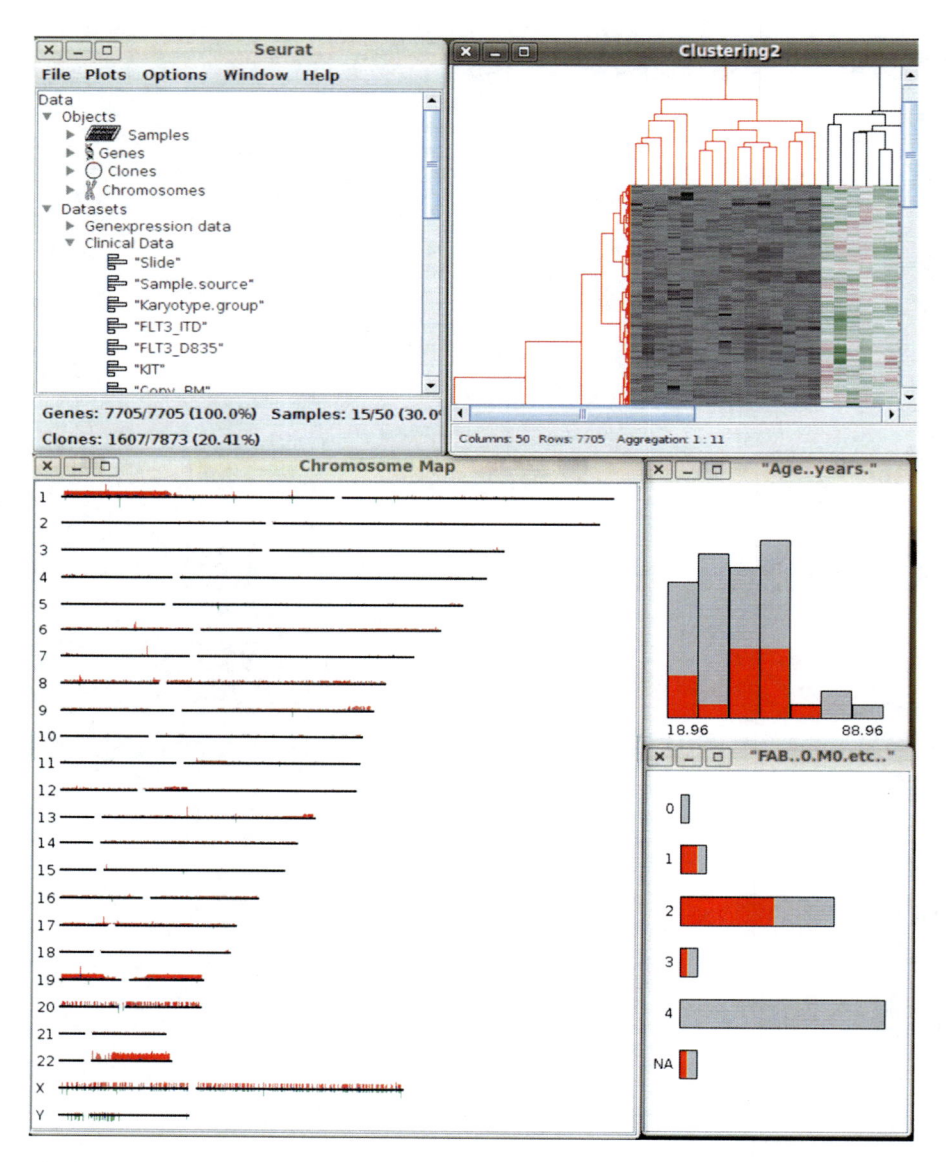

Fig. 8.18 Screenshot of an interactive analysis using SEURAT. The *upper left window* is the main window that organizes the different data sets and clustering results. The *upper right window* shows a result of a hierarchical clustering. The patients corresponding to the left part of the dendrogram were selected, indicated by *gray shading*. The *lower left window* shows a chromosome map that displays the relative frequencies of genetic gains and losses measured by array CGH. The *lower right windows* show a histogram and a *bar chart plot* that display additional clinical data, e.g., the age in years and a classification into disease classes. The *red highlighted parts* correspond to the patients that belong to the left part of the dendrogram in the *upper right plot*

For the different types of data SEURAT provides conventional graphics, e.g., gene expression data can be visualized by heatmaps encompassed by dendrograms while continuous clinical variables are displayed using histograms. Moreover, CGH data and SNP array data holding information about cytogenetic gains and losses can be explored with a chromosome browser graphic. This chromosome browser provides a global view of all chromosomes of the human genome. The relative frequencies of gains and losses are visualized by bar charts along the chromosomes. In addition, each chromosome can be explored in a larger individual plot in which cytoband information is also displayed interactively. Furthermore, time-to-event data can be visualized by so-called event charts (Goldman 1992). Unlike commonly used survival plots such as the Kaplan–Meier curve that shows aggregate information in the form of estimated survival probabilities over time, the event chart displays each individual observation by horizontal lines. This representation is more suited for the framework of interconnected graphics. When dealing with time-to-event data, possible censoring has to be considered. To display observed events within the event charts, a small vertical bar is drawn at the end of the horizontal line. A missing bar indicates that the event of interest has not been observed and thus the observation time is censored.

For exploratory data analysis the software provides different clustering algorithms such as hierarchical clustering, k-means clustering, and biclustering algorithms. To perform clustering and seriation algorithms SEURAT establishes a connection to the statistical software R.

Figures 8.17 and 8.18 are based on data from AML patients as described in Gribov et al. (2010).

8.5　A.1 Appendix

Table A.1 Subsets of genes identified in 5th principal component. Gene titles are truncated to show at most 40 characters

Subset	Affymetrix ID	Gene symbol	Gene title	Entrez ID	Cytoband
2	214218_s_at	XIST	X (inactive)-specific transcript (non-pr…	7503	Xq13.2
2	221728_x_at	XIST	X (inactive)-specific transcript (non-pr…	7503	Xq13.2
2	224588_at	XIST	X (inactive)-specific transcript (non-pr…	7503	Xq13.2
2	224589_at	XIST	X (inactive)-specific transcript (non-pr…	7503	Xq13.2
2	224590_at	XIST	X (inactive)-specific transcript (non-pr…	7503	Xq13.2
2	227671_at	XIST	X (inactive)-specific transcript (non-pr…	7503	Xq13.2
3	201909_at	RPS4Y1	Ribosomal protein S4, Y-linked 1	6192	Yp11.3
3	204409_s_at	EIF1AY	Eukaryotic translation initiation factor…	9086	Yq11.223
3	204410_at	EIF1AY	Eukaryotic translation initiation factor…	9086	Yq11.223
3	205000_at	DDX3Y	DEAD (Asp-Glu-Ala-Asp) box polypeptide 3…	8653	Yq11
3	206624_at	USP9Y	Ubiquitin specific peptidase 9, Y-linked…	8287	Yq11.2
3	206700_s_at	JARID1D	Jumonji, AT rich interactive domain 1D	8284	Yq11
3	214131_at	CYorf15B	Chromosome Y open reading frame 15B	84663	Yq11.222
3	223646_s_at	CYorf15B	Chromosome Y open reading frame 15B	84663	Yq11.222
3	230760_at	ZFY	Zinc finger protein, Y-linked	7544	Yp11.3
3	232618_at	CYorf15A	Chromosome Y open reading frame 15A	246126	Yq11.222
3	236694_at	CYorf15A	Chromosome Y open reading frame 15A	246126	Yq11.222

8 Graphical Displays for Biomarker Data   165

Table A.2 Subset of genes identified in first principal component. Gene titles are truncated to show at most 40 characters

Subset	Affymetrix ID	Gene symbol	Gene title	Entrez ID	Cytoband
1	1561937_x_at	IGHM	Immunoglobulin heavy constant mu	3507	
1	200762_at	DPYSL2	Dihydropyrimidinase-like 2	1808	8p22-p21
1	201110_s_at	THBS1	Thrombospondin 1	7057	15q15
1	202018_s_at	LTF	Lactotransferrin	4057	3p21.31
1	203382_s_at	APOE	Apolipoprotein E	348	19q13.2
1	203535_at	S100A9	S100 calcium binding protein A9	6280	1q21
1	205863_at	S100A12	S100 calcium binding protein A12	6283	1q21
1	206390_x_at	PF4	Platelet factor 4	5196	4q12-q21
1	209374_s_at	IGHM	Immunoglobulin heavy constant mu	3507	14q32.33
1	209687_at	CXCL12	Chemokine (C-X-C motif) ligand 12 (strom...	6387	10q11.1
1	210933_s_at	FSCN1	Fascin homolog 1, actin-bundling protein...	6624	7p22
1	211633_x_at	IGHG1	Immunoglobulin lambda heavy chain	3500	14q32.33
1	211634_x_at	LOC100133862	Similar to hCG1773549	3507	
1	211635_x_at	LOC642131	Similar to hCG1812074	642131	
1	211637_x_at	VSIG6	V-set and immunoglobulin domain containi...	652128	
1	211639_x_at	VSIG6	V-set and immunoglobulin domain containi...	652128	
1	211640_x_at	LOC100133862	Similar to hCG1773549	3500	
1	211642_at	IGHG1	Immunoglobulin lambda heavy chain	83695	
1	211643_x_at	LOC440871	Similar to hCG2043206	50802	
1	211644_x_at	LOC440871	Similar to hCG2043206	50802	
1	211645_x_at				
1	211648_at	LOC100133862	Similar to hCG1773549	3500	
1	211649_x_at	LOC642131	Similar to hCG1812074	642131	
1	211650_x_at	LOC100126583	Hypothetical LOC100126583	3509	
1	212998_x_at	HLA-DQB1	Major histocompatibility complex, class ...	3119	6p21.3
1	212999_x_at	HLA-DQB1	Major histocompatibility complex, class ...	3119	6p21.3
1	213674_x_at	IGHD	Immunoglobulin heavy constant delta	3495	
1	214146_s_at	PPBP	Pro-platelet basic protein (chemokine (C...	5473	4q12-q13
1	214370_at	S100A8	Calcium-binding protein in macrophages (...	6279	1q21
1	214768_x_at	FAM20B	(Clone TR1.6VL) Anti-thyroid peroxidase ...	9917	2p12

(continued)

Table A.2 (continued)

Subset	Affymetrix ID	Gene symbol	Gene title	Entrez ID	Cytoband
1	214777_at	IGKV4-1	Immunoglobulin kappa variable 4-1	28908	2p12
1	215049_x_at	CD163	CD163 molecule	9332	12p13.3
1	215621_s_at	IGHD	Immunoglobulin heavy constant delta	3495	
1	216401_x_at	LOC652493	Similar to Ig kappa chain V-I region HK1…	652493	
1	216412_x_at	IGL@	Immunoglobulin lambda locus	3535	22q11.2
1	216430_x_at	IGL@	Immunoglobulin lambda locus	3535	
1	216491_x_at	IGHM	Immunoglobulin heavy constant mu	3507	14q32.33
1	216510_x_at	IGHV@	Immunoglobulin heavy variable group	3509	
1	216541_x_at	LOC100133862	Similar to hCG1773549	3500	
1	216557_x_at	LOC100133739	Similar to hCG2038920	3509	
1	216560_x_at	IGL@	Immunoglobulin lambda locus	3535	22q11.2
1	216576_x_at	LOC652694	Similar to Ig kappa chain V-I region HK1…	652694	
1	216829_at	LOC652694	Similar to Ig kappa chain V-I region HK1…	652694	
1	216984_x_at	IGLV2-23	Immunoglobulin lambda variable 2-23	28816	
1	217084_at	IGHV@	Immunoglobulin heavy variable group	3509	
1	217227_x_at	IGL@	Immunoglobulin lambda locus	3535	
1	217235_x_at	IGLV2-23	Immunoglobulin lambda variable 2-23	3537	
1	217258_x_at	IGL@	Immunoglobulin lambda locus	3535	
1	217281_x_at	LOC652494	Similar to Ig heavy chain V-III region V…	652494	
1	217360_x_at	LOC652494	Similar to Ig heavy chain V-III region V…	652494	
1	217378_x_at	LOC100130100	Similar to hCG26659	100130100	
1	217384_x_at	LOC647224	Hypothetical LOC647224	647224	
1	217757_at	A2M	Alpha-2-macroglobulin	2	12p13.31
1	218232_at	C1QA	Complement component 1, q subcomponent, …	712	1p36.12
1	218559_s_at	MAFB	V-maf musculoaponeurotic fibrosarcoma on…	9935	20q11.2-q13.1
1	219607_s_at	MS4A4A	Membrane-spanning 4-domains, subfamily A…	51338	11q12

(continued)

Table A.2 (continued)

Subset	Affymetrix ID	Gene symbol	Gene title	Entrez ID	Cytoband
1	222717_at	SDPR	Serum deprivation response (phosphatidyl...	8436	2q32-q33
1	225353_s_at	C1QC	Complement component 1, q subcomponent, ...	714	1p36.11
1	233969_at	IGL@	Immunoglobulin lambda light chain	3535	22q11.2
1	234419_x_at	LOC100133739	Similar to hCG2038920	3509	
1	234792_x_at	LOC100131845	Similar to hCG1742309	3493	
1	240336_at	HBM	Hemoglobin, mu	3042	16p13.3

References

Arabie P, Hubert L (1996) An overview of combinatorial data analysis. In: Arabie P, Hubert L, Soete G (eds) Clustering and classification. World Scientific, River Edge, NJ, pp 5–63

Barlogie B, Tricot G, Anaissie E, Shaughnessy J, Rasmussen E, van Rhee F, Fassas A, Zangari M, Hollmig K, Pineda-Roman M, Lee C, Talamo G, Thertulien R, Kiwan E, Krishna S, Fox M, Crowley J (2006) Thalidomide and hematopoietic-cell transplantation for multiple myeloma. New Engl J Med 354(10):1021–1030

Benjamini Y, Hochberg Y (1995) Controlling the false discovery rate: a practical and powerful approach to multiple testing. J Roy Stat Soc B 57(1):289–300

Binder H, Schumacher M (2008) Adapting prediction error estimates for biased complexity selection in high-dimensional bootstrap samples. Stat Appl Genet Mol Biol 7: Article 1

Biomarkers Definitions Working Group (2001) Biomarkers and surrogate endpoints: preferred definitions and conceptual framework. Clin Pharmacol Therapeut 69:89–95

Eisen M, Spellman P, Brown P, Botstein D (1998) Cluster analysis and display of genome-wide expression patterns. Proc Natl Acad Sci USA 95(25):14863–14868

Friedman J, Hastie T, Tibshirani R (2010) Regularization paths for generalized linear models via coordinate descent. J Stat Softw 33(1):1–22. URL http://www.jstatsoft.org/v33/i01/

Gelman A, Unwin A (2012) Infovis and statistical graphics: different goals, different looks. Unpublished manuscript. URL http://www.stat.columbia.edu/~gelman/research/published/vis14.pdf

Goeman JJ, Bühlmann P (2007) Analyzing gene expression data in terms of gene sets: methodological issues. Bioinformatics 23(8):980–987

Goeman JJ, van de Geer SA, de Kort F, van Houwelingen HC (2004) A global test for groups of genes: testing association with a clinical outcome. Bioinformatics 20:93–99

Goldman AI (1992) EVENTCHARTS: visualizing survival and other timed-events data. American Statistician 46(1):13–18. URL http://www.jstor.org/stable/2684402

Gribov A, Sill M, Lück S, Rücker F, Döhner K, Bullinger L, Benner A, Unwin A (2010) SEURAT: visual analytics for the integrated analysis of microarray data. BMC Med Genom 3:21. doi:10.1186/1755-8794-3-21, URL http://dx.doi.org/10.1186/1755-8794-3-21

Jolliffe I (2002) Principal components analysis, 2nd edn. Springer series in statistics. Springer, New York

Madeira SC, Oliveira AL (2004) Biclustering algorithms for biological data analysis: a survey. IEEE/ACM Trans Comput Biol Bioinform 1:24–45. doi:http://dx.doi.org/10.1109/TCBB.2004.2, URL http://dx.doi.org/10.1109/TCBB.2004.2

Mechelen IV, Bock HH, Boeck PD (2004) Two-mode clustering methods: a structured overview. Stat Methods Med Res 13(5):363–394

Meinshausen N (2008) Hierarchical testing of variable importance. Biometrika 95(2):265–278

Merico D, Isserlin R, Stueker O, Emili A, Bader G (2010) Enrichment map: a network-based method for gene-set enrichment visualization and interpretation. PLoS One 5(11):e13,984

Monti S, Tamayo P, Mesirov J, Golub T (2003) Consensus clustering – a resampling-based method for class discovery and visualization of gene expression microarray data. In: Machine Learning 52(1–2):91–118

Pittelkow Y, Wilson S (2003) Visualisation of gene expression data - the GE-biplot, the chip-plot and the gene-plot. Stat Appl Genet Mol Biol 2: Article 6

Porzelius C, Binder H, Schumacher M (2009) Parallelized prediction error estimation for evaluation of high-dimensional models. Bioinformatics 25(6):827–829

Rousseeuw P (1987) Silhouettes: a graphical aid to the interpretation and validation of cluster analysis. Comput Appl Math 20:53–65

Shi L, Campbell G, Jones WD et al (2010) The MicroArray quality control (MAQC)-II study of common practices for the development and validation of microarray-based predictive models. Nat Biotechnol 28(8):827–838

Sill M, Kaiser S, Benner A, Kopp-Schneider A (2011) Robust biclustering by sparse singular value decomposition incorporating stability selection. Bioinformatics 27(15):2089–2097 doi:10.1093/bioinformatics/btr322, URL http://dx.doi.org/10.1093/bioinformatics/btr322

Smyth GK (2004) Linear models and empirical bayes methods for assessing differential expression in microarray experiments. Stat Appl Genet Mol Biol 3: Article 1

Smyth GK (2005) Limma: linear models for microarray data. In: Gentleman R, Carey V, Dudoit S, Irizarry R, Huber W (eds) Bioinformatics and computational biology solutions using R and Bioconductor. Springer, New York, pp 397–420

Subramanian A, Tamayo VK P Mootha, Mukherjee S, Ebert BL, Gillette MA, Paulovich A, Pomeroy SL, Golub TR, Lander ES, Mesirov JP (2005) A knowledge-based approach for interpreting genome-wide expression profiles. PNAS 102(43):15545–15550

Tibshirani R (1996) Regression shrinkage and selection via the lasso. J Roy Stat Soc B Methodological 58:267–288

Turner HL, Bailey TC, Krzanowski WJ, Hemingway CA (2005) Biclustering models for structured microarray data. IEEE/ACM Trans Comput Biol Bioinform 2(4):316–329. doi:10.1109/TCBB.2005.49, URL http://dx.doi.org/10.1109/TCBB.2005.49

Walker R, Barlogie B, Haessler J, Tricot G, Anaissie E, Shaughnessy J, Epstein J, Hemert R, Erdem E, Hoering A, Crowley J, Ferris E, Hollmig K, Rhee F, Zangari M, Pineda-Roman M, Mohiuddin A, Yaccoby S, Sawyer J, Angtuaco E (2007) Magnetic resonance imaging in multiple myeloma: diagnostic and clinical implications. J Clin Oncol 25(9):1121–1128

Wang X, Terfve C, Rose J, Markowetz F (2011) HTSanalyzeR: An R/Bioconductor package for integrated network analysis of high-throughput screens. Bioinformatics 27(6):879–880

Part III
Clinical Trial Graphics

Chapter 9
Statistical Graphics in Clinical Oncology

Kye Gilder

This chapter is dedicated to my parents, Gerald and Carole Gilder, for their courageous battles against cancer.

Abstract Oncology clinical trials are often complex leading to years of research and generation of vast amounts of data. Statistical graphics play an invaluable role in transforming these multifaceted data into crisp and simplified visuals that assist researchers to quickly and accurately study the results, detect data trends and patterns, and suggest hypotheses. In other words, "Excellence in statistical graphics consists of complex ideas communicated with clarity, precision, and efficiency." The focus of this chapter is to provide a sampling of useful statistical graphics routinely used in clinical oncology research and their utility in communicating information clearly and more efficiently than solely reviewing tables of numerical output. The graphics presented are commonly used during trial design planning, interim analyses, and final analyses of clinical efficacy data.

9.1 Introduction

Cancer is a group of diseases characterized by unregulated cell growth and spread of abnormal cells. If the spread is not controlled, it can ultimately result in death (National Cancer Institute 2012). According to the American Cancer Society, in the United States an estimated 1,638,910 new cases are expected to be diagnosed in 2012, with approximately 577,190 expected deaths (American Cancer Society 2012). Cancer is the second most common cause of death in the United States, exceeded only by heart disease. An estimated 200 types of cancer exist that affect at least 60 different body organs (CancerHelp UK 2012).

K. Gilder (✉)
Biostatistics Department, NuVasive, Inc., San Diego, CA, USA
e-mail: kgilder@nuvasive.com

Cancer can be caused by external factors (e.g., tobacco, infectious organisms, chemicals, lack of physical activity, poor diet, obesity, environmental pollutants, radiation) and internal factors (e.g., inherited genetic mutations, hormones, immune conditions) (Anand et al. 2008). These causal factors may act together or in sequence to initiate or promote cancer development. Cancers are typically treated with a combination of one or more therapeutic approaches including surgery, radiation, chemotherapy, hormone therapy, and biological therapy.

9.2 The Oncology Drug Development Process

Oncology drug development has historically been relatively consistent in approach. Current estimates suggest drug development takes on average 10–15 years from discovery to regulatory approval at a cost exceeding $1 billion (Tufts Center for the Study of Drug Development 2006). According to CenterWatch, 4 new oncology drugs have been approved so far in 2012 and 77 were approved during the period 2000–2011, with an average of 6.4 new oncology drugs per year (minimum of 2 in 2005 and maximum of 12 in 2011) (CenterWatch 2012). According to IMS Health, annual United States prescription sales in 2008 were $291 billion (Seeking Alpha 2012) with the sales of oncology products exceeding $48 billion, contributing nearly 17 % of the pharmaceutical sales that year (Reuters 2012).

Although a common development approach is generally followed, clinical trials can be as diverse and complex as cancer itself. Regulatory approval is based on the demonstration of substantial evidence of safety, efficacy, and risk-benefit obtained from well-controlled clinical trials. While clinical development paths may vary for every candidate compound or drug, the general process is similar and comprises the following stages:

- Nonclinical Research
  - Identify promising ("lead") compound, with favorable properties of administration
  - Identify promising molecular or genetic targets
  - Evaluate in vitro activity
  - Evaluate in vivo activity (animal models)
- Phase 1 Clinical Trials (includes first in human trials)
  - Evaluate the side-effects of increased dosing (adverse events and toxicity profile)
  - Estimate the maximum tolerated dose (MTD) and identify any dose-limiting toxicities (DLT)
  - Evaluate dose(s) (dose-ranging)
  - Evaluate dose schedule(s)
  - Evaluate routes of administration
  - Evaluate clinical pharmacology (food effects, drug-drug interactions)
  - Evaluate QT/QTc interval prolongations (thorough QT studies)
  - Characterize the pharmacokinetic (PK) and pharmacodynamic (PD) profiles

- Phase 2 Clinical Trials (therapeutic exploratory trials, often nonrandomized)
  - Evaluate antitumor activity (signal of clinically meaningful effect or clinical response)
  - Evaluate limited number of doses/schedules to further characterize dose schedule
  - Evaluate combination treatments and/or various disease indications
  - Evaluate clinical pharmacology (food effects, drug-drug interactions)
  - Characterize short-term safety profiles
  - Characterize short-term pharmacokinetic (PK) profiles
  - Phase 2a (pilot, proof of concept, dose-ranging in one or more combination regimens)
  - Phase 2b (well-controlled to evaluate safety and efficacy)
- Phase 3 Clinical Trials (registration or confirmatory trials, generally randomized)
  - Compare efficacy to "standard of care" treatment/regimen
  - Characterize safety profile
  - Confirm antitumor activity translates into prolonged clinical benefit
  - Confirm safety and efficacy (risk-benefit) with reasonable confidence for product labeling
- Phase 4 Clinical Trials (post-approval and post-marketing trials)
  - Evaluate optimal use, including expanded indications and different patient populations
  - Evaluate long-term safety profile

9.3 Oncology Clinical Endpoints

Traditional efficacy oncology endpoints include time-to-event endpoints such as overall survival (OS), progression-free survival (PFS), time-to-progression (TTP), event-free survival (EFS), relapse-free survival (RFS), time-to-response, response classification, and duration of response (DR) (Food and Drug Administration 2007). In addition, response classification, such as complete response (CR), partial response (PR), stable disease (SD), and progressive disease (PD) is often evaluated according to pre-specified criteria for the particular cancer being studied. Also, quality of life (QoL) measures, patient reported outcome (PRO) measures, and change from baseline for various endpoints are commonly used in clinical trials.

9.4 Oncology Statistical Graphics

With the massive amount of data collected in oncology clinical development, there is an acute need for useful, informative, and visually appealing statistical graphics. While it is impossible to present every type of statistical graph used in oncology clinical development, the focus of this chapter is to illustrate graphics routinely used. Specifically, this chapter shows selected graphics used during trial design, interim analyses, and final analyses of clinical data in an oncology clinical trial, with an emphasis on displaying efficacy endpoints clearly, precisely, and efficiently. "Excellence in statistical graphics consists of complex ideas communicated with clarity, precision, and efficiency" (Tufte 1983). With this idea in mind, a variety of statistical graphics are presented that follow the principles of Tufte and Cleveland, while limiting the theoretical or technical discussion of the statistical methods (Tufte 1983; Cleveland 1993).

Graphics related to safety, pharmacokinetic, pharmacodynamic, and biomarker analyses are not included in this chapter. However, they are discussed in select chapters in the book.

The R software package (version 2.14.1) was used exclusively for the generation of all graphics in this chapter. However, many of these statistical graphics can also be generated using other comprehensive statistical graphics packages (e.g., SAS™, JMP™, S-PLUS™). The data used and presented throughout the chapter are either R datasets (e.g., ovarian, lung) or simulated data. In addition, the following R packages were used: `Hmisc`, `rms`, `survival`, `graphics`, `lattice`, `grid`, and `RColorBrewer`. The R code can be obtained from the book's companion website or directly from the author.

9.4.1 Trial Design

Statistical graphics are useful during trial design planning and interim data monitoring, particularly when clinical development teams, medical research scientists, clinical pharmacologists, and investigators are collaborating on clinical research projects and programs.

9.4.1.1 Phase 1 Trial Design

The goal of a phase 1 cancer clinical trial is frequently to estimate the highest dose of an experimental agent associated with a tolerable level of toxicity. Although there are numerous trial designs for phase 1 oncology clinical trials, such as the continual reassessment method (CRM) (O'Quigley et al. 1990; O'Quigley and Reiner 1998), accelerated titration designs (ATD) (Simon et al. 1997), and dose escalation with overdose control (EWOC) (Tighiouart et al. 2005; Babb et al. 1998), the traditional algorithm-based designs are still widely used because of their

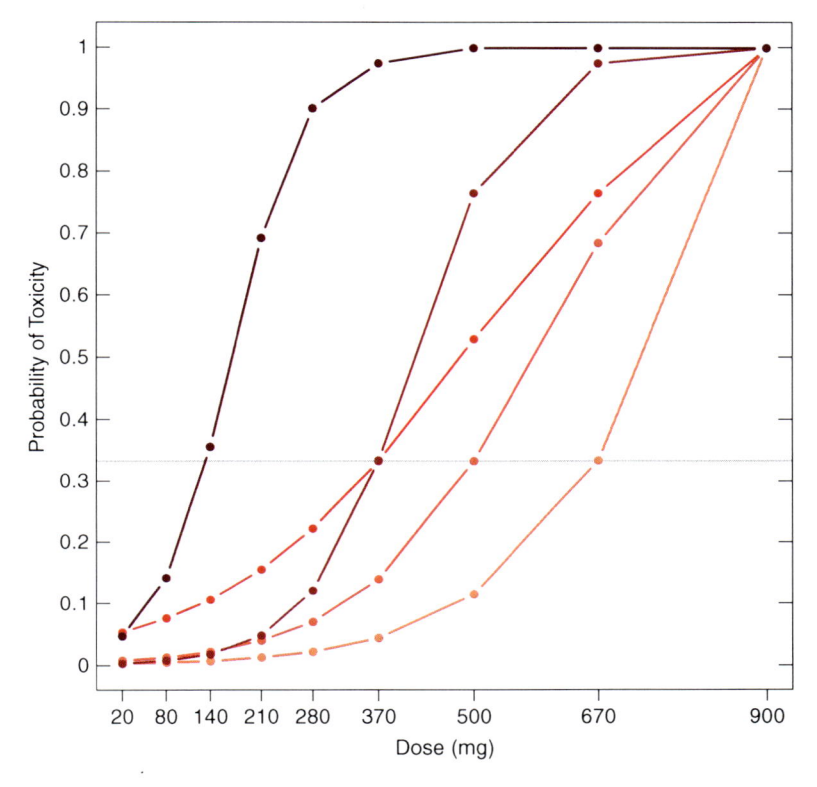

Fig. 9.1 Line plot illustrating simple dose-toxicity profiles for a new agent based on the logistic function p(dose) = 1/(1 + exp(−α − β × dose)) with varying model parameters α and β

practical and operational simplicity for the sponsor and familiarity and comfort by regulatory authorities (Reiner et al. 1999; Storer 1989). The "A + B" design (e.g., "3 + 3"), with or without de-escalation, is a commonly used rule-based design (Babb et al. 1998). As with any trial design, operational design characteristics (e.g., maximum sample size, average sample size, average number of DLT), are important for sponsors, clinical development teams, statisticians, clinical pharmacologists, and investigators to gain an understanding of the trial design.

Dose-response and dose-toxicity profile graphics are routinely generated based on either empirical data or theoretical models (e.g., logistic, sigmoid, quadratic, hyperbolic tangent, Lyman, Emax). Figure 9.1 illustrates potential dose-toxicity profiles for a new agent based on a generalized logistic function with varying model parameters.

Figure 9.2 displays various operating characteristics of a "3 + 3" design within one single figure. These include the assumed dose-toxicity profile, the probability of a dose being selected as the MTD, the expected number of patients treated at each dose level, the expected number of total patients treated, the expected DLT incidence at each dose level, and the total number of expected DLTs

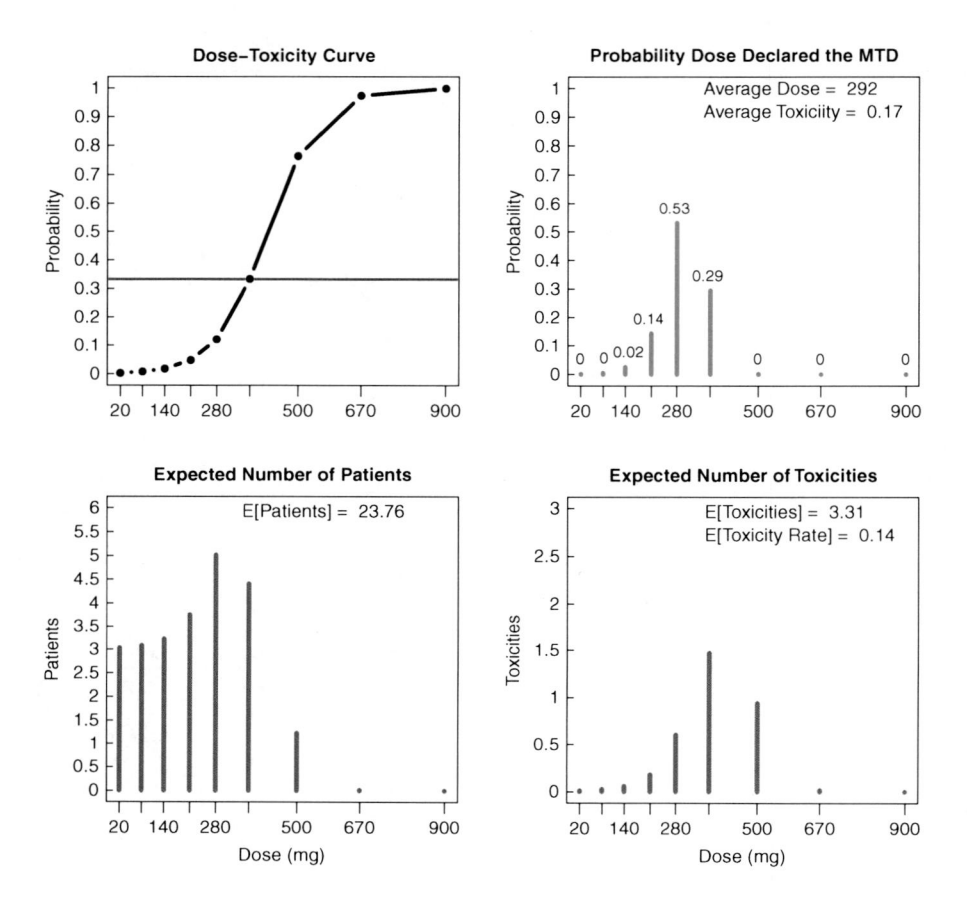

Fig. 9.2 Multi-panel plot illustrating key operating characteristics of a "3+3" phase 1 trial design with de-escalation. Assumed dose-toxicity profile (*upper left*), the probability that a dose is declared the MTD (*upper right*), the expected number of patients treated at each dose (*lower left*), and the expected number of dose-limiting toxicities (DLTs) at each dose (*lower right*). In addition, the expected number of total patients treated and the expected total number of DLTs are annotated on the *lower left* and *lower right panels*, respectively

(Lin and Shih 2001). A multi-panel graphic can be prepared under various assumed "true" dose-toxicity profiles to evaluate the performance characteristics of the trial design over a range of plausible scenarios. Undoubtedly, a graphic of this complexity will require additional review time to clearly understand the data being presented. However, this additional time and approach may highlight potential scenarios where the trial design may be inadequate. Based on specific design assumptions, a dash-board graphic such as this can be helpful to project teams as it summarizes several important components of the operating characteristics of the design such as the expected number of patients at each dose, the expected total number of patients, and the expected MTD. In addition, trial cost and duration can be inferred from preliminary trial design graphics.

9.4.1.2 Phase 2 Trial Design

Multi-stage designs are widely used in Phase 2 oncology clinical trials to minimize the number of patients treated with ineffective experimental agents and the amount of clinical expenses. Multi-stage designs allow for early clinical trial termination when experimental agents are ineffective. In addition, if carefully planned and properly executed, multi-stage designs are practically and operationally simplistic (Schlesselman and Reis 2006). As an example, Fig. 9.3 displays the stopping criteria for two multi-stage designs from a lung cancer clinical trial (Lee and Liu 2008): Simon's two-stage design (Simon 1989) and the Bayesian Predicted Probability design (Lee and Liu 2008).

With the commonly used Simon's two-stage design, project teams can clearly visualize the number of responders that must be observed for the trial to proceed to the second stage (more than 3 responders out of the first 17 patients). Once in the second stage, at least 11 total patients out of 37 must respond to ultimately reject the trial's null hypothesis, which is $H_0: p_0 \leq 0.20$ (objective response rate). As an alternative, the Predicted Probability design (for maximizing power) is a Bayesian multi-stage design that allows the trial to be monitored continuously or by cohorts. The trial will be stopped and the treatment is considered ineffective when the number of responses first falls into the rejection region.

In this example, the Predicted Probability design requires at least 1 responder out of the first 10 patients to proceed into the second stage. The subsequent stage's rejection regions (numbers of responses/n) are 1/17, 2/21, 3/24, 4/27, 5/29, 6/31, 7/33, 8/34, 9/35, and 10/36. Compared to Simon's optimal two-stage design, the trial is monitored more frequently in the Predicted Probability design, which also has a larger probability of early termination and a slightly larger expected sample size in the null case. Both trial designs have a Type I error of 0.10, statistical power of 90%, and a maximum of 37 treated patients. Project teams can easily compare and contrast the use of these two multi-stage designs in terms of operational budgets, expected sample size, time to complete the trial, probability of early termination, etc. In addition, they can compare multi-stage designs with a fixed (traditional) design that requires approximately 38 patients. Simple graphics can provide a visual approach for the clinical trial management team, statisticians, and investigators to understanding the trial design and evaluate trial assumptions.

9.4.1.3 Phase 3 Trial Design

Phase 3 oncology clinical trials are primarily comparative trials aimed to support regulatory approval with regard to the safety, efficacy, and the overall risk-benefit

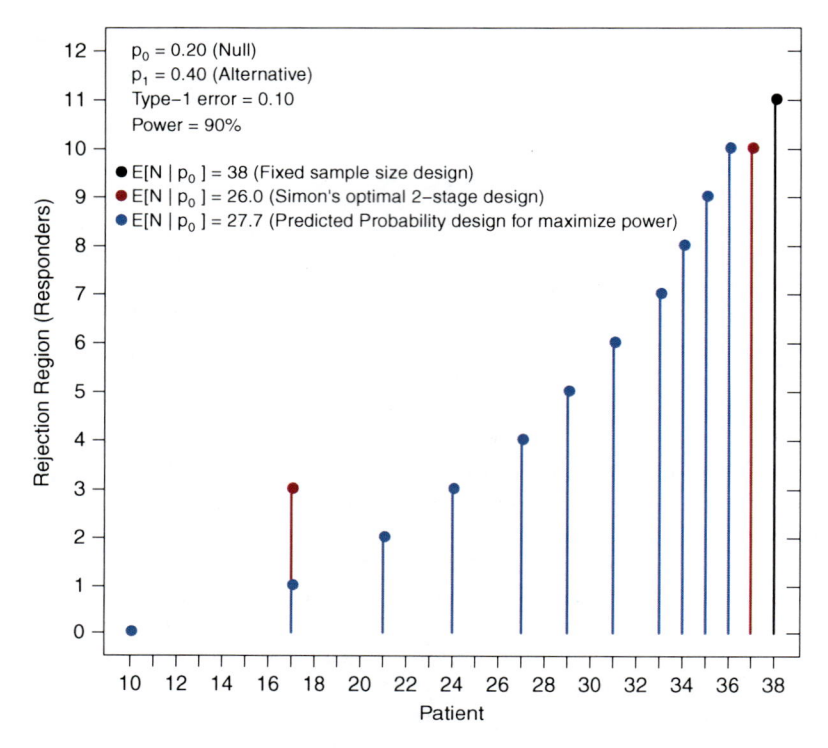

Fig. 9.3 Operating characteristics comparing two multi-stage Phase 2 trial designs evaluating objective response rate. Both trial designs have Type I error of 0.10, statistical power of 90%, and a maximum of 37 treated patients. Simon's two-stage design has analysis milestones that occur after the 17th and 37th patients (*vertical red bars*), while the Predicted Probability design is assessed at 11 different stages. For comparison, a fixed (traditional) sample size trial is included using the same design assumptions

relationship of the investigational agent. Phase 3 trials are both costly and time-consuming. Hence, simulations are often used to assess the trial design assumptions and gain insight into the trial's operating characteristics before commencement of the actual clinical trial. Figure 9.4 is a graphical representation of a simulated clinical trial under assumed treatment effect using the triangular test group sequential trial design that allows early stopping at interim analysis (Whitehead 1997; Jennison and Turnbull 2000). For each individual trial simulation, the trial was terminated when either a boundary was crossed or the maximum sample size of 200 enrolled patients was achieved. In the upper panel of the plot, the solid blue line is the superiority boundary (values above this boundary indicate experimental treatment significantly better), the dashed red line is the inferiority boundary (values below this boundary indicate experimental treatment is significantly worse), and the solid red line is the

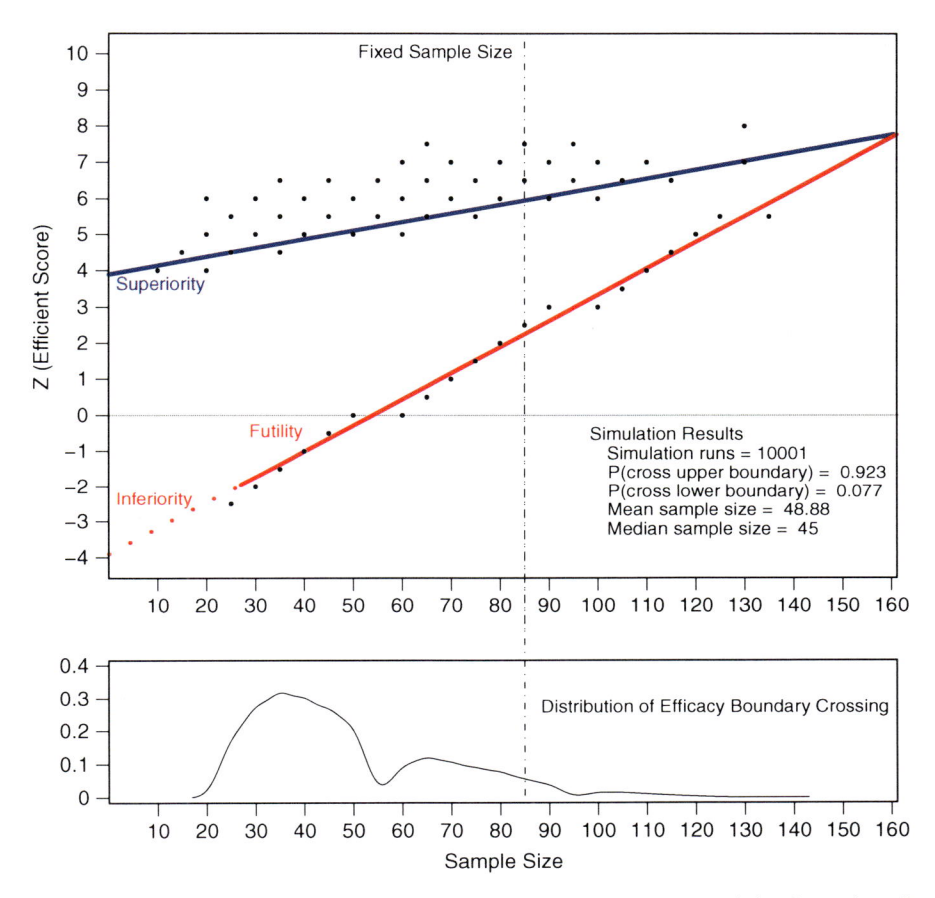

Fig. 9.4 Multi-panel plot displaying simulated data and operating characteristics for a phase 3 trial using the triangular test design. In the *top panel*, the *solid blue line* is the superiority boundary, the *dashed red line* is the inferiority boundary, and the *solid red line* is the futility boundary. The *black dots* represent the Z (efficient score) test statistic at the termination of each of simulation run. The *bottom panel* displays the empirical density plot of the final sample sizes from the simulation. The graphic is annotated with simulation summary statistics, along with a *dashed vertical line* across both panels that represents the fixed sample size from a traditional one-look trial design with 85 patients

futility boundary (values below this boundary indicate no significant difference and continuation of trial unlikely to establish experimental treatment superiority). The black dots represent the Z (efficient score) test statistic at the termination of each simulation run.

Continuous monitoring is an abstract mathematical concept (Brownian motion with drift) in which the plot of Z against sample size develops as an unbroken path (Whitehead 1997). Continuous monitoring is nearly impossible to apply in practice because data inspection even after every patient response causes sample size to increase in small increments. Thus, adjustments to the triangular test boundaries are implemented to account for discrete monitoring that are termed "Christmas that tree boundaries" because of their shape (Whitehead 1997). When assessing the individual simulated trial termination, the Christmas tree boundary corrections were used. As a result, some simulations terminated with a Z test statistic that was inside the continuous triangular boundaries (displayed) but outside the Christmas tree boundaries (not displayed). The bottom panel displays the empirical density plot of the final sample sizes from the simulation. Across both panels is a vertical line that represents the fixed sample size from a traditional one-look trial requiring 85 patients. Figure 9.4 is also annotated with simulated results, such as Type I error, statistical power, and average sample size. Based on the assumptions in this simulated clinical trial, the benefits of the group sequential design in terms of a smaller expected sample size compared to a fixed sample size design are visually and numerically presented. Using various design assumptions and enrollment rates, a trial design graphic can clearly and efficiently communicate a simulated trial's operating characteristics as well as illustrate the impact of trial design assumptions such as treatment effects and enrollment rates.

9.4.2 Interim Analyses

An interim analysis is an assessment of data performed during the patient enrollment or follow-up stages of a trial, but before completion of the trial, for the purpose of assessing such aspects as center performance, the quality of the data collected, sample size assumptions, or treatment effects. With adaptive designs, mid-trial data analyses are used for a variety of reasons, including stopping early for efficacy or futility, increasing sample size, dropping treatment groups, or modifying randomization schemes. Figure 9.5 illustrates an example of a dash-board style graphic used in a Phase 2 Bayesian adaptively randomized clinical trial comparing 3 treatment groups with regard to PFS in ovarian cancer. In this clinical trial, after a period of equal randomization (e.g., 15:15:15), patients were subsequently randomized to the 3 treatment groups in blocks of 9, with the randomization probability based on their respective posterior probabilities. For instance, if the 3 treatment groups had comparable performance, 3 patients each would be randomized to treatment groups A, B, and C, respectively. If a treatment group is demonstrating better performance (i.e., better mean PFS), the treatment group will receive a larger allocation of the 9 patients. The trial was continued to a maximum number of treated patients or when pre-specified termination criteria were met. During planned interim analyses, the Kaplan-Meier plot for the 3 treatment groups is presented, along with the current estimate of each group's posterior distribution for mean PFS, the current total number of patients randomized per group, and the randomization allocation of the next 9 patients.

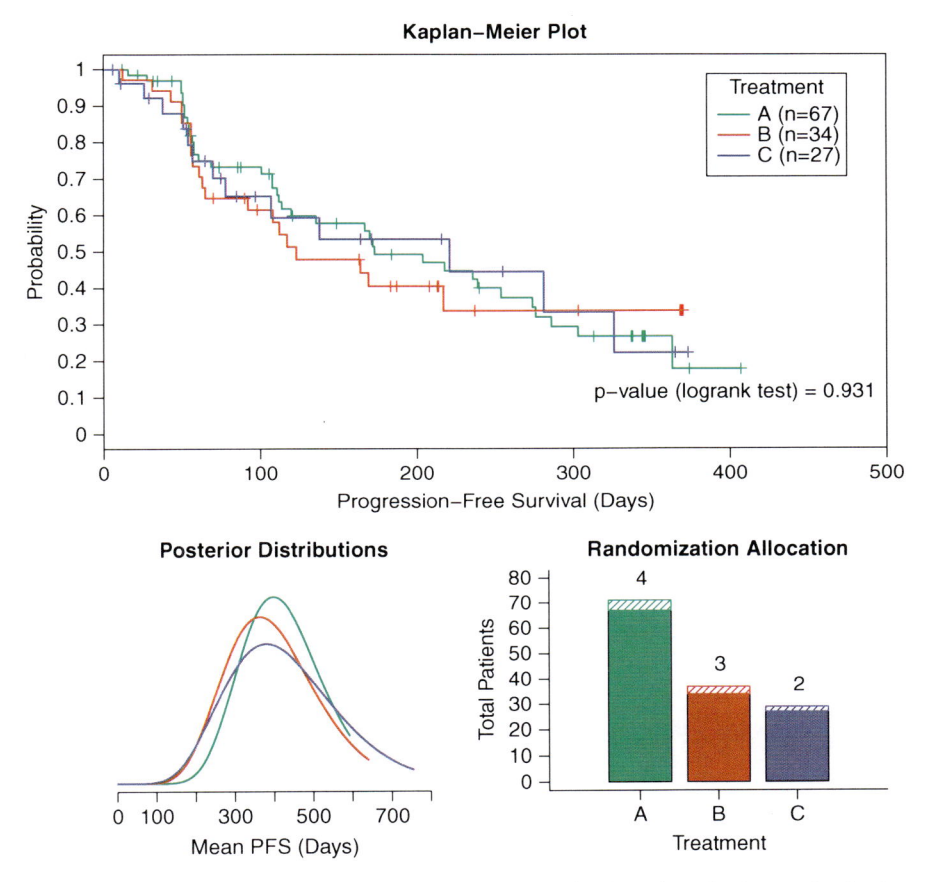

Fig. 9.5 Multi-panel plot illustrating an interim analysis of a Bayesian adaptive randomization trial. The *top panel* illustrates the Kaplan-Meier plot comparing progression-free survival by treatment group. The *bottom panel* is comprised of 2 plots which include a line plot of the posterior distributions of the mean progression-free survival (*bottom left*) for each treatment group, the total number of patients treated in each group (*bottom right, solid bars*), and the allocation of next 9 patients (*bottom right, shaded bars*)

9.4.3 Final Analyses

Graphics for oncology efficacy data are primarily used to show treatment effects on final efficacy endpoints. Traditionally, efficacy graphics, such as bar charts, may use more ink than other graphics because the endpoints are simple measures of treatment effect such as mean change from baseline or a patient's maximum reduction in tumor size. Several commonly used statistical graphics for the presentation of clinical results of efficacy data are presented.

9.4.3.1 Waterfall Plots

Waterfall plots are a useful method of illustrating the antitumor activity of investigational agents. The plots are also sometimes referred to as Delta plots (Chuang-Stein et al. 2001). The waterfall plot is not a minimalist use of ink as suggested in Tufte's principles (National Cancer Institute 2012); however, the ink is well used in clearly showing one aspect of the treatment effect. The plots are termed waterfall because if the patient data are ordered least to largest (or vice versa), it gives the appearance of a waterfall. Waterfall plots can display patterns or distributional location shifts in comparative settings, but caution should be used if the randomization is not approximately 1:1. Antitumor activity, such as the change in tumor growth or shrinkage from baseline, is often reported longitudinally at study time points for each patient. Alternatively, each patient's maximum (i.e., best) change in tumor growth or shrinkage is often reported. The waterfall plot can be used to assess treatment effects at the population level as well as to highlight individual patients. Unlike an average of patient response, or a response rate without information about the magnitude of response, the waterfall plot demonstrates the variability of an experimental agent. The waterfall plot, however, does not provide information on the duration of the response.

Figure 9.6 presents the maximum percentage change in the sum of the longest diameter (SLD) from baseline for 60 patients (ordered from largest increase to largest decrease). The individual patients are colored to differentiate their best overall response determinations.

Figure 9.7 illustrates the same results by treatment group. If the majority of tumor shrinkage (negative change from baseline) were to occur in one treatment group, this pattern would be readily apparent from the plot. As seen in Fig. 9.7 there is no apparent grouping of Treatment A or Treatment B in either the tumor growth (increasing tumor burden) or tumor shrinkage (decrease in tumor burden).

9.4.3.2 Forest Plots

Forest plots are routinely used when combining results from multiple studies or conducting subgroup analyses. When combining results from multiple studies (i.e., a meta-analysis), the individual trial results (e.g., means, proportions, odds ratios, hazard ratios) are represented by a vertical tick mark surrounded by a square, typically with 95% confidence intervals (CI) represented by a horizontal line. The size of the square is proportional to the weight of the trial in the meta-analysis (i.e., the weight applied to the individual trial result to estimate the pooled overall result). The estimate of the overall result and its associated 95% confidence interval are often represented by a filled circle with confidence bars or an elongated diamond where the center of the diamond is the overall pooled estimate and the ends of the diamond are the 95% confidence interval (Cappelleri et al. 2000). Figure 9.8 illustrates a comprehensive

9 Statistical Graphics in Clinical Oncology

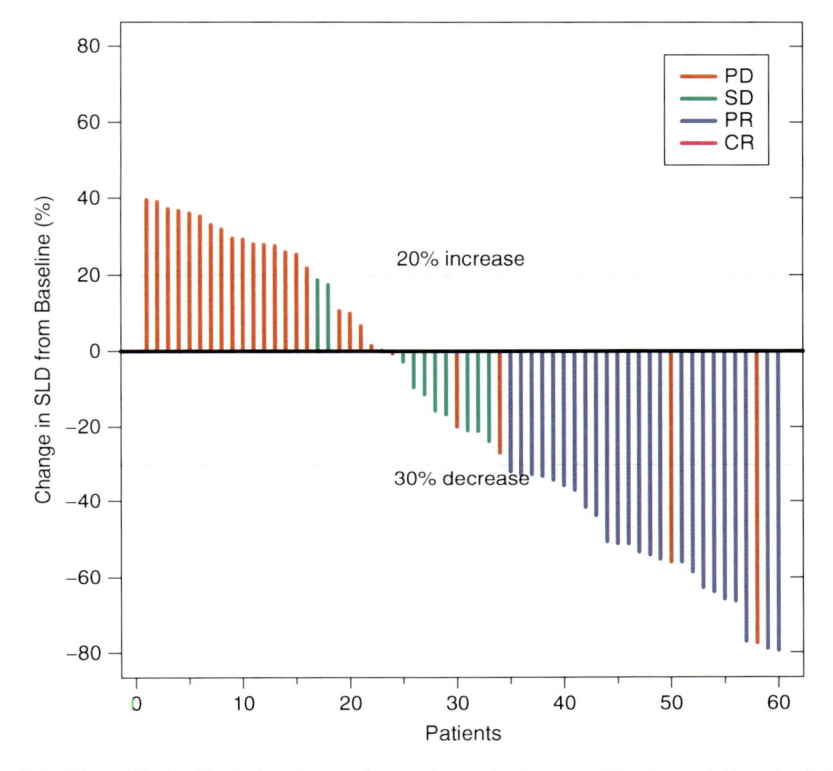

Fig. 9.6 Waterfall plot displaying the maximum change in the sum of the longest diameter (SLD) from baseline. The *vertical lines* represent the maximum change in each patient's SLD from baseline. Additionally, the *vertical lines* are colored by the patient's final response status, where progressive disease (PD) is *orange*, stable disease (SD) is *green*, partial response (PR) is *purple*, and complete response (CR) is *pink*. There are no patients with a complete response (CR)

forest plot for a random-effects meta-analysis of 4 pivotal randomized clinical trials. In addition to the forest plot, the graphic includes summary statistics, measures of heterogeneity, a vertical line for the odds ratio (OR) of 1.0 (no treatment effect), and a vertical line at the overall treatment effect level, making it easy to evaluate if a confidence individual from an individual trial differed significantly from the overall effect and to assess the consistency and robustness of the data.

Forest plots have become a useful tool to conveniently and graphically summarize the relative treatment effect of numerous separate analyses in one figure. Figure 9.9 illustrates a basic subgroup analysis forest plot, with the hazard ratio (HR) and 95% confidence intervals based on a Cox proportional hazards regression model. A useful aspect of the forest plot is to assess whether the effect size for different subgroups differs significantly from the main effect.

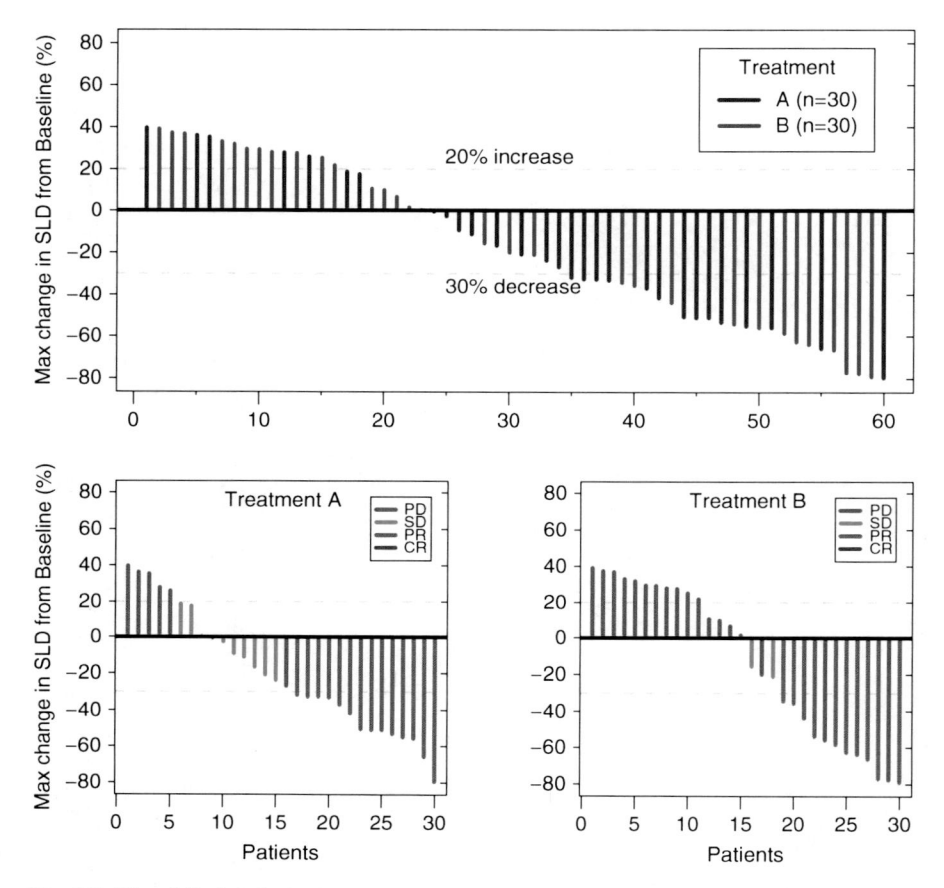

Fig. 9.7 Waterfall plot displaying the maximum change in the SLD from baseline by treatment group. The *vertical lines* (*top*) represent the maximum change in each patient's SLD from baseline and are colored by treatment group. Additionally, with the results separated by treatment group (*bottom left* and *bottom right*), the *vertical lines* are colored by the patient's final response status, where progressive disease (PD) is *orange*, stable disease (SD) is *green*, partial response (PR) is *purple*, and complete response (CR) is *pink*. There are no patients with a complete response (CR)

9.4.3.3 Change from Baseline Plots

"Change from baseline" is a useful and routinely used estimate of interest for both continuous safety and efficacy measures in oncology clinical trials. Raw change, absolute change, and percentage change are variations often reported. Depending on the sample size and normality of the data, the appropriate graphic should be selected. The dot plot with 95% confidence interval, box plot, empirical cumulative distribution plot, and density plot are attractive plots for presenting continuous data. Figure 9.10 shows examples of 4 different plots of change from baseline date from a visual analog scale (VAS) for pain for 2 treatment groups. Simple statistical

9 Statistical Graphics in Clinical Oncology

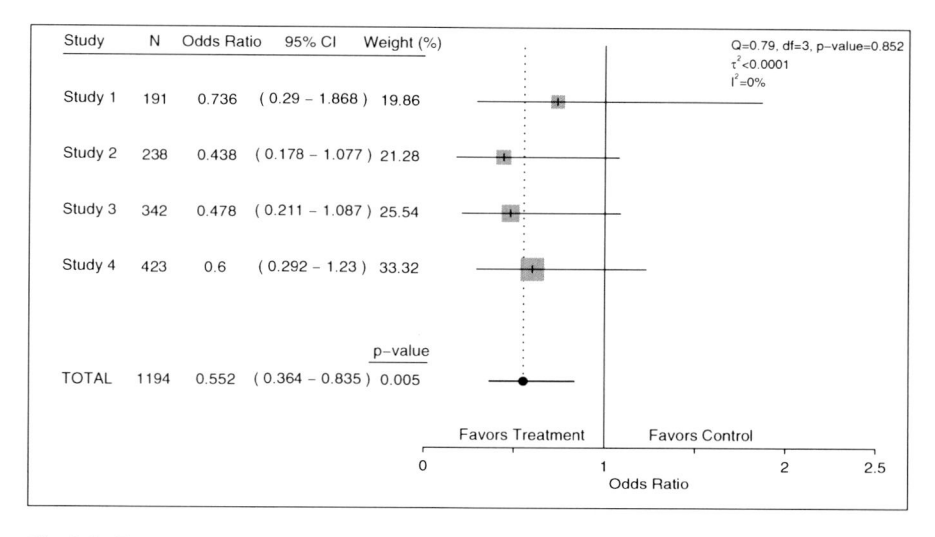

Study	N	Odds Ratio	95% CI	Weight (%)
Study 1	191	0.736	(0.29 – 1.868)	19.86
Study 2	238	0.438	(0.178 – 1.077)	21.28
Study 3	342	0.478	(0.211 – 1.087)	25.54
Study 4	423	0.6	(0.292 – 1.23)	33.32
				p-value
TOTAL	1194	0.552	(0.364 – 0.835)	0.005

Q=0.79, df=3, p-value=0.852
$\tau^2 < 0.0001$
$I^2 = 0\%$

Favors Treatment | Favors Control
Odds Ratio

Fig. 9.8 Forest plot that displays both numerical and graphics results from a meta-analysis. Each trial's estimated odds ratio (OR) is represented by a *vertical tick mark* surrounded by a *square*. The individual trial's 95% confidence interval for the OR is represented by a *horizontal line*. The size of the square is proportional to the weight of the trial in the meta-analysis (i.e., the weight applied to the individual trial OR to estimate the pooled OR). The estimate of the pooled OR and its associated 95% confidence interval is represented by a *filled circle* and *horizontal line*, respectively. In addition, a *dashed line* is vertically extended from the estimate of the pooled OR. The OR = 1.0 (no treatment effect) is illustrated with a *solid vertical line*. The plot is further annotated with the heterogeneity statistics Q, t^2, and I^2

graphics such as these, particularly when used together, provide useful visual summaries and aid in identifying distributional properties of the data. In addition, they provide insight into choosing the appropriate statistical methodology for comparing treatment groups.

9.4.3.4 Event Charts

For both randomized and nonrandomized trials, in particular those with a small number of patients, events charts are a convenient graphical approach to illustrating time-to-event outcomes and other key events (e.g., randomization, treatment, response). Using the overall survival data from a randomized trial comparing two treatments for ovarian cancer (R dataset ovarian) (Therneau 2012), treatment dates over a 2-year period were randomly generated and assigned to 26 patients. Based on the randomly generated treatment date and the duration of the patients overall survival, their event date was derived. The event was either death or censored, where the patient was either lost-to-follow-up or still alive at the time of the analysis. Figure 9.11 plots the patients by calendar time (i.e., chronologically as they were treated), while Fig. 9.12 plots the patients by survival time (shortest survival time to longest survival time). Figure 9.11 provides a nice overview of the enrollment rate

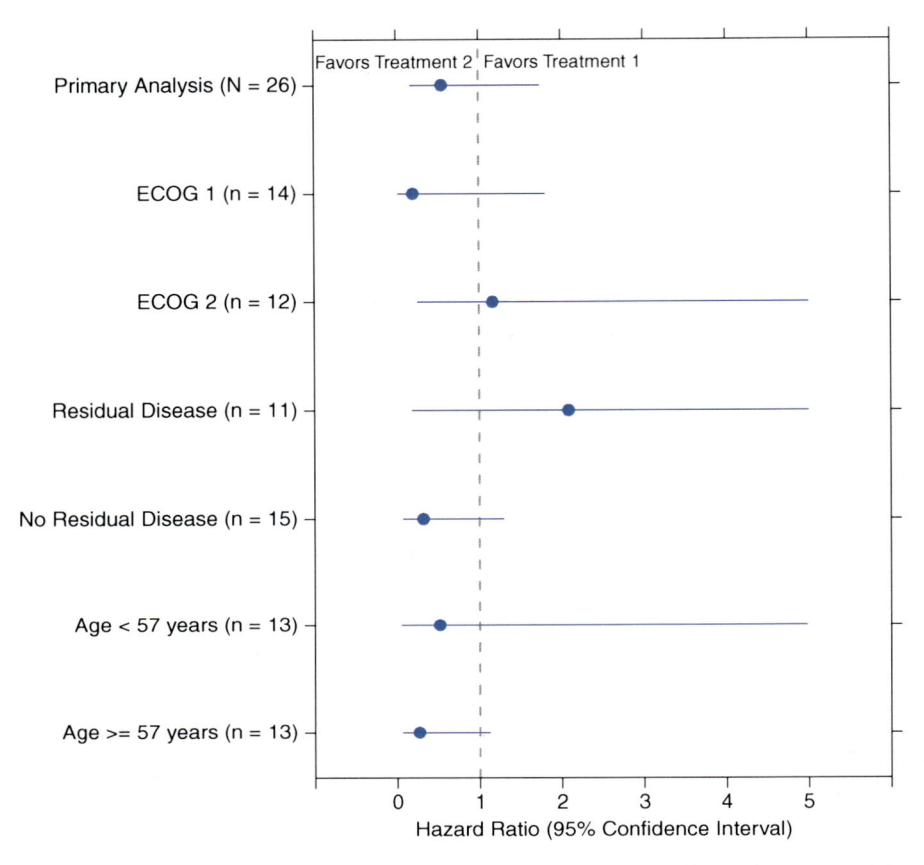

Fig. 9.9 Forest plot summarizing the treatment effect for the primary analysis and numerous subgroups. The *blue dot* represents the hazard ratio (HR) based on a Cox proportional hazards regression model, while the *horizontal lines* represent the 95% confidence interval for the hazard ratio. The HR = 1.0 (no treatment effect) is illustrated with a *dashed vertical line*

and overall trial duration. In this example, patient 25 was still on-going at the time of data cut (March 26, 2012), while the other censored patients could be considered lost-to-follow-up. Figure 9.12 provides some insight into the overall survival, while quickly illustrating which patients have died or are censored. Both styles of the event chart can provide sponsors and project teams with useful trial information, particularly in exploratory studies with a small number of patients where all the patients can be displayed clearly on a single graphic. Conversely, for studies with a large number of patients, the display may be difficult to interpret when attempting to display in a single graphic.

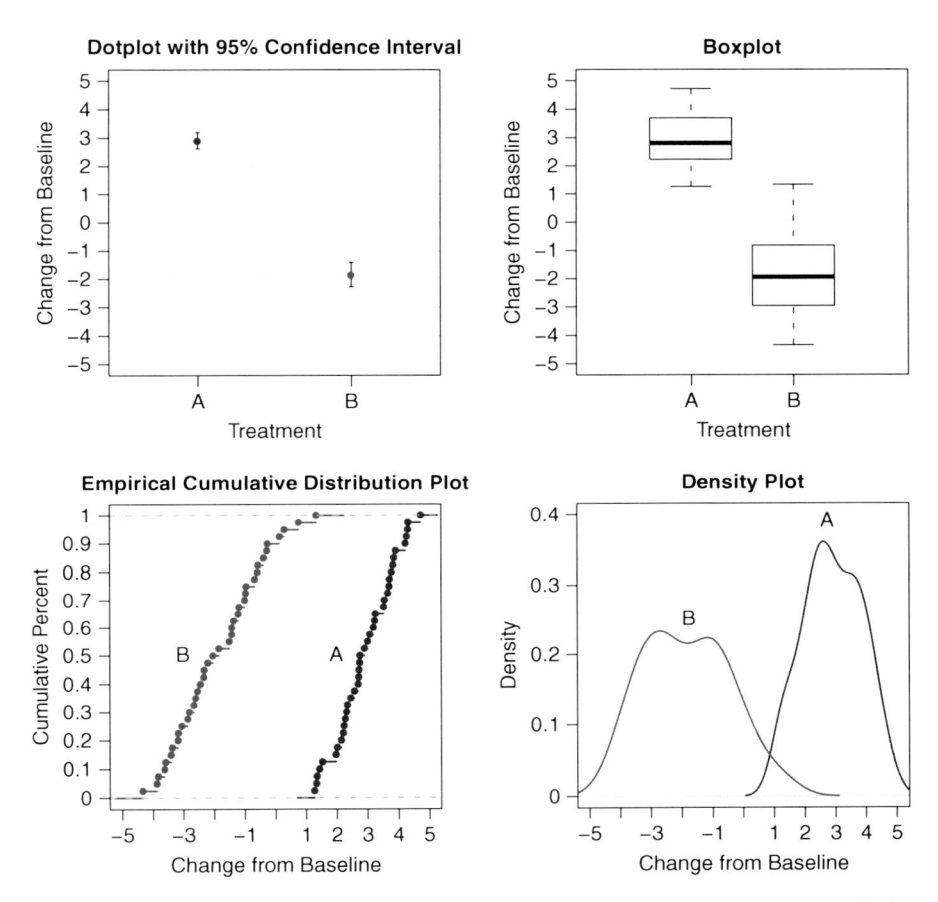

Fig. 9.10 Selected plots illustrating the change from baseline in visual analog scale (VAS) for pain by treatment group. The point estimate for the change from baseline in VAS is represented by a *dot*, with the associated 95% confidence interval (*upper left*), *box plots* (*upper right*), empirical cumulative distribution functions (*bottom left*), and probability density plots (*bottom right*)

9.4.3.5 Time-to-Event Plots

Graphical summaries of time-to-event data are cornerstone in the analysis and reporting of results from oncology clinical trials. Time-to-event analyses examine and model the time it takes for an event to occur. Time-to-event endpoints include overall survival (OS), PFS, TTP, EFS, RFS, time-to-response, and duration of response (DR).

The most frequently used graphical technique for time-to-event data is the Kaplan-Meier plot. The nonparametric Kaplan-Meier estimator, also known as the product limit estimator, is an unbiased estimator for estimating the survival function for time-to-event data (Kaplan and Meier 1958; Parmar and Machin 1995).

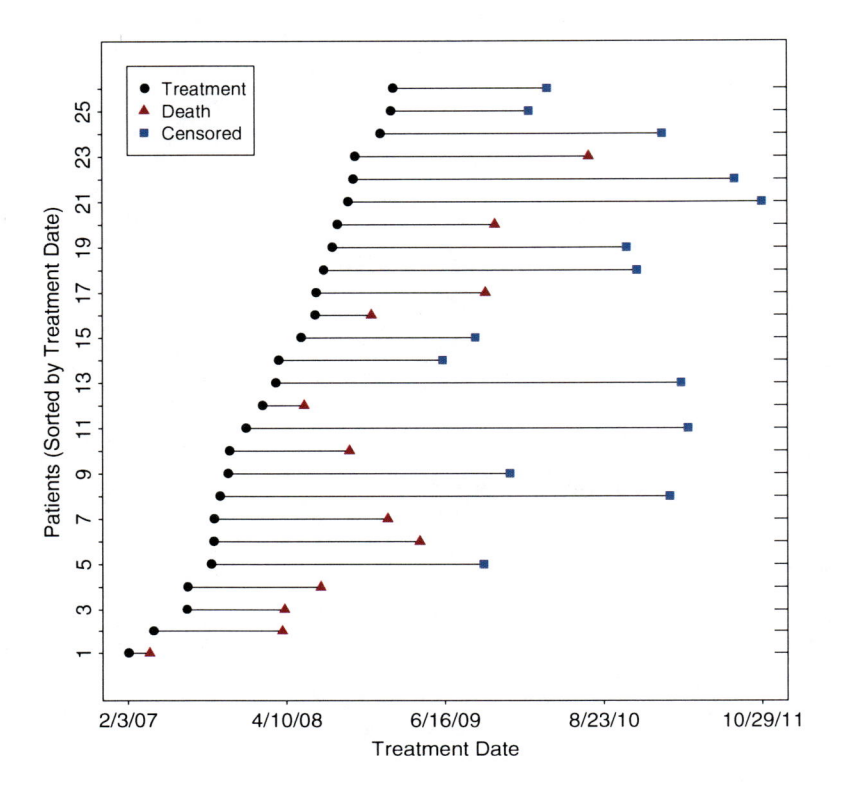

Fig. 9.11 Calendar event chart. Each of the 26 treated patients is represented chronologically (i.e., patient 1 was treated first and patient 26 was treated last patient). The y-axis measures actual calendar dates. Each patient's treatment date is denoted with a *black dot*. With regard to the event, either the date of death is denoted with a *red triangle* or the censored event is denoted with a *blue square*. The length of the *black horizontal line* indicates the patient's overall survival in days

An important advantage of the Kaplan-Meier method is that it takes into account censored data, particularly right-censoring, which occurs if a patient withdraws or is lost from the trial before the final outcome is observed or has not experienced the event at the time of the analysis.

A plot of the Kaplan-Meier estimate of the survival function is a series of horizontal steps of declining magnitude which approaches the true survival function when the sample size is large. To illustrate the variability in the Kaplan-Meier estimator, plots often include confidence intervals, confidence bands, or confidence bars. Confidence intervals and bands can provide an attractive display when a single survival function or multiple well-separated survival functions are presented. However, if the multiple survival functions are not well separated, the confidence intervals and confidence bands are generally confusing and difficult to interpret. As an alternative, confidence bars at selected time points offer a clean assessment of variability. Other useful features often employed in Kaplan-Meier plots include

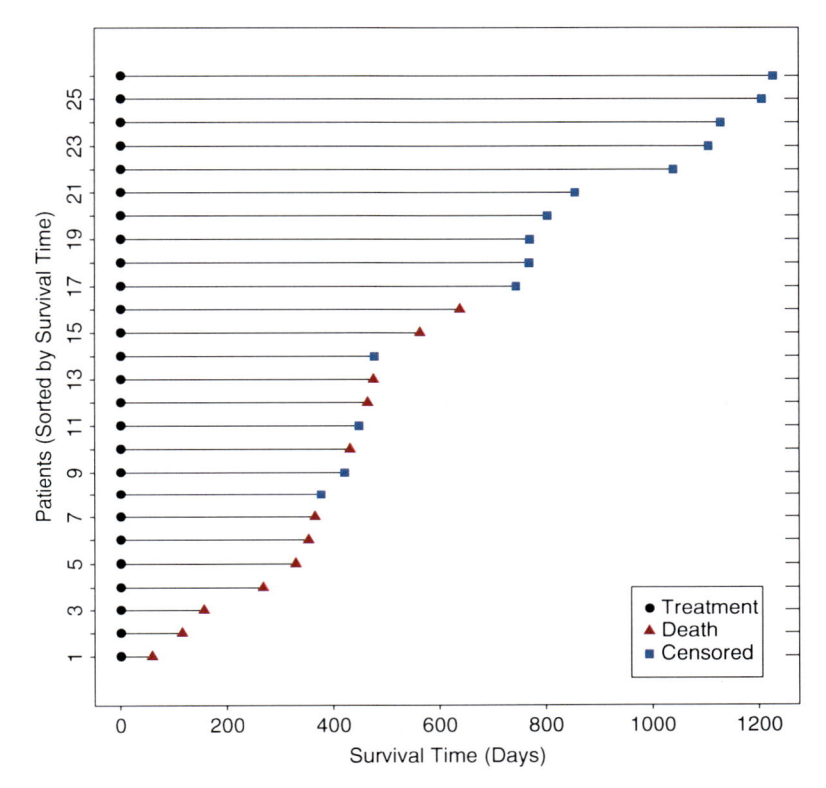

Fig. 9.12 Event chart. Each of the 26 treated patients is represented in increasing order of overall survival (i.e., patient 1 has the shortest overall survival and patient 26 has the longest overall survival). The *y*-axis measures overall survival in days. Each patient's treatment is denoted with a *black dot* at Day 0. With regard to the events, a death is denoted with a *red triangle* and a censored event is denoted with a *blue square*

annotations regarding the number of patients "at risk" over time and censored observations. Censored observations are typically represented by small tick marks. Both of these features are important for proper interpretation of survival functions, particularly as the pool of patients "at risk" decreased as time progresses.

Figure 9.13 shows a simple Kaplan-Meier plot of overall survival in patients with advanced lung cancer (R dataset lung) (Loprinzi et al. 1994). The solid line in the Kaplan-Meier survival function is surrounded by the 95% error band. The median survival, along with the 95% confidence interval, is annotated on the graphic.

Cumulative incidence plots are also often used to present time-to-event data. A cumulative incidence plot illustrates the cumulative proportion of patients that experience the event, while the Kaplan-Meier plot illustrates the proportion of patients free of the event. Figure 9.14 illustrates the cumulative incidence plot in the patients

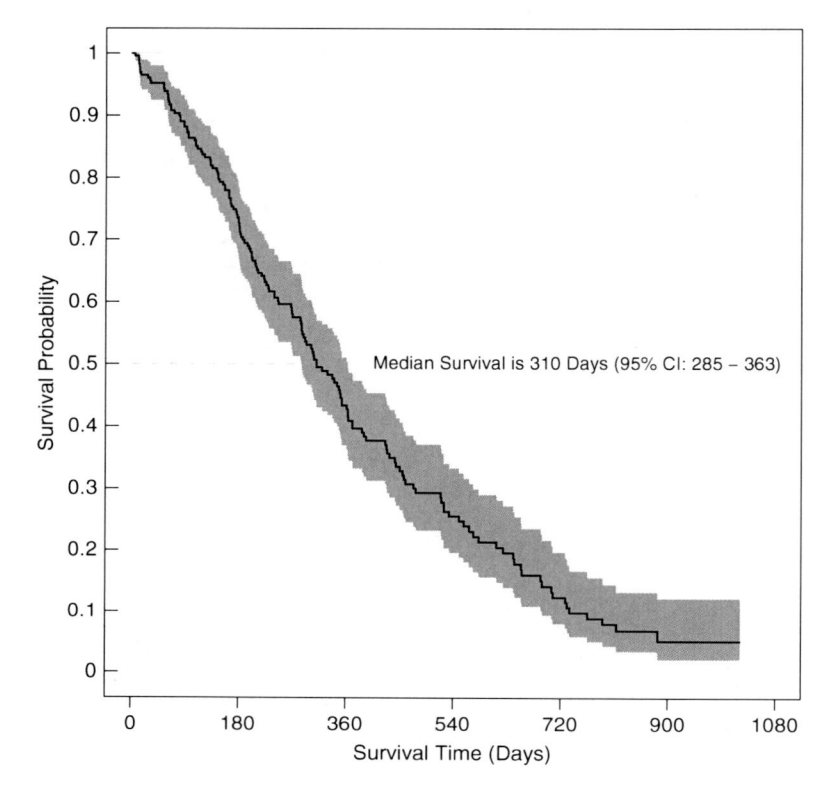

Fig. 9.13 Kaplan-Meier plot of overall survival. The *solid line* is the Kaplan-Meier survival function estimate and the *solid gray band* is the 95% confidence band

with advanced lung cancer. The solid line is the Kaplan-Meier cumulative incidence function estimate, surrounded by the 95% confidence interval. Note that censored observations are denoted on the cumulative incidence plot by small tick marks. The Kaplan-Meier and cumulative incidence plots display comparative results, provided they are banked to 45° (i.e., the aspect ratio such that the average absolute angle in curves is 45°). A detailed discussion of Kaplan-Meier and cumulative incidence plots is provided by O'Connell and Treder (O'Connell and Treder 2009).

Figure 9.15 shows a more advanced Kaplan-Meier plot of overall survival by sex. The solid purple line is the Kaplan-Meier survival function for males, while the solid green line is the survival function for females. At approximately 6-month intervals (180 days), 95% confidence bars are presented around the survival estimate and the number of "at risk" patients are presented for both groups. Finally, the graphic is annotated with faint dotted references lines on the y-axis and the p-value from the nonparametric logrank test, a commonly used test for comparing survival functions of two groups (Parmar and Machin 1995).

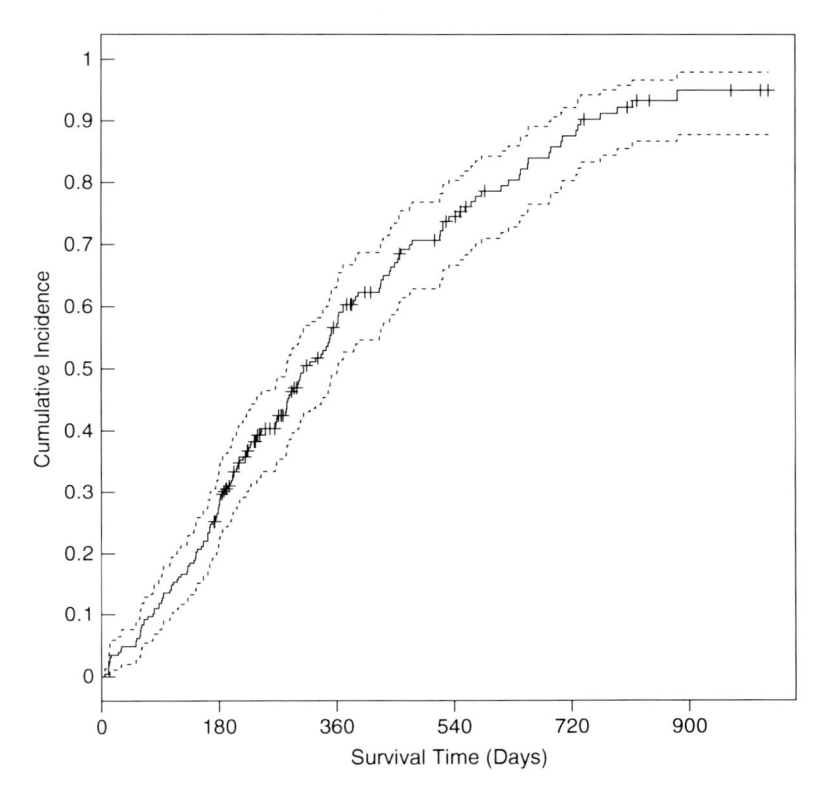

Fig. 9.14 Kaplan-Meier cumulative incidence plot of cumulative death (1—survival). The *solid line* is the Kaplan-Meier cumulative incidence function estimate and the *dashed lines* are the 95% confidence interval. Censored observations are denoted by *small tick marks*

An alternative to Figs. 9.15 and 9.16 is a plot of the difference between the survival functions for the male and female groups with an associated 95% confidence band. The confidence bands could be replaced by confidence intervals or selected confidence bars to minimize the amount of ink used. The *p*-value from the nonparametric logrank test is also included. The fact that the solid line lies below difference=0 indicates that the estimate of the survival probability for females is always larger than that of the males. In many cases, if follow-up is long enough, the two survival curves will approach 0%. As a result, two survival curves that come together at the end of the follow-up do not necessarily indicate a lack of difference between the two curves.

In addition to the Kaplan-Meier methodology, the semi-parametric Cox proportional hazards model is extensively used in the analysis of time-to-event data (Cox 1972). It is important that the proportional hazards assumption holds for valid interpretation of regression coefficients in a Cox proportional hazards model. Figure 9.17 contains an example of 4 diagnostic plots used with the Cox model (Fox 2002). Both, Fig. 9.17a and b are graphical diagnostic tests for the proportional

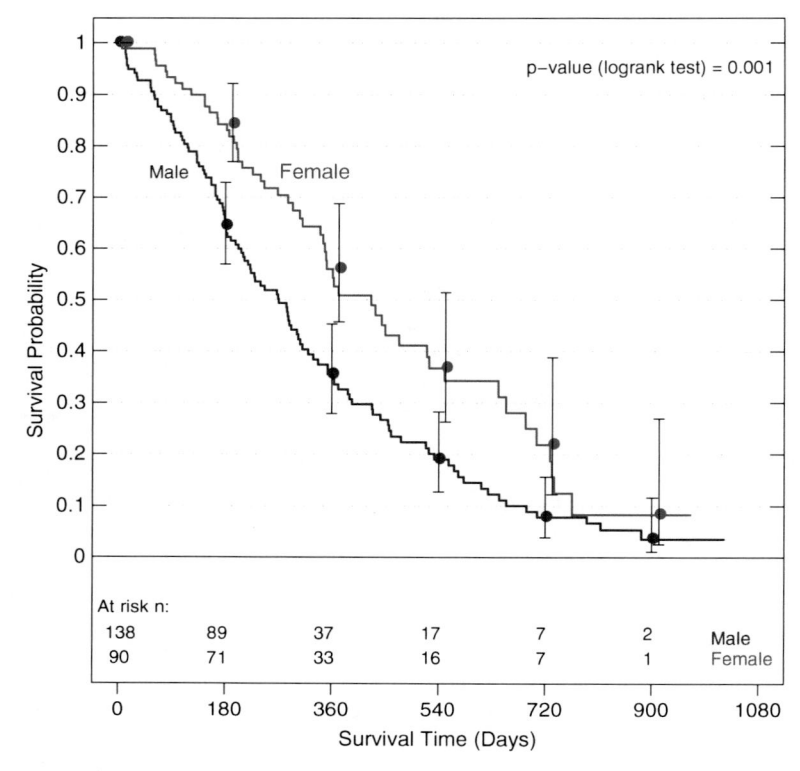

Fig. 9.15 Kaplan-Meier plot comparing overall survival by sex. The *solid lines* are the Kaplan-Meier survival function estimates, with females denoted in *green* and males denoted in *purple*. In addition, 95% confidence intervals are displayed at approximately 6-month intervals (180 days), along with the number of patients "at risk"

hazards assumption. Figure 9.17a shows the plot of log(survival time) vs. log(−log(survival probability)). In this example, the curves appear to be approximately parallel and the proportional hazards property appears to hold. Figure 9.17b shows the scaled Schoenfeld residuals vs. survival time, where the solid line is the smoothing-spline fit, the dotted line is a ±2-standard-error band around the smoothing-spline fit, and the blue dashed line is the least squares regression line. Systematic departures from a horizontal line are indicative of nonproportional hazards. Although the solid line has a slight upward trend, a formal test of the proportional hazards assumption holds ($p=0.117$). Figure 9.17c is an index plot for assessing influential observations by comparing the magnitudes of the largest dfbeta values to the regression coefficients. Finally, Fig. 9.17d is a plot where the martingale residuals

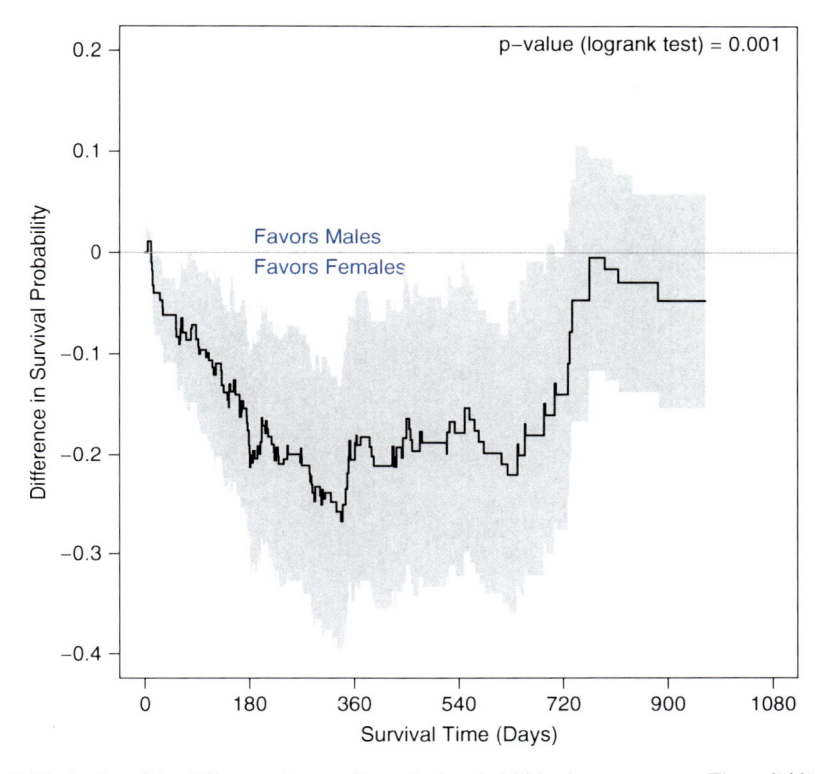

Fig. 9.16 A plot of the difference in overall survival probabilities between sexes. The *solid line* is the difference between sexes in Kaplan-Meier survival probability estimates, and the *solid gray band* is the 95% confidence band for the difference

are plotted against covariates (e.g., age) to detect nonlinearity, an incorrectly specified functional form in the parametric part of the Cox model. The solid lines are fit by local linear regression (lowess). There appears to be slight nonlinearity.

9.5 Conclusion

Statistical graphics are an invaluable tool for the efficient and accurate analysis and interpretation of clinical data. A graphic can stand alone when provided with sufficient data and information. In fact, a few graphics can provide information equivalent to numerous tables. Efficacy graphics may be relatively basic and easy to comprehend or they may be quite complex and require careful review. Better graphics simply lead to better clinical research and more effective communication of

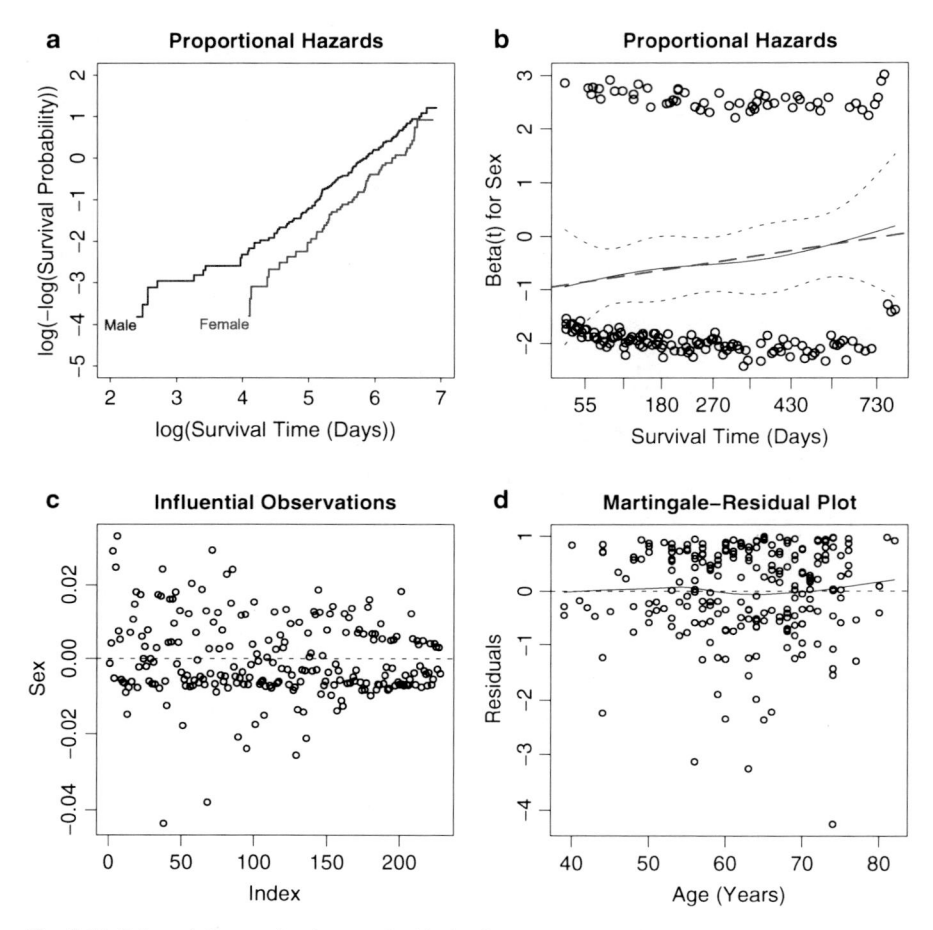

Fig. 9.17 Selected diagnostic plots used with the Cox proportional hazards model. (**a**) Is a plot of log(survival time) vs. log(−log(survival probability)); (**b**) illustrates the scaled Schoenfeld residuals vs. survival time, where the *solid line* is the smoothing-spline fit, the *dotted line* is a ±2-standard-error band around the smoothing-spline fit, and the *blue dashed line* is the least squares regression line; (**c**) is an index plot for assessing influential observation whereby comparing the magnitudes of the largest dfbeta values to the regression coefficients; and (**d**) is a plot where the martingale residuals are plotted against covariates. The *solid lines* are fit by local linear regression (lowess)

information. The use of concise, compelling, and common graphical analysis in oncology clinical research contributes to sound and scientific analysis and interpretation of clinical data, which ultimately results in more efficient use of patient data, cost savings, and faster approval of oncology drugs and improved patient care.

Acknowledgments The author thanks Kara Choquette, Thian Kheoh, Byron McKinney, Sandeep Menon, Michael O'Connell, and Denise Trone for their many insightful comments and recommendations that led to the improvement in the structure, content, and focus of the chapter.

References

American Cancer Society (2012) Cancer facts & figures 2012. American Cancer Society, Atlanta
Anand P, Kunnumakkara A et al (2008) Cancer is a preventable disease that requires major lifestyle changes. Pharm Res 25(9):2097–2116
Babb J, Rogatko A, Zacks S (1998) Cancer phase I clinical trials: efficient dose escalation with overdose control. Stat Med 17:1103–1120
CancerHelp UK (2012) How many different types of cancer are there? http://cancerhelp.cancerresearchuk.org/about-cancer/cancer-questions/how-many-different-types-of-cancer-are-there. Accessed 01 March 2012
Cappelleri J, Ioannidis J, Lau J (2000) Meta-analysis of therapeutic trials. In: Chow S-C (ed) Encyclopedia of biopharmaceutical statistics. Marcel Dekker, Inc., New York, NY, pp 307–316
CenterWatch (2012) FDA Approved Drugs for Oncology. http://www.centerwatch.com/druginformation/fda-approvals/drug-areas.aspx?AreaID=12. Accessed 10 Apr 2012
Chuang-Stein C, Le V, Chen W (2001) Recent advancements in the analysis and presentation of safety data. Drug Inf J 35:377–397
Cleveland W (1993) Visualizing data. Hobart Press, Summit, NJ
Cox D (1972) Regression models and life-tables (with discussion). J R Stat Soc 34:187–220
Food and Drug Administration (2007) Guidance for industry: Clinical Trial Endpoints for the Approval of cancer drugs and biologics, [WWW] http://www.fda.gov/downloads/Drugs/GuidanceComplianceRegulatoryInformation/Guidances/ucm071590.pdf (cited 01 September 2012. 10.22), Rockville, MD, U.S. Department of Health and Human Services, Food and Drug Administration
Fox J (2002) An R, S-PLUS Companion to Applied Regression: Cox Proportional-Hazards Regression for Survival Data, Sage Publications, Inc., Thousand Oaks, CA
Jennison C, Turnbull B (2000) Group sequential tests with Applications to Clinical Trials. Chapman and Hall/CRC, Boca Rotan, FL
Kaplan E, Meier P (1958) Nonparametric estimation from incomplete observations. J Am Stat Assoc 53:457–481
Lee J, Liu D (2008) A predictive probability design for phase II cancer clinical trials. Clin Trials 5(2):93–106
Lin Y, Shih W (2001) Statistical properties of the traditional algorithm-based designs for phase I cancer clinical trials. Biostatistics 2(2):203–215
Loprinzi C et al (1994) Prospective evaluation of prognostic variables from patient-completed questionnaires. North Central Cancer Treatment Group. J Clin Oncol 12(3):601–607
National Cancer Institute (2012) Defining Cancer. http://www.cancer.gov/cancertopics/cancerlibrary/what-is-cancer. Accessed 01 March 2012
O'Connell M, Treder B (2009) Overview of Descriptive and Graphical Methods for Time-to-Event Data. In: Design and analysis of clinical trials with time-to-event endpoints. Chapman and Hall/CRC, Boca Rotan, FL, 227–239
O'Quigley J, Pepe M, Fisher L (1990) Continual reassessment method: a practical design for phase I clinical studies in cancer. Biometrics 46:33–48
O'Quigley J, Reiner E (1998) A stopping rule for the continual reassessment method. Biometrika 85:741–748
Parmar M, Machin D (1995) Survival analysis: a practical approach. Wiley, New York
Reiner E, Paoletti X, O'Quigley J (1999) Operating characteristics of the standard phase I clinical trial design. Comput Stat Data Anal 30:303–315
Reuters (2012) Cancer drug sales could hit $80 billion by 2011: IMS. http://www.reuters.com/article/2008/05/15/us-cancer-forecast-idUSN1453543620080515. Accessed 10 Apr 2012
Schlesselman J, Reis I (2006) Phase II clinical trials in oncology: strengths and limitations of two-stage designs. Cancer Invest 24:404–412
Seeking Alpha (2012) U.S. Prescription Drug Sales Grow Slowly; Hydrocodone Most Prescribed. http://seekingalpha.com/article/128003-u-s-prescription-drug-sales-grow-slowly-hydrocodone-most-prescribed. Accessed 10 Apr 2012

Simon R (1989) Optimal two-stage designs for Phase II clinical trials. Control Clin Trials 10:1–10
Simon R et al (1997) Accelerated titration designs for phase I clinical trials in oncology. J Natl Cancer Inst 89:15
Storer B (1989) Design and analysis of phase I clinical trials. Biometrics 45:925–937
Therneau T (2012) <therneau.terry@mayo.edu> Survival analysis, including penalised likelihood. R package version 2.36-12. http://CRAN.R-project.org/package=survival, Accessed September 1, 2012
Tighiouart M, Rogatko A, Babb J (2005) Flexible Bayesian methods for cancer phase I clinical trials: dose escalation with overdose control. Stat Med 24:2183–2196
Tufte E (1983) The visual display of quantitative information. Graphics Press, Cheshire, CT
Tufts Center for the Study of Drug Development (2006) Average Cost to Develop a New Biotechnology Product Is $1.2 Billion, According to the Tufts Center for the Study of Drug Development, 9 November 2006. http://csdd.tufts.edu/NewsEvents/NewsArticle.asp?newsid=69. Accessed 18 Dec 2006
Whitehead J (1997) The design and analysis of sequential clinical trials (revised 2nd edition). Wiley, New York

Chapter 10
Efficient and Effective Review of Clinical Trial Safety Data Using Interactive Graphs and Tables

Harry Southworth

Abstract Clinical trials collect a great deal of data relating to the safety of the trial participants. The data are complex in nature and traditional approaches to data review involve using summary tables and listed data. An alternative approach to the review of clinical trial safety data is presented, allowing reviewers to access individual subject data via hyperlinked plots and tables. Examples of presentations of data for very large studies, and of the inclusion outputs from modern statistical methods are demonstrated.

10.1 Introduction

A typical clinical trial collects a great deal of data from the trial subjects (healthy volunteers or patients), and it is usually the case that most of the data are related to the safety, not efficacy, of the study drug. These data relate to adverse events, laboratory investigations, vital signs, possibly electrocardiograms, medical history, demography, concomitant medications, urinalysis and often more.

The data are of various types: adverse event data are usually time to event data (though often presented as though they are binomial) and have associated data relating to severity, assessed causality and other associated things; the laboratory data are usually continuous and the variables are often highly skewed and subject to outliers; urinalysis data are often ordered categorical data.

The large amount of data and their complicated structure makes interpretation difficult. Historically, the approach to the presentation of such data has been to produce some summary tables (usually frequencies of adverse events and means,

H. Southworth (✉)
AstraZeneca, Alderley Park, Macclesfield SK10 4TG, UK
e-mail: harry.southworth@astrazeneca.com

standard deviations and other statistics for continuous data) and to list the data by domain. The result is often thousands of pages of output that clinical reviewers have to then work their way though, attempting to find patterns and signals of adverse effects. The process is slow, difficult and probably error prone.

This chapter describes an alternative approach that makes use of modern technology, and which has been found to be valuable and popular with clinical reviewers.

10.2 Principles

Experience illustrates that humans are able to gain greater and more rapid understanding of data when the data are presented graphically than when they are presented in tabular or listed format, so it follows that any modern data review system ought to rely heavily on graphical presentations. However, some kinds of data do not lend themselves well to graphical presentation, and sometimes it is useful to be able to quickly see exact numbers and not have to read an approximate value off an axis. Therefore, when appropriate, we opt to provide tabular presentations of data and seek to give reviewers the ability to rapidly pick out the important pieces of information from tables by use of conditional formatting—i.e., colour-coding table cells according to the value they contain—rather than coerce data into ill-suited graphical presentations.

A further observation is that reviewers very often require the ability to view multiple pieces of data relating to a single clinical trial subject. For example, if it is noticed that a particular subject had an unusually large value of alanine aminotransferase (a laboratory variable that might indicate liver damage), the reviewer will immediately want to know how other liver-related laboratory variables behaved (typically aspartate aminotransferase, bilirubin and alkaline phosphatase, but possibly others as well). But the reviewer's needs do not stop at the laboratory variables. It is also necessary to establish if there were any adverse events that were liver-related, if there were concomitant medications that could be implicated, if the subject had a medical history of liver problems, and so on. Therefore, it is necessary for the reviewer to always be able to view all the data collected on a subject, as soon as the subject becomes of interest.

Thus, 3 principles have emerged: present most of the data graphically; provide tabular data when it makes most sense; always enable reviewers to quickly access a report containing all of any individual subject's data.

10.3 Graphical Presentations

Other chapters in this book contain a great deal of advice on the construction of good graphs for clinical trial safety data. At the risk of repetition, good boxplots and shiftplots of study level data are indispensible, specific plots of certain variables

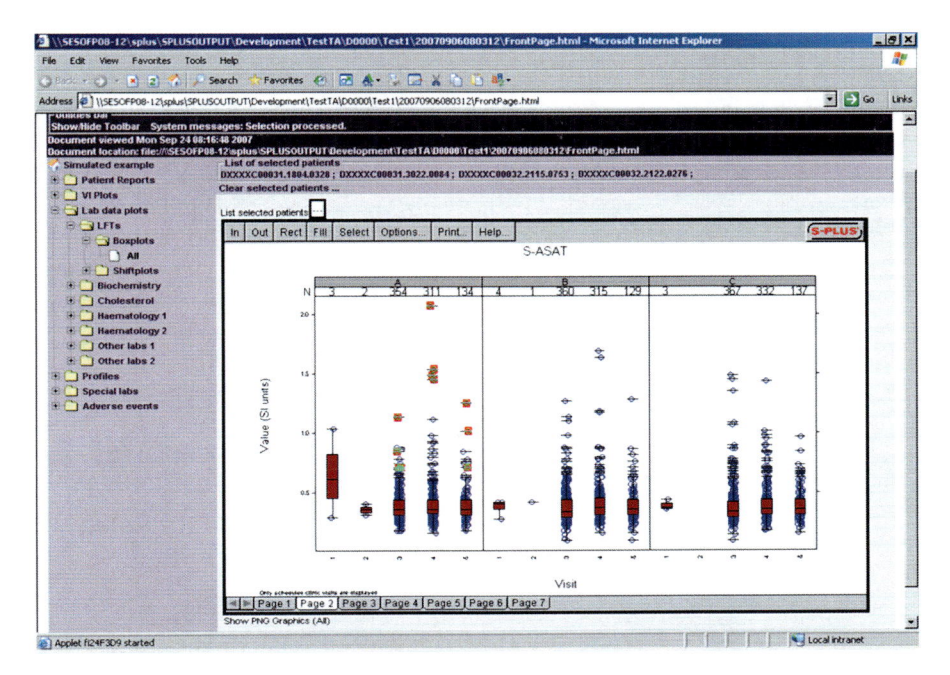

Fig. 10.1 A typical data review tool. Boxplots are superimposed over a plot of the individual observations, jittered to separate points. The reviewer selects points by clicking them and can then navigate to reports on the individual study subjects by clicking their identifiers. Upon clicking, points are highlighted across boxplots, and also across the tabbed pages of the plot

such as alanine aminotransferase versus bilirubin should be produced as these can quickly highlight potential liver injury, and plots of patient profiles are also valuable. Typically, it is the outliers that are the most important values, so these should never be excluded from plots.

In order to make the individual patient data available to a reviewer upon seeing, say, an outlier on a shiftplot, hyperlinking can be used. (Sect. 10.6.1) contains further discussion of software. The basic functionality used in our implementation involves the reviewer using an ordinary Web browser to view a collection of graphs and tables that are easily navigated. The reviewer can click on a point on a plot (a shiftplot or boxplot, say). That point is then highlighted, as are other observations (if any) that come from the same clinical study subject, and the subject identifier is displayed on the screen. The reviewer can then click the subject identifier to bring up a report containing all the data available on that subject. Alternatively, the reviewer can select several points at once and then browse a list of patient identifiers.

Figure 10.1 shows a screenshot of a typical data review tool. The various graphs, tables and patient reports that are contained in the tool can all be accessed by expanding and collapsing elements of the tree menu on the left of the screen. In the particular screenshot, the item Lab data plots→LFTs→Boxplots→All has been selected ("LFT" is short for "liver function test" although not all the variables actually relate to function). Breadcrumbs at the top left of the screenshot allow the user to navigate backwards and forwards within the data review tool.

The boxplots in Fig. 10.1 are of a particular laboratory variable, *S-ASAT* (serum aspartate aminotransferase). The clinical visit number is on the horizontal axis and the plot has been conditioned (trellised) on the treatment group (labelled A, B and C here). Along the top of each panel, the numbers of observations in the boxplots are shown.

When the user moves the mouse pointer over an observation in a plot, the related subject identifier appears on the plot (not shown). To select one or more points, the user clicks on the point. In panel A of Fig. 10.1 several observations are highlighted with red squares as a result of the user having selected these points. Immediately a point is selected, observations relating to the same subject are highlighted in the plots at other visits. The user can then click on List selected patients... above the plot to have the identifiers of the selected subjects listed as shown. These identifiers are hyperlinked to reports on the individual subjects (see Sect. 10.5).

10.3.1 Graphical Presentations for Large Studies

Some clinical studies are notably larger than others: the HPS study (Heart Protection Study Collaborative Group 2002) followed 20,500 patients for 5 years; the JUPITER study (Ridker et al. 2008) followed approximately 17,800 patients for 1.9 years; the PLATO study (Wallentin et al. 2009) followed approximately 18,600 patients for 1 year. For such studies, plotting all of the data can result in plots that are cluttered and packed with too many points.

10.3.1.1 Scatter Plots for Large Amounts of Data

One way to successfully produce a scatter plot for a large amount of data is to use hexagonal binning—essentially to display a two-dimensional histogram of the data. Since the data being reviewed are usually related to safety, it is the outliers that are of interest. As such, the subjects whose data fall into hexagons in the central region of the plot are unlikely to be of interest. This suggests that hyperlinks to patient reports only need to be included in the hexagons around the periphery of the plot, and these are only a small minority of all patients, thus overcoming issues due to shortage of memory (if all patient identifiers are hyperlinked, the need of the browser to hold all the URLs in memory can cause it to fail).

Figure 10.2 shows an example of a hexagonal bin shiftplot. The data are for a particular variable, *Creat* (creatinine), with baseline values on the horizontal axis and post-baseline values on the vertical axis. The axes cover the same range as each other, the aspect ratio has been set to 1, and a diagonal (0, 1) reference line appears on the plot. The vertical reference lines are at the lower and upper limits of normal.

The plot is conditioned on the treatment group, and on the clinical visit number. In Fig. 10.2 the data from visit 4 are displayed, as indicated by the strip label above

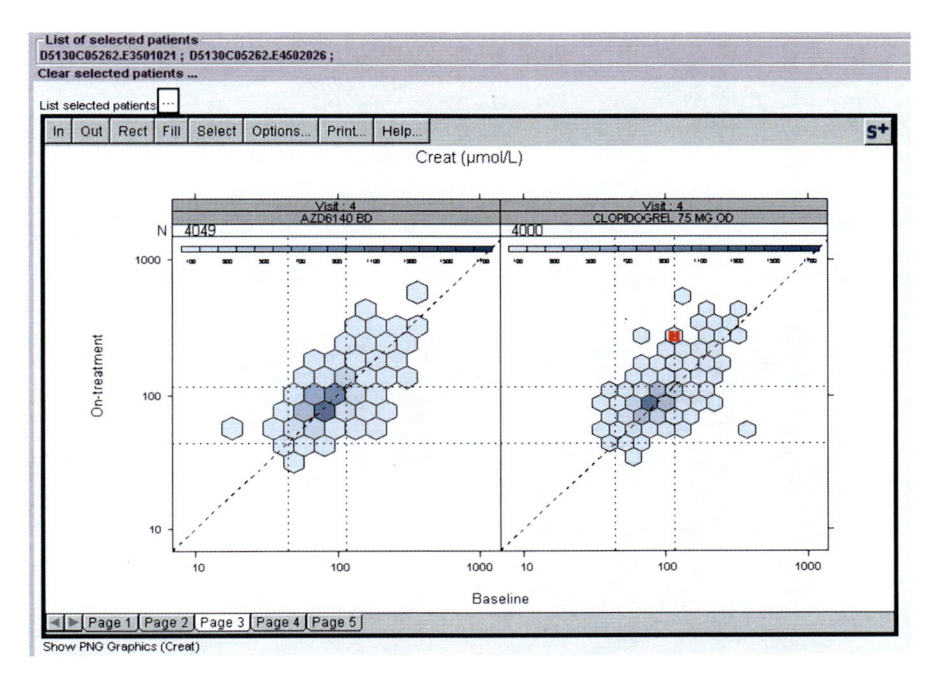

Fig. 10.2 A hexagonal bin shiftplot with patient identifiers included only in peripheral cells. Selecting a cell gives access to the hyperlinked list of patient identifiers

the plots, and the data from the other visits appear in pages accessed via the tabs beneath the plot. The key at the top of each panel gives an indication of how many patients are in each hexagonal bin. In this example, the user has selected a hexagon in the rightmost panel and listed all patient identifiers associated with data in that hexagonal bin. The listed patient identifiers are hyperlinked to patient reports (Sect. 10.5).

10.3.1.2 Boxplots for Large Amounts of Data

Boxplots display a summary of the majority of the data (the central region in the box, and the region to the ends of the whiskers) and the outliers. Again, it is chiefly the outliers that are of interest, so if only the outliers are hyperlinked to patient reports, memory usage is minimal.

However, it is important to be able to see what happened across time, and across different laboratory variables, for any subject who generates an outlying laboratory observation. Therefore, the information that links the observations in time and across variables needs to be contained in the graph so that once a reviewer selects an outlier by clicking it, the subject's data across time and other variables are also highlighted.

10.4 Tabular Presentations

Whilst graphical displays are generally preferred, tabular presentations of data can be useful, so long as reviewers can easily and quickly find what they are looking for. For frequencies or rates of adverse events, we typically produce both, the graphical representation described by Amit et al. (2007) and the simple tables of the same.

A typical adverse event summary table has a column for each treatment group. In cases where there are only 2 treatment groups, a column for relative risk can

Names	A	B	C
Nasopharyngitis	25 (06.56%)	31 (08.07%)	36 (09.25%)
Upper respiratory tract infection	26 (06.82%)	42 (10.94%)	31 (07.97%)
Headache	18 (04.72%)	39 (10.16%)	23 (05.91%)
Dizziness	21 (05.51%)	17 (04.43%)	21 (05.40%)
Influenza	16 (04.20%)	12 (03.13%)	20 (05.14%)
Cough	22 (05.77%)	21 (05.47%)	17 (04.37%)
Diarrhoea	18 (04.72%)	27 (07.03%)	16 (04.11%)
Back pain	12 (03.15%)	21 (05.47%)	15 (03.86%)
Arthralgia	16 (04.20%)	11 (02.86%)	15 (03.86%)
Hyperhidrosis	4 (01.05%)	9 (02.34%)	12 (03.08%)
Oedema peripheral	12 (03.15%)	20 (05.21%)	12 (03.08%)
Nausea	13 (03.41%)	11 (02.86%)	12 (03.08%)
Pain in extremity	17 (04.46%)	12 (03.13%)	10 (02.57%)
Shoulder pain	5 (01.31%)	3 (00.78%)	9 (02.31%)
Urinary tract infection	8 (02.10%)	9 (02.34%)	9 (02.31%)
Abdominal pain upper	11 (02.89%)	6 (01.56%)	9 (02.31%)
Lethargy	1 (00.26%)	4 (01.04%)	8 (02.06%)
Constipation	5 (01.31%)	7 (01.82%)	8 (02.06%)
Bronchitis	8 (02.10%)	9 (02.34%)	8 (02.06%)
Tremor	8 (02.10%)	6 (01.56%)	8 (02.06%)
Fatigue	9 (02.36%)	10 (02.60%)	8 (02.06%)
Dyspepsia	8 (02.10%)	10 (02.60%)	7 (01.80%)
Hypertension	8 (02.10%)	11 (02.86%)	7 (01.80%)
Flatulence	2 (00.52%)	1 (00.26%)	6 (01.54%)
Neck pain	3 (00.79%)	3 (00.78%)	6 (01.54%)
Pharyngolaryngeal pain	7 (01.84%)	8 (02.08%)	6 (01.54%)

Fig. 10.3 A frequency table of adverse event preferred terms from a study with 3 treatment groups, labelled A, B and C. If the user clicks a column heading, the table is sorted according to the frequencies in that column (in this case, the table has been sorted on column C). If the user clicks a cell in the table, a list of all the patient identifiers associated with that cell appears at the top of the page, and the identifiers are hyperlinked to their patient reports

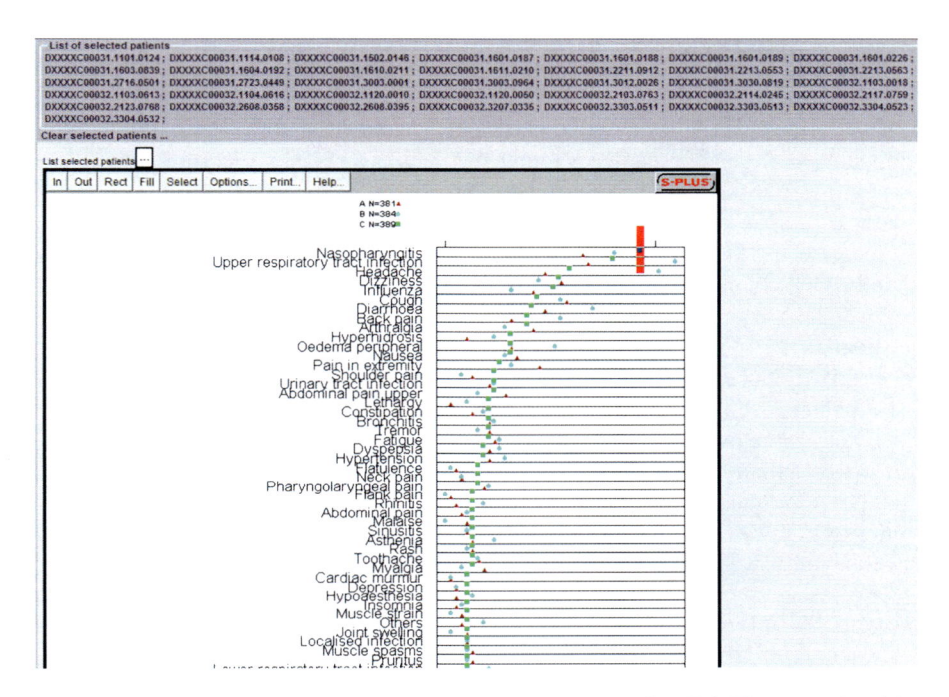

Fig. 10.4 A dotplot of the same adverse event data that appears in Fig. 10.3. The sort-order of the events can be changed by moving the mouse pointer over the vertical axis and clicking. In this case, the events have been sorted according to the percentage of patients reporting the event in group C, and the hyperlinked list of patients reporting nasopharyngitis in group C has been generated

easily be added, or perhaps a measure of the strength of evidence for a treatment difference such as a χ^2 statistic or deviance. In computing relative risks, it is desirable to avoid zero and infinite values. In the case of (almost) equal sample sizes, this can be achieved by considering the data as being a 2 ×2 table and adding 1/2 to each cell (equivalently, treating the data as coming from 2 Beta distributions and using the Jeffreys prior). In any case, when there are small numbers of events, interpretation of such statistics needs account for the fact.

To make the table easy for reviewers to use, the table can be sorted according to the values in any particular column by clicking on the header for that column, and reverse sorted by clicking again. If the reviewer is interested in the patients in a particular treatment group who reported a particular adverse event, they can click the appropriate table cell to bring up a hyperlinked list of the patient identifiers, and then access the patient reports by clicking in that list.

Figure 10.3 shows an example of a sortable frequency table from a study with 3 treatment groups. The user has moved the mouse pointer over the row labelled *Nasopharyngitis* and clicked on one of the cells in that row to bring up the hyperlinked list of patient identifiers associated with that event in one of the treatment groups (A, B or C). The same information is presented graphically in Fig. 10.4.

10.5 Patient Reports

When it comes to reports containing all the relevant and available information for an individual trial subject, we have found that tabular presentation, with conditional formatting, is often preferable to graphical presentation. Whilst it is possible to render all of the data graphically, doing so often makes some information more difficult to see. For example, showing each lab value over time as a symbol that changes depending on whether the value is above, beneath, or within the normal range, hides the actual values of the data and forces the reviewer to do more work to see them.

In order to enable a clinical reviewer to skip up and down the report quickly, a list of tables that appear in the report is displayed in a small font above each table. Each item in the list is anchored to the relevant table so that clicking the item causes the report to jump to that table.

It is common for an adverse event to be coded as "serious" if it led to hospitalization, death, congenital deformity, disability or was some other important medical event. Naturally, such events require more attention from reviewers than non-serious events. As such, if an adverse event is recorded as "serious," the table cell that contains the information is formatted bright red so that it will stand out to the reviewer. Similarly, it is common for the severity of each adverse event to be recorded (often on a scale of mild, medium, severe) and formatting is used to make the cell background yellow for the mildest level, deep red for the most severe level, and shades of orange and red in between.

Laboratory values that are outside of the normal range are of more interest to clinical reviewers than those within the normal range, so it makes sense to draw the reviewer's attention to these. Using a pale blue cell background to indicate that a value is beneath the lower limit of normal, and an orange cell background to indicate that it is above the upper limit or normal achieves this, but immediately raises the question of what the normal range actually was. Rather than having a separate table containing the normal ranges, and rather than increasing the amount of information in each table 3-fold by including the normal ranges, our implementation is such that when the reviewer moves the mouse over the table cell, a tool-tip displaying the normal range appears. Figure 10.5 displays a fragment of a typical patient report in which the conditional highlighting and a tooltip can be seen.

10.5.1 Profile Graphs

In some studies, it is known that certain laboratory values are of special interest. For example, if earlier work with the experimental compound indicated a possible adverse effect on the liver, plots of alanine aminotransferase, aspartate aminotransferase, total bilirubin and alkaline phosphatase over time will be of interest. It is possible to create such patient profile graphs, including indicators of adverse events

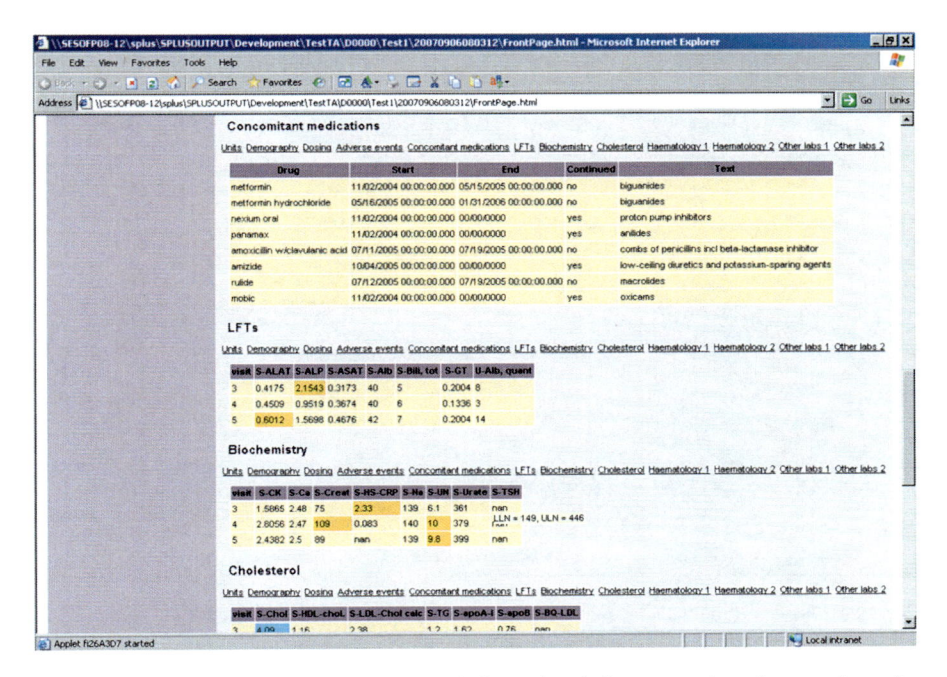

Fig. 10.5 A patient report containing tabular information. Laboratory values that are above the upper limit of normal are highlighted orange, those that are beneath the lower limit of normal are highlighted *blue*, and a tooltip displays the normal range on mouseover

of interest and the dosing interval, and to embed them directly in the patient reports. Alternatively, the patient profile graphs can be trellised by patient and hyperlinked to the patient reports.

Figure 10.6 shows plots of 3 liver-related variables, scaled by their upper limits of normal so that they display well together on the same vertical scale. The horizontal axis is the number of days since the start of the study. The shaded background region in each panel indicates the interval during which the patient was receiving treatment. Clicking into one of the panels enables access to the individual's patient report.

10.5.2 Static Renditions

The interactivity described above (the use of anchors to navigate the report, and the use of mouse-over to see the normal range) is natural when the report is displayed in a Web browser, but sometimes it is useful for reviewers to take static versions of the reports, perhaps to review off-line, or to discuss in meetings with colleagues. Such static renditions lose the interactivity and alternative ways of communicating

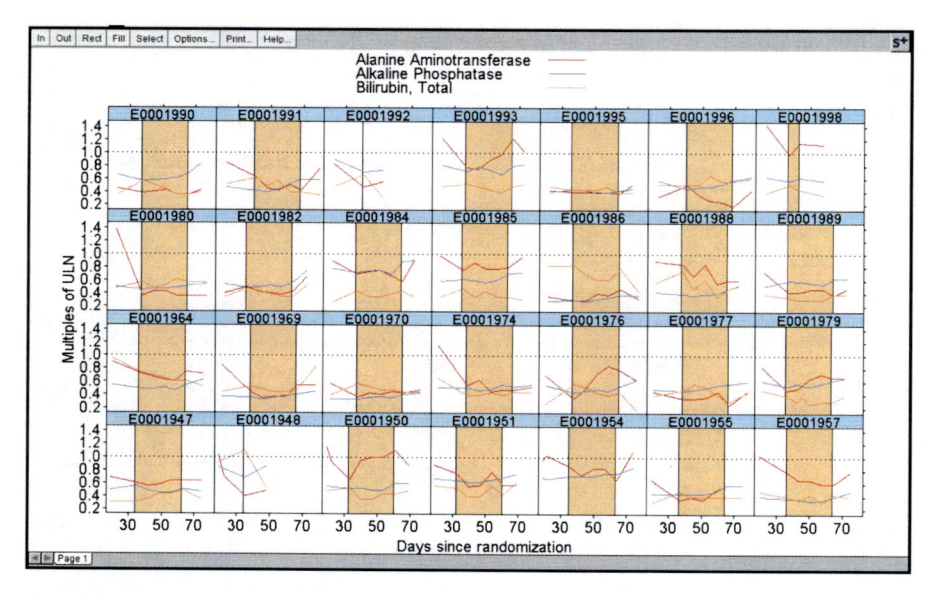

Fig. 10.6 Plots of three liver related laboratory variables over time. The *shaded regions* indicate the periods during which the patients were receiving treatment

the information are required. For displaying the normal ranges, we have found that including the values as subscripts (lower limit of normal) and superscripts (upper limit of normal) in a small font in each cell is acceptable.

10.6 Experience with the System

10.6.1 Implementation

The implementation at AstraZeneca is a bespoke system based on the S-PLUS™ statistical software system. Much of the output is HTML, together with XML, XSL, Javascript and related technologies that any modern Web browser can interpret. Proprietary off-the-shelf systems also exist, including one based on the Spotfire graphical software system, and we are currently considering moving to a hybrid system involving S-PLUS™ and Spotfire. An open source approach to implementation could use Scalable Vector Graphics, so the implementation of an interactive data review system need not be expensive.

In our implementation, the patient reports are XML files, and rendering into HTML in the browser is done with XSL. A separate XSL stylesheet is used to create PDF renditions that can be easily taken off-line and which print more nicely onto paper than the HTML versions.

10.6.2 Review Processes

In a typical clinical trial, there will often be more than one timepoint at which it is necessary or desirable to review the data. Whilst our system was primarily designed for the review of completed trials, it is natural that users would want to be able to access the same functionality at other times during the execution of the study.

When the clinical trial has completed and the blind has been broken, the primary reviewers are medics and drug safety personnel looking for any indications that the experimental compound has any unexpected adverse effects. However, medical writers and auditors also use the system to quickly access information and to check that other outputs from the study are consistent with the data.

It is common for a blinded review of the data to be performed prior to the final data being made available. At such a blinded review it is common for there to be errors in the data, such as lab or ECG data that have been recorded in the wrong units, and such errors are readily identifiable using an interactive data review system. Any errors discovered can then be reported and corrected prior to the final database lock. The blind review also makes it possible to cross check the data with other planned outputs from the trial to ensure that the programming has been carried out correctly, reducing the scope for errors and delays later in the reporting process.

Some clinical trials employ a safety monitoring committee that performs regular blinded or unblinded reviews of the safety data in order to identify and advise on any emerging safety concerns. It is always necessary to interpret any possible safety signal in context, and being able to easily browse the data interactively enables rapid access to all relevant information collected during the study to date.

In Phase I studies, it is often necessary to closely monitor the emerging data in order to identify any unexpected adverse events, or any adverse events that are occurring with a higher than expected frequency. In this setting, the treatment will often be blinded and the data prone to errors, and it is necessary to frequently update the data review tool. We are currently piloting a system based on a slimmed down version of our S-PLUS™ tool together with Spotfire. The system is updated with new data every 24 h so that reviewers are always in possession of the most up-to-date information.

10.6.3 Establishing if the System adds Value

In order to establish whether the system was proving useful with clinical reviewers in practice, feedback was sought. When formal questionnaires were sent out, the response rate was approximately 50% (46 people) and 93% of the feedback from clinical data reviewers was positive. It was concluded that the system saved time and money, and led to a much deeper understanding of clinical trial data than was obtained using static tables and listings. Anecdotal evidence in the form of comments made verbally and in writing supports this strong positive conclusion.

It is difficult to quantify value when the main benefit comes from ease of use and ability to understand data and find information. Perhaps the best evidence of such value is observed in cases in which the clinical study data had been reviewed in the old-fashioned way, and then in a data review system at a later date. In 2 separate instances, a data review system was created for a study that had already closed and had its final report written. When reviewers went back over the data using the review system, they found significant potential safety issues that had not previously been noticed. Conversely, we have had situations in which reviewers raised concerns over safety having seen a small number of outliers in listings of laboratory data. When the same data were reviewed in one of our data review systems, the context allowed a deeper understanding of the data and dispelled much of the concern.

Another aspect of the value added that is difficult to quantify is the reduction in the number of requests for additional presentations of data. When following the old manual approach to data review, it is common for reviewers to see something that looks interesting, and to then request additional summary tables, listings and plots; 62% of clinical reviewer questionnaire respondents said that they found they requested fewer additional outputs as a result of having the data review system, and only 5% responded that they needed more.

In summary, although benefits are difficult to quantify, there is evidence that an interactive data review system can enable expert reviewers to identify potential safety issues earlier in clinical development, put potential safety issues into context so that they can be managed accordingly, and reduce the amount of time and resource needed to properly review the data.

10.7 Additional Outputs

Since the system can display tables and graphs, there are very few limitations on what kinds of outputs can be included. Whilst good graphs of the data and useful summary tables are a good starting point, it is possible to include the results of more sophisticated analyses.

10.7.1 *Data Mining Adverse Events*

Southworth and O'Connell (2009) describe several methods for ordering adverse events according to the strength of evidence for a treatment effect. The authors make a distinction between formal hypothesis testing and data mining and then describe a general approach of using the adverse event frequencies to classify patients to treatment group, and to then find out which features (adverse events) are the most important for performing the classification. The classification methods considered include penalized logistic regression (Firth 1993), random forests (Breiman 2001) and gradient boosted models (Friedman 2001). When these classification methods are compared with a modified version of the approach for the analysis of frequency tables developed by Berry and Berry (2004), the authors find the various approaches work similarly or better on the example datasets that they use.

Simple graphical displays of the outputs of the various methods, such as variable importance or relative influence plots, can be used to help clinical reviewers identify the adverse events most likely to be of interest in terms of strength of evidence for a treatment difference. If variable importance or relative influence is used to order the events, it is useful to also present estimates of relative risk or risk difference, together with interval estimates.

A "drill-down" approach is also possible, in which selecting an adverse event on a variable importance plot produces the list of patients who reported the event, and these patient identifiers are linked to individual patient reports. To take the idea a step further, it is possible to create a report for each adverse event which, in turn, links to the individual patient reports. Such a report can contain frequency tables of the event split by treatment group, sex and race (for example), cumulative frequency plots of the time to first onset, or plots of predicted probabilities (or relative hazard) of the event across the observed range of ages or BMIs, as obtained from a generalized additive model.

Figure 10.7 displays a hyperlinked plot of adverse event terms sorted according to their relative influence as estimated by a random forest. The 25 events with the largest values of relative influence are displayed (it is possible to make the plot longer when displayed in the browser, but for purposes of displaying it on a fixed page, only a subset of events is included). Since relative influence or variable importance from a random forest will always put *some* order on the predictors, it is important to investigate further. The tabbed pages at the bottom of Fig. 10.7 contain plots of relative risks and risk differences for the events, in the same order. Figure 10.8 shows the risk difference plot (i.e., the difference between the observed probabilities of having the event in the two treatment groups). The interval estimates are 90% posterior intervals simulated from beta distributions resulting from updating Jeffreys priors. It is clear from Fig. 10.8 that most of the events identified by the random forest have only weak evidence of a difference between treatment groups.

10.7.2 Outlier Detection

Penny and Jolliffe (2001) and Southworth (2008) describe methods for outlier detection in clinical laboratory data, and Lin et al. (2011) refine the approach by excluding clinically irrelevant outliers. In general, these methods proceed by robustly estimating the Mahalanobis distance for each multivariate observation and then labelling values above a (somewhat arbitrary) threshold as being outliers.

Having reduced the data to their robust Mahalanobis distances, it is not difficult to produce two-dimensional plots that will fit naturally into the data review system. These include plots of robust distances against robust leverage values, or perhaps

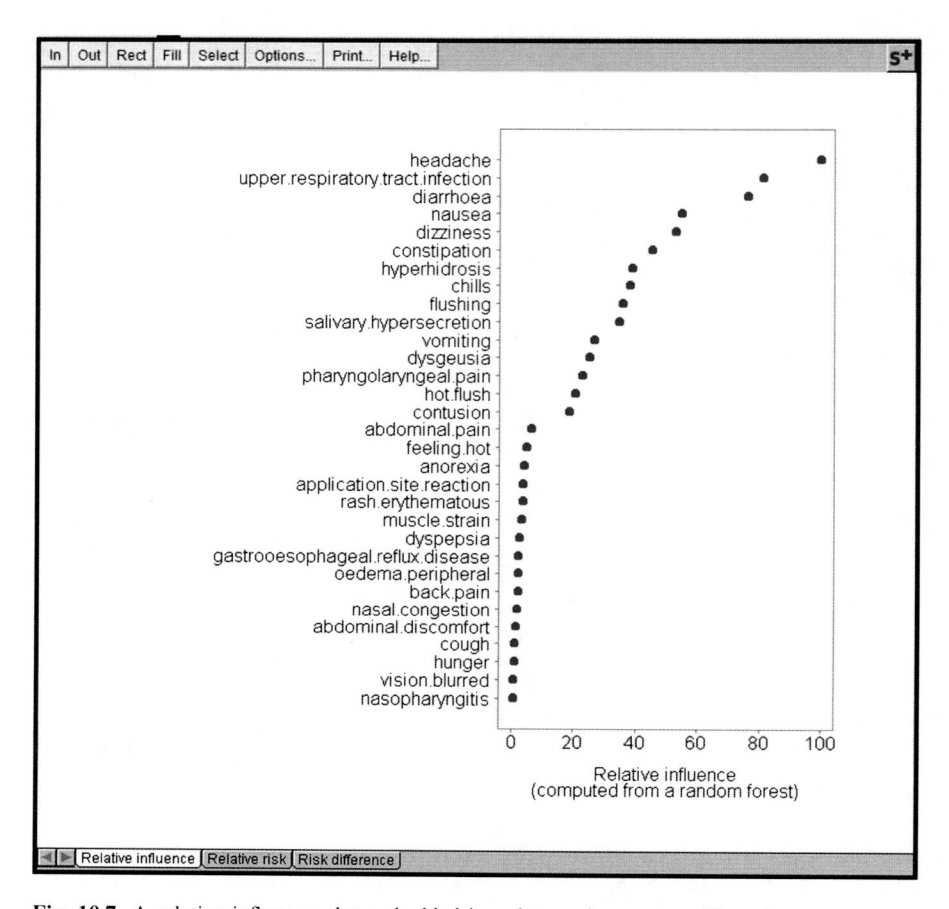

Fig. 10.7 A relative influence plot embedded in a data review system. The adverse events are sorted such that those with the most evidence for a treatment difference appear at the top. The points on the plot are hyperlinked to reports displaying more information about the event, including summaries by sex, age and race

boxplots of the robust distances by treatment group. In order to establish why a multivariate outlier has been identified as such, it is useful to then consider univariate outlier detection plots.

Figures 10.9 and 10.10 show examples of outlier detection plots. Four liver-related laboratory variables were included in the analysis (alanine aminotrasferase, aspartate aminotransferase, alkaline phosphatase and total bilirubin, abbreviated in Fig. 10.10 as ALT, AST, ALP and TBL, respectively). The multivariate robust distances in Fig. 10.9 are computed by using robust regression to eliminate baseline effects, and then finding a robust estimate of Mahalanobis distance for each

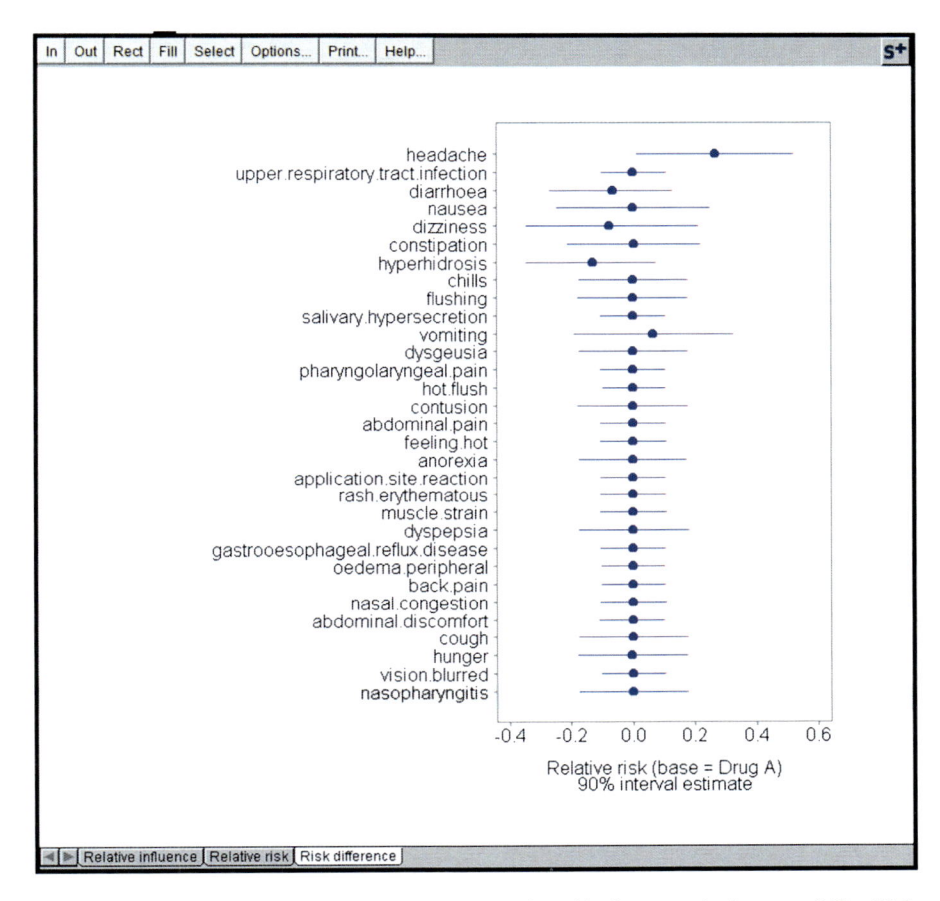

Fig. 10.8 A risk difference plot. This is the plot on the tabbed page at the bottom of Fig. 10.7. It can be seen that differences between the treatment groups are mostly small at the interval estimates are quite wide

four-dimensional observations. These are then plotted against robust Mahalanobis distances of the baseline values. There appears to be some evidence of a dose response effect. More insight can be gained by looking at univariate outlier detection plots, available on the tabbed page, displayed in Fig. 10.10. These are plots of robustly scaled residuals for each variable. The user has selected the largest residual for ALT and the associated residuals for AST, TBL and ALP have been highlighted as a result, making clear that the patient was an outlier in 3 of the 4 variables simultaneously. As ever, the patient reports can be accessed via hyperlinks.

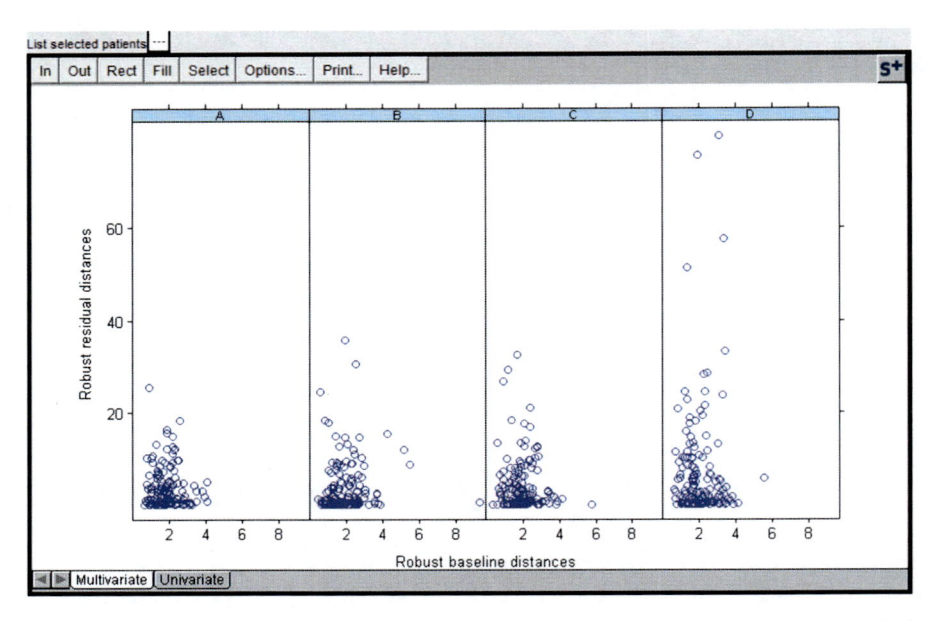

Fig. 10.9 Multivariate outlier detection plot. Any observation with a large value on the vertical axis can be considered to be a multivariate outlier. To gain more insight, the tabbed page for univariate outliers can be selected

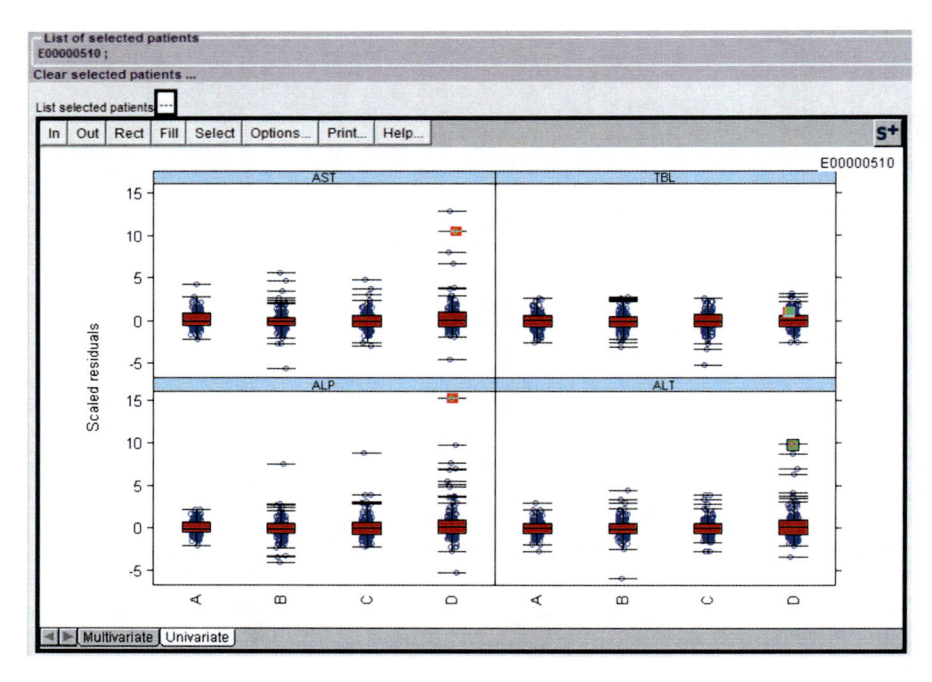

Fig. 10.10 Boxplots of robustly scaled residuals for each of the variables considered in the multivariate outlier detection. Selecting the largest scaled residual for ALT highlights the related values in the other panels

10.8 Closing Comments

Clinical review of clinical trial data using listings and summary tables is outmoded and better approaches are available. There is plentiful technology, both proprietary and open source, for creating graphs and tables that are hyperlinked to related documents. The available evidence suggests that clinical reviewers find value in having such an interactive data review system, in that they gain deeper understanding of the data and are able to more effectively find and interpret potential safety issues in clinical trials.

References

Amit O, Heiberger RM, Lane PW (2007) Graphical approaches to the analysis of safety data from clinical trials. Pharmaceut Stat 7:20–35
Berry SM, Berry DA (2004) Accounting for multiplicities in assessing drug safety: a three-level hierarchical mixture model. Biometrics 60:418–426
Breiman L (2001) Random forests. Mach Learn 45:5–32
Firth D (1993) Bias reduction of maximum likleihood estimates. Biometrika 80:27–38
Friedman J (2001) Greedy function approximation: a gradient boosting machine. Ann Stat 29: 1180–1232
Heart Protection Study Collaborative Group (2002) MRC/BHF Heart Protection Study of cholesterol lowering with simvastatin in 20 536 high-risk individuals: a randomised placebo-controlled trial. Lancet 360:7–22
Lin X, Parks D, Zho L, Curtis L, Stell H, Rut A, Mooser V, Cardon L, Menius A, Lee K. Truncated robust distance for clinical laboratory safety data monitoring and assessment. J Biopharm Stat (to appear)
Penny KI, Jolliffe IT (2001) A comparison of multivariate outlier detection methods for clinical laboratory safety data. Statistician 50:295–308
Ridker P, Danielson E, Fonseca FAH, Genest J, Gotto AM, Kastelein JJP, Koenig W, Libby P, Lorenzatti AJ, MacFadyen JG, Børge BA, Nordestgaard G, Shepherd J, Willerson JT, Glynn RJ (2008) Rosuvastatin to prevent vascular events in men and women with. New Engl J Med 359:2195–2207
Southworth H (2008) Detecting outliers in multivariate laboratory data. J Biopharmaceut Stat 18:1178–1183
Southworth H, O'Connell M (2009) Data mining and statistically guided clinical review of adverse event data in clinical trials. J Biopharm Stat 19:803–817
Wallentin L, Becker RC, Budaj A, Cannon CP, Emanuelsson H, Held C, Horrow J, Husted S, James S, Katus H, Mahaffey KW, Scirica BM, Skene A, Steg PG, Storey RF, Harrington RA (2009) Ticagrelor versus clopidogrel in patients with acute coronary syndromes. New Engl J Med 361:1045–1057

Chapter 11
Visualizing Dose–Response When the Signal to Noise Ratio Is Low: The Bronchodilatory Response in Chronic Obstructive Pulmonary Disease

Michael Looby and Didier Renard

Abstract Spirometry is a safe, cheap, and easy-to-use methodology for the assessment of lung function. Spirometry biomarkers such as the forced expiratory volume in 1 second (FEV_1) and the forced vital capacity (FVC) are commonly used in the diagnosis of conditions such as asthma and chronic obstructive pulmonary disease. In recent years, FEV_1 in particular has also been used to support dose selection in bronchodilator drug development programs. Despite its convenience and objectivity as a measure of pulmonary function, FEV_1 has a very low signal to noise ratio (SNR) as a marker of bronchodilator response. This problem is exacerbated when the biomarker is analyzed using traditional dose ranging study designs that do not provide an explicit and precise estimate of the dose response relationship. The combination of low SNR and imprecise methodology means that traditional dose finding activities for bronchodilators are inefficient and may lead to the selection of suboptimal doses. Using graphics produced during the development of the novel long-acting β_2-agonist, indacaterol, the issues outlined above are described and an alternative approach, built on a model-based characterization of the bronchodilatory dose response relationship is presented.

11.1 Introduction

Bronchodilators are drugs that dilate the bronchi and bronchioles, decreasing resistance in the respiratory airway and increasing airflow to the lungs. Typically, bronchodilators are used to treat obstructive lung diseases of which asthma and chronic obstructive pulmonary disease (COPD) are the most common. Bronchodilators are either short-acting or long-acting. Short-acting medications provide quick or "rescue"

M. Looby (✉) • D. Renard
Modeling and Simulation, Novartis Pharma AG, Basel, Switzerland
e-mail: Michael.Looby@novartis.com

relief from acute bronchoconstriction. Long-acting bronchodilators help control and prevent symptoms. The main types of prescription bronchodilating drugs are β_2-agonists (short- and long-acting) and anticholinergics (short- and long-acting). These drugs are administered directly into the lungs by inhalation. The β_2-agonists are used to treat both asthma (in combination with anti-inflammatory medication) and COPD; the anticholinergics are used in COPD. While the short-acting drugs are used when needed, to provide fast temporary relief from symptoms and flare-ups for periods of 4–6 h, the long-acting drugs are taken at fixed intervals to control and prevent bronchoconstriction. Typically, the long-acting medication takes longer to begin working and provides relief for up to 12 h. Indacaterol is a new long-acting β_2-agonist which provides both a rapid onset of action and 24 h control (EMA 2009a, b) in the maintenance treatment of airflow obstruction in patients with COPD.

Spirometry is a pulmonary function test used to diagnose respiratory disorders and assess the efficacy of bronchodilators. It measures the volume and velocity of air that can be inhaled or exhaled. These quantities are captured in the metrics forced vital capacity (FVC) and forced expiratory volume at 1 second (FEV_1). Figure 11.1 shows the spirometric time course used to derive these key metrics. Results are usually given in both raw data (liters, liters per second) and percent predicted—the test result as a percent of the "predicted values" for the patients of similar characteristics (height, age, sex, and sometimes race and weight). The ratio of FEV_1/FVC is a measure of airflow limitation. A post-bronchodilator FEV_1 <80% of the predicted value in combination with an FEV_1/FVC <70% confirms the presence of airflow obstruction—either COPD or asthma. An improvement of FEV_1 of ≥12% (either spontaneously or after inhalation of a bronchodilator and/or following a 2-week course of oral corticosteroids) indicates reversibility, and therefore suggests a diagnosis of asthma.

The FEV_1 and FVC thresholds used for diagnostic purposes are sufficiently far apart to allow a reasonable chance of correct diagnosis given the inherent variability in these measures. In recent years, FEV_1 has also become the metric of choice to support dose selection of new bronchodilators. In contrast to their use as diagnostics, less attention has been paid to the performance of FEV_1 as a differential marker of the bronchodilatory response.

This chapter uses graphics to explore the properties of the FEV_1 signal as a basis for the assessment of the dose response of a bronchodilator. The following issues are explored:

- The inherent variability of FEV_1 and its consequences as a biomarker of the bronchodilatory response
- The traditional approach to dose ranging based on pairwise comparisons of active dose and placebo
- Model-based characterization of the bronchodilatory dose response

Many of the thoughts captured in this chapter evolved from experience gathered during the development of the novel bronchodilator, indacaterol. The graphics presented hereafter were originally used at various stages of the development program and for this reason are not always uniform with respect to appearance or display.

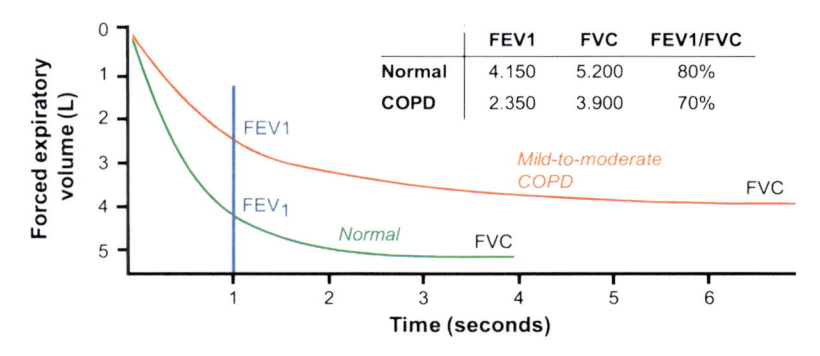

11 Visualizing Dose–Response When the Signal... 219

	FEV1	FVC	FEV1/FVC
Normal	4.150	5.200	80%
COPD	2.350	3.900	70%

Fig. 11.1 Spirometry measures used to assess lung function

11.2 FEV$_1$ As a Marker for Dose Response Assessment

FEV$_1$ is a non-invasive, objective, quick and cheap measure of bronchodilator activity. Despite these appealing characteristics, it suffers from an important drawback; it has a low signal to noise ratio (SNR), i.e., the drug-related change in FEV$_1$ is small relative to the background within- and between-patient variability in the measure.

To compare the degree of variability to the magnitude of the dose response relationship of a bronchodilator, Fig. 11.2 presents pooled raw trough FEV$_1$ data (black dots, left plot) from two dose ranging trials in indacaterol. The red line is a smooth curve to capture the average change in response with dose. A reader unfamiliar with these measures might be forgiven for missing any evidence of a dose response. The right plot zooms in on the dose response curve to present it on the scale it is typically presented on to give an impression of the SNR. While the raw trough FEV$_1$ data ranges between about 0.5 and 3 L, the drug-induced changes range between about 1.25 and 1.45 L. In other words, the approximately 0.2 L drug-related signal is located in 3 L of noise.

An initial conclusion from these figures is that it is very difficult to depict the dose response based on FEV$_1$ for a bronchodilator using even large quantities of *raw* data alone due to the low SNR. In order to tease out a more useful signal for the purposes of graphical representation of the data, it is necessary to improve the signal by removing sources of variability.

Figure 11.3 presents the estimated SNR in FEV$_1$ based on an analysis of covariance (ANCOVA) of individual steady-state trough FEV$_1$ measurements taken from 5,558 patients and 34,615 observations collected in 9 studies across the indacaterol development program. Patient characteristics, defined by baseline FEV$_1$ and reversibility, account for much of the total variability. Having taken all covariates into account, about 210 mL of variability remains unexplained. This provides the denominator estimate for the SNR estimates in Fig. 11.3. Placing this noise in the context of the indacaterol dose response indicates that the SNR for differentiation of most doses is less than 15%. For example, the signal in the differentiation of the 75 and 150 μg

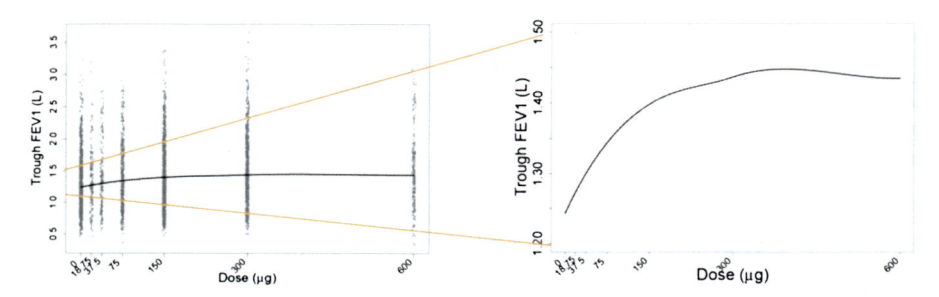

Fig. 11.2 Raw trough FEV_1 data vs. dose on original scale (*left*). Zoom of dose response (*right*)

is only about 1/10th of the noise. Even the largest dose of indacaterol, 600 µg, when compared to placebo has a SNR of less than 1. As we shall see below this has consequences for the precision with which active doses can be differentiated in reasonably sized trials.

Baseline FEV_1, i.e., FEV_1 measured prior to treatment, accounts for a major part of total variability, and for this reason, it must be accounted for when assessing the response to a bronchodilator. Figures 11.4 and 11.5 show the impact of baseline normalization. Figure 11.4 shows the individual 24 h profiles (black lines) for a placebo and indacaterol dose, respectively, in a COPD population. The response in the indacaterol cohort is hardly discernable from the placebo cohort at the population mean level (red line) due to the large variability. Figure 11.5 shows the impact on these profiles of subtracting the baseline FEV_1 value for each patient. Having now accounted for the biggest source of between-patient variability, the drug-induced average increase in FEV_1 becomes more apparent.

The next step to increasing the SNR is the correction for systematic deviations due to circadian variability or other extraneous systematic study-related variability. In a review of 4,756 pulmonary function tests from individuals who required one for any reason, Medarov et al. (2008) reported a diurnal variation of 17.6% between the lowest (1.80 L) and highest (2.12 L) mean FEV_1 values. The circadian variability represents the natural daily change in the time course of broncho-responsiveness over the course of the day. Typically, the FEV_1 nadir is reached shortly before awaking and subsequently approaches a peak by mid morning, from where it descends slowly again to its nadir again over the course of the remaining day. To account for these systematic changes, the baseline corrected curves are adjusted by subtracting the placebo profile. Figure 11.6 presents the impact of placebo correction on the FEV_1 profiles on a range of indacaterol doses. The biggest placebo-induced changes occur in the morning where the curve rises rapidly from the nadir to the peak. This is also the time where trough FEV_1 is assessed. Any discrepancies in the time of dose relative to the sampling time or in sampling times across patients will inflate variability in the corrected FEV_1 measures.

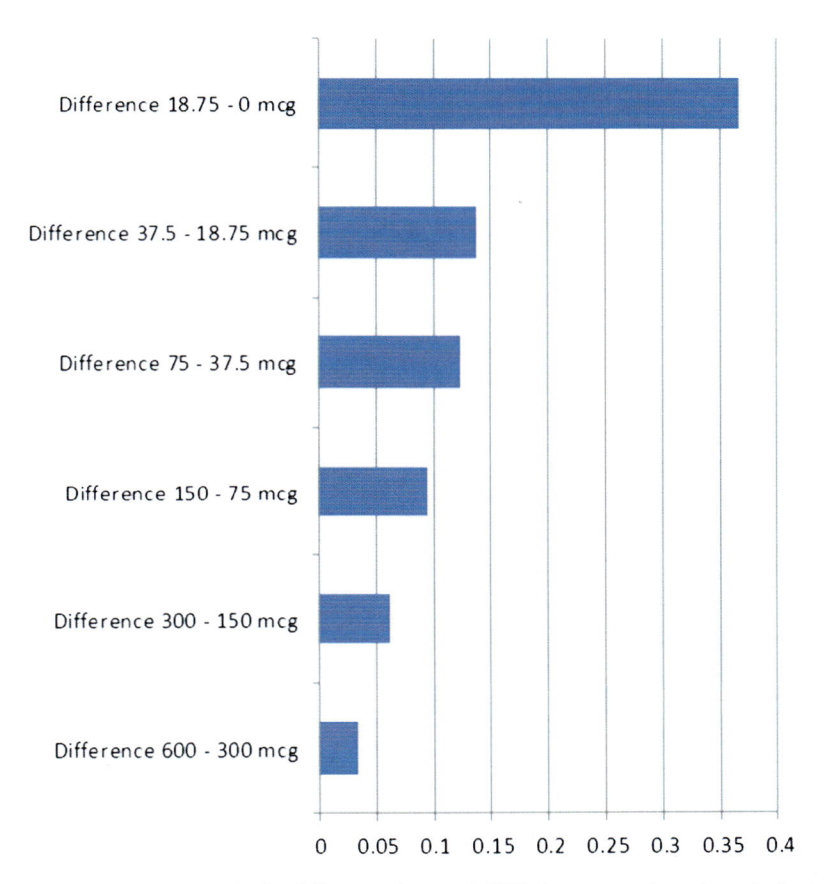

Fig. 11.3 Signal to noise ratio for differences in trough FEV$_1$ between various doses in the indacaterol dose response

For the purpose of dose selection in the development of new bronchodilators, the FEV$_1$ response is assessed for a range of doses at steady state, over the course of a dosage interval. Typically, several metrics are considered: peak FEV$_1$ to represent the maximum response expected in a dosage interval; area under the curve of serial FEV$_1$ measurements within a dosage interval to give the average bronchodilation over the course of the day; and trough FEV$_1$ measured just prior to the next administration, to give the minimum degree of bronchodilation.

Figure 11.7 presents the corrected mean trough FEV$_1$ response to 150 µg indacaterol from 6 trials (one in each panel) with between 92 and 420 COPD patients at each study visit (blue dots). This analysis has taken into account the important covariates, such as baseline and reversibility, measured in each trial and accounts for the known major differences in patient characteristics. Despite this, considerable unaccounted within- and between- study variability remains. The median responses (red

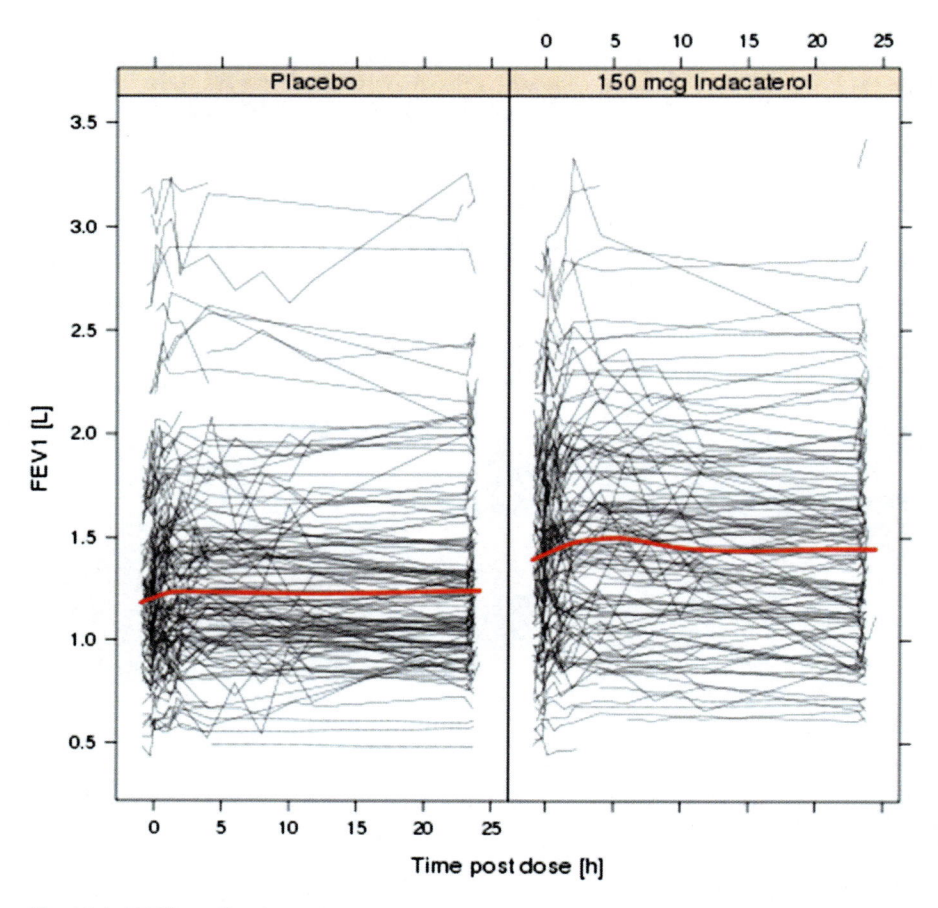

Fig. 11.4 FEV1 profile after a single dose of placebo (*left*) and indacaterol (*right*)

lines) across the 6 trials span 60 mL. Within trials, the mean trough FEV_1 spans a maximum of about 50 mL (gray areas).

Given that baseline accounts for a major portion of difference between patients and is necessary to tease out treatment signals, it is important to understand its behavior over time when interpreting baseline corrected data. Figure 11.8 presents the change in trough FEV_1 relative to baseline in 385 COPD patients treated with placebo on study days 1, 8, 15, and 29. Note: all FEV_1 values are the mean of two adjacent measures 15 min. apart within each subject. The black lines join the observations in each patient to give an impression of the within-patient variability. The box and whisker plots show the distribution of values on the various study days. Several trends are evident:

- The FEV_1 values within any one patient can vary considerably from day to day in the absence of any drug. This is apparent from the varied trajectories of the black lines connecting the observations on the respective study days

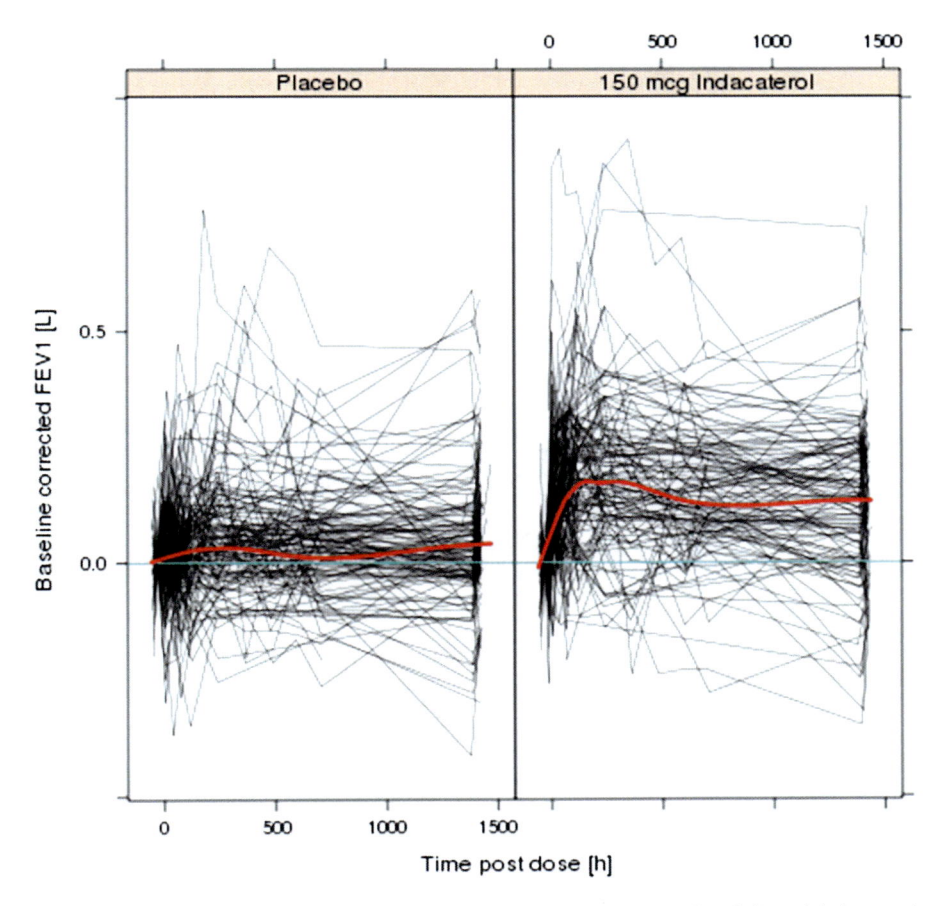

Fig. 11.5 Baseline corrected FEV$_1$ profile after a single dose of placebo (*left*) and indacaterol (*right*)

- Baseline correction is most effective for the trough value one day after commencement of treatment. However, with increasing time, the impact of baseline correction becomes diluted. This is apparent from the increasing width of the boxes over time
- The dilution of baseline correction is noticeable beyond the first dose and increases continually over the course of the study

This simple exercise demonstrates that the imprecision of baseline corrected values can increase 2- to 3-fold relative to the response observed immediately after the first dose. This effect is mainly driven by unexplained within-patient variability in baseline. In the relatively large sample presented in Fig. 11.8, the confidence interval for the fractional difference from baseline will be over 5% by day 30 which adds about 60 mL uncertainty to the drug-related response. This effect explains the common observation that a dose response relationship, presented as its baseline

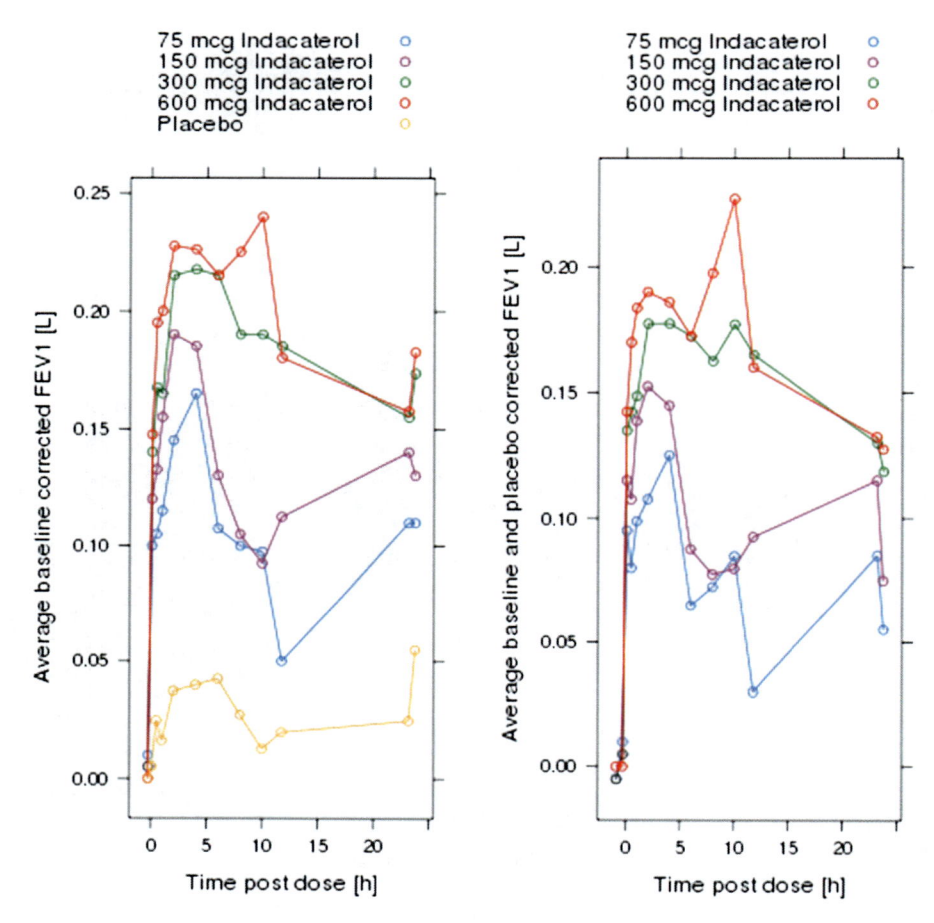

Fig. 11.6 Baseline corrected mean FEV_1 profiles before (*left*) and after (*right*) placebo correction

corrected value, is almost always more pronounced following the first dose compared to subsequent doses. Figures 11.10, 11.11, and 11.12 bear out this point: while a clear dose response is observed on Day 2, the greater imprecision on subsequent days confounds dose differentiation on subsequent days.

In summary, this section shows that bronchodilator-induced changes in FEV_1 are small relative to the natural between- and within-patient variability. In order to tease out the bronchodilator signal from the raw data, it is necessary to remove the major sources of variability such as baseline and circadian effects. However, even when the known sources of variability have been accounted for, large uncertainty in the underlying true bronchodilatory response remains. In particular baseline correction is most effective for accounting for variability following the first dose. Within-patient variability will dilute its corrective power as the duration of the study increases.

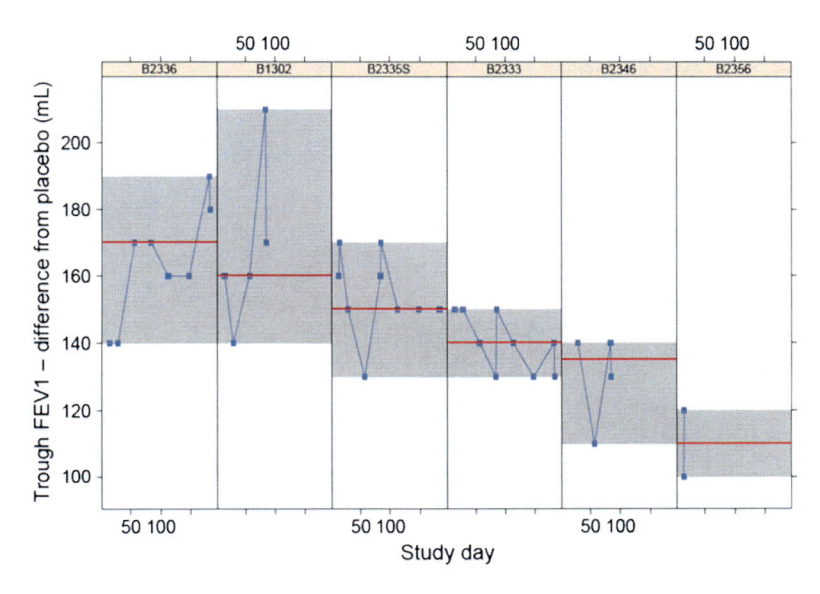

Fig. 11.7 Steady-state FEV_1 response to 150 μg indacaterol across visits in six different studies

In order to provide a robust assessment of the underlying dose response of a bronchodilator, the analysis methods used will have to have adequate precision to tease out the low drug-related signal while appropriately accounting for the various sources of high variability.

11.3 Traditional Dose Ranging

As shown above, characterizing the bronchodilatory response requires significant manipulation of raw FEV_1 data in order to tease out the drug-related signal. The traditional approach uses ANCOVA to account for the major sources of variability. For example, a typical ANCOVA analysis will use a statistical model to account for covariates such as treatment (dose), baseline FEV_1, FEV_1 reversibility to one or more bronchodilators, smoking status, country, and center in which the trial was carried out. By accounting for the known differences between patients and measurements, the SNR is improved.

The traditional ANCOVA approach to dose ranging compares active doses and placebo (contrast statistics) to determine the existence of a dose response and, if so, to select a target dose. If the difference between placebo and at least one of the doses is statistically significant, the presence of a dose–response relationship has been confirmed: i.e., there is a positive treatment effect. The target dose is then estimated as the smallest statistically significant dose which has an average effect that is clinically relevant according to a pre-specified value of clinical relevance (Bornkamp et al. 2007). Although the simplicity of this approach is appealing, it has the following issues:

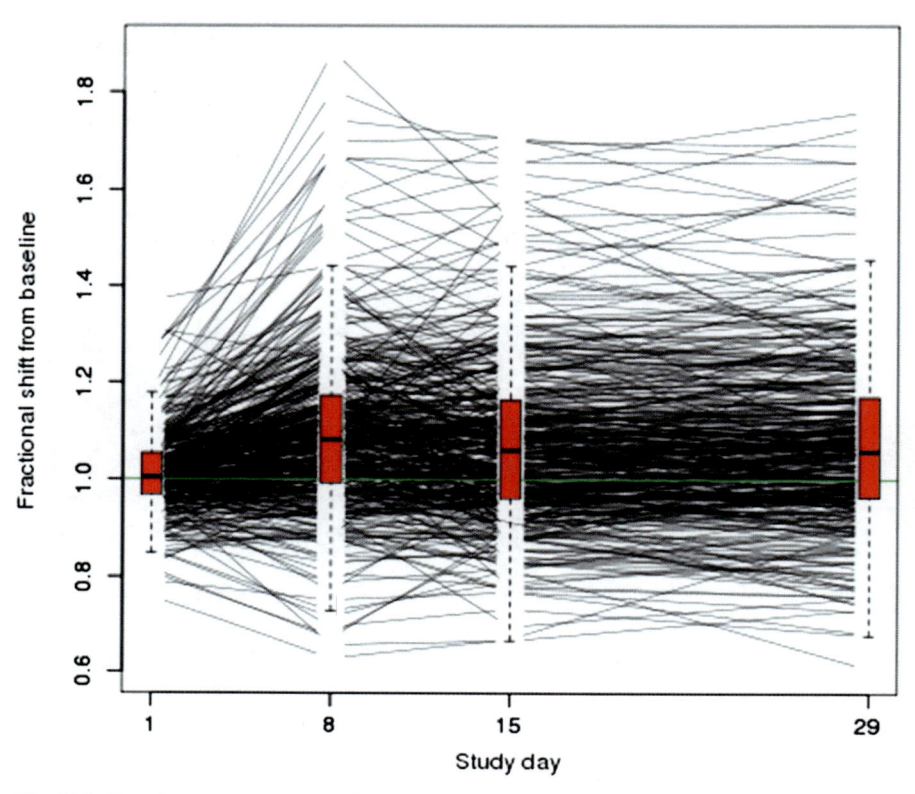

Fig. 11.8 Fractional change in trough FEV_1 relative to baseline in placebo-treated COPD patients over 28 days

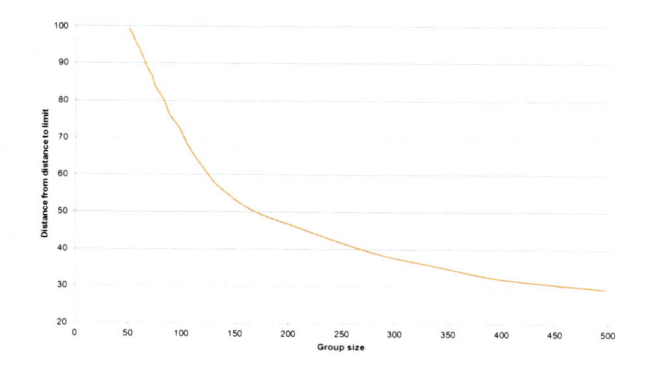

Fig. 11.9 Distance (in mL) from point estimate to limit of 95% CI vs. sample size (per group)

1. It requires the definition of a pre-specified value of clinical relevance. However, the FDA has not determined a minimal clinically important dose (MCID) for FEV_1 use in regulatory submissions (Michele 2011)

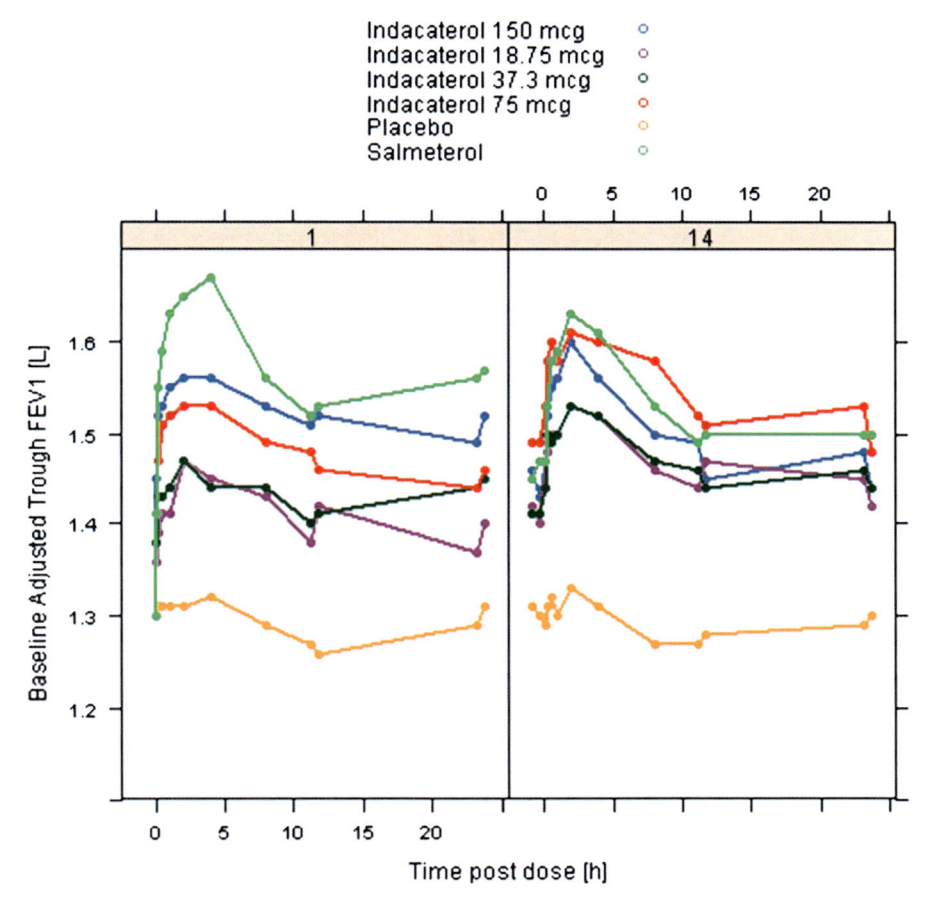

Fig. 11.10 FEV_1 profile after first (1) and last (14) dose from a traditional dose ranging trial for indacaterol in COPD

2. The traditional trial is primarily designed to detect the presence of a dose response, i.e., to detect a significant treatment effect, but not to characterize it. The traditional trial is not designed to differentiate active doses from each other. Hence sample sizes are chosen to allow statistical differentiation from placebo only
3. The low SNR associated with FEV_1 in the bronchodilatory response means that the estimated responses at each dose level will lack the precision to differentiate active doses from each other and cannot easily be benchmarked against any target level of response

The third issue is a question of sample size. Therefore, it is instructive to consider the relationship between precision and group size for typical bronchodilator studies as presented in Fig. 11.9. Precision here is defined as the distance from the point estimate to the limit of the 95% CI (i.e., half the length of the interval). The sample

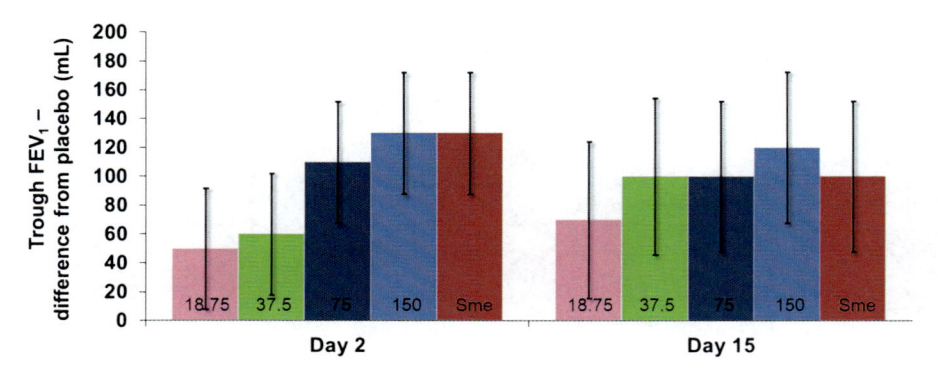

Fig. 11.11 Trough FEV$_1$ least square mean estimate (placebo contrasts) after first and last dose from a traditional dose ranging trial for indacaterol in approximately 86 COPD patients per treatment group

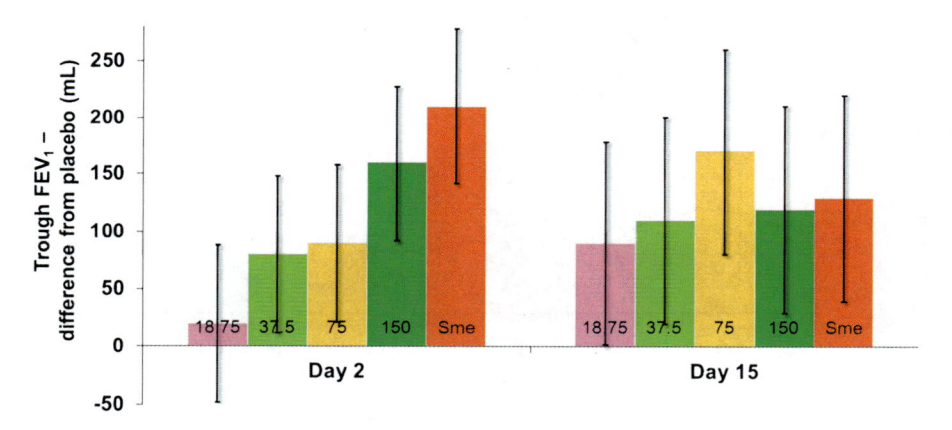

Fig. 11.12 Trough FEV$_1$ least square mean estimate (placebo contrasts) after first and last dose from a traditional dose ranging trial for indacaterol in 86 asthma patients per treatment group

size is determined such that the precision is obtained with 95% coverage probability, i.e., in 95% of the studies one would expect to see a precision less than what is reported on the graph.

For example, a sample size of 100 patients per group would result in a 95% CI that extends no further than 68 mL (half the interval) from the observed point estimate with 95% chance. While this precision is acceptable for discriminating a particular dose level from placebo, it lacks the precision to reliably construct a graded dose response relationship. To discriminate doses with the precision necessary to reconstruct a graded dose response, much larger sample sizes would be required. For example, Fig. 11.9 shows that 500 patients per group would provide a precision of no more than 30 mL (half the interval) with 95% chance.

The precision with which the response to adjacent doses can be estimated is the measure by which we can reliably reconstruct the graded response over the dose range.

The higher the precision, the easier it will be to robustly identify the threshold to the plateau of the dose response relationship. By the same token, the lower the precision, the higher the likelihood that too low a dose will be identified as being significantly different from the plateau of the dose response.

Figure 11.10 is a typical representation of the results of a traditional dose ranging trial in COPD, with about 86 patients per treatment arm. It depicts the baseline corrected longitudinal response after the first dose and last dose for a range of indacaterol doses, a positive control salmeterol and placebo. A figure very similar to this was recently published (Chowdhury et al. 2011) to discuss the dose response relationship of indacaterol. Unfortunately, this graphical presentation that focuses on the point estimates ignores the uncertainty in their estimation. Without this information, it is impossible to draw any statistically meaningful conclusions about possible treatment differences, never mind any discussion of the evolution of treatment effects across study days.

A clearer presentation of the summary data is achieved by focusing on a cross-sectional assessment of a particular metric, such as trough FEV_1, providing the point estimates together with confidence intervals. Figure 11.11 presents the ANCOVA derived placebo contrasts for trough values from the same study shown in the previous figure, whereby the Day 2 values are the trough observations 24 h after the first dose and the Day 15 values are the corresponding troughs after the 14th (last) dose. The bars with numbers correspond to the tested indacaterol doses, SME is the response to the positive control salmeterol.

Treatment-related changes in response by dose can now be interpreted in the context of the uncertainty in the estimates. The confidence intervals for all responses on Day 15 are larger than the confidence intervals on Day 2, and they all overlap. The fluctuations in all indacaterol treatments are of similar magnitude to the fluctuations in the positive control. In other words, the least squares mean estimates at steady state lack the necessary precision to allow any statistically meaningful differentiation between the active doses. On comparing the responses on both study days, it is apparent that the responses of some doses increase and others decrease, and all are within the range of the confidence intervals of approximately ±50 mL.

In the belief that the greater responsiveness of asthma patients to β_2-agonists would allow better differentiation of doses, the FDA requested dose exploration in this population (Chowdhury et al. 2011). Figure 11.12 presents the results of a traditional dose ranging trial carried out in asthma patients. Note that this trial is identical in design to the one shown in the previous figure except for the patient population.

While the absolute response in asthma patients is greater, as expected, the associated confidence intervals are disproportionately wider. Hence, whatever might be gained in terms of signal is lost in terms of noise. Performing dose ranging in asthma patients confers little or no advantage whatsoever in terms of dose finding. The performance of the positive control (salmeterol) across study days highlights the difficulty in interpreting the data; between Day 2 and Day 15, there is an 80 mL reduction in response which is more than half the maximum response noted for indacaterol in this trial; this is not a statistically significant difference. It is worth noting that given the width of the confidence intervals, it is perfectly possible that the dose of 18.75 μg could have produced a response larger than all the other doses simply by random chance. This underlines the fact that the traditional approach to the assessment of dose response lacks the precision to adequately differentiate doses of a bronchodilator.

As predicted from Fig. 11.8, the confidence intervals are consistently wider on Day 15 compared to Day 2. This is most likely due to the impact of within-patient variability which dilutes the corrective power of the baseline. The poorer precision on Day 15 further confounds the assessment of the response estimates.

The sobering conclusion of this brief graphical presentation is that dose differentiation based on this endpoint using traditional methodologies is not robust. The traditional trials are designed to differentiate active doses from placebo and not active doses from one another. Furthermore, it is not feasible to adequately expand the trial size for the differentiation of a sufficient number of doses necessary to estimate a graded dose response. The inherent variability in FEV_1 relative to the treatment difference is of a magnitude that makes precise treatment differentiation nearly impossible using traditional methodology. Replacing COPD patients with asthma patients does not provide a solution.

In summary, these examples demonstrate that the traditional approaches to dose ranging and dose selection of bronchodilators preferred by some regulatory authorities do not provide a rational basis for the differentiation and selection of doses. In the examples presented, any of the doses tested could have produced the numerically largest response given the imprecision of the chosen approach. In the search for differentiation, it is likely that the traditional approach will be biased toward selecting too low a dose for the simple reason that the bigger the difference in the doses, the more likely that differentiation will be apparent.

11.4 Indacaterol Dose Selection

The indacaterol program, following regulatory precedent and guidance, relied heavily on traditional empirical approaches to selecting doses. The emphasis of the initial approach was to identify the lowest doses that were superior to the comparators formoterol and tiotropium in terms of trough and peak FEV_1. Note that the inclusion of peak was encouraged by the FDA. The selection rule aimed to identify the minimum dose from the range of 75, 150, 300, and 600 μg that had a response numerically superior to the comparators. Both the minimum and next highest doses were tested together in a longer extension of the trial.

Note that the highest doses of 300 and 600 μg were independently tested in a year-long trial in order to establish the safety of the highest available doses, irrespective of the ultimate clinical doses.

Figure 11.13 depicts the outcome of the primary dose selection trial. Based on the criterion of numerically exceeding the trough values of the comparators, the doses of 75 and 150 μg would have qualified, however, on the basis of the peak metric, the doses of 150 and 300 μg were selected. These were the doses subsequently approved in most countries around the world.

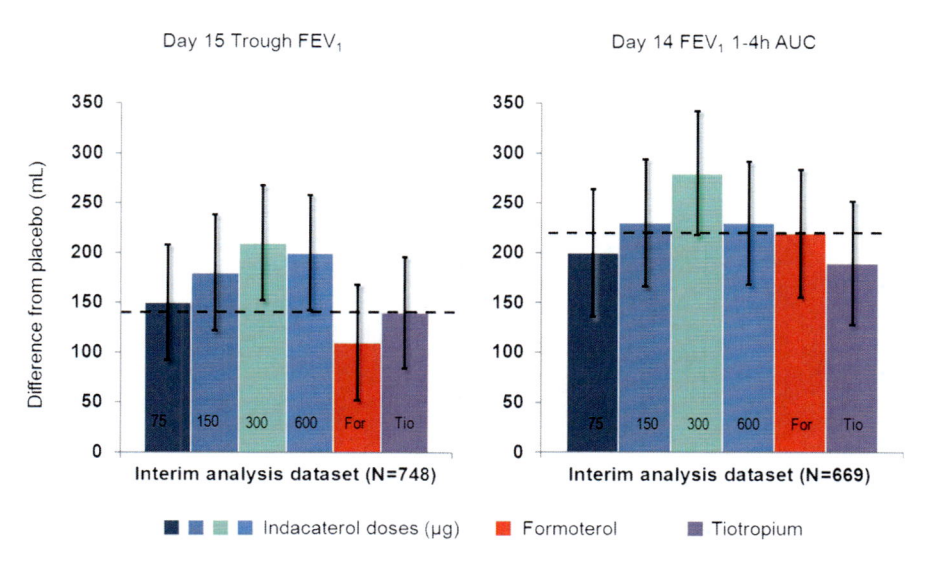

Fig. 11.13 Dose selection trial for indacaterol in COPD

On submission of these data to the FDA, a request was made to better explore efficacy at lower doses. For this purpose, the dose ranging trials depicted in Figs. 11.11 and 11.12 were carried out and were central to the FDA decision for approval of the 75 μg dose (Chowdhury et al. 2011).

In the course of the development program, the shortcomings of the traditional approach to dose selection of bronchodilators started to become more apparent. For this reason, alternative approaches were explored to provide a more rational and robust means of dose selection (Renard et al. 2011), prior to the submission to the FDA. The remainder of this article provides a graphical exploration of this approach.

11.5 Estimation of the Dose Response Relationship

The traditional approach to dose ranging assesses the response to each dose independently and tests whether each dose is different from placebo. In other words, the traditional approach simply detects whether a dose response is present, but does not explicitly characterize it (Bornkamp et al. 2007). However, if sufficiently many doses have been tested it is not uncommon that a dose response relationship is inferred from such an analysis, even though no such relationship has been or was ever intended to be estimated. Chowdhury et al. (2011), for instance inferred a dose response relationship from a study that was not designed to have adequate precision to support such inference (see Figs. 11.11 and 11.12).

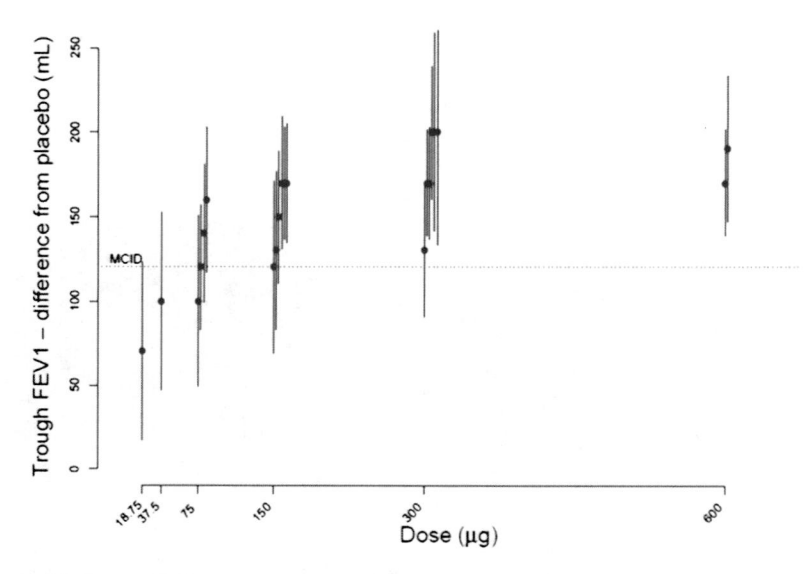

Fig. 11.14 Trough FEV_1 placebo contrasts for the primary statistical analysis at each doses level in 12 studies in COPD patients

A more rational approach is to use our knowledge of the underlying pharmacology of bronchodilators to determine the most likely shape of the dose response relationship and use a mathematical model to approximate this functional form based on the available data across all dose levels. It is known that direct acting bronchodilators cause increased bronchodilation with increasing dose until a maximum threshold is achieved.

Figure 11.14 depicts the primary analysis results for the 12 major COPD efficacy trials carried out in the indacaterol development program. Each point represents the ANCOVA derived least squares mean contrast trough FEV_1 at each dose level assessed across the trials; this figure summarizes data from 8,111 patients studied for up to 6 months. The horizontal line represents the minimum clinically important difference (MCID) as defined by some authors (Cazzola et al. 2008). The vertical lines represent the 95% confidence intervals of each estimate and indicate the large uncertainty in even the largest trials. In contrast to the raw data in Fig. 11.3, a dose response signal is apparent; whereby; with increasing dose, trough FEV_1 appears to increase to a plateau.

On fitting a mathematical model to such data, it is possible to provide an explicit estimation of the graded dose response relationship which allows the relative potency of each dose to be calculated. Once an explicit estimate of the dose response is available, it is straightforward to identify the dose that is on the threshold of the plateau of the dose response. Typically, this is the dose that attains up to 80–90% of the maximum response (or some other clinically relevant target). In other words, a model-based approach turns the dose selection process in to a calibration problem—it allows efficient and robust identification of the minimum dose that attains optimal efficacy, given an adequate safety threshold.

11.6 Application of Model-Based Approaches to Indacaterol Dose Selection

Two retrospective approaches were taken for the model-based analysis of the indacaterol dose response (Renard et al. 2011):

- A Bayesian study-level meta-analysis using summary level study data
- A nonlinear mixed effects analysis of patient level data

Readers are referred to the original article (Renard et al. 2011) for precise details. For the purpose of this graphical exploration, only the study-level analysis will be considered here.

Typically, model-based approaches use the raw patient level data to derive an estimate of the dose response relationship. However, in the case of the indacaterol program, the large number of long trials provided a rich data base of study level data which allowed estimation of the dose response from the summary level data while accounting for both within- and between-trial variability.

Specifically, the least square mean placebo contrast estimates for each dose level (between 18.75 and 600 µg), visit and trial were included in an E_{max} dose response model. The complete analysis included data from 12 trials which had data from 8,111 patients. Positive control data was also included to allow benchmarking against important comparators.

Since all trials had similar inclusion/exclusion criteria and the ANCOVA analysis used to derive the least squares mean estimates accounted for major known sources of variability, a level playing field was created for the purpose of the model-based meta-analysis. Figure 11.15 provides an overview of the study-level analysis data set. This figure is essentially an expansion of Fig. 11.14, that now captures all visits in each trial. Box plot summaries of the data have been superimposed on the points to give an impression of the shape of the underlying dose response relationship. It would be safe to conclude that with increasing dose, the response approaches a plateau; this is expected for the dose response relationship for a beta-agonist. The dotted line marked MCID indicates the minimally clinically important difference, a threshold of 120 mL believed by some experts to be the degree of bronchodilation that should be exceeded to achieve optimal clinical benefit (Cazzola et al. 2008).

The final analysis of these data, including the comparators formoterol (12 µg), salmeterol (50 µg), and tiotropium (18 µg), is presented in Fig. 11.16. This analysis allows the dose response relationship to be explicitly estimated while accounting for the considerable within- and between-study variability. It is apparent that 150 µg is located on the threshold of the plateau and that it is the minimum dose that provides a consistent advantage over the comparators.

Given the explicit characterization of the dose response, it is possible to rank the doses on the effective dose scale that calculates the percentage of the maximum response achieved by a given dose. Table 11.1 presents the results of the analysis. Key to dose selection is the identification of the threshold to the plateau of the dose response; below this point, the response will drop off very rapidly, and above this

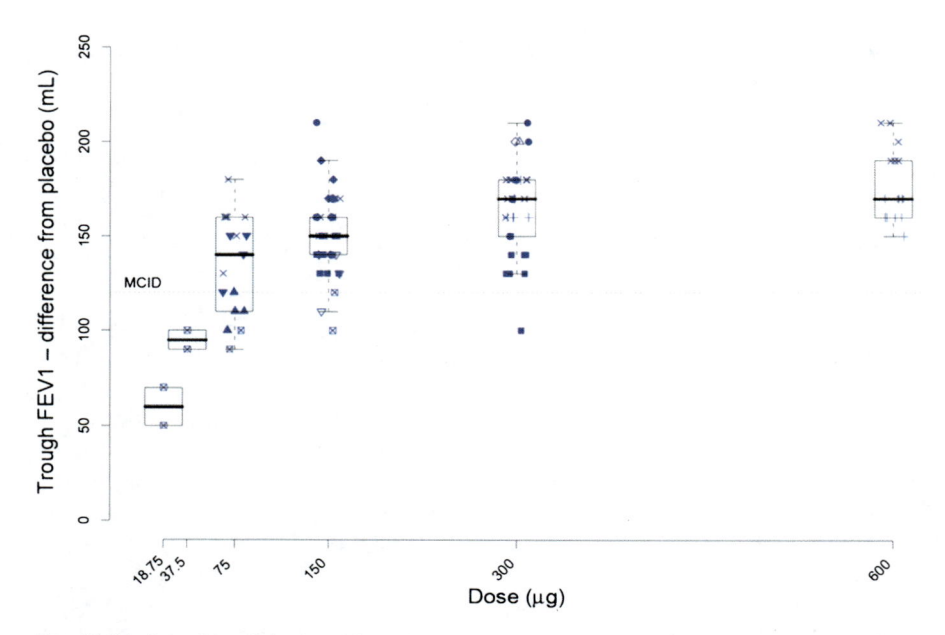

Fig. 11.15 Overview of data used in study level dose response analysis

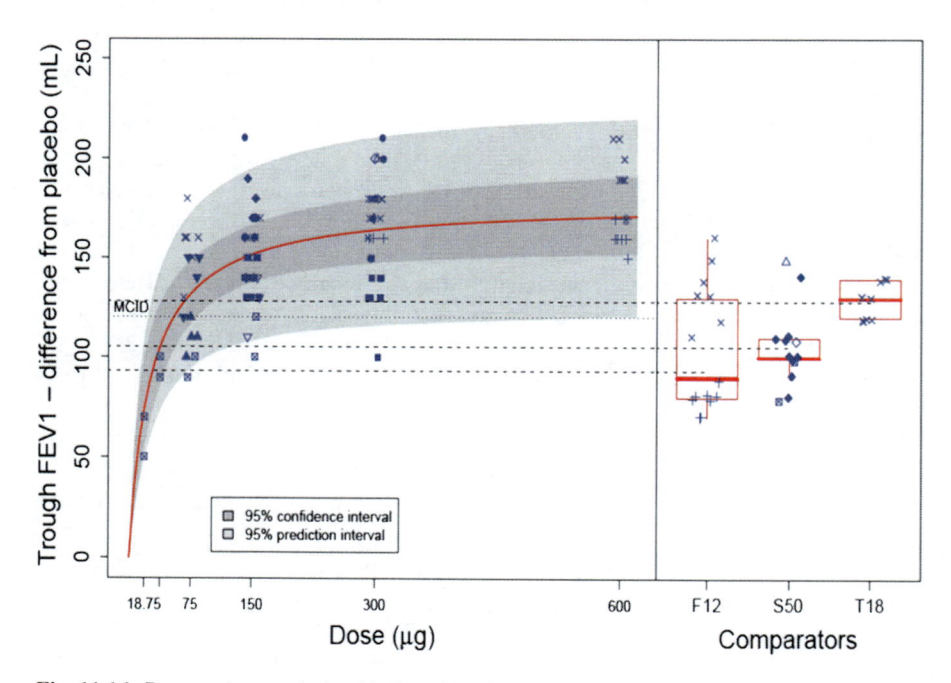

Fig. 11.16 Dose response relationship based on the totality of the study level data

Table 11.1 Results of the study level dose response analysis

Parameter	Mean	SD	Q2.5	Q50 (median)	Q97.5
Model parameters					
E_{max} (L)	0.179	0.012	0.156	0.178	0.204
ED_{50} (µg)	27	9	12	26	46
Derived parameters					
% Max effect at 18.75 µg	43	8	29	42	60
% Max effect at 37.5 µg	59	8	45	59	75
% Max effect at 75 µg	74	6	62	74	86
% Max effect at 150 µg	85	4	77	85	92
% Max effect at 300 µg	92	2	87	92	96
% Max effect at 600 µg	96	1	93	96	98

point little additional benefit is achieved. From the table, we see that 75 µg represents the ED_{74}, 150 µg the ED_{85}, and 300 µg the ED_{92}. In other words, these doses span the transition to the plateau of the dose response relationship.

Placing the response predictions in the context of a graded dose response, the MCID and the comparator data, provides a clear means of assessing the relative benefit of the various doses. Figure 11.17 presents a ranking of the doses according to their predicted response. In contrast to the traditional approach presented in Figs. 11.11, 11.12, and 11.13, the model based estimates are much more precise and allow differentiation of doses. This increase in precision has been attained by pooling information across doses to estimate an overall dose response relationship. Based on the study level analysis presented in this figure, it is possible to state the following:

- There is a 92% probability that 37.5 µg is below the MCID of 120 mL
- There is a 95% probability that 75 µg exceeds the MCID
- 150 µg has an incremental benefit over 75 µg and is the lowest indacaterol dose that exceeds the average bronchodilation observed for the comparators
- 150 µg is located mid-way between the MCID and the maximum response and exceeds all comparators
- 300 µg intersects the maximum response

The ability to characterize the indacaterol dose response in this precise manner bears testimony to the power of the study level dose response analysis. It would not be possible to derive information of this quality and precision in any single trial of practical size.

It is important to stress, the purpose of this presentation is not to question the choice of the 75 µg dose of indacaterol by the FDA which is based on the Agency's risk benefit assessment (Chowdhury et al. 2011). The model based analysis shows that 75 µg is the lowest tested dose that both exceeds the MCID and is as good as the best comparator tested. For this reason, Novartis proposed 75 µg as the lowest dose in its submission to the FDA, while claiming that incremental benefit can be achieved at higher doses. However, it is the aim of this publication to highlight the

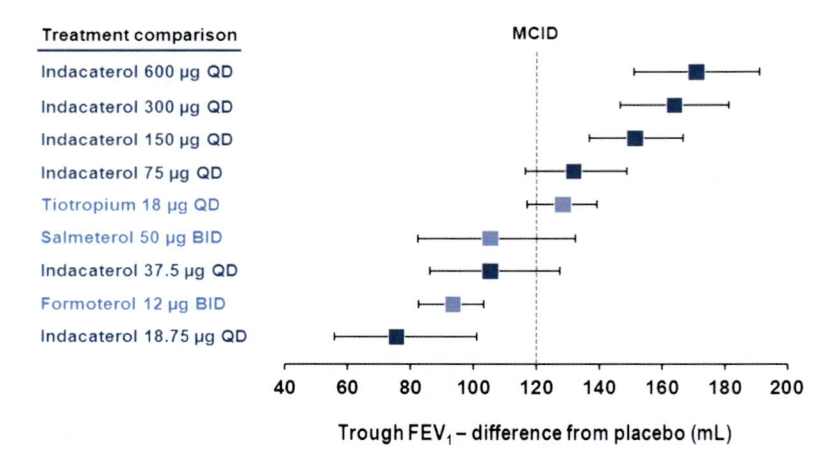

Fig. 11.17 Ranking of trough FEV_1 responses based on dose response analysis

shortcomings of the traditional approach to dose ranging favored by the Agency. It is apparent that the dose response data as presented in Figs. 11.11, 11.12, and 11.13 do not allow any statistical meaningful differentiation of doses. Given the confidence intervals, it is perfectly possible that doses of 18.75 or 37.5 μg could *randomly* produce responses greater than any of the higher doses in the traditional dose ranging trials. It is not clear how the agency would have reacted had this possibility actually arisen. Indeed, the large differences observed between the Day 2 and Day 15 responses for the positive control, salmeterol, in both the asthma and COPD trials (Figs. 11.11 and 11.12) suggest that such spurious results are likely to occur in the traditional design and analysis. To avoid confounding the dose selection process with spurious results, it is necessary to design studies and analysis methods that are capable of appropriately handling the low SNR in FEV_1. Placing the process of dose selection in the context of explicit dose response assessment greatly increases the probability of efficiently selecting the optimal dose.

11.7 Sensitivity Analyses

Despite the advantages of the study-level analysis presented above, it is nevertheless a retrospective meta-analysis and hence requires qualification. Such qualification or sensitivity analyses assess the robustness of the results to various assumptions or natural constraints of the analysis. For example, in a meta-analysis, it is typical to test the sensitivity of the results to the sample of included studies.

The history of the program provided a natural means for such a sensitivity analysis. Prior to the first submission to the FDA, the available efficacy trials were included in a study level meta-analysis of the dose response. After the first submission, the agency requested studies with lower doses. By the time of resubmission, a further 6

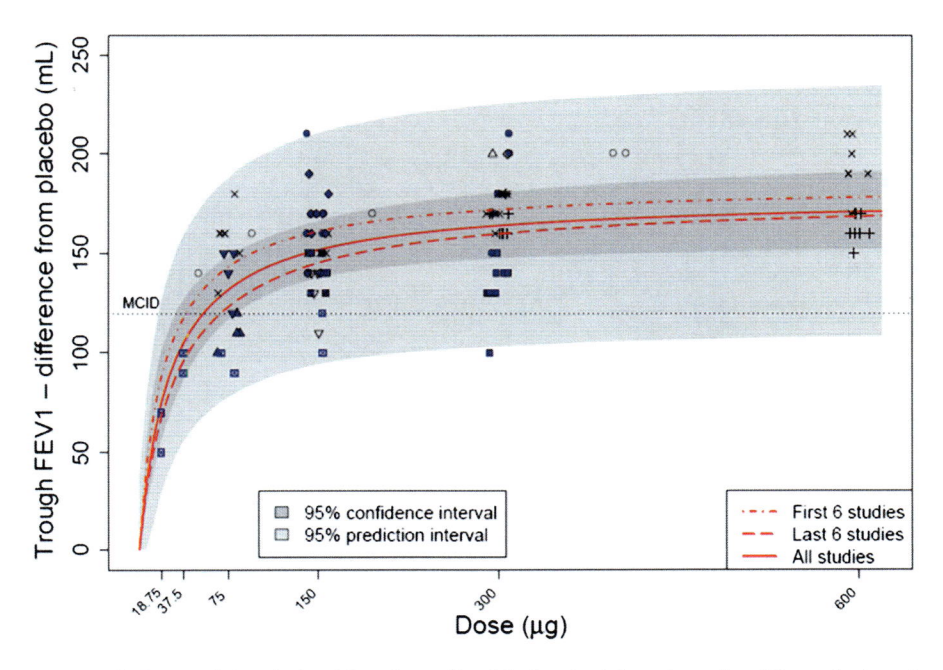

Fig. 11.18 Sequential analysis of first six studies (*black points*), last six studies (*blue points*), and all 12 studies together

studies had become available with more data at lower doses and were analyzed independently using the same methodology. Given the excellent correspondence between both independent analyses, all studies were pooled in the final analysis as presented in Fig. 11.16.

Figure 11.18 shows the graphical comparison of all 3 analyses. Although no doses lower than 75 μg were available for the first analysis, it nevertheless allows an estimate of the dose response that is consistent with the second analysis that mainly included studies with lower doses. As expected, the analysis of all 12 trials lies between the previous 2 and its confidence interval includes the mean predictions of both.

Given that only one study included doses lower than 75 μg, it was considered important to assess the impact of this trial on the overall analysis if the results of the low dose study had been different. For this purpose, a sensitivity analysis was carried out whereby the responses of all doses in low dose study were adjusted relative to the response of the 150 μg dose. So for example, the responses of the low dose study were all adjusted corresponding to the fractional deviation for the 150 μg response from the 25th, 50th, and 75th percentile of the population estimate for the response to this dose. Figure 11.19 presents the results of the analysis where the responses of the low dose study were adjusted upward by the fractional deviation of the 150 μg dose from the third quartile of the population response for this dose. As expected, significantly increasing all the responses of the low dose trial caused the

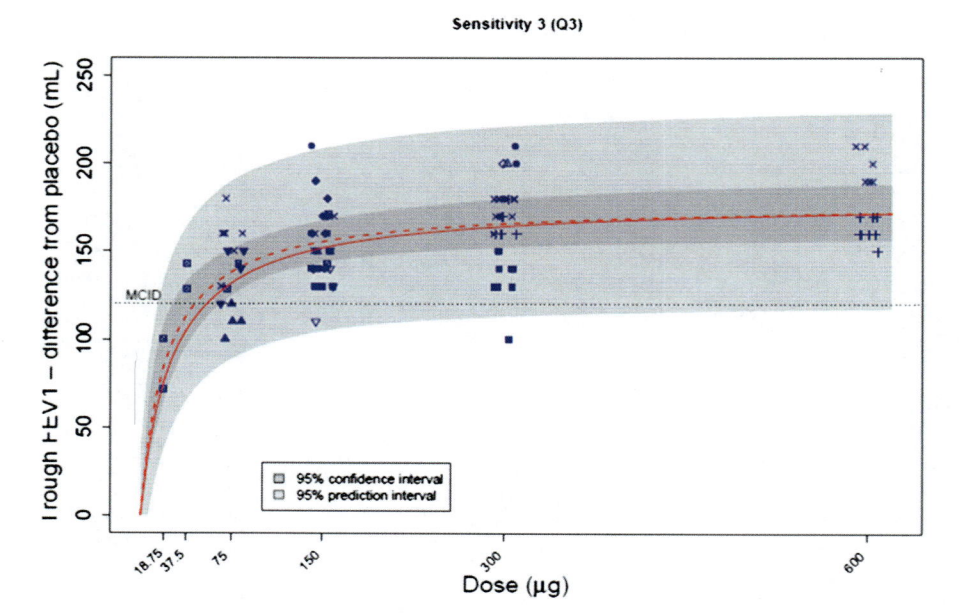

Fig. 11.19 Sensitivity analysis: the response of the low dose study is adjusted to correspond to 3 quartiles of the population estimate

estimated dose response to increase, but not by a margin that would alter the conclusions of the original analysis. This analysis provided confidence that the rest of the data set (without the lowest doses) has adequate information to provide a robust estimate of the dose response.

11.8 Summary and Conclusions

FEV_1 is commonly used as a diagnostic tool for respiratory disorders. It is simple to interpret and has adequate SNR to assist the diagnosis of conditions such as asthma or COPD. Using graphical analysis, it has been shown that the low SNR of FEV_1 as a marker of the bronchodilatory response becomes problematic when assessing dose response relationships

In the assessment of bronchodilator efficacy, the problem of the low SNR is compounded by the poor precision of the traditional approaches to dose ranging trials advocated by some regulatory authorities. While these trials are adequate for detecting a dose response, they lack the precision to adequately differentiate active doses. Simply increasing the trial size to compensate for the poor precision is not a viable option given the number of patients that would be required to cover all doses necessary to characterize a dose response relationship.

In the case of indacaterol, it was shown that there was considerable variability among the indacaterol doses and positive controls on the respective trial days and the confidence intervals of all doses tested by the traditional approaches overlapped. It is concluded that the traditional approach cannot distinguish active doses from one another in any statistically meaningful manner. Given the large between- and within-trial variability, it is difficult to conclude the dose selection decisions based on such methodology alone can be robust.

A model-based approach that provides an explicit estimation of the dose response was presented to support dose selection. This approach used the totality of the data from all major trials to estimate the dose response relationship while accounting for the within- and between-trial variability. Pooling information across the program in this manner allowed precise estimation of the population dose response and provided a robust basis to support dose selection.

The study design and analysis methods used to support dose selection must be tailored to account for the properties of the underlying data. Placing the dose selection process of bronchodilators in the context of an explicit estimation of dose response greatly increases the chances of robustly and efficiently identifying the optimal dose.

References

Bornkamp B, Bretz F, Dmitrienko A, Enas G, Gaydos B, Hsu CH, König F, Krams M, Liu Q, Neuenschwander B, Parke T, Pinheiro J, Roy A, Sax R, Shen F (2007) Innovative approaches for designing and analyzing adaptive dose-ranging trials. J Biopharm Stat 17(6):965–995

Cazzola M, MacNee W, Martinez FJ, Rabe KF, Franciosi LG, Barnes PJ, Brusasco V, Burge PS, Calverley PM, Celli BR, Jones PW, Mahler DA, Make B, Miravitlles M, Page CP, Palange P, Parr D, Pistolesi M, Rennard SI, Rutten-van Molken MP, Stockley R, Sullivan SD, Wedzicha JA, Wouters EF (2008) Outcomes for COPD pharmacological trials: from lung function to biomarkers. Eur Respir J 31:416–469

Chowdhury BA, Seymour SM, Michele TM, Durmowicy AG, Dongmei L, Rosebraugh CJ (2011) The risks and benefits of indacaterol—the FDA's review. N Engl J Med 365(24):2247–2249

European Medicines Agency (2009a) Assessment report for Onbrez Breezhaler. p 8. http://www.ema.europa.eu/docs/en_GB/document_library/EPAR_-_Public_assessment_report/human/001114/WC500053735.pdf

European Medicines Agency (2009b) EPAR summary for the public. http://www.emea.europa.eu/docs/en_GB/document_library/EPAR_-_Summary_for_the_public/human/001114/WC500053733.pdf

Medarov BI, Pavlov VA, Rossoff L (2008) Diurnal variations in human pulmonary function. Int J Clin Exp Med 1:267–273

Michele TM (2011) NDA 22-38: Cross-Discipline Team Leader Review Addendum. p 16. http://www.accessdata.fda.gov/drugsatfda_docs/nda/2011/022383Orig1s000CrossR.pdf

Renard D, Looby M, Kramer B, Lawrence D, Morris D, Stanksi DR (2011) Characterization of the bronchodilatory dose response to indacaterol in patients with chronic obstructive pulmonary disease using modelbased approaches. Respir Res 12(54):1–9

Chapter 12
Statistical Graphics in Late Stage Drug Development

Julia Wang and Surya Mohanty

Abstract Well-structured statistical graphics help us understand subject characteristics and behavior, and their relationships to efficacy and safety of study treatment. They are increasingly being used to extract and communicate information from the clinical trial data in a lucid and succinct manner. Rapid advancements have been made in recent years in statistical visualization techniques. Statistical software such as S-PLUS™ and R provide an extensive set of tools to construct graphics to aid in the interpretation and presentation of clinical data. In this chapter, several examples are presented to demonstrate the uniquely illuminating and enlightening roles of graphics in clinical trial data analysis. Some of these graphics have played significant roles in regulatory submissions and FDA advisory committee meetings.

12.1 Introduction

Human eyes are ill-equipped to discern patterns from statistical output presented in numeric or tabular forms. The spatial qualities of these numbers vanish once they are listed or tabulated on paper.

Graphical techniques allow numbers to be displayed pictorially and easily convey a sizable amount of information at a glance. Exploratory data analysis relies heavily on statistical graphics to provide insight into one or more aspects of the

---

J. Wang (✉)
Janssen Pharmaceutical Companies of Johnson and Johnson,
920 Route 202 South, Raritan, NJ 08869, USA
e-mail: JWANG@its.jnj.com

S. Mohanty
Johnson and Johnson Pharmaceutical Research and Development, LLC,
Advanced Modeling and Simulation Group, Raritan, NJ, USA
e-mail: SMohanty@its.jnj.com

underlying structure of the data and guidance into the appropriate statistical analysis.

In addition, thoughtfully designed statistical graphics provide a convincing means of communicating the essential messages hidden in the data in a clear, precise, and efficient manner.

The saying that "A picture is worth a thousand words" illustrates the fact that complex ideas can easily be conveyed with graphics. Furthermore, these thousand-word pictures rarely require a thousand words to explain.

Every picture tells a story. It time and again tells the story better than pages of tabulated numbers and descriptive text. The more than one-dimensional spatial quality of pictures effortlessly transcends the inherent limitation of sequential numbers and words stringed linearly together.

A series of real-world examples of statistical graphics in action produced at Janssen Pharmaceutical Companies of Johnson & Johnson are presented below.

12.2 Does the New Analgesic Work?

A new analgesic was studied in a phase II, acute pain, multiple-dose, bunionectomy study for up to 4 days. The study drug was allowed to be taken once every 4–6 h. Since placebo would not provide much pain relief, it was expected that subjects randomized to placebo would repeat dosing closer to 4 h on average. On the other hand, if the new analgesic was effective, the actively treated subjects would repeat dosing closer to 6 h on average.

In Fig. 12.1, subjects' dosing history were plotted over time, one line per subject, with the actual dosing time presented as pink open dots, for both the placebo group and the treated group separately. To our surprise, the dosing patterns painted by the dots looked very similar between the two groups without any apparent visual difference. The placebo subjects were not repeating treatment more frequently than the treated group. Most subjects completed the study without early termination.

What had happened was that in this study, in order to minimize early termination, subjects were allowed therapeutic doses of rescue pain medication Tylenol when the pain was not sufficiently controlled by the blinded study treatment. This is a commonly adopted approach to minimize missing pain relief and pain intensity measurement scores inevitably will be caused by early study drug termination. However, such non-missing scores collected with the aid of rescue pain medication no longer reflect the pure therapeutic effect of the study drug, but a combined effect with the rescue medication. In this study, the efficacy effect represented by the pain relief and pain intensity measurement scores also became indistinguishable between the two groups, because with adequate amount of Tylenol at hand, all subjects self-titrated to an adequate level of pain control, which is the same between the two groups distributionally.

Does the new analgesic work? Yes it does.

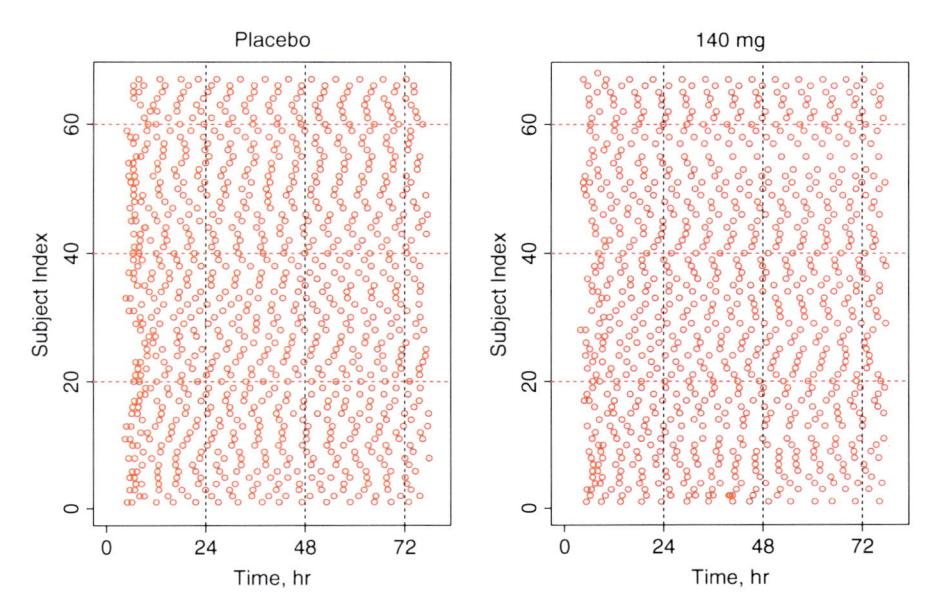

12 Statistical Graphics in Late Stage Drug Development

Fig. 12.1 Subject-level study drug dosing information

In Fig. 12.2, subjects' rescue pain medication dosing history was superposed, as blue solid dots, on top of the pink open dots. Clearly the placebo subjects were taking more rescue pain medication than the actively treated group, in order to reach the same level of pain relief. What's more, for each group, the rescue intake was the highest on the first day after surgery and decreased over time, as the acute bunionectomy pain was self-limiting. These 2 graphs provided straight-forward views into what had happened in dosing patterns that was very hard to convey using summary statistics. They also provided us a very comfortable and convincing feeling that the active drug worked, without looking at the usual "hard" statistics.

This study was analyzed exactly as specified by the FDA guidelines on the analysis of analgesic studies at that time and it failed, because there were no differences in pain relief and pain intensity scores due to the rescue medication use. However, this graph provided reassuring evidence supporting analgesic efficacy as well as insight on supplemental rescue medication use. It was very appealing to the clinicians.

12.3 How Did Subjects Take Rescue Pain Medication?

The same analgesic as in the previous section was studied in a phase II, osteoarthritis study for up to 30 days. Subjects were also allowed rescue pain medication on an as-needed basis to reduce early termination of study drug.

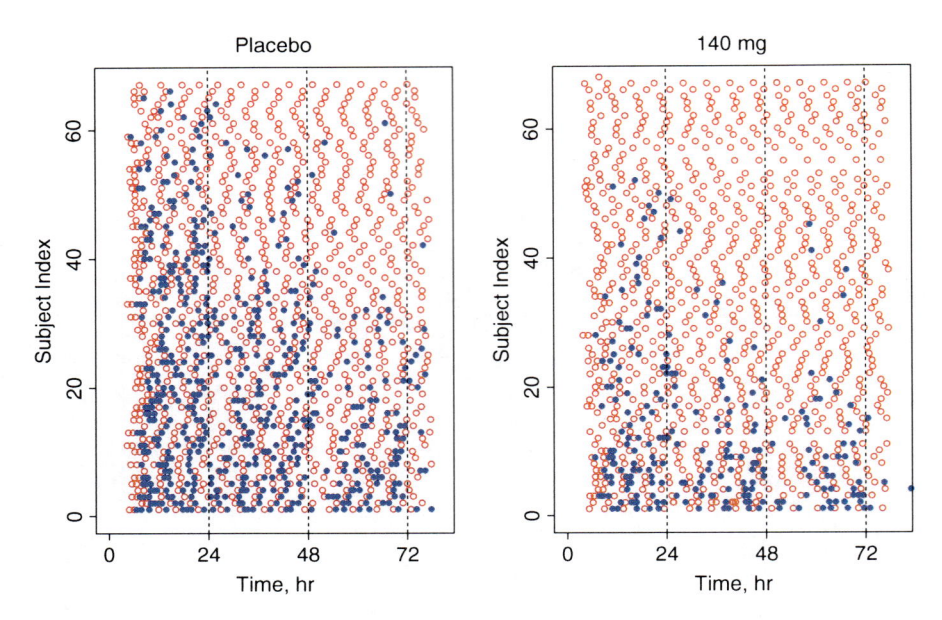

Fig. 12.2 Subject-level study drug (*pink open dots*) and rescue pain medication (*blue solid dots*) dosing information

During the analysis of this study, it became necessary to investigate whether indeed subjects took rescue pain medication as intended for additional needed pain relief. To this end each individual subject's study drug and rescue medication dosing profiles were examined over time in relation to average daily pain relief. Well, most of them did take rescue when the pain was insufficiently controlled by study drug. However, a few surprises showed up and they are illustrated in Fig. 12.3.

In this study, subjects were instructed to take the study drug two times a day, once in the morning between 7 and 8 am, and once in the evening between 7 and 8 pm. If pain was not sufficiently controlled by this regimen, subjects were allowed to take a designated rescue medication up to 4 times a day.

In Fig. 12.3, each individual subject's dosing profile was plotted over time. The times of study drug administration were plotted as black dots. They were around 7 to 8 am and 7 to 8 pm for all subjects. The times of rescue pain medication were plotted as red dots. In addition, the average daily pain intensity scores were plotted as blue line segments. The height of the blue line indicates the pain intensity.

The last subject featured in Fig. 12.3, Subject 517, consistently took the rescue pain medication only on days when the pain intensity was highest, indicating a serious pain flare. This was the pattern observed with most of the subjects in the study.

But, some unexpected dosing patterns were also observed.

Subject 501 took rescue pain medication 3 times a day regardless of daily pain scores. Subject 505 initially only took rescue pain medication when needed. However after about a 2-week period of high pain scores with the help of rescue pain medication, he decided to keep taking frequent daily rescue medication as a form of insurance even after his pain was well controlled. Lastly, subject 510 took a

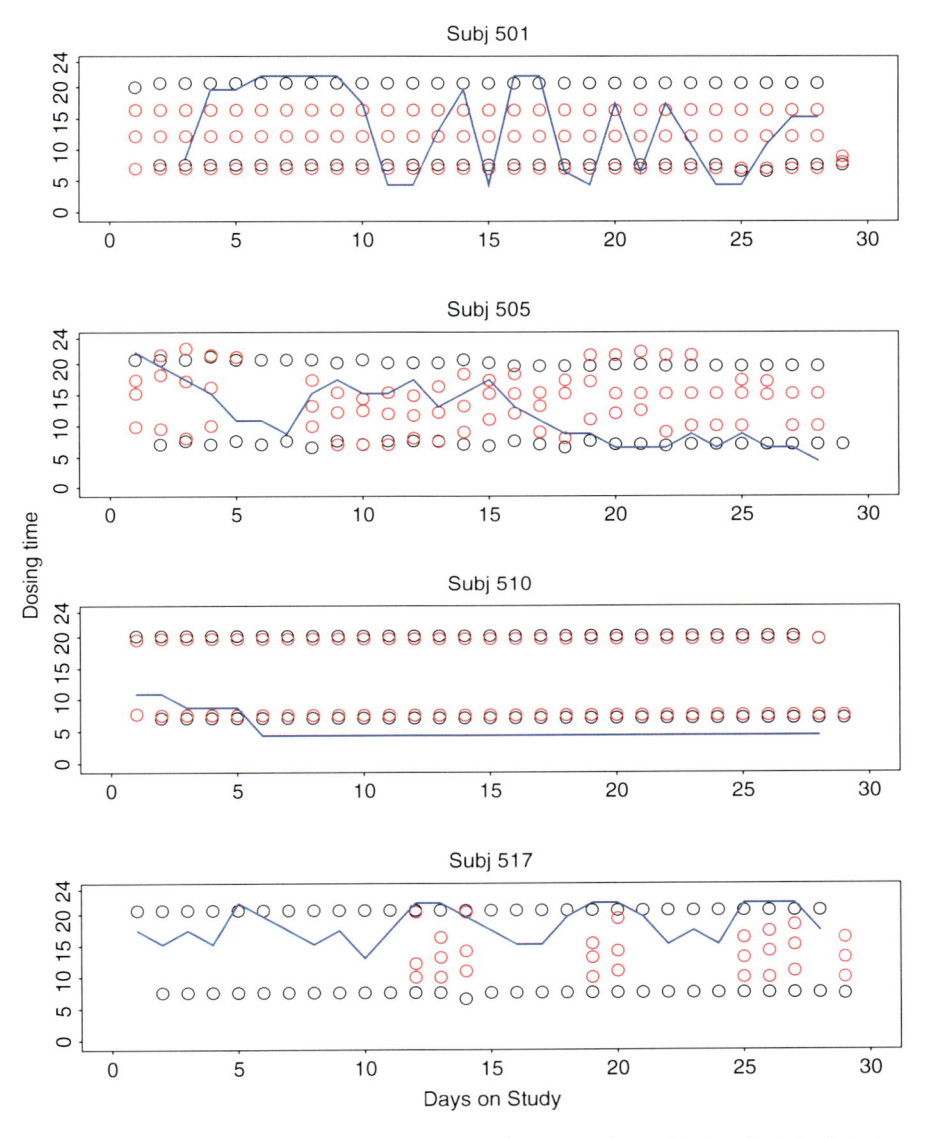

Fig. 12.3 Subject-level study drug (*black dots*) and rescue pain medication (*red dots*) dosing information for selected subjects

combined dose of study drug with rescue medication 2 times a day even though the pain was well controlled.

The insight gained from these graphs, which would normally escape our attention, helped the clinical trial team to give better instructions on the rescue medication intake during a similar phase III osteoarthritis study. Study sites were requested to clearly explain to the subjects the purpose of rescue pain medication.

12.4 Why Did Subjects Drop Out?

In an 8 h single-dose study, many subjects dropped out of the study during the 8 h observation period, which was a bit of a surprise.

There were various speculations as to why this was the case, some more plausible than others but no definitive conclusions can be made. This mystery was solved when individual subjects' efficacy scores as measured by variable Z were plotted over time stratified by the time of dropout as presented in Fig. 12.4, the famous spaghetti plot.

Because lower Z scores indicated better efficacy, the figure shows that for subjects who dropped out by hour 1, no treatment effect was observed. In fact, for the 2 subjects on the graph, their condition continued to worsen. For subjects who dropped out between hours 2 and 7, some efficacy effect was observed and then lost, just before dropout. For subjects that completed the 8 h study, sustained efficacy was obtained except for a few subjects who showed signs of losing it at the end.

In Fig. 12.5, the mean Z scores were plotted over time stratified by the time of dropout and same observations as above can be reached. It clearly illustrated the cause of subject dropout as efficacy-related.

12.5 What is the Appropriate Dose Range?

Many years ago an effective oral analgesic in tablet form was studied in a phase II double-blind study in children 7–16 years of age before the liquid formulation was available. The intended liquid dosage regimen had it been available at that time would have been in a milligram by kilogram (mg/kg) form. Based on safety and efficacy data already collected on adults, it was expected that the effective dose range for children would be 1–2 mg/kg.

In the study, the children were randomly assigned to the "A=approximately 1 mg/kg" group and the "B=approximately 2 mg/kg" using the dosing table based on body weight (Table 12.1).

This was a morphine sparing study. Children were allowed to use a morphine pump as needed to supplement the pain relief. The primary efficacy variable was the amount of morphine used to achieve a comparable pain relief profile between the 2 randomized groups.

There was a statistically significant difference between the amount of morphine used between the "Approximately 1 mg/kg" group and the "Approximately 2 mg/kg" group.

Due to variation in weight in each group and because the mg/kg was approximated using different amounts of 25 mg tablets, when the actual weight adjusted dose was calculated for each child in each randomized group, we obtained a range instead of a single point of actual mg/kg doses taken for each group, as shown in

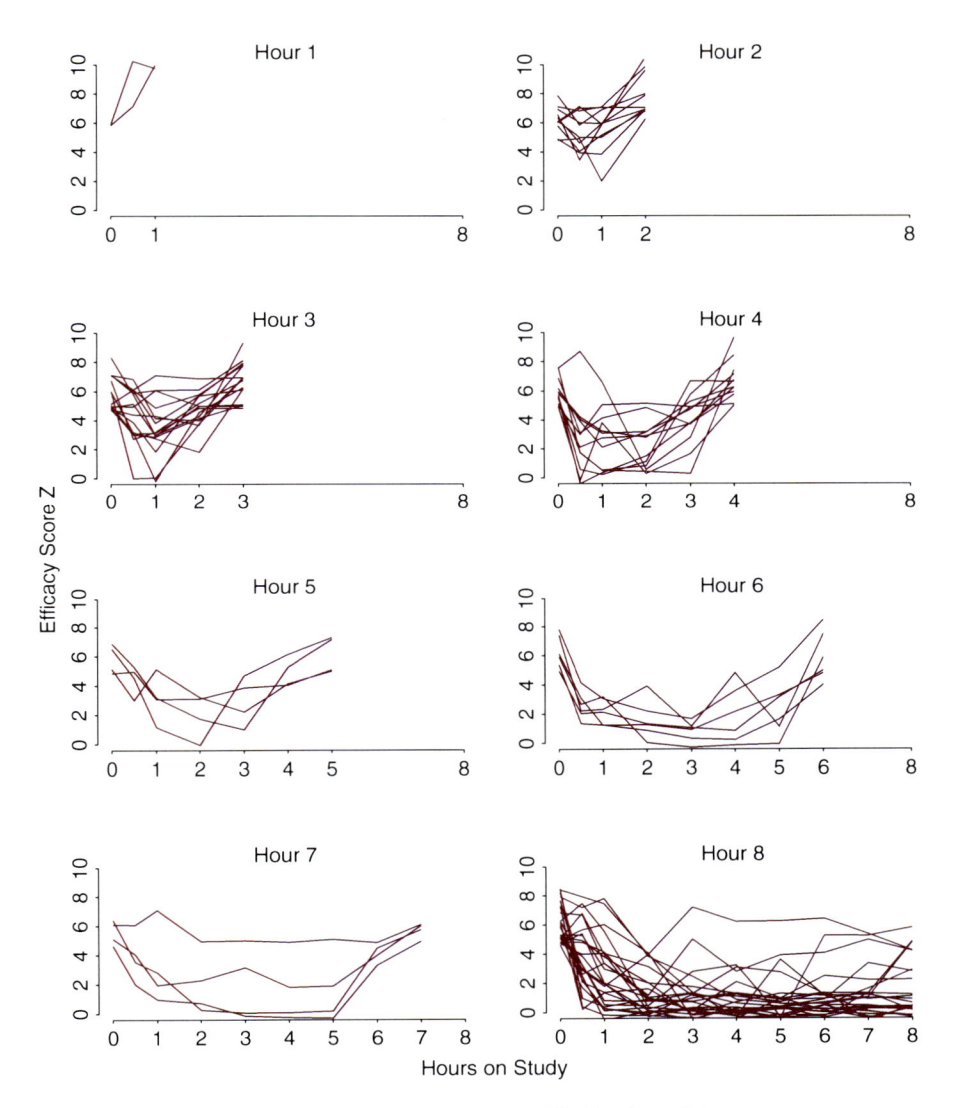

Fig. 12.4 Individual profile of variable Z over time stratified by time of dropout

Fig. 12.6. When the two groups were combined together, a larger mg/kg dose range spanning from 0.4 to 2.5 mg/kg was obtained.

In Fig. 12.6, the primary efficacy variable was plotted against the actual mg/kg dose taken. A lowess smoother was added to illustrate the trend. Even though children were not randomized to these mg/kg doses, this graph strongly suggested that the 1–2 mg/kg dose range to be efficacious in children, which was later confirmed by other studies.

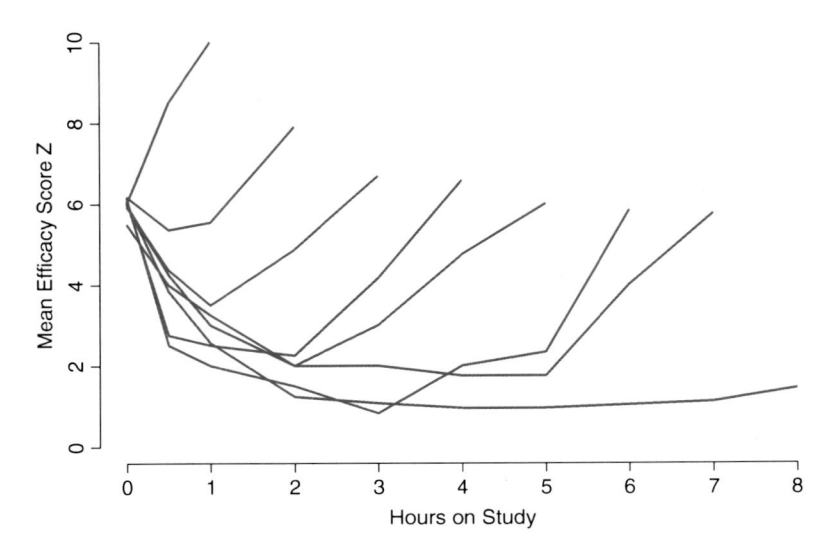

Fig. 12.5 Mean profile of variable Z over time stratified by time of dropout

Table 12.1 Dosing chart: Tablet dosage by body weight

Weight (kg/mg)	25 mg (1 tablet)	50 mg (2 tablets)	75 mg (3 tablets)	100 mg (4 tablets)
20–29.9	A	B		
30–34.9	A		B	
35–44.9		A	B	
>45		A		B

A denotes a dosing of approximately 1 mg/kg, B of approximately 2 mg/kg

12.6 Does Efficacy Depend on INR Control?

In a phase III study for the prevention of stroke and systemic embolism in patients with non-valvular atrial fibrillation, the primary hypothesis that the study drug is non-inferior to warfarin in the prevention of the composite endpoint of stroke and non-CNS systemic embolism was statistically demonstrated. However, the adequacy of the warfarin management in terms of INR (International Normalized Ratio) control was questioned. Compared to similar contemporary studies that had study-wise TTRs (%) (Percentage of time in therapeutic range) in the mid 60s, this study has a study-wise warfarin TTR (%) of 55%, which was considered much lower numerically.

Adjusted-dose anticoagulation with warfarin has been the most effective intervention to mitigate the risk of thromboembolic events in subjects with atrial fibrillation, as this clinical condition significantly increases the risk of stroke. The intensity of anticoagulation by warfarin has been measured by the international normalized ratio (INR). Maintaining subjects in the narrow therapeutic range of INR between 2 and 3 has been considered a critical aspect in warfarin dose management. The benefits of

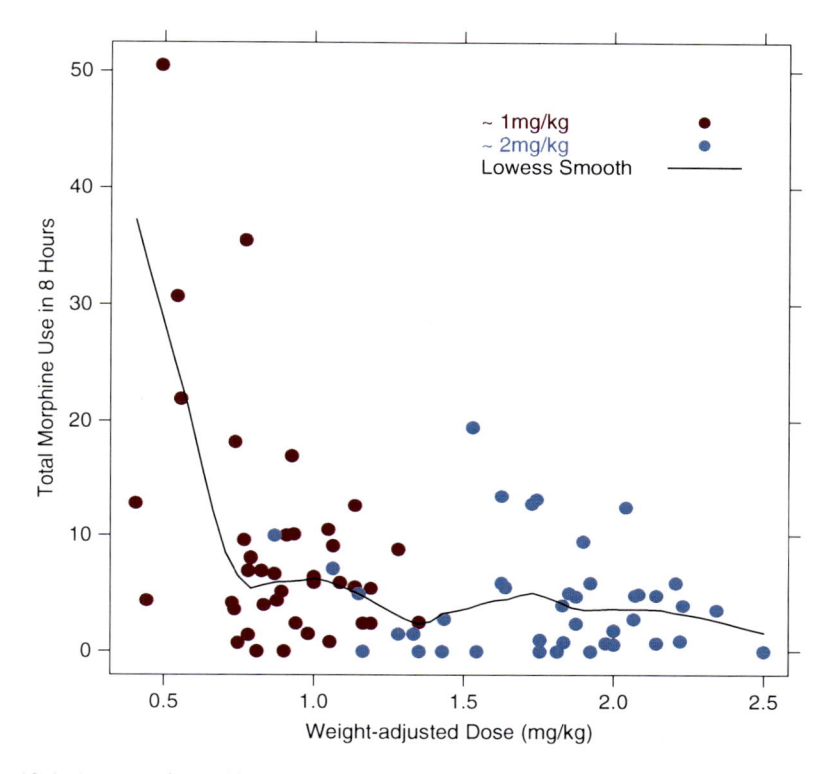

Fig. 12.6 Amount of morphine used vs. actual mg/kg dose

warfarin in thromboembolic events prevention and potential harm of bleeding are inversely related to INR. There is an increased risk of thromboembolic events if INR is less than 2.0 and an increased risk of bleeding if INR is larger than 3.

In a clinical study using warfarin as a comparator, the percentage of time a warfarin subject is maintained within the therapeutic range of INR between 2 and 3 can be calculated as the subject-level TTR (%). The study-wise TTR (%) is calculated as the average of the warfarin subject-level TTRs. This study-wise TTR (%) is used as a measure of how well warfarin treatment is managed in the study. The higher the TTR, the better is the warfarin dose management considered to be, which logically should lead to better efficacy and safety of the warfarin treatment in the study.

Because the study-wise TTR at 55% was lower than what was achieved in other similar contemporary studies, and the majority of the time the subject was out of the 2–3 INR range was in the INR <2 region, where the thromboembolic events in the warfarin arm would be higher, the question naturally arose: would the study drug still be non-inferior to warfarin efficacy-wise had the study-wise warfarin INR control been better?

This study enrolled subjects from many study centers across 4 continents. Just as study-wise TTR is considered as a measure of quality-of-care in the study, the center TTR, calculated as the average of warfarin subject-level TTRs in that center, is considered as a measure of quality-of-care in the center.

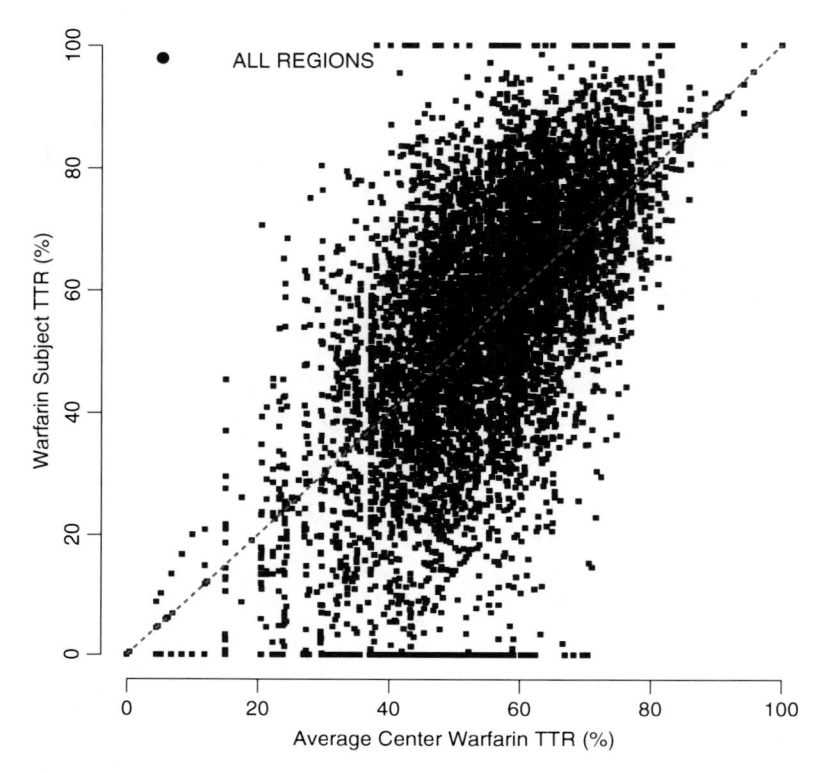

Fig. 12.7 Subject TTR vs. average center TTR for warfarin subjects

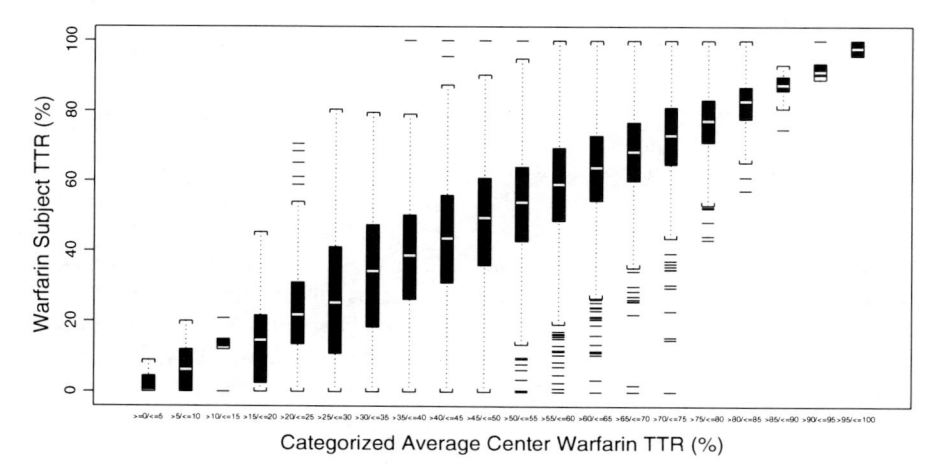

Fig. 12.8 Box plot of subject TTR vs. categorized average center TTR

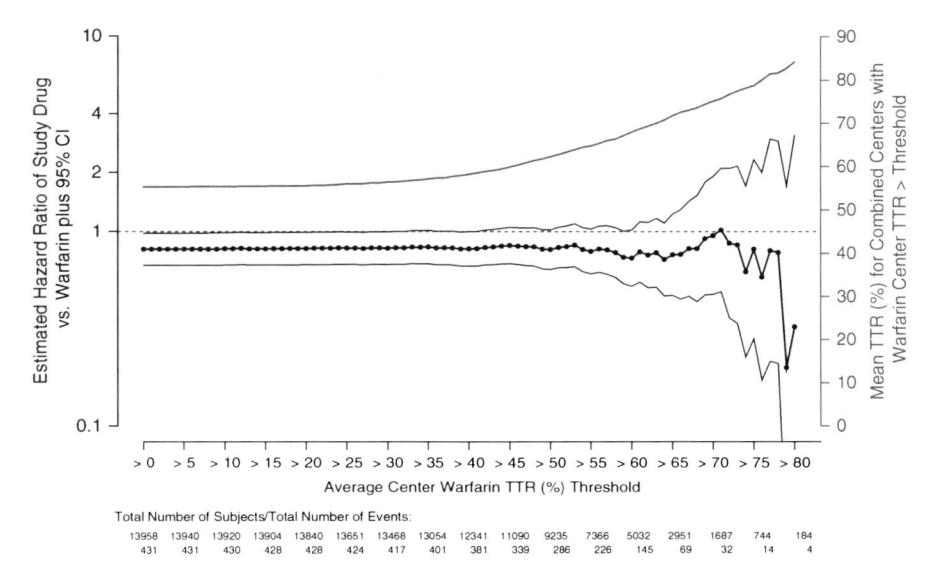

12 Statistical Graphics in Late Stage Drug Development

Fig. 12.9 Estimated treatment effect for sliding populations of combined centers with center average warfarin TTR > threshold on a log scale

There was a large variation in warfarin subject-level TTRs, ranging from 0 to 100%. In Fig. 12.7, individual subject-level TTRs were plotted against the center TTRs of the centers they are in as a scatter plot. In Fig. 12.8, box plots of subject-level TTRs were plotted for categorized center TTR groups in increments of 5%. Both graphs showed that warfarin subjects in centers with higher center TTRs also have higher individual subject-level TTRs.

If we discard the centers with lower center TTRs from the study population, subjects with lower subject-level TTRs will also be deleted. The remaining population can be treated as a smaller double-blind, randomized study, and its "study-wise" TTR will be higher.

Therefore, to answer the question of what the efficacy of this study would have looked like had we had better study-wise warfarin TTR than the observed TTR of 55%, centers with smaller average warfarin TTRs were progressively dropped (using cutpoints ranging from 0 to 100% in 1% increments) out of the study population. This resulted in remaining subpopulations that can be considered as randomized studies on their own, having increasingly higher and higher average TTRs. As shown in Fig. 12.9, the efficacy results for these smaller "studies" remain stable and consistent with the whole-study results until their TTRs reach the mid 60s to 70%, at which point the estimation breaks down and becomes unreliable due to the much smaller sample sizes left.

In Fig. 12.9, the analysis results of these 100 substudies were plotted. The visual impact of the stability of the treatment effect as the substudy-wise TTR steadily increases is beyond expression.

This simple analysis is more appropriate than the usual regression analysis that explores the association between efficacy and TTR because association is not equal to causation supportable by a randomization argument. In general, those subjects not being able to be managed to the higher TTRs are not comparable to those who can be managed to the same high TTRs, even at the same center with the same quality-of-care. The two groups of subjects are not comparable, with one being generally much sicker and with more complications than the other. Any differences we see in either efficacy or safety at the same TTR on a regression curve can very likely be due to the difference in subject populations rather than to the treatment group difference. The two contributing factors, population difference and treatment difference, are confounded and non-separable.

12.7 Would the Study Still Be Positive?

In a phase III study of a new antiplatelet therapy in acute coronary syndrome (ACS), the primary efficacy variable was the time to cardiovascular death, myocardial infarction, or stroke. The study was successful in statistically demonstrating the superiority of the study drug vs. placebo. The primary efficacy variable was analyzed using the Cox proportional hazards model under the non-informative censoring assumption. However, more than 10% of the subjects discontinued the study before the trial end date without experiencing any event, and the impact of this amount of missingness on the efficacy conclusion was questioned.

To evaluate the robustness of the primary efficacy results with respect to missing data caused by early withdrawals, a sensitivity analysis was carried out using model-based simulation from an exponential model built from the observed data. The individually fitted hazard rate in actively treated subjects was inflated from 0 to 100% while such fitted hazard rate in placebo subjects remained as observed.

For all censored early withdrawal subjects, virtual events and durations were imputed through random sampling from the fitted exponential distribution and the study was re-analyzed. This process was repeated 1,000 times for each sensitivity scenario.

The distributions of simulated hazard ratios, as brown dots, and 95% upper confidence limits, as blue dots, were plotted in Fig. 12.10.

Each vertical bar comprised 1,000 tiny dots representing 1,000 hazard ratios or 95% upper confidence limits. These dots were so numerous that they were stacked up on top of each other leaving almost no space among them except at the two extreme ends where points get sparse, hence the appearance of solid bars.

The dots in the bars were binned and the percentage of points in each bin was plotted proportionally to the right of the bars to indicate the shapes of distribution of the dots that closely resemble normal distributions.

The key message from this graph comes from the percentage of simulated trials with 95% upper confidence limits equal to or larger than 1. When this happens, the simulated trial would have failed to achieve superiority claim.

This percentage remains at zero up to 30% inflation. It reaches 1% at 50% inflation, and becomes noticeable at 70% inflation. When the assumed hazard rates

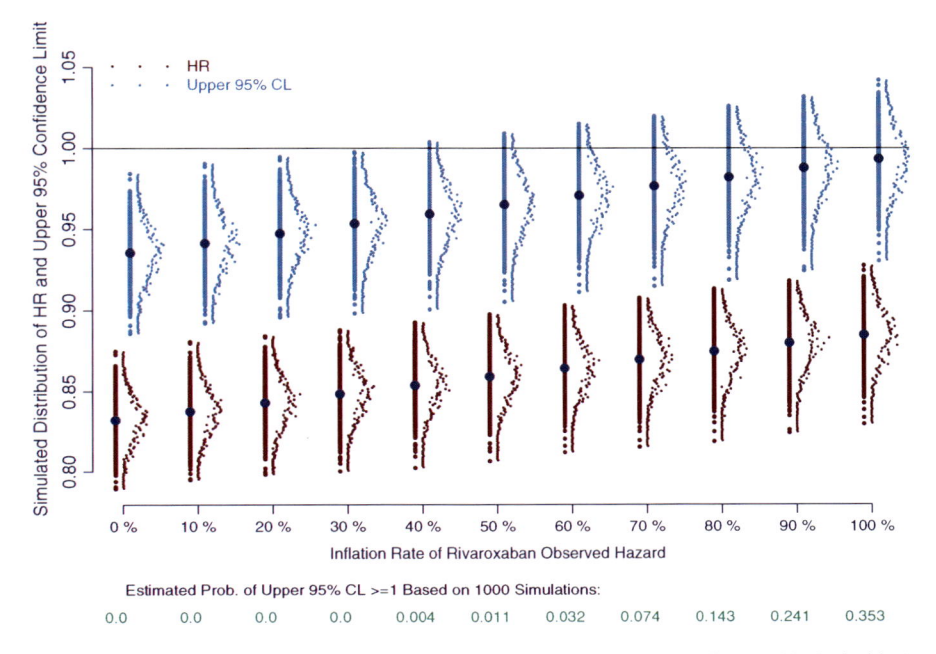

Fig. 12.10 Distributions of Simulated HR (in *brown*) and 95% Upper Confidence Limit (in *blue*)

in actively treated group are twice as bad as those observed, it becomes 35%. However, the majority of the simulated trials, 65%, would still have had superior efficacy results in favor of the study drug.

This analysis supports the robustness of our primary efficacy analysis with respect to missing data caused by early withdrawal.

12.8 Concluding Remarks

Simple graphs such as histograms and scatter plots have been routinely generated and proven useful. Custom-designed graphs for a special problem at hand require more effort. Some graphs (such as Fig. 12.9) packed so much information into one plot that it had been considered mind-boggling in the beginning. However, once properly understood, they can be much more effective than tables of numbers to get across the messages that would have gone unnoticed or distorted otherwise.

As human consciousness evolves upwards and the logical linear left brain is increasingly integrated with the intuitive spatial right brain, the future tools of communication will be more pictorial than linguistical. Until the days come when instantaneous thought transfer between human minds is possible, creative graphical statistical representation of data will be an enormously useful aid to our information sharing and decision making process.

Chapter 13
Graphical Data Exploration in QT Model Building and Cardiovascular Drug Safety

Ihab G. Girgis and Surya Mohanty

Abstract Graphical data exploration of clinical trial results is an imperative step prior to any model-based analyses. Thorough understanding of the raw data and the biological and statistical significances will certainly increase the likelihood of constructing a useful model and evade excessive complex data representation and overfitting. In this work, we use graphical data exploration to assess the cardiovascular safety of Drug X by estimating the propensity for the drug to alter the duration of the QT interval. We also identify model building strategies and the potential models that may be tested incrementally. Insights gained from this exercise will improve the efficiency of the model building process, communicate a clear and simple representation for complex data, and provide a useful decision making instrument for the drug development program.

13.1 Introduction

Advancements in mathematical modeling and simulation in drug development can be attributed to much of what has been accomplished in engineering and applied physics. The concept of modeling a real life scenario through a computer prior to building a working prototype has been the cornerstone for much of the advancement in engineering. For example, in aerospace industry, a major part of the development of airplanes, missiles, and spacecrafts relies on aerodynamics 3D modeling (computational fluid dynamics, CFD) through solutions to Navier–Stokes equations, the basic governing equations, written down in the nineteenth century, for fluid mechanics (Girgis et al. 2006).

Unlike some of the physics-based models, biological modeling does not possess a unique set of differential equations to describe or link different drugs to the efficacy

---

I.G. Girgis(✉) • S. Mohanty
Johnson and Johnson Pharmaceutical Research and Development, LLC,
Advanced Modeling and Simulation Group, Raritan, NJ, USA
e-mail: IGirgis@its.jnj.com

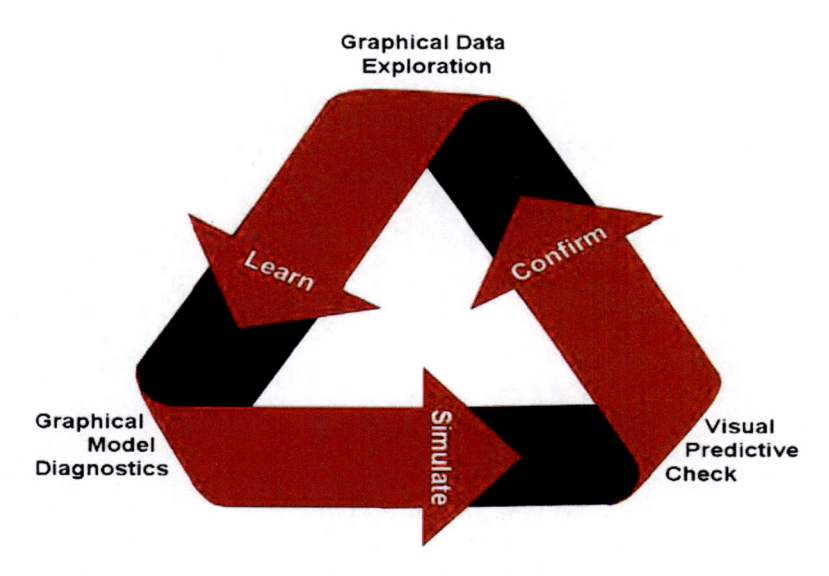

Fig. 13.1 Graphical data presentation cycle in modeling

and safety endpoints. Consequently, pharmacometrics (drug and disease modeling) requires a high level of art and proficiency in combining experimental evidences, biological fundamentals, and scientific reasoning in order to analyze various observations. This is particularly true in modeling of biological phenomena since it usually has a high level of complexity, large variability, and numerous degrees of freedom. Therefore, graphical visualization and exploration is a crucial element of biological model building. It provides a powerful tool for scientific data exploration as well as for communicating quantitative information. In addition, it uncovers many quantitative and qualitative relationships and confounding information inherent in the data.

13.2 Graphical Data Exploration

Graphical data presentation is the backbone of pharmacometrics in all its different phases (data exploration, modeling building, model validation, and simulation scenarios) (Dykstra et al. 2010). These modeling phases (Fig. 13.1) mirror, in a local level, the Learn and Confirm Paradigm in drug development (Sheiner 1997). While graphical data presentation plays a different role in each of these phases, graphical data exploration is the first essential step, and perhaps the most important step, to better understand the data and identify potential models. It provides a meaningful visual view of multidimensional data, clues for the clinical significance of various variables and its trends, key relationships, and sanity check on the quality of the data and trial conduct. We provide one such case study where graphical data exploration was utilized to build a useful model.

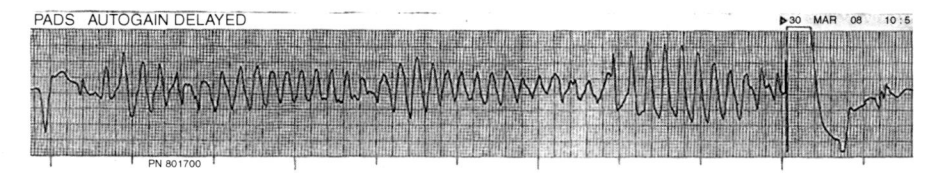

Fig. 13.2 Lead II ECG of TdP and cardioverter-defibrillator shock at end of the strip

13.3 Objective

Detecting drug-induced effects on cardiac repolarization, measured by the length of the QT interval on an ECG, is a closely monitored safety element in drug development. Drug-induced prolongation of the QT interval has been linked to cardiotoxic risk and the occurrence of torsades de pointes (TdP), a polymorphous ventricular life-threatening arrhythmia. An ECG of a patient with TdP is shown in Fig. 13.2. Every compound is required to demonstrate absence of QT prolongation beyond a predefined safety margin. A regulatory guidance document (U.S. Department of Health and Human Services 2005) provides justification for a thorough QT (TQT) study and suggests a level of rigor required for conducting such a trial to support a regulatory submission in the US. A TQT study is designed to determine the drug cardiovascular safety by detecting the magnitude of the QT changes caused by the drug. A mean time-matched QT prolongation of 5 ms, with the upper bound of the 95% one-sided confidence interval excluding 10 ms, is considered a threshold level by regulatory bodies (U.S. Department of Health and Human Services 2005).

QT intervals can be influenced by a number of factors, such as heart rate, administration of placebo, gender, and natural circadian rhythm (Piotrovsky 2005; Girgis et al. 2007). The objective of this work is to provide an example in utilizing graphical visualization to understand and help the model building in pharmacokinetics and pharmacodynamics (PK and PD); namely, the change of the individual QT interval following administration of Drug X, a noncardiac drug, or Moxifloxacin (PK/PD relationship) in connection with different covariate effects, such as gender, placebo effect, RR interval (RR = 60/HR, HR is the heart rate), and circadian rhythm. This approach is valuable for future nonlinear mixed-effects population modeling development, if needed. A schematic presentation of the durations and intervals of a typical ECG is shown in Fig. 13.3.

13.4 Data Used

To ensure a thorough assessment of the potential electrocardiographic effects of Drug X (half-life of 10–15 h), precise measurements of ECGs in healthy adults, with particular attention to the QT interval duration, were collected when Drug X is administered twice a day at therapeutic and supratherapeutic doses (4 times the therapeutic dose).

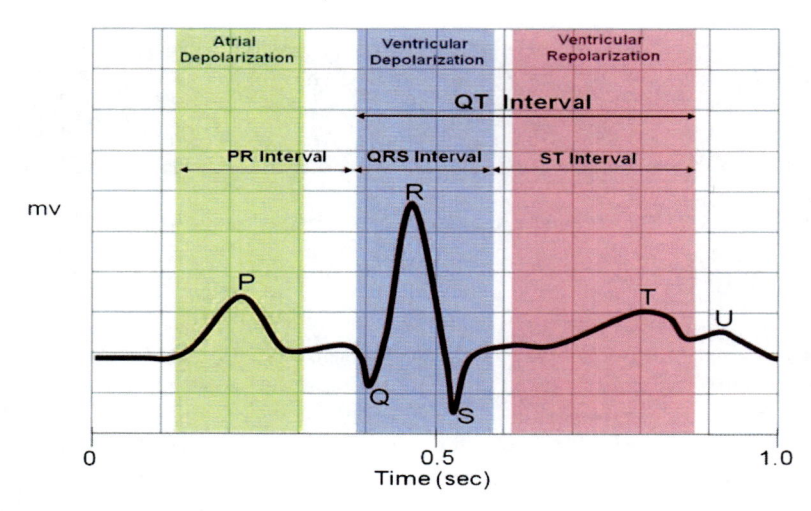

Fig. 13.3 Schematic for the ECG trace and the QT interval

This TQT study is a double-blind, randomized, placebo- and positive-controlled, 3-way crossover study in 40 healthy subjects (22 M, 18 F, age 30±8 years, BMI 24±2) between 18 and 50 years of age. The study consisted of 3 phases: a screening phase of up to 21 days, a double-blind treatment phase with three 11-day (days −2 to 9) treatment periods, and end of study/early withdrawal assessments. During the double-blind phase, each treatment period was separated by a washout period of at least 10 days, but not more than 14 days, between the last dose in a treatment period and the first dose in the next treatment period. A placebo control was used to establish the frequency and magnitude of changes in clinical endpoints that may occur in the absence of active treatment. Subjects received all 3 treatments, one during each of the 3 treatment periods. Moxifloxacin (400 mg single oral dose) was used as a positive control for evaluation of the sensitivity to detect changes in the QT interval.

The study is also designed to evaluate the steady-state of the Drug X effect (Day 7), and the assay sensitivity (Day 8, moxifloxacin). On Days −1, 7, and 8 in each treatment period, 13 h continuous 12-lead ECG recordings are obtained for each subject. Each of these continuous collection periods initiated 30 min before the time of the morning dose (including days when no drug is given). On Days 7 and 8 in each treatment period, blood samples are collected predose and within 5 min after each time point to measure the plasma concentrations of Drug X and moxifloxacin (Table 13.1).

Subject time point measurements, for each treatment period, are obtained at the same time point on the baseline day (Day −1). The dataset used in this analysis contains 2,350 time points for baseline, 738 time points for placebo (Day 8 placebo observations are not included due to residual drug effect), 557 time points for Moxifloxacin, and 1,318 time points for Drug X. The clock time data, 7 A.M. to 9 P.M., was imputed from the time relative to dosing and dosing window, where the

Table 13.1 Overview of study design

Arm	Period 1		Period 2		Period 3
1	D(−1): Baseline D1–7: Placebo D8: Moxifloxacin		D(−1): Baseline D1–7: Low dose Drug X D8: Placebo		D(−1): Baseline D1–7: High dose Drug X D8: Placebo
2	D(−1): Baseline D1–7: Placebo D8: Moxifloxacin		D(−1): Baseline D1–7: High dose Drug X D8: Placebo		D(−1): Baseline D1–7: Low dose Drug X D8: Placebo
3	D(−1): Baseline D1–7: Low dose Drug X D8: Placebo	Washout 10–14 days	D(−1): Baseline D1–7: High dose Drug X D8: Placebo	Washout 10–14 days	D(−1): Baseline D1–7: Placebo D8: Moxifloxacin
4	D(−1): Baseline D1–7: Low dose Drug X D8: Placebo		D(−1): Baseline D1–7: Placebo D8: Moxifloxacin		D(−1): Baseline D1–7: High dose Drug X D8: Placebo
5	D(−1): Baseline D1–7: High dose Drug X D8: Placebo		D(−1): Baseline D1–7: Placebo D8: Moxifloxacin		D(−1): Baseline D1–7: Low dose Drug X D8: Placebo
6	D(−1): Baseline D1–7: High dose Drug X D8: Placebo		D(−1): Baseline D1–7: Low dose Drug X D8: Placebo		D(−1): Baseline D1–7: Placebo D8: Moxifloxacin

morning dose was taken between 8:00 A.M. and 10:00 A.M., and the corresponding evening doses, which were taken 12 h later.

13.5 Data Overview

Data set preparation is performed using S-PLUS™ 8.0 for Windows (Tibco Software Inc, Palo Alto, CA). Data exploration and data visualization are carried out by Prism™ 5.01 (GraphPad Software, Inc, La Jolla, CA). There are many factors that make the data highly variable across population. Analysis of such data is complex. A boxplot of the data stratified by the treatment group (Fig. 13.4) is presented to give first insights about the data quality, variability, and trend. The horizontal line in the interior of the box is located at the median of the data. The height of the box is equal to the interquartile distance, or IQD, which is the difference between the 3rd and 1st quartiles of the data. Approximately 95 percent of the data fall inside the whiskers (the lines extending from the top to the bottom of the box). The outliers are presented in a staggered format. Compared to the median of the placebo arm (407 ms), Drug X low dose shows a QT shortening of 4 ms, while the high dose shows a shortening of 10 ms. Additionally, the positive control (Moxifloxacin) has

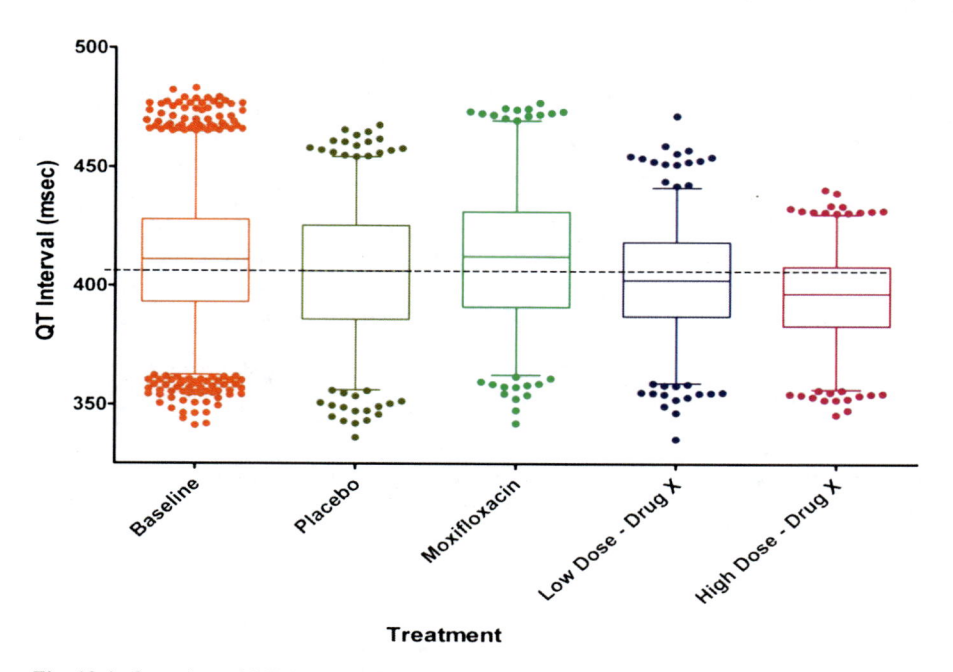

Fig. 13.4 Overview of QT data stratified by treatment group

a QT prolongation effect of 6 ms. This naive estimate agrees with previous results from literature and shows a preliminary evidence to establish the ability of the study to detect the effect of the study drug.

The lowess smoother curve is one of many useful exploratory graphical tools. It follows the trend of the data, using an algorithm developed by Chambers et al. (1983) and implemented by Prism™. It is a robust, local smooth regression of scatterplot data using weighted linear least squares without computationally expensive methods. In this work, coarse lowess curves, with 5 points smoothing window, are used. Figure 13.5 shows a temporal overview of all the data stratified by the treatment group. Significant circadian rhythm variation can be seen with a similar pattern along all treatment groups. Such an effect is important to be taken into consideration in the modeling since it may interfere with the drug effect.

13.6 Baseline Data

Accurate modeling and understanding of baseline QT is the initial step in evaluating effects of drugs. Changes to this baseline model after the administration of the investigational drug will reflect the effect on the QT interval. Data variability has mixed-effects (random and fixed) components. Random (unexplained) variability is difficult to relate to controllable variables. However, it may be identified on different levels; namely, intra-individual variability (IIV, i.e., measurement error),

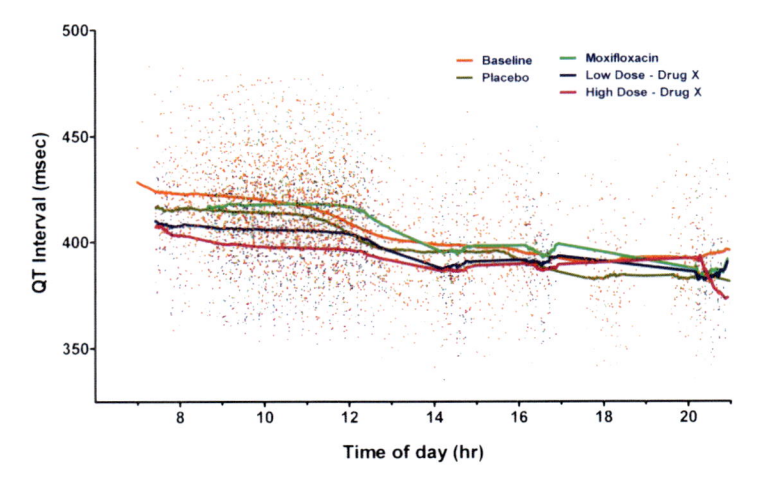

13 Graphical Data Exploration in QT Model Building and Cardiovascular Drug Safety

Fig. 13.5 Temporal overview of QT data stratified by treatment group

between-individual variability (BIV, i.e., individual difference), and between-occasion variability (BOV, i.e., different day). On the other hand, fixed effects (variability) are associated with changes in known covariates (variables). Among the most important fixed effects on QT, besides the drug effect, are placebo effect, heart rate (presented with RR interval), gender effect, and time effect (circadian rhythm) (Dykstra et al. 2010).

13.6.1 Placebo Effect

Since, in some cases, placebo effect (response) is substantial and can vary substantially, understanding and correcting this effect is important to qualify the drug effect. Figure 13.6 shows a column scatter plot of the baseline and placebo data where all the individual data points are shown and stacked into different columns according to their values. The scatter plot shows a thorough view of the data density, its distribution, median, and the IQD range. Figure 13.7 shows the placebo effect across time of day, and a slight shortening of the QT interval by approximately 3 ms, during placebo treatment compared to baseline. Both baseline and placebo curves demonstrate similar daily temporal variation, consequence of a circadian rhythm, suggesting an additive placebo correlation model.

13.6.2 RR Effect

As shown in Figs. 13.8 and 13.9, the QT interval is highly correlated with heart rate (or RR interval). The most conventionally used correction method for the heart rate

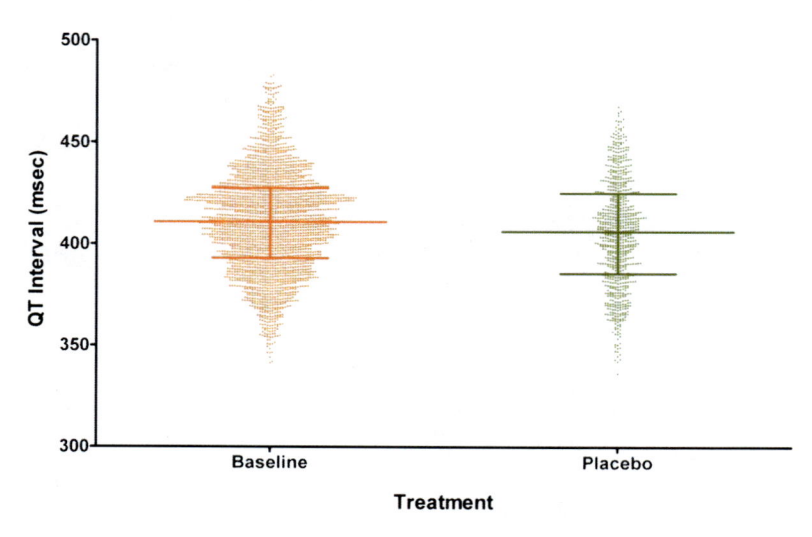

Fig. 13.6 Column scatter plot of baseline and placebo QT interval data

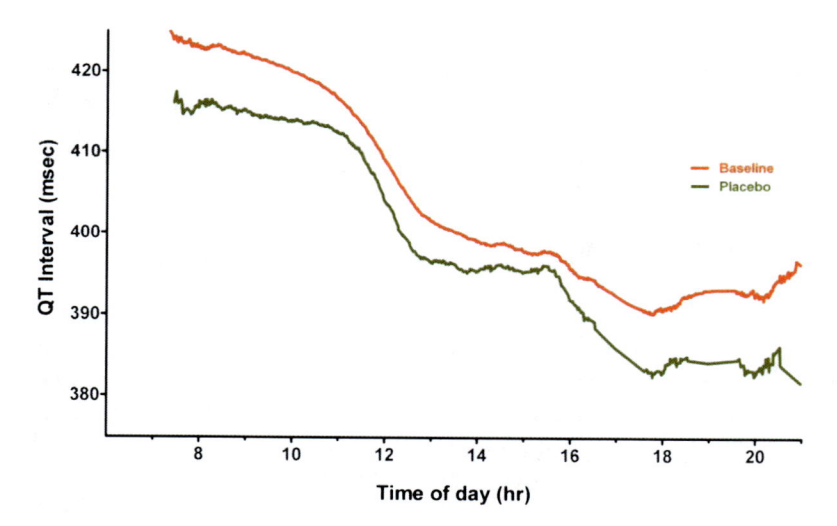

Fig. 13.7 Lowess smoother curves of the daily temporal variation of the baseline and placebo QT interval data

effect on the QT interval has a general power formula of $QTc = QT/RR^n$, where the RR interval is in seconds. The value of the exponent (n) may be estimated based on the pool or subject-specific data. For the fixed correction method, $n = 0.5$ (Bazett's correction, QTcB) or $n = 0.333$ (Fridericia's correction, QTcF). Figure 13.10 shows that the power formula provides an adequate correction for changes in RR with a baseline RR-corrected QT (QTc) of 417 ms and n of 0.375 (using naive fitting with no covariate effect).

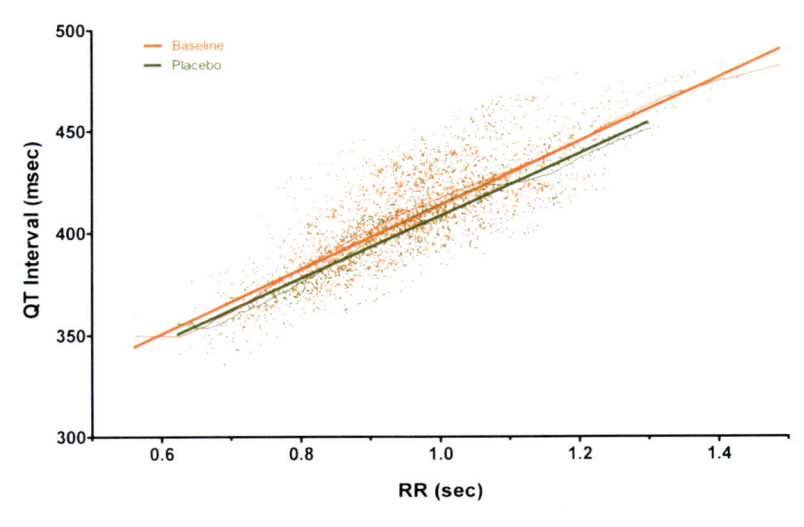

Fig. 13.8 Linear fit and lowess smoother curves of QT interval vs. RR interval for baseline and placebo data

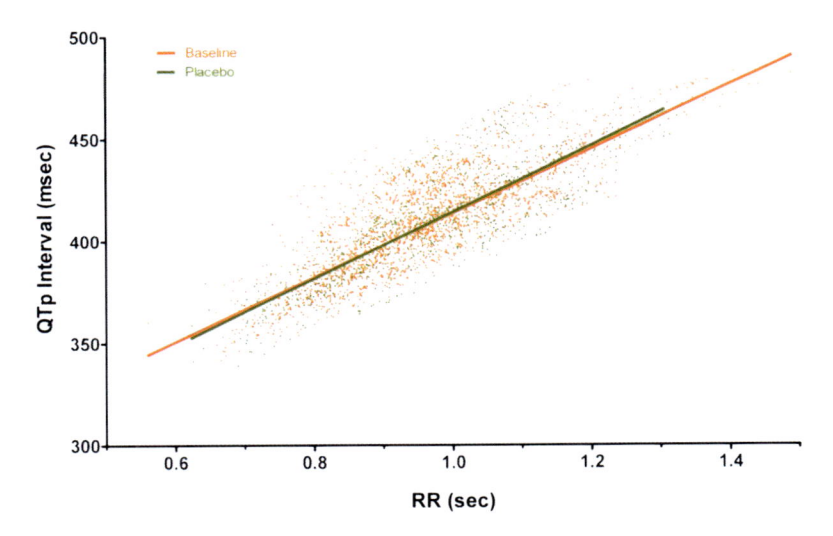

Fig. 13.9 Linear fit of placebo-corrected QT (QTp) compared to the baseline QT

13.6.3 Gender Effect

One of the important demographic factors that impacts the QT interval is the subject's gender. Females usually exhibit a longer QT interval than males. Figure 13.11 illustrates the gender effect on the baseline QT. As shown, females consistently exhibit a significantly longer QT interval than males by about 15 ms, which implies

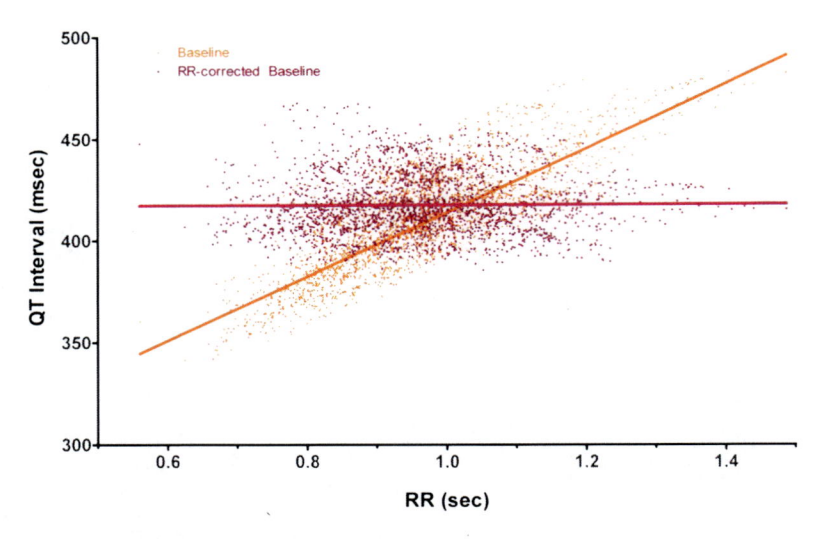

Fig. 13.10 RR-corrected baseline QT (QTc) compared to the uncorrected QT

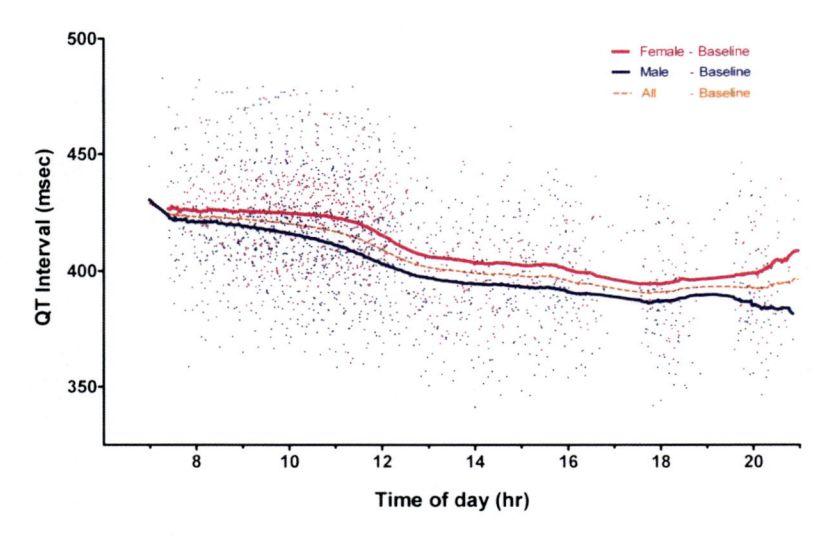

Fig. 13.11 Temporal distribution of baseline QT and QTc intervals for males and females

that either an additive or a multiplicative simple gender effect model will be adequate to capture this difference.

It is intriguing to gain significant insights using a simple graphical presentation of the pooled data. Based on the above discussion, the effects of heart rate, placebo, and gender could be adequately captured by simple expressions, as follows:

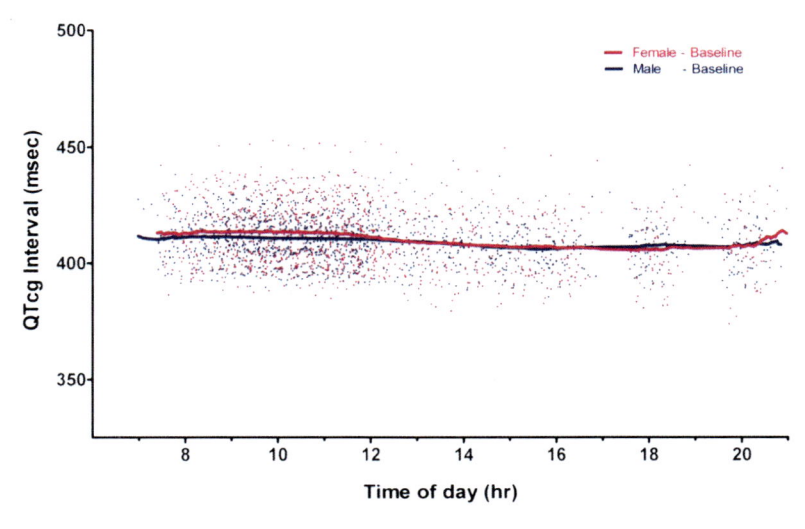

Fig. 13.12 Temporal view of QTcg interval for the baseline data by gender

Corrected QT for heart rate (QTc):

$$QT_c = QT / RR^n$$

Corrected QT for heart rate and gender effect (QTcg):

$$QTcg = QT / RR^n + \Delta Gen$$

Corrected QT for heart rate, placebo effect, and gender (QTcgp):

$$QTcpg = QT / RR^n + \Delta Gen + \Delta Pla$$

where, as discussed earlier, $n = 0.375$, $\Delta Gen = -15$ ms, and $\Delta Pla = +3$ ms. Figure 13.12 shows the calculated QTcg of baseline data for males and females. Compared to Fig. 13.11, the used fixed-effects model for QTcg significantly reduces and explains the baseline data variability by deducing the relationships with heart rate and gender covariates. Similarly, Fig. 13.13 shows the calculated QTcg and QTcgp of baseline and placebo data. Figure 13.14, a box and whisker plot, contrasts the raw and corrected baseline and placebo data.

13.6.4 Circadian Rhythm

Circadian Rhythm is an internally driven, 24 h cycle biological clock. It is influenced by a number of effects, such as light–dark cycles, timing of food

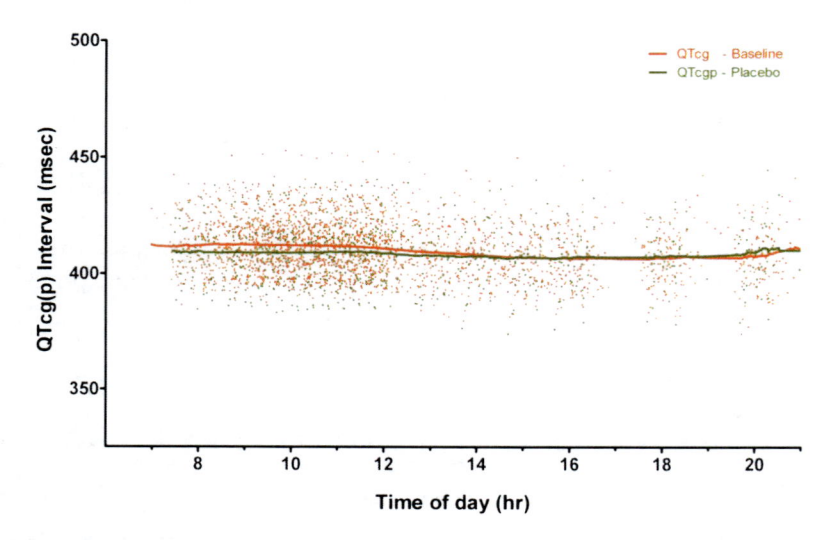

Fig. 13.13 Temporal view of QTcg(p) interval of the baseline and placebo treatment groups

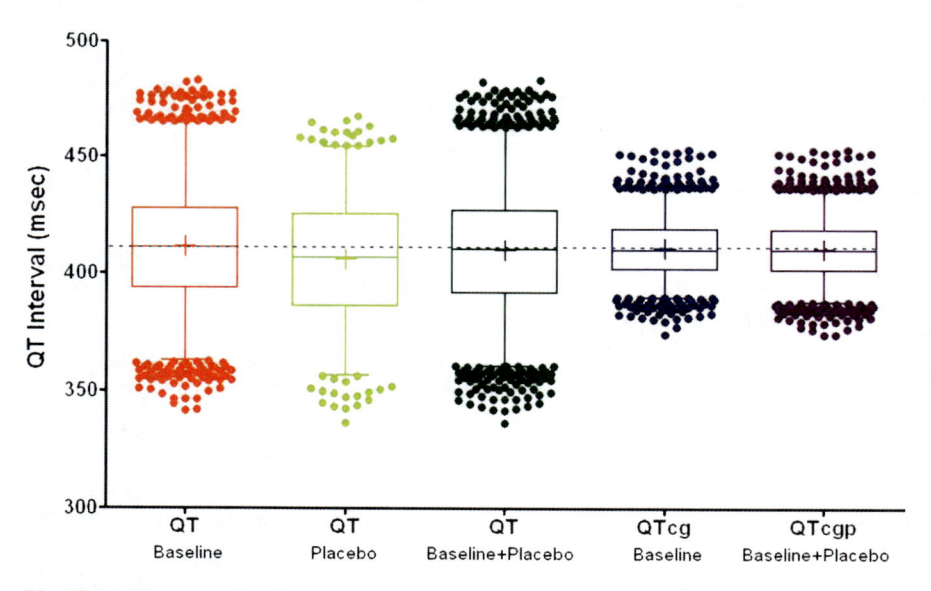

Fig. 13.14 Uncorrected and corrected QT interval of the baseline and placebo treatment groups

intake, and temperature. Therefore, one key fixed effect on QT is day time. The impact of the day time on QT is critical to be quantified in order to be able to draw a conclusion about drug effects. The corrected QT (QTcg and QTcgp) interval for pooled baseline and placebo data show a circadian variation of about 5 ms, peaking during morning hours (Fig. 13.15). The same pattern is observed at the

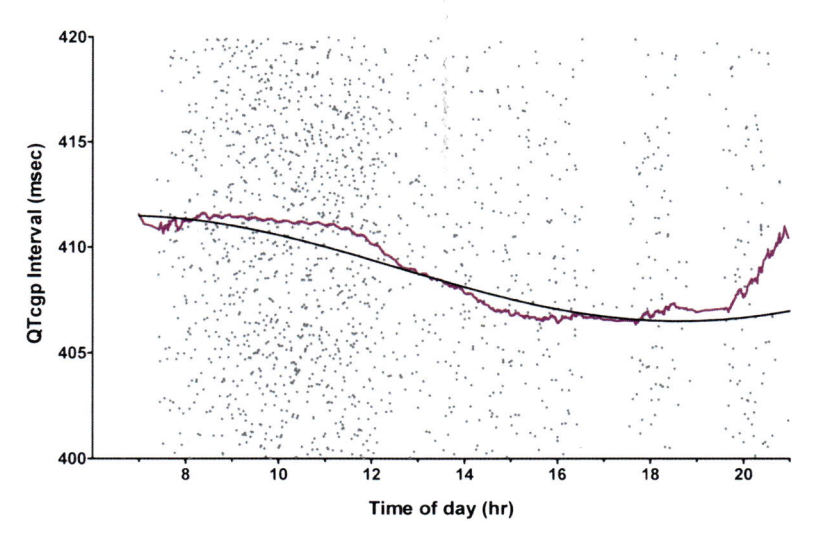

Fig. 13.15 Circadian rhythm trend line and cosine model for baseline and placebo pooled data

individual level (Girgis et al. 2007). For our purpose, a simple, 24 h harmonic cosine function

$$(A \cos(2\pi / 24(\text{Time} - \varphi)))$$

is used to describe the circadian rhythm (diurnal fluctuations) effect, where A is the amplitude parameter and φ is the phase shift parameter. The parameter values are estimated by trial and error. The corrected QT for heart rate, placebo effect, gender, and circadian rhythm (QTcgpt) can then be expressed as follows:

$$\text{QTcgpt} = \text{QT} / \text{RR}^n + \Delta\text{Gen} + \Delta\text{Pla} + A \cos(2\pi / 24(\text{Time} - \varphi))$$

where $\Delta\text{Gen} = -15$ ms, $\Delta\text{Pla} = +3$ ms, $n = 0.375$, $A = 2.5$ ms, and $\varphi = 6.6$ h.

13.7 Drug Effect

13.7.1 Moxifloxacin Drug Effect

Since the effect of moxifloxacin is well known, it was administrated and included in this study as a positive (active) control. Active control is necessary in order to show the validity of the trial conduct and to help assess any false QT liability. Using pooled data from 20 studies, Florian et al. (2011) described the moxifloxacin concentration–QTc relationship by a linear model with a mean slope of 3.1 (2.8–3.3)

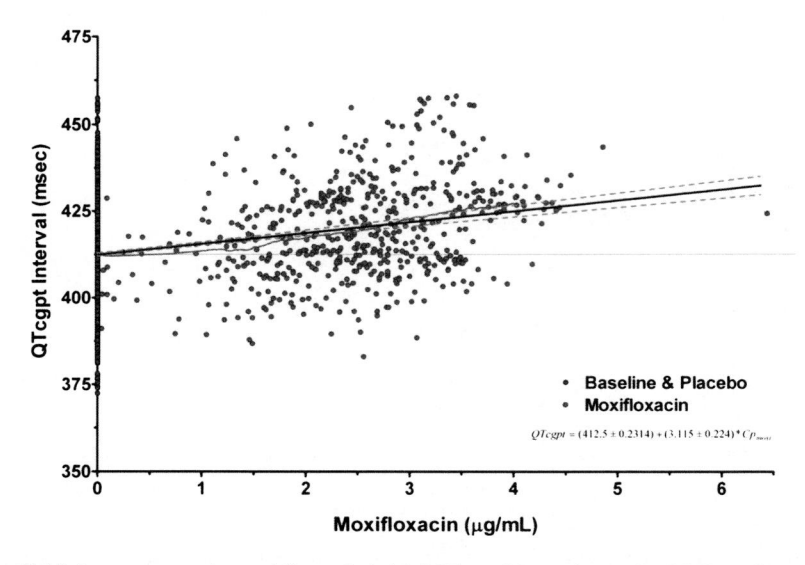

Fig. 13.16 Lowess smoother and linear fit (with 95% confidence interval) of QTcgpt interval vs. moxifloxacin concentration

milliseconds per μg/mL of moxifloxacin. For the current data, corrected baseline and placebo data were combined with corrected moxifloxacin (QTcgpt) data to present the zero moxifloxacin concentration effect (model intercept). As shown in Fig. 13.16, the predicted mean slope is 3.12 ms per μg/mL. Thus, the result validates and gives credibility for the graphical data exploration process and technique used.

13.7.2 Drug X Effect

Figure 13.17 shows that as the Drug X concentration increases, the QT interval decreases. The linear fit predicts a QT shortening with a negative slope of −0.45 ms per μg/mL. While a prolonged QT interval is linked to the risk for life-threatening events, little is known about shortened QT intervals. Nevertheless, QT interval shortening has previously been associated with sudden death (Gaita et al. 2003).

13.8 Modeling Results

Based on the above rationale, Girgis et al. (2007) used a hierarchical Bayesian approach to establish a population model for baseline and placebo data. The final structure model was similar to the current proposed corrections for the QT interval (QTcgpt). Figure 13.18 illustrates an example of the model fit for different subjects.

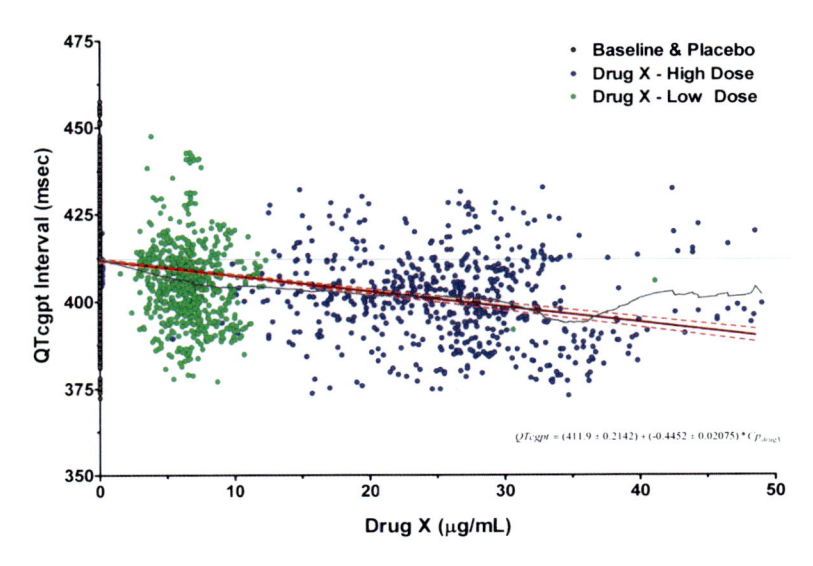

13 Graphical Data Exploration in QT Model Building and Cardiovascular Drug Safety 269

Fig. 13.17 Lowess smoother and linear fit (with 95% confidence interval) of QTcgpt interval vs. Drug X concentration

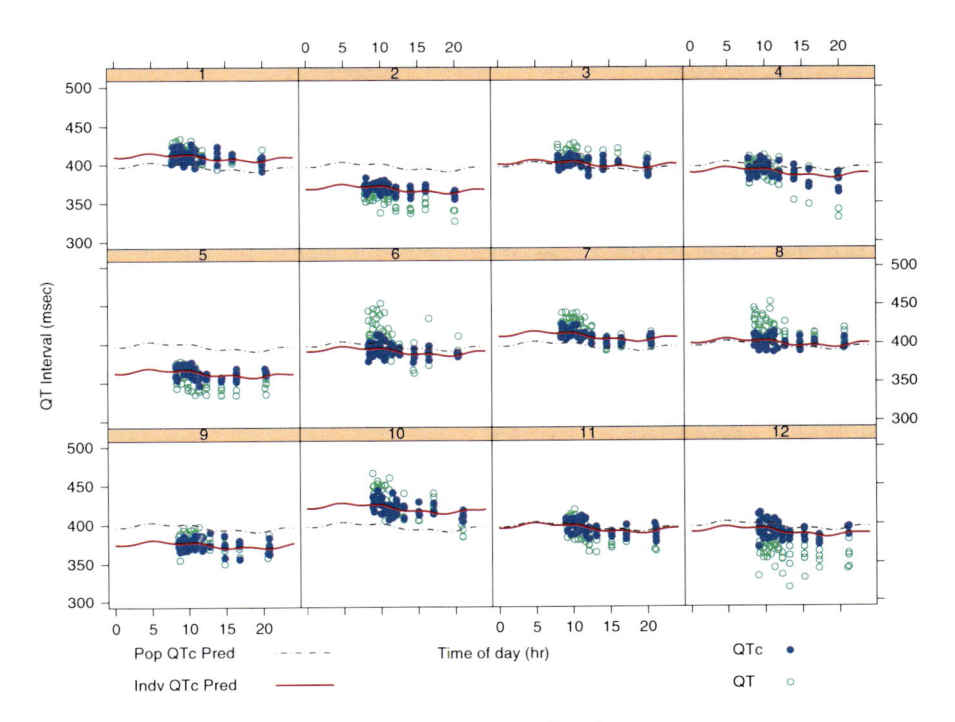

Fig. 13.18 Example of model fit for individual QT baseline data

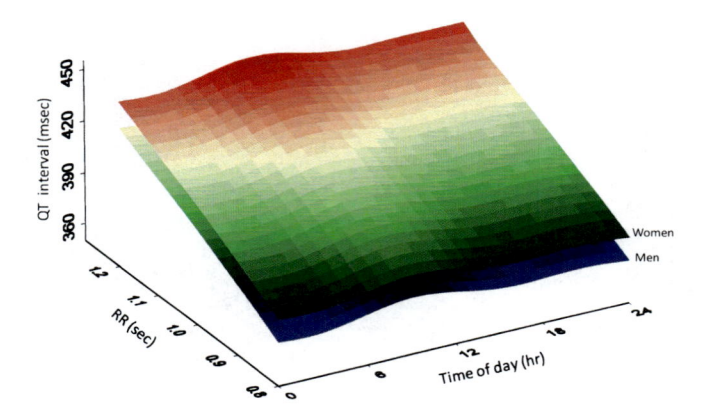

Fig. 13.19 Three dimensional plot of the modeled QT interval

It shows the time course of the measured QT (open green circles), individually corrected QT, (QTic = QT/RRni, filled blue circles), QTic individual predictions (red solid line), and QTic population predictions for a typical male (dashed gray line) for 12 subjects. As shown, QT measurements are well described by the final model. Figure 13.19 shows the 3-dimensional plot of the QT Interval and its relationship to heart rate, gender, and circadian rhythm based on the Girgis et al. baseline model (Girgis et al. 2007).

13.9 Summary

Data from a thorough QT study were used as an example to illustrate the benefits of graphical data exploration to assess the cardiovascular safety of a Drug X. As shown, graphical data exploration of clinical trial results greatly helps to better understand the data and increase the likelihood of constructing a useful model. It also could improve the efficiency of the model building process and provide a useful decision making instrument for the drug development program.

Acknowledgments The authors thank Dr. Douglas A. Marsteller for his help during his summer internship at J&J.

References

Chambers JM, Cleveland WS, Tukey PA, Kleiner B (1983) Graphical methods for data analysis. Wadsworth and Brooks, Pacific Grove

Dykstra K, Pugh R, Krause A (2010) Visualization concepts to enhance quantitative decision making in drug development. J Clin Pharmacol 50:130S–139S

Florian JA, Tornøe CW, Brundage R, Parekh A, Garnett CE (2011) Population pharmacokinetic and concentration–QTc models for moxifloxacin pooled analysis of 20 thorough QT studies. J Clin Pharmacol 51(8):1152–1162

Gaita F, Giustetto C, Bianchi F, Wolpert C, Schimpf R, Riccardi R, Grossi S, Richiardi E, Borggrefe M (2003) Short QT syndrome: a familial cause of sudden death. Circulation 108:965–970

Girgis IG, Shneider MN, Macheret SO, Brown GL, Miles RB (2006) Creation of steering moments in supersonic flow by off-axis plasma heat addition. J Spacecraft Rockets 43(3):May–June

Girgis IG, Schaible T, Nandy P, De Ridder F, Mathers J, Mohanty S (2007) Parallel Bayesian methodology for population analysis. PAGE 16:Abstr 1105. www.page-meeting.org/?abstract=1105

Guidance for industry: e14 clinical evaluation of QT/QTc interval prolongation and proarrhythmic potential for non-antiarrhythmic drugs. U.S. Department of Health and Human Services, Food and Drug Administration, Oct 2005

Piotrovsky V (2005) Pharmacokinetic-pharmacodynamic modeling in the data analysis and interpretation of drug-induced QT/QTc prolongation. AAPS J 7(3):E609–E624

Sheiner LB (1997) Learning versus confirming in clinical drug development. Clin Pharmacol Ther 61:275–291

Chapter 14
Data Visualization at the Individual Patient Level

Matthew Austin and Alicia Zhang

Abstract Although displays of clinical trial data in clinical research primarily focus on group level data, researchers often find themselves focusing on individual patients in a clinical trial. Data displays of individual patient data in clinical research are needed for many reasons, such as individual case review to understand potential outliers or important adverse events. When reviewing individual patient data, understanding how the individual is different from or similar to reference populations is very important. Without the correct context, individual data can be very difficult to interpret. Following good principles of graphics, we can build helpful displays that allow for faster and more accurate decision making.

Keywords Patient profiles • Axis scales • Layering • Sorting

14.1 Introduction

A single patient in a clinical trial provides an amazing volume of information. Each individual patient may provide information on safety, efficacy, quality of life, demographic characteristics, study conduct and a variety of other endpoints. Laboratory assessments alone may result in hundreds of observations on an individual patient in a relatively short clinical trial. In many situations, we need to review data at the individual level. We may want to study individual patients who report specific adverse events, review individual responses to a drug, or review individual cases for data cleaning. Depending on the purpose of the review, we may want to review all available data or selected data from an individual patient. Graphical representations of patient profiles have become more popular with the creation of software that

M. Austin (✉) • A. Zhang
One Amgen Center Drive, Global Biostatistical Science,
Amgen Inc., Thousand Oaks, MS 24-2-C, CA, USA
e-mail: maustin@amgen.com; aliciaz@amgen.com

makes displaying an individual's data easier; however, many data reviews are still based on tabular data that are spread across multiple documents. This is inefficient in the best case and may lead to an incorrect interpretation in the worst case.

The advantage of graphical displays over tabular displays is the ability to make compact, data-dense displays that provide a more complete context for interpretation. A properly designed display of individual patient data will provide all the information needed to make decisions and eliminate *page thrashing* during the data review process. Page thrashing is the process of flipping through multiple documents to retrieve all the data that are needed to make a decision. When the data are not organized and are not presented in a comprehensive manner, it may take many iterations across multiple functions (clinical, statistics, data management, etc.) to get the necessary information to make a decision. This is an inefficient use of resources and it can also disrupt the flow of thought and lead to misinterpretations of the data. When all of the necessary information is collected and organized in a single, well-designed display, decisions can be made quickly, efficiently and accurately.

The data from an individual need to be put into context to be most effective. It is relatively simple to create a plot of laboratory measurements over time in a clinical trial (Fig. 14.1); however, to create an informative and insightful data display all relevant data need to be presented (Fig. 14.2). For example, to monitor renal function we may be interested to see if a patient's creatinine values are within the normal range throughout the entire trial. A display of creatinine values should therefore include some indicator of the normal range. If the study is in patients with low renal function, it might be more helpful to put an individual's creatinine levels in the context of the specific trial population because we would expect creatinine levels outside the normal range. Throughout this chapter, we will revisit concepts for contextualizing the information from an individual patient.

Review of individual efficacy data is usually limited to a small number of parameters of interest. In contrast, when an unusual adverse event or an adverse event of special interest is observed in an individual, the scope of the information needed for this individual patient can be very broad. To help understand more about the patient who had the adverse event, we may be interested in all the other adverse events the patient experienced as well as all concomitant medications, complete medical history, laboratory assessments, and demographics. To highlight the different types of displays that could be used depending on the specific situation, 3 examples are used to demonstrate principles of design presented in the next section:

- Example 1 (Sect. 14.3.1) presents a review of a single endpoint of interest
- Example 2 (Sect. 14.3.2) presents a targeted review of a small number of endpoints
- Example 3 (Sect. 14.3.3) presents a general review of a large number of parameters for an individual patient

The graphs from these examples are referred to throughout this chapter, so it may be helpful to review the example overviews in Sect. 14.3 to have a better understanding of what is presented in the graphs. Section 14.3.4 presents an interactive display that allows the user to move from aggregate level adverse event data to individual data.

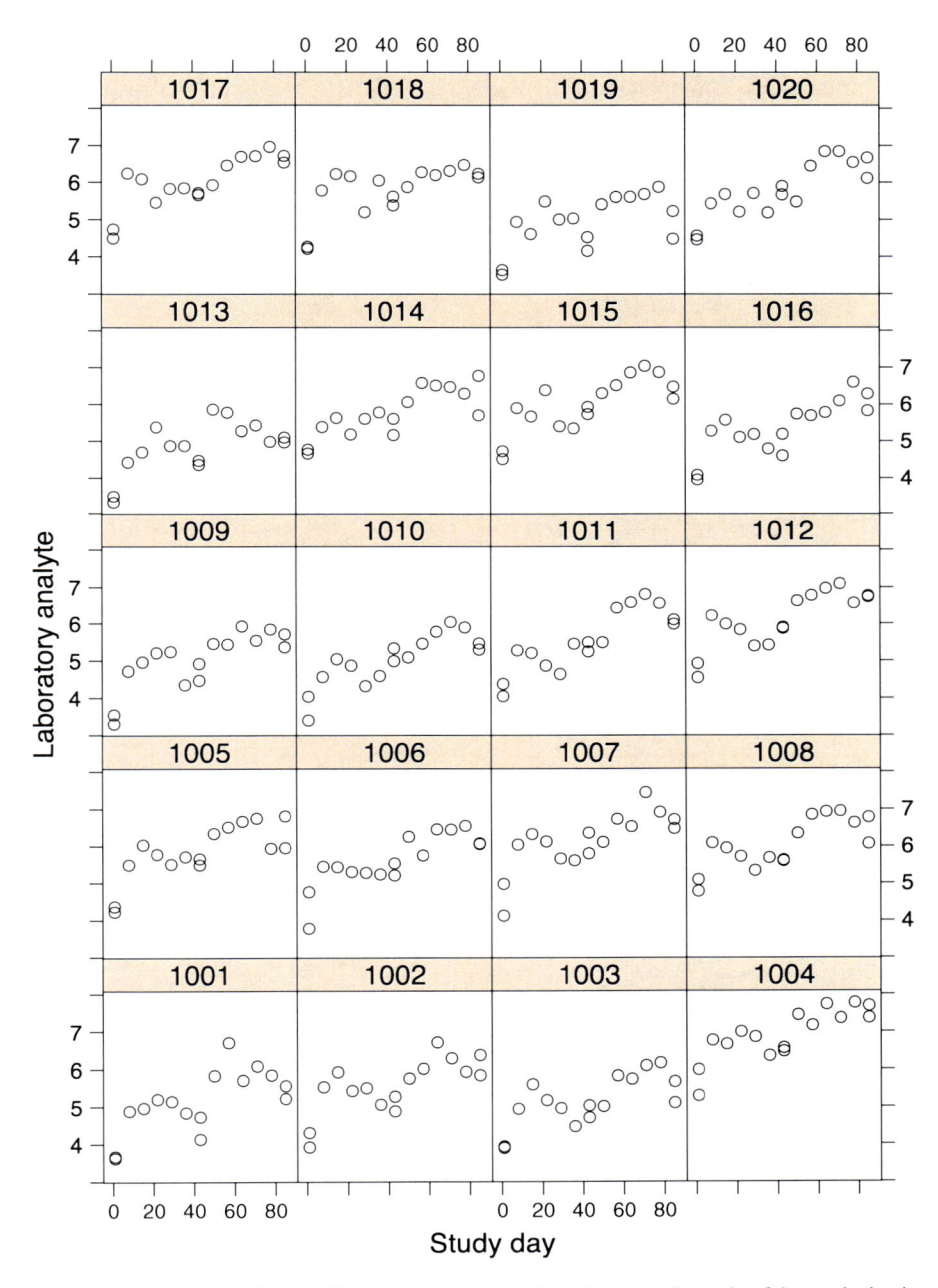

Fig. 14.1 Individual patient profiles of a laboratory analyte. Note that the scale of the vertical axis is the same for each patient. The sorting order is by patient identification number which does not encode any information. The primary hypothesis is that this investigational medication will increase the level of the analyte within the target range of 6–8 units. Without any reference lines, it is difficult to assess how many patients are actually reaching the target range. Also, we do not know if patients have received their doses of the investigational product, what doses they received (high or low) or if they received the concomitant medication that could impact the analyte

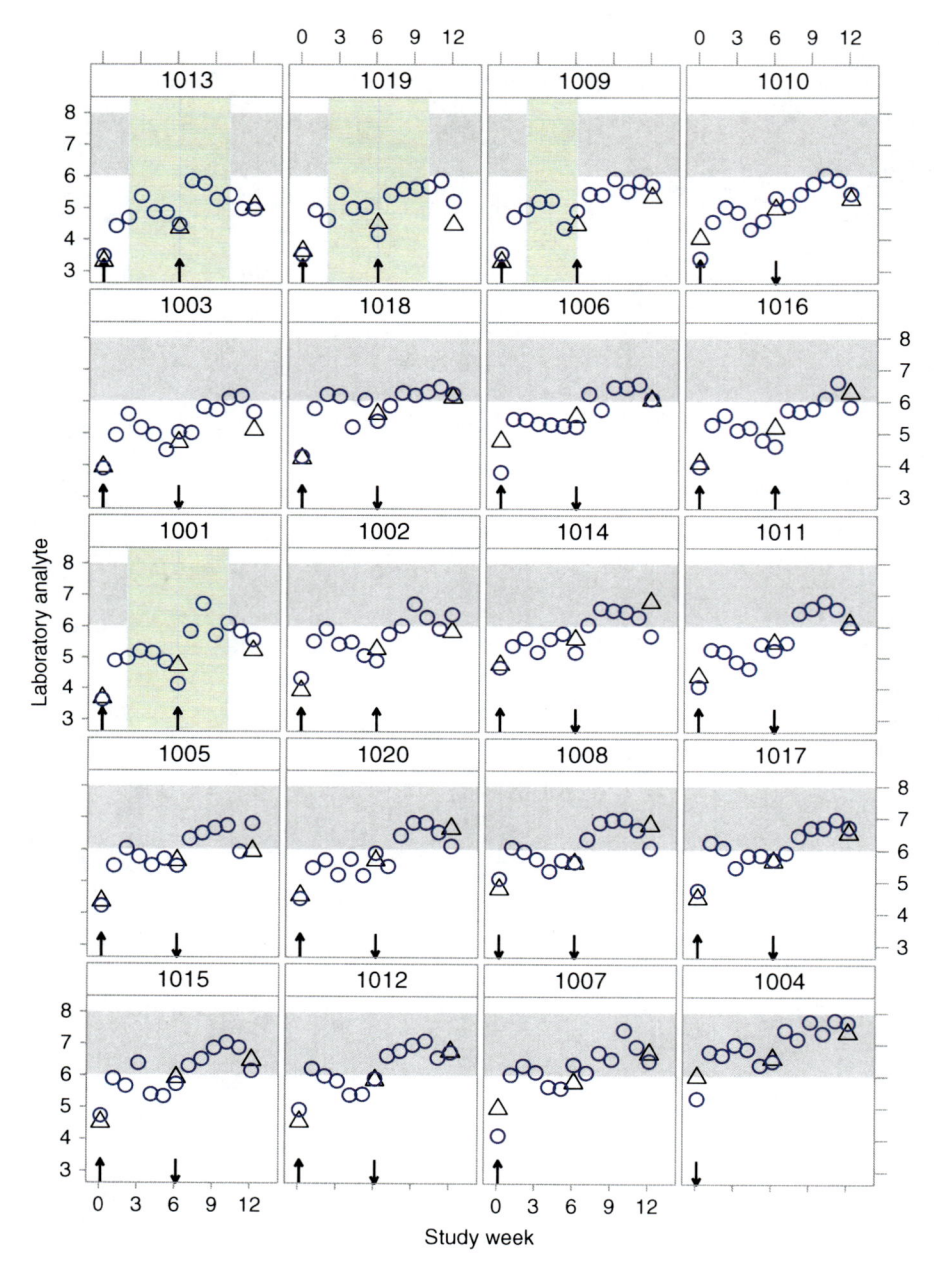

Fig. 14.2 Individual patient profiles of a laboratory analyte with layered data representing types of measurements (*blue circles* = local laboratory and *black triangles* = central laboratory), administration of concomitant medication known to impact the laboratory analyte (*yellow highlighted regions* represent the 28 day impact of this medication), and dosing information (*up arrows* = high dose and *down arrows* = low dose). The scale of the axes is the same for all patients and the display is sorted so that the patients with the worst response (lowest maximum on-study value) are at the *top-left*, while the best responders are at the *bottom-right*. Additionally, the target range of 6–8 units is displayed as a gray shaded region

Graphs in this chapter were created using S-PLUS™ (TIBCO Software Inc. 2012) and R (R Development Core Team 2011), including the Hmisc (Alzola and Harrell 2006) and Design (Alzola and Harrell 2006) libraries in S-PLUS™ and the lattice (Sarkar 2008) and latticeExtra (Sarkar and Andrews 2011) packages in R.

14.2 Principles of Design

14.2.1 Overview

Data visualizations need to be tailored to their specific purpose and audience. For example, visualizations used in presentations during meetings are generally shown for seconds and need to immediately convey a very specific point. These visualizations are usually very simple and straightforward. Graphical review of individual patient data falls on the other end of the spectrum. Displays of individual patient information are usually created for detailed study. These types of displays are generally very data dense and require more orientation for the reviewer. Caution should be taken to only present data that are relevant to the question. Presenting unnecessary data can add noise to the overall presentation that distracts from the important information. Few (2006) states

> By removing any information that is not really necessary, you automatically increase focus on the information that remains.

Few (2004) presents the following objectives for communicating quantitative information that are helpful when designing individual patient displays:

- Organize the data
  - Group the data
  - Prioritize the data
  - Sequence the data
- Highlight the data
  - Reduce the non-data ink
  - Enhance the data ink

These objectives are an extension of Tufte's principles (Tufte 1983) as summarized in Chap. 3.

Organizing the data is an important concept when producing data dense displays. A disorganized display is confusing and potentially misleading. When presenting individual patient data, it is common to group all the data from an individual in the same panel in a trellis format where each individual panel contains a single patient's information (Figs. 14.1 and 14.2). When examining an individual patient's response across a small number of endpoints, we can group the data by endpoint where the individual patient of interest is highlighted in a similar manner across the different panels (Figs. 14.5 and 14.6). When the question of interest requires looking at all the information from a single patient, we can group the information by types of data such as dosing information, adverse events, and laboratory assessments (Fig. 14.9). Highlighting the important data is key to an effective display. In the effort to include all necessary data to understand the endpoint of interest, it is crucial not to overwhelm the data that are most important. Using bold saturated colors of primary hues for the key data and using less emphasis on data presented for contextualization highlights the most important features of graph.

14.2.2 Sorting

Organizing the display of data through sorting is one of the easiest ways to increase the amount of information a reader gains from the display. Once the reviewer understands the sorting order, the interpretation should be easy and intuitive. Figures 14.1 and 14.2 display the same data. Figure 14.2 sorts the panels by a function of the response. The sorting order in this example is from lowest maximum on-study value to highest maximum on-study value. The patient with the lowest maximum on-study value is at the upper left and the highest is at the bottom right. If the reviewer is interested in patients with the lowest values on-study, they can simply focus on the top row. Alternatively, if the maximum response is of primary interest, the bottom row is most useful (although, in this case, sorting in the reverse order might be more helpful so that the most important information is at the top). Sorting by patient number is rarely useful, because the patient numbers in most clinical trials are arbitrary.

Besides sorting by a function of the response variable, other study parameters can be useful. Figures that organize information by country and/or investigative site can help spot trends and sources of variability in the data. However, remember that presenting the countries by alphabetical order or the study sites by numeric order is probably not helpful. Consider sorting the outer grouping parameter by some useful function of the response. For example, if you are interested in identifying sources of variability, sort the countries by the variance or interquartile range.

14.2.3 Layering Data

Layering is combining different types of related data in a single display to give a more complete picture of the endpoint of interest. Layering information should increase the efficiency and accuracy of the interpretation as compared to reviewing the data in different sources. Figure 14.2 uses layering to combine different factors to display a complete picture of the components that may influence the endpoint of interest. In this example, the primary interest is the impact of treatment on a laboratory analyte. However, several other factors also influence the measurement and need to be taken into account for proper interpretation (Sect. 14.3.1 for more detail on this example).

Layering data is also key to contextualizing the individual patient data to some larger set of data. Presenting normal ranges of laboratory data or summaries of the distribution of the individual patient's treatment group within the trial is a form of layering information.

14.2.4 Axes

14.2.4.1 Scaling

The scale of the axes greatly influences the interpretation of data. When the axes are determined only by values of a single patient, small, unimportant changes can appear very large. The information needs to be put into context. When the data are measured as a continuous variable, we have several choices to help contextualize the information from an individual patient:

- Normal ranges (when available)
- Within study data
- Clinical opinion

If normal ranges are available for the measurement, the axes should usually contain at least the upper and lower normal limits. Reference lines for the normal range should be added to the individual patient data so it is easy to see any departures from the normal range. An exception to this advice is when displaying information where the disease state is related to the parameter of interest. In patients with renal failure, having both upper and lower limits for parameters measuring renal function may not be necessary because all patients in this special population may be above or below the normal range depending on the laboratory parameter being measured.

Superimposing group level summary statistics with the individual patient data can also be helpful for contextualization. For example, present the mean ± standard deviation or medians with interquartile range from the patient's treatment group or control group along with the individual patient information (Fig. 14.4). Note that the summary statistics selected should summarize the distribution of the data (e.g., standard deviation) and not the precision of measuring central tendency (e.g., standard error). Displaying standard errors or confidence intervals are not helpful in this context because we are generally interested in where the individual patient falls in the overall distribution and not specifically if they are near the parameter estimate. Using standard errors or confidence intervals potentially leads to misinterpretation of the individual data. For example, in large studies most of the individual data do not fall within a single standard error of the mean or the 95% confidence interval. Non-statisticians may interpret individual data not falling within this interval as outliers if the mean ± standard error or confidence interval is used as a reference range.

When presenting data as change from baseline, consider adding reference lines for the least significant change if available. With some laboratory parameters there are specific thresholds of change that trigger clinical review and are useful references for contextualizing the size of the change in an individual patient.

Superimposing the actual data (as opposed to summary statistics) from other patients in the same treatment group or control group is another way to put the individual patient data into context. When displaying all the individual patient data in the same graph, there are some special considerations. One or more of the patients used as reference data may have very high or very low values compared to other patients. Although these values may be of particular importance, they may to hamper the interpretation of the patient of interest in the plot. One potential solution is to only show the individual data from the reference group that falls between the 5th, 95th percentile. If you select this method, this should be clearly stated in the title, footnote or axis label.

When adding reference lines for the normal range, summary statistics or actual data from a reference group, the data for the patient of interest must be more prominent. Use of more diluted colors (gray) for reference information and alpha transparency is very helpful. Alpha transparency gives the appearance of overlap which helps when over-plotting is an issue. The line thickness, symbol size, and color of the individual patient of interest should be thicker, larger and more saturated than the reference information.

When both the sample size and the number of parameters of interest are small, all of the individual patient profiles can be placed on a single trellis display leveraging the value of small multiples (Cleveland 1994). The scale of the vertical axis is usually determined by the range of the data from all the individuals; however, having at least the upper and lower limits of normal if available is still very helpful for the overall context.

14.2.4.2 Aspect Ratio

The aspect ratio of a graph is the height of the plotting region divided by the width of the plotting region. The plotting region does not include titles, axis labels, or footnotes. Generally, the plotting region is defined by the 4 axes of a graph– although not all graphs actually display axes.

One commonly misused method for determining the aspect ratio is matching the size of the plot to the available space on a page or presentation slide. *Determination of the aspect ratio should not be based on the size of the paper or display device.*

There are good algorithms for determining aspect ratio, such as the banking algorithm suggested by Cleveland (1993). However, these algorithms do not account for axis scaling based on adding external data such as laboratory normal ranges or group level summary statistics to individual patient data. A good rule of thumb, when banking does not apply, is to use an aspect ratio of 1. An aspect ratio of 1 is not ideal for every circumstance; hence, it is a rule of thumb and not a law. For example, using an aspect ratio of 1 does not work for a single boxplot. The key is to actively think about the aspect ratio, because it is of second importance to the actual data and can easily mask trends in the data if used incorrectly.

14.2.4.3 Axis Lines

The lines representing the axes of a plot denote the plotting region and contain the tick marks and labels needed for interpreting the information. The lines themselves contain very little information—perhaps none when you consider the tick marks as separate entities from the axis lines. The axes can actually distract from the actual data being plotted. This is particularly apparent in trellis plots which have many axes on a single graph (Fig. 14.1). The more these lines can be put into the visual background of the graph and the actual data can be moved to the foreground, the better the impact of the graph. This can most easily be achieved by using less saturated color such as gray for the axis lines (Fig. 14.2). In some graphs, the axis lines can be removed completely leaving only the tick marks and labels (Fig. 14.5).

14.2.4.4 Tick Marks and Labeling

Placements of tick marks on the axes is an important aspect in making a graph easier to interpret. Too many tick marks can be distracting, while too few can leave the reader without appropriate reference points. This is particularly important when using trellis displays because the size of each panel is relatively small. The number of tick marks should be proportional to the size of the individual panel.

The tick marks and labels should be placed at relevant values. For example, if a study collects data at baseline and weeks 3, 6, 9, and 12, do not label the axis with "Study day" and place tick marks at days 20, 40, 60, and 80 (Figs. 14.1 and 14.2). The axis should be consistent with the study design. This is particularly important with longer duration studies. If a study is 3 years long, it is more appropriate to place the tick marks and label as months than to annotate with weeks or days.

14.2.4.5 Axis Breaks

When time data are presented on the horizontal axis, it is generally appropriate to use a straightforward linear scale. However, it may be helpful to utilize axis breaks to account for differing frequency of data collection. Study designs commonly specify more frequent measures at the beginning of a trial and may have time periods of intense sampling for pharmacokinetic parameters. Displaying information on a constant scale can cause some areas to have overlapping information while others have nothing at all. Breaks in time (horizontal) axis can separate areas where time is measured in different units (e.g., hours, days, weeks, months) as in Fig. 14.9. Axis breaks are generally discouraged for the response variable (vertical axis); however, the study design may provide good reasoning for breaks on a the horizontal time axis.

14.2.4.6 Multiple Vertical Axes

Multiple continuous variables can be presented in separate graphs on the same page or all combined in a single graph. When combining many parameters into a single graph, it is tempting to present multiple vertical axes. Selection of the scales of the different axes can greatly impact the interpretation of the data. One potential solution when working with laboratory data is to scale the different axes by the lower and upper normal ranges for each parameter. Under this scaling, the limits would all be aligned vertically. This has a direct benefit of having a clinically meaningful scaling and also reduces the number of horizontal reference lines representing the normal ranges of the parameters. If changes from baseline are being displayed, another option is to scale the data by the standard deviation or some other measurement of variability. Displaying changes on the standardized scale is familiar to clinicians.

14.2.5 Shapes, Colors, and Line Types

Bold, familiar colors and shapes can encode information (red, orange, yellow, and green or circles, triangles, octagons). Few states that *full saturated, bright versions of just about any primary hue tend to demand attention* (Few 2004). Combining bold colors with familiar shapes can also trigger natural associations. Red octagons immediately make a reader think of stop signs and yellow triangles signal a warning. Care should be taken when using combinations that will be reviewed by people in different regions where they can have different meanings; however, if most of the intended reviewers are from a specific region these combinations can be very useful.

Reference information should not overwhelm the primary data of interest. The primary data should be visually perceived as being in the foreground while the reference data appears to be in the background. Reference lines and symbols should be smaller in size and less saturated in color. Solid, black, thick lines might represent for the primary data while dashed, gray, thin lines are used for reference (Fig. 14.6). Similarly, large, black symbols filled with a saturated primary hue might represent the primary data while smaller, gray, non-filled symbols are used for reference (Fig. 14.5). Alpha transparency can be a very useful tool to create this effect. Too many symbols or colors can saturate the senses and potentially reduce the interpretability. The focus should always be on the primary data. Any other information should only enhance the interpretability of the primary data.

14.3 Examples

14.3.1 Single Outcome of Interest

In this hypothetical example of a 20-patient single-arm phase I study, a single laboratory analyte is the primary endpoint. The endpoint is assessed at a local laboratory

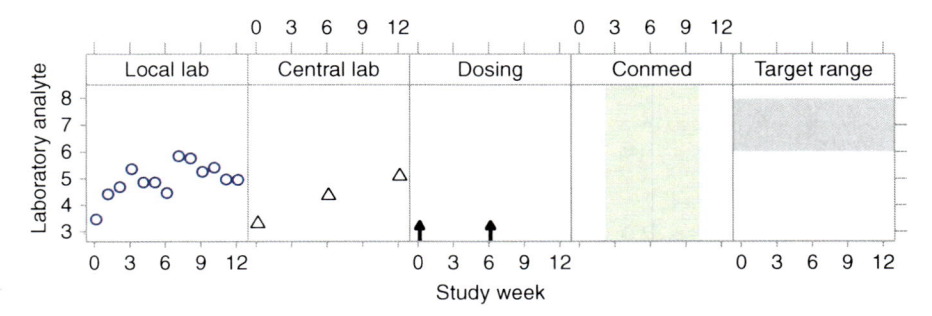

Fig. 14.3 Separation of layered data for patient 1,013 from Fig. 14.2. The data includes five layers: local laboratory values, central laboratory values, dosing level, concomitant medication, and target range. All of these components are necessary for interpreting the response profile of the individual patients

at all time points and additionally at a central laboratory at 3 time points during the study. The treatment under investigation is expected to increase and, through dose titration, maintain the level of the primary endpoint within a target range of 6–8 units. There is a concomitant medication some patients receive that also impacts the primary endpoint. The impact of a single dose of this concomitant medication lasts 28 days.

Figure 14.1 displays the endpoint data for each individual in the study. Figure 14.2 presents the same data as Fig. 14.1, but the sort order is by increasing maximum on-study value of the laboratory analyte. Additionally, Fig. 14.2 adds information about dosing, the concomitant medication, the target range, or whether the analyte was assessed at a local laboratory or the central laboratory. Whether or not a patient received a high or low dose can be derived from the up or down arrows (up = high, down = low). Because the impact of the concomitant medication lasts 28 days, a shaded region is shown that highlights the 28 day window starting at the time the patient begins the medication. If a patient, such as patient 1,013, receives the concomitant medication more than once, there will be multiple shaded regions.

Figure 14.3 displays a single patient from the study and separates out the 5 layers of data presented in Fig. 14.2. Figure 14.4 separates out the 4 patients who received the concomitant medication that is known to impact the endpoint in this example. To put the changes of these 4 individuals into the context of the overall study, the mean and standard deviation of the endpoint for the entire study cohort are added as reference.

14.3.2 Multiple Outcomes of Interest

In this hypothetical example, we have 5 variables (Var1–Var5) measured on 100 patients (Fig. 14.5). These 5 variables measure different attributes of the same condition as we may encounter in clinical research such as analytes measuring liver

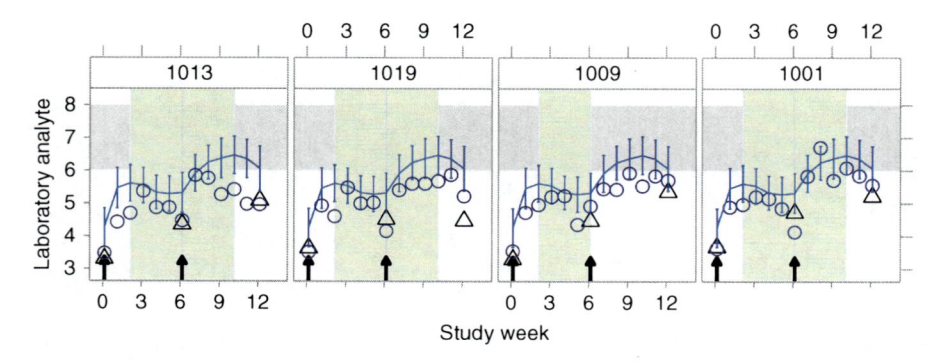

Fig. 14.4 Displaying only the 4 patients from Fig. 14.2 who received the concomitant medication that impacts the outcome of interest. Means and standard deviations are presented for all 20 patients to add context to the response profiles of these 4 patients

Fig. 14.5 Boxplots displaying the distribution of data from 5 variables with the same patient highlighted across all variables

function or renal function. When reviewing the values of $Var1$, we notice that an individual patient has a value above 2.0 which is a prespecified threshold and will need further investigation. We now want to investigate the response of this patient on the other 4 related variables to see if this patient is also different from the rest of the patients in this study. If $Var1$ is the only variable where this patient differs from the rest of the patients, we may think that the individual laboratory measurement could have been erroneous. However, if we see a trend across multiple measurements we would be more sure of the potential uniqueness of this individual.

Figure 14.5 simply highlights data from the patient with the highest value of $Var1$ in the distributions of the other variables. Note that the symbol size is larger than other symbols used in the plot and has a colored (green) fill while the other data are unfilled. This graph shows that the patient with the highest value of $Var1$ also has high values for $Var2$ and $Var3$ while the patient is in the middle of the distribution

Fig. 14.6 Parallel coordinate plot from 5 variables with an individual patient highlighted

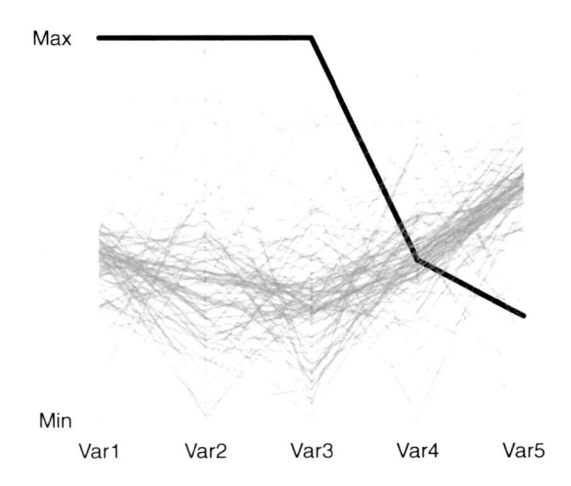

for *Var4* and in the lower 25% of *Var5*. Figure 14.6 presents the same data in a parallel coordinate plot. The parallel coordinate plot standardizes the variables so that they have the same scale. The data points from each individual are connected which allows the visual inspection of the trend among the variables. As with the boxplots, we can clearly see that this patient has high values for the first three variables and a low value for the 5th variable. Figure 14.7 presents the same data by using density plots. While boxplots are extremely useful in many circumstances, if any of the variables have more than one mode they do not accurately depict the distribution of the data. Using density plots allows a visualization of more complex distributions. The graphs are aligned vertically so that it is easier to compare where the individual of interest is located in distributions of the five variables. If the graphs were in a single row or in a grid, the comparison would be less efficient. Figure 14.8 presents the same data by using pairwise scatter plots. The previous displays of these data concentrated on univariate descriptions of the 5 variables. The scatter plots allow us to examine the bivariate relationships between the variables and visualize where the individual patient of interest falls in these distributions.

The graphs shown in this example are static; however, several software packages, such as Spotfire™, JMP™, and Mondrian, allow dynamic exploration of individual patients through *brushing* (Unwin et al. 2006) and the ability to drill down through multiple levels of data. Turning a static display into an interactive display by adding the capability to drill down is an amazingly useful tool (Sect. 14.3.4). The ability to link information among different endpoints, including both continuous and categorical data, using multiple types of displays can speed review and lead to insights that might be missed otherwise.

14.3.3 Patient Profile

Many situations provide reasons for looking at all the data available for an individual patient. When a patient experiences an adverse event of interest during a clinical

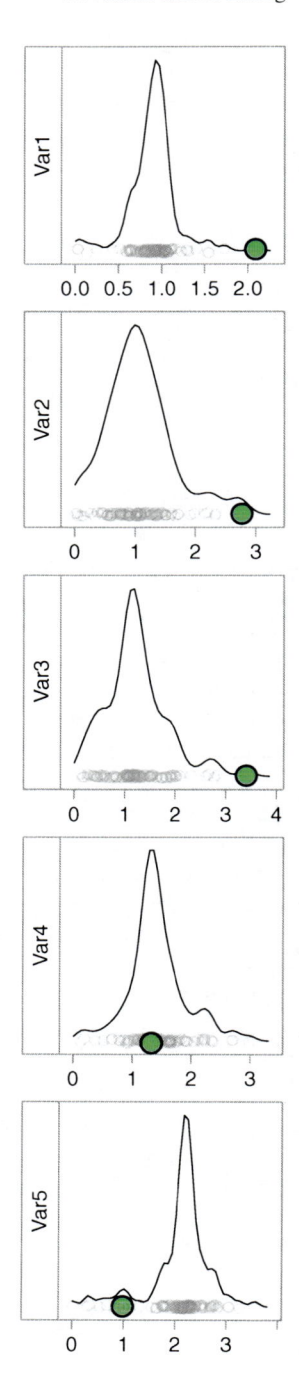

Fig. 14.7 Density plots for 5 variables with an individual patient highlighted

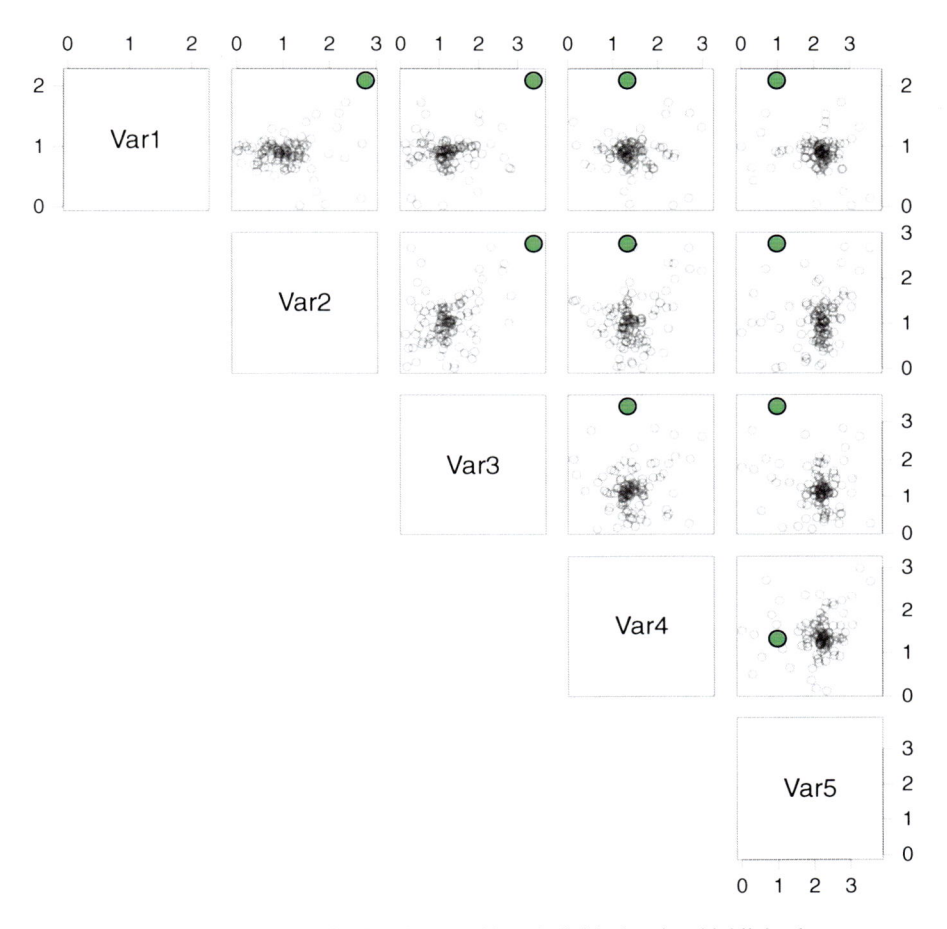

Fig. 14.8 Pairwise scatter plots for 5 variables with an individual patient highlighted

trial, there is often a need to examine most—if not all—of the data from this patient. A full patient profile is needed to give context to the event and help determine what important factors may have contributed to likelihood of this event occurring. We may want to examine the temporal relationship of the event to the last dose of investigational drug or determine the number of doses given prior to the event. Also of interest is the medical history including previous on-study adverse events and concomitant medications. Laboratory data may also be helpful to understand the event.

In these circumstances it is difficult to prespecify exactly what data are needed, so providing as much relevant data as possible on the individual patient can prevent too many back-and-forth iterations. The most common solution is to provide patient listings that contains as many parameters as possible; however, tabular formats are not as effective as graphical displays because they cannot achieve the information density and interpretability of a graph. Providing a single, well-designed graphical patient profile is a good alternative to multiple listings.

Powsner and Tufte (1994) proposed a very compact graphical representation of laboratory data by presenting each laboratory analyte in separate panels on a consistent time axis. They also provide narratives of patient information on the display to complete the picture for the reviewing clinician.

An alternative, as presented in Fig. 14.9, is a display that mixes tabular and graphical elements. This display provides information on demographics, dosing, adverse events, concomitant medications, laboratory measurements, and medical history on a consistent time axis.

Dosing information is presented at the top which gives the reviewer access to the timing and amount (numbers in plotting symbol indicate the dose of drug) of the dose received. This data are presented at the top of the display and in red to emphasize the importance of this information.

Laboratory information is presented by placing the results from a single laboratory analyte in a row. Values outside the normal ranges are given in bold font superimposed on colored rectangles that indicate the CTC (Common Toxicity Criteria) grade. A clinician can quickly scan all the laboratory analytes to see what values are aberrant as well as look at any analytes of interest to see if any are decreased or elevated. The laboratory analytes are sorted by type (chemistry, hematology, etc.) to create logical groupings of information.

Next are adverse events and concomitant medications. Within adverse events, the preferred (dictionary coded) terms are sorted by system organ class so that similar events are grouped together. Adverse events are presented by using plotting symbols with the severity of the adverse event indicated by numerical code within the plotting symbol and color (1 = mild/green, 2 = moderate/yellow, 3 = severe/orange, 4 = life-threatening/red, 5 = fatal/red). Whether or not an adverse event is serious (events that require hospitalization, are life-threatening or fatal) is indicated by the type of plotting symbol (circle = non-serious, triangle = serious). A line extends from the plotting symbol indicating the duration of the adverse event. An arrow on the right indicates that the adverse event is continuing at the time of analysis. Concomitant medications are presented in a similar style as the adverse events. An arrow on the left indicates that the patient was on medication at the beginning of the study.

Medical history is presented by text at the bottom of the display aligned along the left margin with the start date of the condition and an indicator of whether or not the event was present at the start of the study.

14.3.4 Aggregate to Individual

In small trials, examining each individual patient in detail may be feasible. However, in large trials, specific events usually trigger individual patient review. For example, an imbalance in the occurrence of an adverse event between treatment and control groups may trigger an investigation of the patients who have the event.

Figure 14.10 presents comparisons of adverse event incidences between two treatment groups. For each figure, the vertical axis represents the nominal p-value

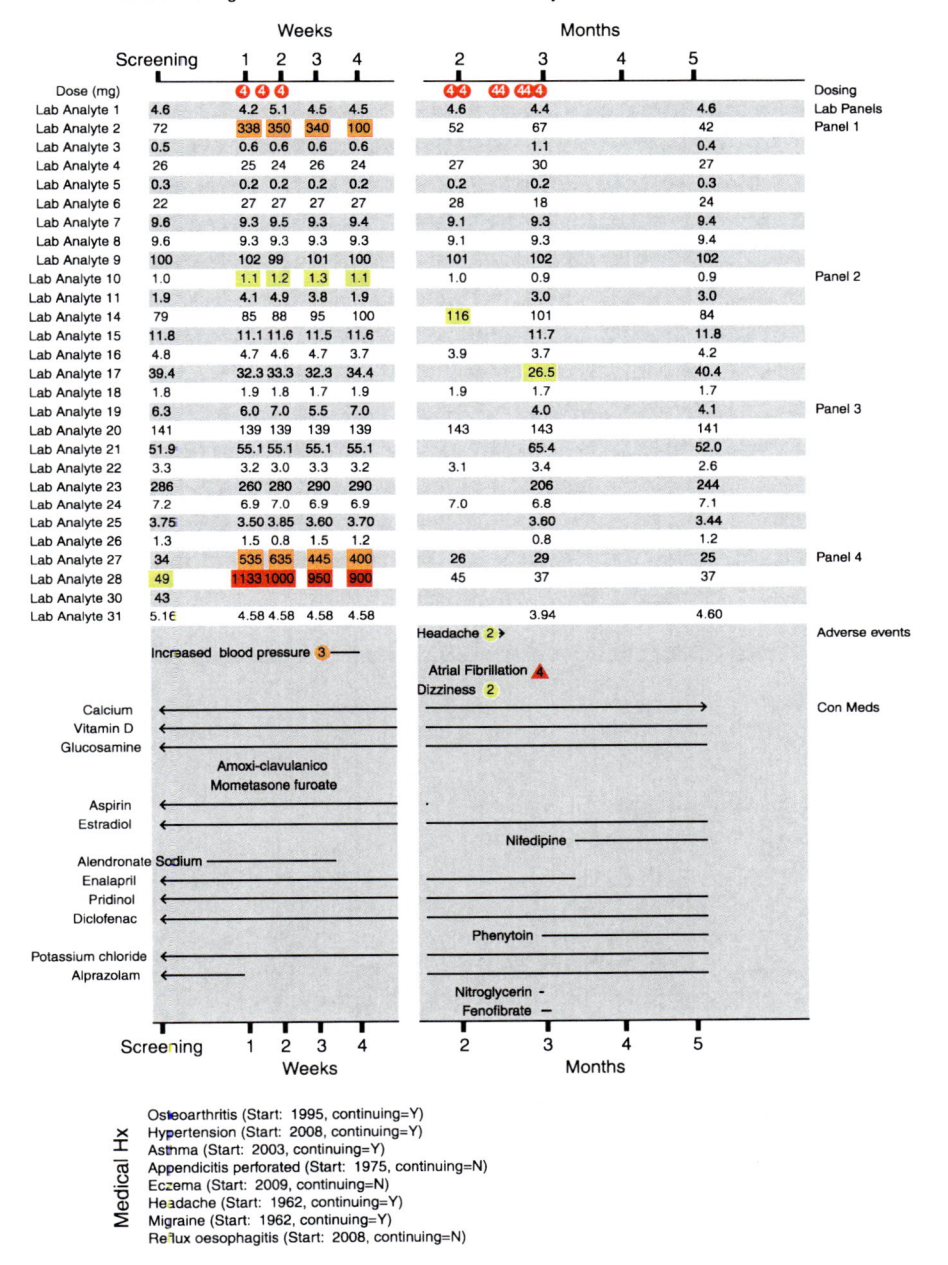

Fig. 14.9 Individual patient profile utilizing a compact tabular design. The horizontal axis represents study week or month (randomization at the beginning of week 1). A break in the horizontal axis is due to higher sampling frequency in the first 4 weeks. Categories of data are listed in the right margin. Dosing times are represented by *red circles* with the actual dose level given as a number inside the circle. For laboratory data, *bold, italicized* numbers superimposed on colored *rectangles* represent measurements outside the normal range (*green* = CTC grade 1, *yellow* = CTC grade 2, *orange* = CTC grade 3, *red* = CTC grade 4). For adverse event data, serious and non-serious adverse events are represented by *triangles and circles*, respectively. The severity of the adverse event is given as a number inside the plotting symbol (1 = mild, 2 = moderate, 3 = severe, 4 = life-threatening, 5 = fatal). *Arrows* for adverse events and concomitant medications represent events or medications that started before the study or were continuing at the end of study

Adverse Events

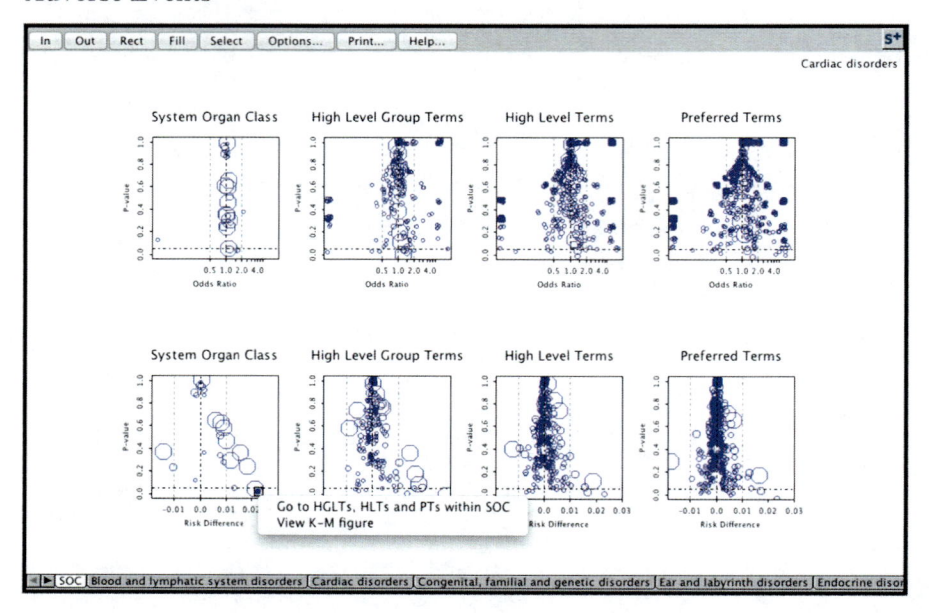

Fig. 14.10 Example of an interactive graphical display designed to help identify imbalances in adverse events. The vertical axis displays the nominal p-value based on a log-rank test and the horizontal axis represents the effect size (odds ratio in top row and risk difference in *bottom row*). Columns represent the hierarchy of the MedDRA® coding dictionary. *Circles* represent an adverse event term and the size is proportional to the overall incidence

and the horizontal axis represents the effect size. Each circle represents an adverse event term and the size of the plotting symbol is proportional to the overall incidence. The top row of figures shows the odds ratio between the treatment groups (values larger than 1 indicate higher risk on treatment compared to control) and the bottom row represents the risk differences (values larger than 1 indicate higher risk on treatment). The columns represent different levels of grouping of the adverse events. From left to right, the 4 columns represent the system organ classes (SOC), high level group terms (HLGT), high level terms (HLT), and preferred terms (PT) as classified in the MedDRA® coding dictionary.[1] The system organ class is the broadest grouping of medical concepts in the coding dictionary which is why it has the fewest terms which are represented by the circles. Mousing over[2] the plotting symbol will display the name of the term in the top-right of the screen. Figure 14.10

---

[1] MedDRA®: the Medical Dictionary for Regulatory Activities terminology is the international medical terminology developed under the auspices of the International Conference on Harmonization of Technical Requirements for Registration of Pharmaceuticals for Human Use (ICH). MedDRA® is a registered trademark of the International Federation of Pharmaceutical Manufacturers and Associations (IFPMA).

[2] Interactive figures created with S+Graphlets®.

Adverse Events

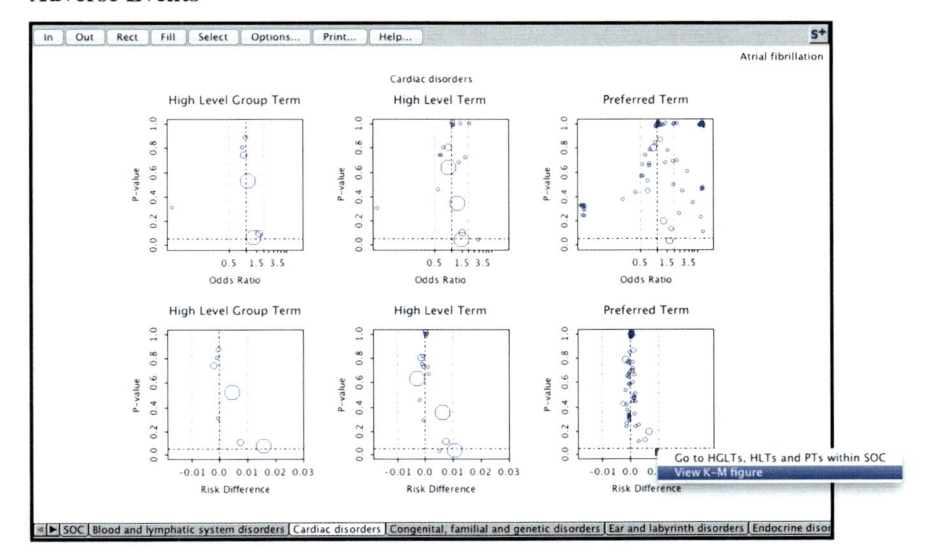

Fig. 14.11 Similar to Fig. 14.10, restricted to Cardiac disorder SOC. The vertical axis displays the nominal *p*-value based on a log-rank test and the horizontal axis represents the effect size (odds ratio in *top row* and risk difference in *bottom row*). Columns represent the hierarchy of the MedDRA® coding dictionary. *Circles* represent an adverse event term and the size is proportional to the overall incidence

has the "Cardiac disorders" term from the SOC level selected which is the largest positive risk difference and significant at the nominal 5% level.

The option "Go to HGLTs, HLTs and PTs within SOC" takes the reviewer to Fig. 14.11 that displays a similar set of figures for odds ratio and risk difference for the HLGTs, HLTs and PTs within the Cardiac disorders SOC.

Mousing over the plotting symbol with the largest positive risk difference in the HLGT, HLT and PT panels indicates that the largest imbalances are for "Cardiac arrhythmias," "Supraventricular arrhythmias," and "Atrial Fibrillation" (shown in figure), respectively. Selecting the "View K-M Figure" option takes the reviewer to Fig. 14.12 that displays the Kaplan–Meier figures for all atrial fibrillation adverse events, serious adverse events, related adverse events and adverse events leading to withdrawal from the study. Additionally, the proportion of patients with events and hazard ratios with 95% confidence intervals are supplied for the different types of adverse events. Patient numbers with hyperlinks to patient profiles similar to Fig. 14.9 are provided for quick access to individual level data.

Although this example was presented by starting at the broadest grouping (SOC) of the coding dictionary, working from the opposite direction can be very helpful. If there is an imbalance in a preferred term, examining the HLT, HLGT, and SOC for that specific preferred term is very helpful to see if similar types of events also show

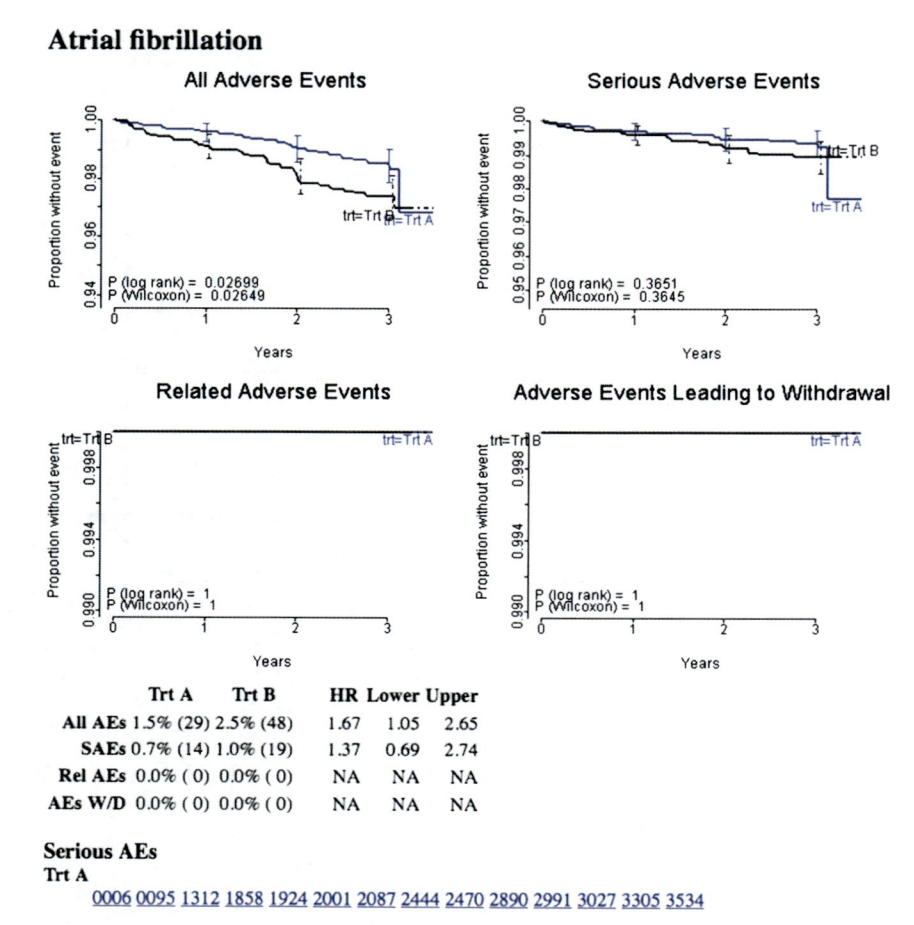

Atrial fibrillation

Fig. 14.12 Summary of atrial fibrillation preferred term. Kaplan–Meier figures for different adverse event definitions with corresponding statistical tests for differences in survival curves. Estimates of incidence and hazard ratio with 95% confidence intervals. Links to patient profiles (Fig. 14.9 for example) for individual patients experiencing serious adverse events coded as atrial fibrillation

	Trt A	Trt B	HR	Lower	Upper
All AEs	1.5% (29)	2.5% (48)	1.67	1.05	2.65
SAEs	0.7% (14)	1.0% (19)	1.37	0.69	2.74
Rel AEs	0.0% (0)	0.0% (0)	NA	NA	NA
AEs W/D	0.0% (0)	0.0% (0)	NA	NA	NA

Serious AEs

Trt A
0006 0095 1312 1858 1924 2001 2087 2444 2470 2890 2991 3027 3305 3534

Trt B
0064 0132 0133 1157 1348 1455 1550 1813 1847 1853 2002 2099 2202 2787 2898 2947 2948 3059 3744

imbalances. Borrowing information from similar medical concepts is what makes hierarchical modeling appealing in this setting.

Figure 14.10 is a good reminder of the multiplicity issues when examining adverse event data. There are thousands of simultaneous comparisons taking place. Imbalances can arise at many levels of the coding dictionary (SOC, HLGT, HLT, and PT) and at any term within a level. It is interesting to note that this example is created by randomly splitting the placebo group from a large clinical trial into 2 groups; therefore, any difference between the groups is, by definition, spurious.

14.4 Summary

While most graphical presentations of clinical trial data focus on group level summary statistics, focusing on individual patient data is necessary to understand many facets of the results from a clinical trial. Potential outliers in important endpoints, unusual outcomes, and/or important adverse events require detailed review of individual patient data that will encompass many important variables measured during the trial. Arranging these data in compact graphical displays can reduce the review time, increase efficiency and lead to more accurate conclusions that may impact the interpretation of safety and efficacy outcomes.

Graphical displays of individual patient data are key to many members of the clinical research team, such as statisticians, clinicians, data managers, and medical writers. Displays of individual patient data while the study is ongoing help in the cleaning of the data and assist in early identification of patients so additional data can be collected prior to the patient ending study. During the analysis, understanding the response of individual patients is key to understanding the full impact of the investigational product for both efficacy and safety. When reporting the data outside the research team, the understanding of data the individual level helps the team answer questions more efficiently and effectively.

When clinical research teams work together to identify important data points for further investigation, they can help each other by identifying what information is needed for the interpretation of individual patient data. With the correct context, individual patient data can be revealing and lead to insightful discoveries.

References

Alzola CF, Harrell FE (2006) An introduction to S and the Hmisc and design libraries. Available from http://biostat.mc.vanderbilt.edu/wiki/pub/Main/RS/sintro.pdf. Electronic book, 310 pages. Accessed on 5 Sept 2012

Cleveland WS (1993) Visualizing data. Hobart Press

Cleveland W (1994) The elements of graphing data. AT&T Bell Laboratories

Few S (2004) Show Me the numbers: designing tables and graphs to enlighten. ISBN 0970601999. Analytics Press

Few S (2006) Information dashboard design: the effective visual communication of data. O'Reilly Series. O'Reilly

Powsner SM, Tufte ER (1994) Lancet 344(8919):386

R Development Core Team (2011) R: A language and environment for statistical computing. R Foundation for Statistical Computing, Vienna, Austria. URL http://www.R-project.org. ISBN 3-900051-07-0

Sarkar D (2008) Lattice: multivariate data visualization with R. Springer, New York. URL http://lmdvr.r-forge.r-project.org. ISBN 978-0-387-75968-5

Sarkar D, Andrews F (2011) latticeExtra: extra graphical utilities based on lattice. URL http://R-Forge.R-project.org/projects/latticeextra/

TIBCO Software Inc., TIBCO Spotfire S+ 8.2 Guide to Graphics

Tufte E (1983) The visual display of quantitative information. Graphics Press

Unwin A, Theus M, Hofmann H (2006) Graphics of large datasets: visualizing a million. Statistics and computing. Springer, New York

Chapter 15
Graphics for Meta-Analysis

Peter W. Lane, Judith Anzures-Cabrera, Steff Lewis, and Jeffrey Tomlinson

Abstract We present 4 types of graphic used in meta-analysis. The commonest is the forest plot, and we discuss important aspects of the basic form of this plot. We present 2 enhanced versions, one displaying the results of subgroup analysis, and the second displaying absolute risks alongside relative risks from a meta-analysis of a binary outcome. The funnel plot is a well-established graph for assessing publication bias. We show some alternative forms, including a recently suggested enhancement using contours. The third type is a bubble plot used to summarize the results of meta-regression. Finally, we show a graphic designed for network meta-analysis, presenting rankings of the treatments that are compared. We prepared programs and graphs using GenStat™, R, RevMan™, SAS™ and Stata™, and these are available from the website.

15.1 Introduction

Meta-analysis is the statistical combination of results from 2 or more studies. It is often part of a wide-ranging "systematic review" of a specific medical intervention, which considers all the scientific evidence and how it may reasonably be summarized to inform patients, physicians, and policy makers. If used appropriately, meta-analysis (abbreviated as MA in this chapter) is a powerful tool to summarize

---

P.W. Lane (✉)
Quantitative Sciences, GlaxoSmithKine, Stevenage, UK
e-mail: peter.w.lane@gsk.com; peterwlane@gmail.com

J. Anzures-Cabrera • J. Tomlinson
Roche, Welwyn Garden City, UK

S. Lewis
Centre for Population Health Sciences, University of Edinburgh,
Edinburgh, UK

quantitative results from multiple studies, and assist in deriving meaningful conclusions. When studies are heterogeneous, the associated technique of meta-regression can provide insight into potential explanations of the heterogeneity.

A major source of meta-analyses, and guidance related to their production, is the Cochrane Collaboration, which is an international, nonprofit, independent organization. We refer in this chapter to some of the resources provided by this organization, which aims to help people make well-informed decisions about healthcare by preparing, maintaining, and promoting the accessibility of systematic reviews of the effects of healthcare interventions (Cochrane reviews). These are updated regularly and published online. The Cochrane Collaboration is committed to involving and supporting people of different skills and backgrounds, reducing barriers to contributing, encouraging diversity, open decision making, and teamwork.

Graphical methods are frequently the most effective way of presenting a MA and communicating the results. Tables and text are needed to provide the background to and detail of an analysis, and to discuss the interpretation, but a good set of graphical displays will provide the key messages. It is essential that the graphs should be constructed to avoid distorting or obscuring the information in a report, because readers will often focus on them more than on the text.

For each graph in this chapter we present a version drawn by one package (stated in the text). The data and code can be found on the website associated with this book, together with code and graphs produced by the other packages we tried.

15.2 Forest Plots and Variations

15.2.1 *Traditional Forest Plots*

Forest plots are the most common graphical displays for presenting results from meta-analyses. In a forest plot the effect estimates from individual studies are displayed together with their confidence intervals (Anzures-Cabrera and Higgins 2010). The individual effect estimates are represented by a symbol whose area is proportional to the weight of the estimate in the meta-analysis. A horizontal line extending either side of the symbol represents the confidence interval. The use of different symbol sizes draws attention to studies with larger weight in the meta-analysis, i.e., with smaller confidence intervals. The summary effect of the meta-analysis is symbolized by a diamond at the bottom of the plot where the width of the diamond represents the 95% confidence interval. It has become conventional to present 95% confidence intervals around the estimates in a forest plot, rather than any other indications of variability such as standard errors. However, for clarity, it should always be stated what the intervals represent.

Any type of effect estimate can be used in a forest plot: odds ratios, risk ratios, hazard ratios, mean differences, standardized mean differences, proportions, etc. Effect estimates measured on a ratio scale are usually displayed on a transformed

15 Graphics for Meta-Analysis

Study	TE	seTE	RR	95%-CI
Ono 2011				
Yamomoto 2011				
Malek 2008	−0.22	1.0681	0.80	[0.10; 6.49]
Yuan 2011	1.25	0.6396	3.48	[0.99; 12.19]
Worrall 2009	0.80	0.6024	2.22	[0.68; 7.23]
Jeong 2011	1.64	0.5305	5.14	[1.82; 14.54]
Trenk 2008	−0.53	0.4993	0.59	[0.22; 1.57]
Campo 2011	0.90	0.4170	2.46	[1.09; 5.57]
Oh 2011	0.95	0.3768	2.58	[1.23; 5.40]
Collet 2009	1.24	0.3731	3.47	[1.67; 7.21]
Giusti 2009	0.82	0.3625	2.28	[1.12; 4.64]
Sawada 2010	0.49	0.3568	1.63	[0.81; 3.28]
Malek 2010	0.60	0.3564	1.83	[0.91; 3.68]
Shuldiner 2009	0.74	0.3363	2.09	[1.08; 4.04]
Harmsze 2011	0.34	0.2533	1.40	[0.85; 2.30]
Komarov 2011	0.21	0.2325	1.23	[0.78; 1.94]
Tello-Montoliu 2011	0.16	0.2054	1.17	[0.78; 1.75]
Pare 2010	0.06	0.1906	1.06	[0.73; 1.54]
Mega 2009	0.40	0.1751	1.49	[1.06; 2.10]
Anderson 2009	0.33	0.1675	1.39	[1.00; 1.93]
Bhatt 2009	0.25	0.1614	1.29	[0.94; 1.77]
Sibbing 2009	0.13	0.1600	1.14	[0.83; 1.56]
Pare 2010	−0.17	0.1558	0.84	[0.62; 1.14]
Tiroch 2010	−0.12	0.1308	0.89	[0.69; 1.15]
Simon 2009	−0.15	0.1256	0.86	[0.67; 1.10]
Wallentin 2010	0.13	0.0938	1.14	[0.95; 1.37]
Fixed effect			**1.18**	**[1.09; 1.28]**
Random effects			**1.34**	**[1.15; 1.56]**

Heterogeneity: I-squared=60.2%, tau-squared=0.0667, p<0.0001

0.1 0.5 1 2 10
Higher risk for A: *1 or *17   Higher risk for B: *2 or *8
Risk ratio and 95% CI

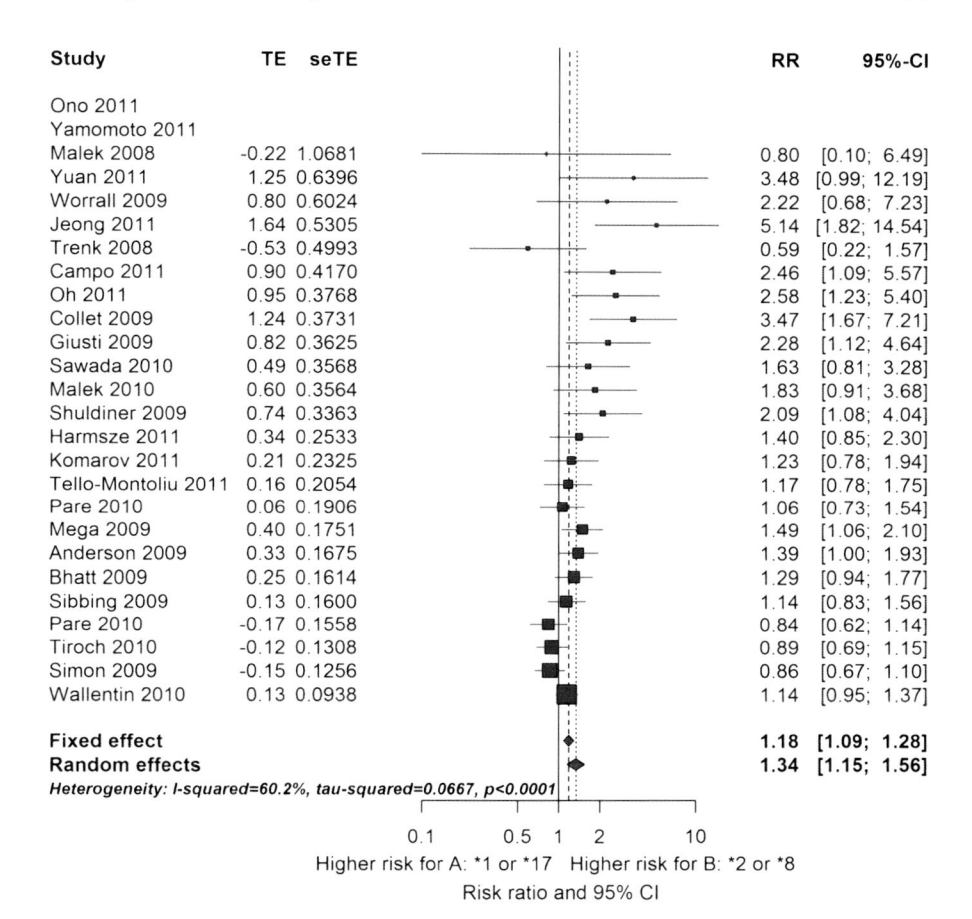

Fig. 15.1 Forest plot of the clopidogrel data, showing risk ratio of cardiovascular events comparing individuals having one or more copies of any CYP2C19 genetic variant associated with reduced enzyme function (i.e., *2, *3, *4, *5, *6, *7, *8) with individuals having none of these alleles (i.e., *1/*1) or having one or more *17 gain-of-function alleles; (higher risk ratio corresponds to higher chance of cardiovascular events with *2–*8). Some counts were not reported (NR). Treatment effect (TE) is listed on the log scale with its standard error (seTE)

scale, the log scale, with the x-axis labels presented on the original scale (Anzures-Cabrera and Higgins 2010).

We illustrate the forest plot in Fig. 15.1, showing a meta-analysis of the risk ratio (RR) of cardiovascular events (all-cause mortality, fatal and nonfatal CHD or stroke, stent thrombosis, target vessel revascularization, and hospitalization for ACS) in trials of clopidogrel (an anti-platelet drug), published by Holmes et al. (2011). The analysis used the standard inverse-variance method (Birge 1932), combining the reported log risk ratios across the 26 selected studies (though 2 of these had no estimate available). The comparison was made on treated subjects only, categorized by genotype associated with reduced enzyme function.

In this example, the individual studies are ordered according to the weight that they have in the meta-analysis, but any other sensible ordering (such as date of study) can be used. At the top of the plot are located the 2 trials for which no estimates were available because there were no events in one of the groups. Other methods of MA, such as logistic regression, include contributions from such studies (though not from studies with no events at all). Confidence intervals were truncated for studies extending beyond the range of 0.2–5 on the risk-ratio scale: these are studies with an effect estimated with low precision. Although the x-axis has a log scale, its labels indicate values on the untransformed scale as this is more readily understood, as mentioned earlier. The forest plot also allows the comparison of different methods in the meta-analysis; here, the results of both fixed-effects and random-effects meta-analysis (using the DerSimonian and Laird 1986 method) are included at the bottom of the graph. The difference between these methods indicates how much heterogeneity there is between the studies.

We drew this graph using the Meta package in R, and also produced versions using GenStat™ and SAS™.

15.2.2 Subgroup Analyses

The results of subgroup analyses are often presented in a forest plot. This is relevant when studies are divided into subsets according to different characteristics, for example types of participants, follow-up duration or treatment. Subgroup analyses are undertaken to investigate heterogeneity or to explore specific questions (Higgins & Green 2008). For each subgroup a meta-analysis is undertaken and the results are presented at the bottom of the group. However, it is the comparison of effects between subgroups that is most important, and the graph allows this to be done visually.

Kirsch et al. (2008) performed a meta-analysis of selective serotonin reuptake inhibitors (SSRI) to establish a relationship between baseline severity of depression and efficacy of antidepressants. The SSRI data was obtained from the FDA; it contained information from clinical trials for efficacy conducted for marketing approval of antidepressants approved between 1987 and 1999.

The aim of the published meta-analysis was to establish a relationship between baseline severity of depression and efficacy of antidepressants. The primary outcome measured in all the trials was a score calculated using the Hamilton Rating Scale for Depression (HRSD). Mean change from baseline was calculated by subtracting the mean baseline HRSD scores from the mean score after treatment. We performed a subgroup analysis by splitting the studies according to the type of SSRI used: fluoxetine, nefazodone, paroxetine, or venlafaxine (Fig. 15.2). The effect measure of the meta-analysis is mean difference (MD). The effect of each drug group is represented by individual diamonds. These diamonds have the same interpretation as the diamond used to represent the result of the overall meta-analysis, i.e., each diamond is centered in the subgroup effect and the width of the diamond

15 Graphics for Meta-Analysis

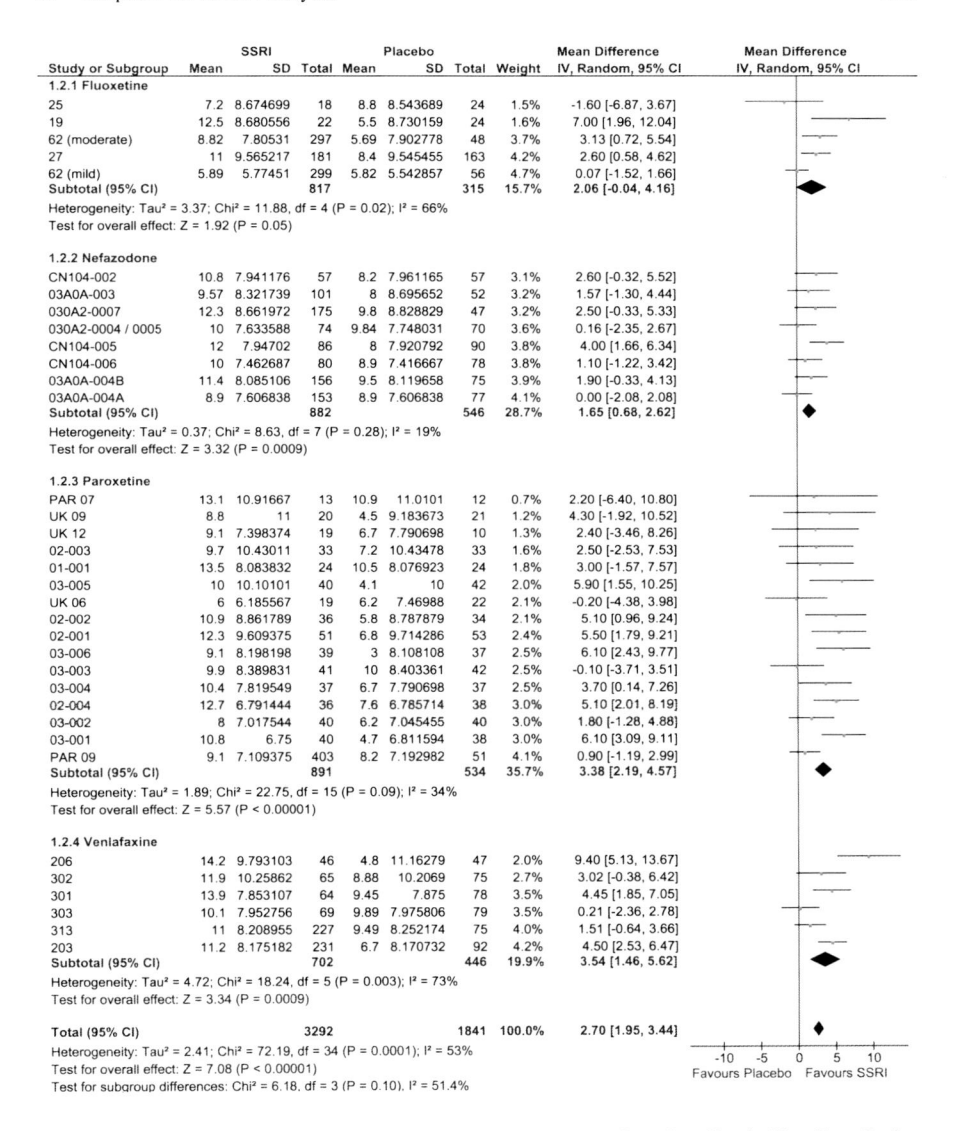

Study or Subgroup	SSRI Mean	SSRI SD	SSRI Total	Placebo Mean	Placebo SD	Placebo Total	Weight	Mean Difference IV, Random, 95% CI
1.2.1 Fluoxetine								
25	7.2	8.674699	18	8.8	8.543689	24	1.5%	-1.60 [-6.87, 3.67]
19	12.5	8.680556	22	5.5	8.730159	24	1.6%	7.00 [1.96, 12.04]
62 (moderate)	8.82	7.80531	297	5.69	7.902778	48	3.7%	3.13 [0.72, 5.54]
27	11	9.565217	181	8.4	9.545455	163	4.2%	2.60 [0.58, 4.62]
62 (mild)	5.89	5.77451	299	5.82	5.542857	56	4.7%	0.07 [-1.52, 1.66]
Subtotal (95% CI)			817			315	15.7%	2.06 [-0.04, 4.16]
Heterogeneity: Tau² = 3.37; Chi² = 11.88, df = 4 (P = 0.02); I² = 66%								
Test for overall effect: Z = 1.92 (P = 0.05)								
1.2.2 Nefazodone								
CN104-002	10.8	7.941176	57	8.2	7.961165	57	3.1%	2.60 [-0.32, 5.52]
03A0A-003	9.57	8.321739	101	8	8.695652	52	3.2%	1.57 [-1.30, 4.44]
030A2-0007	12.3	8.661972	175	9.8	8.828829	47	3.2%	2.50 [-0.33, 5.33]
030A2-0004 / 0005	10	7.633588	74	9.84	7.748031	70	3.6%	0.16 [-2.35, 2.67]
CN104-005	12	7.94702	86	8	7.920792	90	3.8%	4.00 [1.66, 6.34]
CN104-006	10	7.462687	80	8.9	7.416667	78	3.8%	1.10 [-1.22, 3.42]
03A0A-004B	11.4	8.085106	156	9.5	8.119658	75	3.9%	1.90 [-0.33, 4.13]
03A0A-004A	8.9	7.606838	153	8.9	7.606838	77	4.1%	0.00 [-2.08, 2.08]
Subtotal (95% CI)			882			546	28.7%	1.65 [0.68, 2.62]
Heterogeneity: Tau² = 0.37; Chi² = 8.63, df = 7 (P = 0.28); I² = 19%								
Test for overall effect: Z = 3.32 (P = 0.0009)								
1.2.3 Paroxetine								
PAR 07	13.1	10.91667	13	10.9	11.0101	12	0.7%	2.20 [-6.40, 10.80]
UK 09	8.8	11	20	4.5	9.183673	21	1.2%	4.30 [-1.92, 10.52]
UK 12	9.1	7.398374	19	6.7	7.790698	10	1.3%	2.40 [-3.46, 8.26]
02-003	9.7	10.43011	33	7.2	10.43478	33	1.6%	2.50 [-2.53, 7.53]
01-001	13.5	8.083832	24	10.5	8.076923	24	1.8%	3.00 [-1.57, 7.57]
03-005	10	10.10101	40	4.1	10	42	2.0%	5.90 [1.55, 10.25]
UK 06	6	6.185567	19	6.2	7.46988	22	2.1%	-0.20 [-4.38, 3.98]
02-002	10.9	8.861789	36	5.8	8.787879	34	2.1%	5.10 [0.96, 9.24]
02-001	12.3	9.609375	51	6.8	9.714286	53	2.4%	5.50 [1.79, 9.21]
03-006	9.1	8.198198	39	3	8.108108	37	2.5%	6.10 [2.43, 9.77]
03-003	9.9	8.389831	41	10	8.403361	42	2.5%	-0.10 [-3.71, 3.51]
03-004	10.4	7.819549	37	6.7	7.790698	37	2.5%	3.70 [0.14, 7.26]
02-004	12.7	6.791444	36	7.6	6.785714	38	3.0%	5.10 [2.01, 8.19]
03-002	8	7.017544	40	6.2	7.045455	40	3.0%	1.80 [-1.28, 4.88]
03-001	10.8	6.75	40	4.7	6.811594	38	3.0%	6.10 [3.09, 9.11]
PAR 09	9.1	7.109375	403	8.2	7.192982	51	4.1%	0.90 [-1.19, 2.99]
Subtotal (95% CI)			891			534	35.7%	3.38 [2.19, 4.57]
Heterogeneity: Tau² = 1.89; Chi² = 22.75, df = 15 (P = 0.09); I² = 34%								
Test for overall effect: Z = 5.57 (P < 0.00001)								
1.2.4 Venlafaxine								
206	14.2	9.793103	46	4.8	11.16279	47	2.0%	9.40 [5.13, 13.67]
302	11.9	10.25862	65	8.88	10.2069	75	2.7%	3.02 [-0.38, 6.42]
301	13.9	7.853107	64	9.45	7.875	78	3.5%	4.45 [1.85, 7.05]
303	10.1	7.952756	69	9.89	7.975806	79	3.5%	0.21 [-2.36, 2.78]
313	11	8.208955	227	9.49	8.252174	75	4.0%	1.51 [-0.64, 3.66]
203	11.2	8.175182	231	6.7	8.170732	92	4.2%	4.50 [2.53, 6.47]
Subtotal (95% CI)			702			446	19.9%	3.54 [1.46, 5.62]
Heterogeneity: Tau² = 4.72; Chi² = 18.24, df = 5 (P = 0.003); I² = 73%								
Test for overall effect: Z = 3.34 (P = 0.0009)								
Total (95% CI)			3292			1841	100.0%	2.70 [1.95, 3.44]
Heterogeneity: Tau² = 2.41; Chi² = 72.19, df = 34 (P = 0.0001); I² = 53%								
Test for overall effect: Z = 7.08 (P < 0.00001)								
Test for subgroup differences: Chi² = 6.18, df = 3 (P = 0.10), I² = 51.4%								

-10 -5 0 5 10
Favours Placebo   Favours SSRI

Fig. 15.2 Subgroup analysis of the SSRI data, showing change from baseline in Hamilton Rating Scale for Depression (change expressed as amount of improvement, i.e., reduction in HRSD score)

represents the 95% confidence interval. The plot suggests that there may be differences in effect between drug types: the effect of fluoxetine and nefazodone on change from baseline is lower than the effect observed in paroxetine and ventafaxine. These treatment differences by drug type may explain the heterogeneity observed in the overall meta-analysis ($I^2 = 52.9\%$). However, the formal between-subgroup test is not statistically significant (P = 0.10), so the evidence for

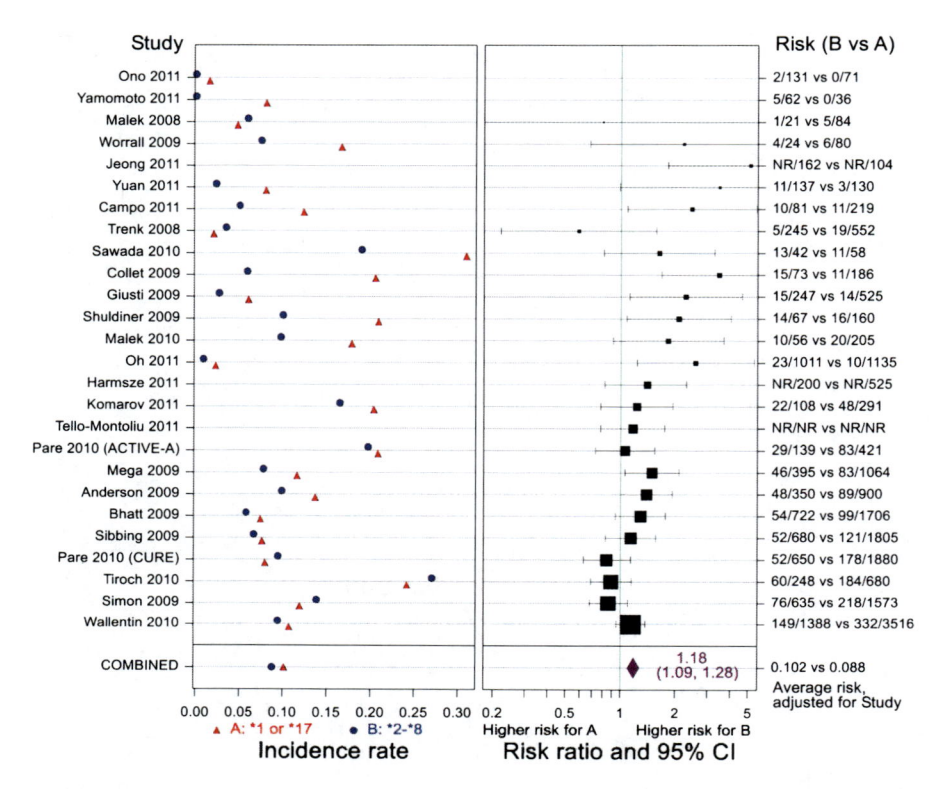

Fig. 15.3 Joint dotplot of the clopidogrel data, showing risks of cardiovascular events in each study alongside risk ratios, and adjusted average risks alongside the combined risk ratio

between-subgroup differences is not very strong and the observed differences should be interpreted with great caution.

We drew this graph using RevMan, and also produced a version using SAS™.

15.2.3 Joint Dotplot

One major problem with the traditional forest plot for binary outcomes is that there is no interpretation of the results on the natural scale of risk. Figure 15.3 shows the forest plot for the clopidogrel data including an additional panel displaying the raw risks in each study graphically, using the idea in Fig. 9 of Amit et al. (2008). The combined fixed-effects estimate of the risk ratio is accompanied by adjusted average risks (across the population of patients in these trials, as explained in Lane 2011), though these exclude the trials with unrecorded counts. This puts into context the headline risk ratio of 1.18, typically reported as an "18% increase in risk." This is, of course, an 18% relative increase, and the absolute increase is actually from 0.088 to 0.102, i.e., an extra 14 CV events per 1,000 patients.

We drew this graph using GenStat™, and did not attempt it with other systems.

15.3 Funnel Plots

A second type of display commonly used in meta-analysis is the funnel plot. This is a straightforward scatterplot of some measure of the precision of each study in the analysis against the estimated effect from the study. The main use of the plot is in the assessment of publication bias. Typically, in a collection of studies affected by publication bias, the pattern of scatter indicates a deficiency of studies with low precision and nonsignificant results, a situation which arises with many collections of published (rather than in-house) trials because of the tendency not to publish the results of trials showing no significance. Several alternative measures of precision have been suggested, such as the sample size, the SE of the estimated effect, 1/SE or 1/variance. Whitehead (2002) used the sample size, whereas Sutton et al. (2000) used 1/SE and Borenstein et al. (2009) used SE. Sterne and Egger (2001) investigated the alternatives and recommend the use of SE; they mention in particular that use of sample size precludes the possibility of adding a funnel constructed from expected confidence limits.

15.3.1 Standard Funnel Plot

Figure 15.4 shows the version with SE plotted against estimate for the SSRI dataset. Note that it is conventional to reverse the y-axis so that the "spout" of the funnel is at the top. The funnel is simply formed from the pointwise 95% confidence limits that would be associated with estimates from hypothetical trials with a range of standard errors, calculated under the null hypothesis of a fixed-effect meta-analysis, i.e., with a common treatment effect across all the trials.

The vertical reference line here is drawn at the position of the combined estimate from a fixed-effect meta-analysis. An alternative choice is to position it corresponding to the null hypothesis of no treatment effect: if there really is any publication bias, then the combined estimate from the collected trials is not reliable. This example of the plot shows no evidence of publication bias. There is one study with a particularly high standard error of the treatment effect, which happens also to have an estimate almost equal to the combined estimate. The other studies show a spread of estimates that appears almost symmetrical about the combined estimate, with 4 estimates outside the funnel and 2 more lying on it. Out of 35 estimates, we would expect to see about 2 outside the funnel, so there is more heterogeneity than expected under the fixed-effect model. For further details of the interpretation of funnel plots, see Sterne et al. (2011).

We drew this graph using SAS™, and also produced versions using GenStat™ and R.

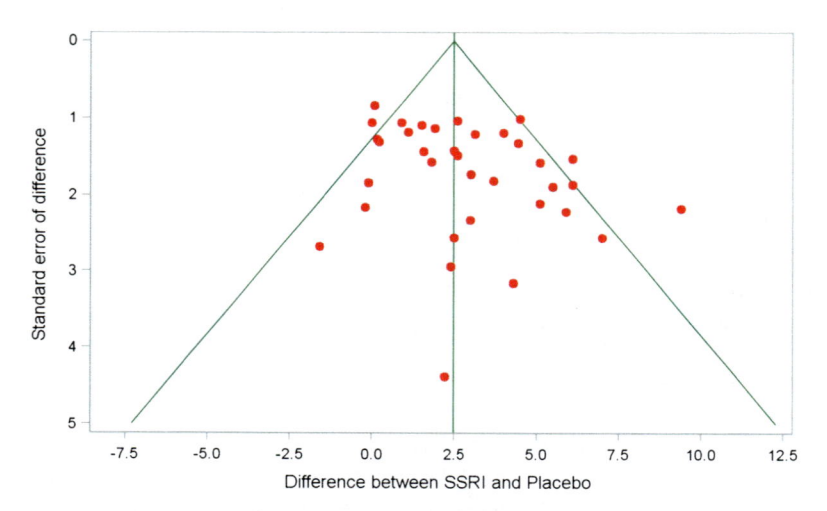

Fig. 15.4 Standard funnel plot of the SSRI data, showing an even distribution of estimates of difference around the combined estimate, regardless of the precision of the estimates

15.3.2 Funnel Plot with Contours, Plotting Reciprocal of Standard Error

An enhancement to the funnel plot was suggested by Peters et al. (2008) to help to differentiate asymmetry due to publication bias from that due to other factors. The enhancement consists of adding contours, at chosen significance levels. The idea of the plot is that asymmetry involving areas of nonsignificance is likely to be associated with publication bias, whereas asymmetry involving areas of significance are more likely to be associated with other causes, such as heterogeneity. A detailed report on the interpretation of funnel plot asymmetry has recently been published by Sterne et al. (2011).

Figure 15.5 shows an enhanced plot of the SSRI data, using significance levels <0.01, <0.05 and <0.1. For variety, we use 1/SE on the y-axis (as in Peters et al.). Note that we have chosen to use lighter shading for less significant and darker for more significant regions, which seems more natural than the opposite scheme used by Peters et al. We also distinguish the different drugs used in the component trials of the meta-analysis: this can be a useful visual check of subgroups in an analysis.

We drew this graph using GenStat™, and also produced versions using R and SAS™. These enhanced funnel plots are provided by the Stata™ system, using the "confunnel" command.

15.3.3 Funnel Plot with Contours, Plotting Standard Error

Figure 15.6 shows a funnel plot of the clopidogrel data, showing a reference line this time corresponding to the null hypothesis of no effect rather than at the reported combined estimate (fixed-effects risk ratio=1.18).

15 Graphics for Meta-Analysis

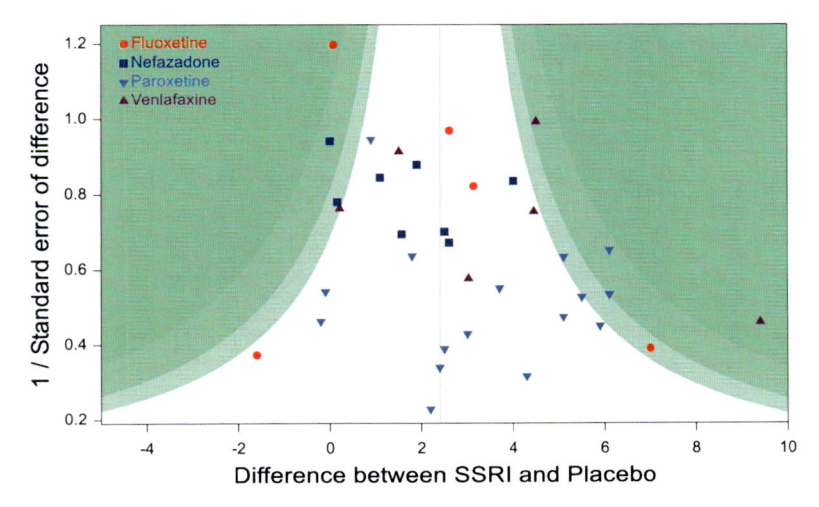

Fig. 15.5 Enhanced funnel plot of the SSRI data, grouping the points according to the drug tested; contours correspond to significance levels 0.01, 0.05, and 0.1

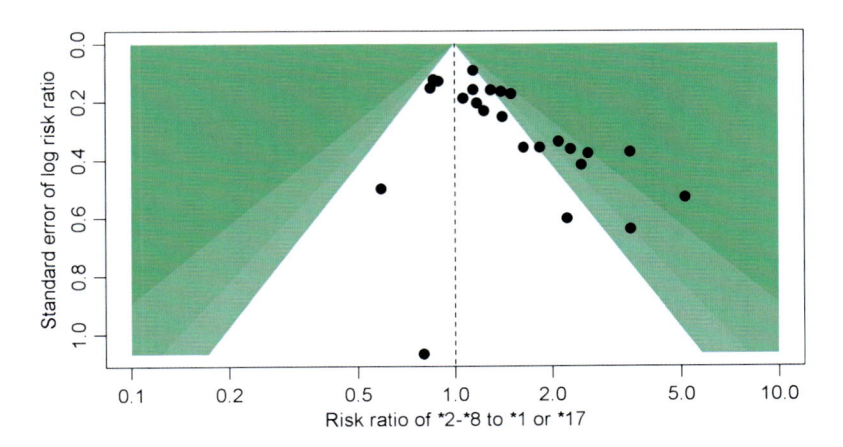

Fig. 15.6 Enhanced funnel plot of the clopidogrel data, showing the spread of the individual estimates about the null value of 1

There is pronounced asymmetry in this plot, with 10 trials with intermediate precision of estimates well above the null line and only one such trial below the line. Most of the 10 trials are in the region of statistical significance of the estimate compared to 1, which is an indication of publication bias according to Peters et al. This plot was shown as e-Figure 2 by Holmes et al., including the result of using a "trim-and-fill" method to attempt to rectify the bias.

We drew this graph using R, and also produced versions using GenStat™ and SAS™.

15.4 Meta-Regression

Meta-regression is a technique that can be used to investigate the relationship between the treatment effect and one or more study-level covariates, although it is only really useful if there are at least 10 studies in the meta-analysis (Borenstein et al. 2009, p. 188). It has been recommended that sources of heterogeneity should be investigated wherever possible (Thompson 1994), and meta-regression can be used to do this. So, for instance, a continuous measure, such as baseline blood pressure, might have been reported in each trial as a mean across all participants. It might be hypothesized that a treatment was most beneficial in patients with higher blood pressures. The treatment effect sizes can be regressed against such study-level mean covariate values, and the analysis weighted so that larger, more precise studies receive more weight that smaller, less precise ones.

Meta-regression involving a single covariate can be represented straightforwardly by a scatter plot of the treatment effect against the covariate, using circles with an area proportional to the weight that the study was given in the analysis (Fig. 15.7). Meta-regression can be based on a fixed-effects or random-effects model, as for meta-analysis: we used the random-effects model described by van Houwelingen et al. (2002). The regression line from the meta-regression analysis can be overlaid, and a confidence interval for the regression line can be added if available. It can also be useful to overlay a line showing the location of zero treatment effect; we have done this, and also included a line showing the NICE clinical significance criterion (see Figure 4 of Kirsch et al. 2008). The data from one Study, 62, were split into those from Mild and Moderate patients by Kirsch et al, and the point from the Mild patients is an outlier in the Baseline dimension, which we have therefore labeled.

This plot was drawn using GenStat™, and we also produced versions using R, SAS™ and the "Metareg" macro in Stata™.

Caution should be used in interpreting these meta-regression plots: using the mean value of a covariate simplifies the information within individual studies. It is possible that a relationship between the covariate of interest and the treatment effect exists within every individual study, but that this is hidden in the between-study means because of some other between-study confounding variable (Thompson and Higgins 2002), or simply because the between-study means do not vary much.

Meta-regression involving the concurrent analysis of 2 covariates can be illustrated using response surfaces (Lau et al. 1998). However, there are rarely enough trials in a meta-analysis to consider this (Thompson and Higgins 2002): you really need at least 10 trials per covariate. To reduce the risk of finding spurious associations, the covariates to be investigated should be kept to a minimum, carefully prespecified and their choice scientifically justified (Higgins and Thompson 2004). According to Thompson et al. (1997), plotting the treatment effect against the average baseline risk is not always recommended as it does not take into consideration the inherited correlation between the baseline and the treatment effect. There are methods available that take this correlation into account, but these are beyond the scope of this article.

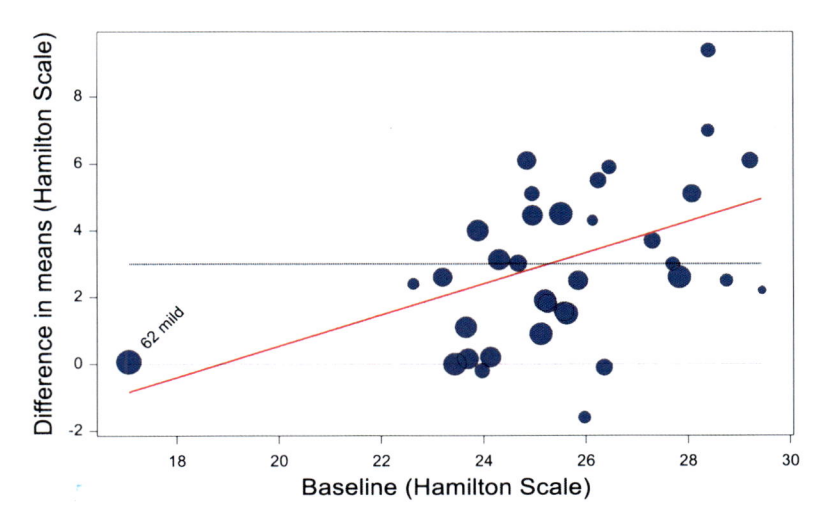

Fig. 15.7 Meta-regression of the SSRI data, showing the relationship between mean treatment differences and mean baseline

15.5 Network Meta-Analysis

Standard methods of meta-analysis deal with just 2 treatments, providing a direct, or head-to-head, comparison between them. When studies provide information on more than 2, meta-analysis can be extended to make comparisons between all the treatments. One special case of this is "indirect meta-analysis," where there are typically 2 sets of studies, one comparing one treatment against a control and another comparing a second treatment against the same control. The term "mixed treatment comparisons" is used when both direct and indirect comparisons can be made, and the 2 types are combined. More generally, however, the term "network meta-analysis" includes both of these special cases, as well as the analysis of any collection of studies involving a set of treatments, with potentially different subsets of the treatments compared in each study (see, for example, Jones et al. 2011).

The sets of studies in a network meta-analysis can be displayed in simple interval plots, showing the estimates and CIs for each study that contribute directly to the estimate of an individual treatment comparison. In a Bayesian framework predictive intervals can be included in the plot (Salanti et al. 2011). However, this does not include the indirect information, and the plots cannot be turned into meaningful forest plots with all contributions shown with an indication of their relative weight.

Network meta-analysis is commonly carried out using a Bayesian hierarchical approach. One of the main drivers for this is the ability to estimate probabilities associated with ordering the treatments that are being compared: for example, the probability of each treatment being the best among those considered. The probabilities are called rank probabilities; they add up to one for each treatment and each ranking. The rank probabilities can then be fed into health economic models to

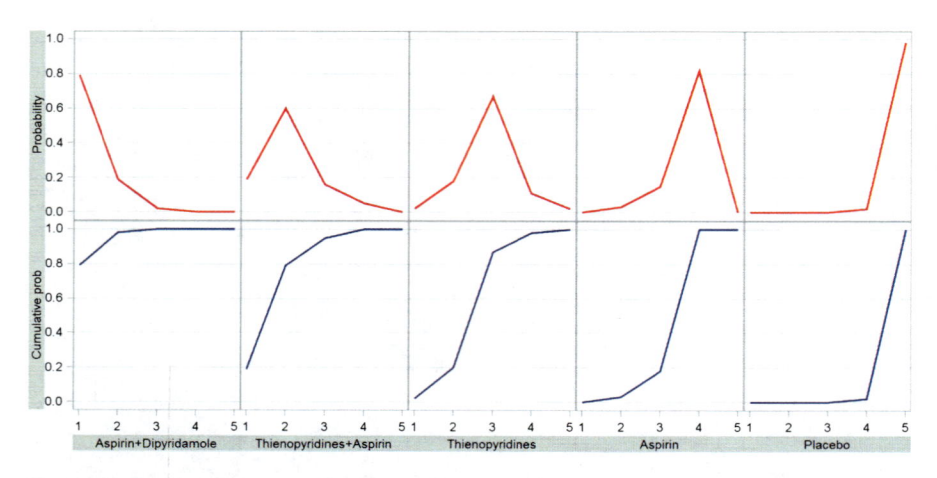

Fig. 15.8 Treatment rankings from a network meta-analysis of 5 anti-platelet regimens. The x-axis shows the 5 possible ranks and the y-axis the probability of achieving that rank

quantify the consequences of selecting treatments in practice. For our purposes, the probabilities provide useful information characterizing the network meta-analysis, which can be readily displayed in graphical form.

Figure 15.8 shows one potential display of this kind. It shows the results of a network meta-analysis of common anti-platelet regimens after transient ischaemic attack or stroke (Thijs et al. 2008), which was re-analyzed using Bayesian methods by Salanti et al. (2011) to illustrate graphical methods for network meta-analysis. Our figure is a composite of 2 of the figures in the latter paper, showing rankograms of the 5 paired regimens in one row of a trellis, and the corresponding cumulative ranking probabilities in the second row. The two plots order treatments from best to worst.

The first point in each panel in the top row shows the probability of one of the treatments being best in terms of the chosen outcome, derived from the posterior distributions of the model parameters in the Bayesian analysis Salanti et al. (2011). The second point shows the probability that this treatment is second best, and so on. By joining the points with a line, we can see a profile of each treatment, which gives a visual representation of how one treatment compares to the others in probabilistic terms. We have ordered the treatments by the position of the peak of the profile. The second row of panels display the cumulative probabilities, which provide the same information but in a different form so that we can also see the probabilities of each drug being in the top two treatments, for example, again with a profile allowing easy visual comparison across treatments. Based on these plots, the best available treatment is the combination of aspirin and dypiridamole.

This plot was drawn using SAS™, and we also produced a version using GenStat™ and R.

15.6 Conclusion

We have shown some of the main types of graph used in meta-analysis, and emphasized some of the important features of them. There are, of course, many more types, and we mention 2 of them here even though we have not included them, because they are frequently used. The L'Abbé plot was introduced by L'Abbé et al. (1987). It is a scatter plot of the contributing summaries from each trial, one arm against the other, usually representing points as bubbles to indicate the amount of information from each trial. It is most often used with binary outcomes, and can be a useful visual summary of the weight of evidence in favor of one treatment compared to another.

The radial plot is another standard way to present the information in a funnel plot, and the Galbraith plot is an enhanced version of it (Galbraith 1988). Starting from a funnel plot using 1/SE as the measure of "study size," the estimates are standardized (divided by their SEs) and the axes are interchanged. The enhanced form includes a circular (or sometimes vertical) axis to calibrate the slope (i.e., the combined estimate) with its confidence interval. This also adds to the visual appreciation of how the individual study estimates contribute.

References

Amit O, Heiberger R, Lane PW (2008) Graphical approaches to the analysis of safety data from clinical trials. Pharm Stat 7:20–35
Anzures-Cabrera J, Higgins JPT (2010) Graphical displays for meta-analysis: an overview with suggestions for practice. Research Synthesis Methods 1:66–80
Birge RT (1932) The calculation of errors by the method of least squares. Physics Reviews 40:207–227
Borenstein M, Hedges LV, Higgins JPT, Rothstein HR (2009) Introduction to Meta-analysis. Wiley, Chichester
DerSimonian R, Laird N (1986) Meta-analysis in clinical trials. Control Clin Trials 7:177–188
Galbraith RF (1988) A note on graphical presentation of estimated odds ratios from several trials. Stat Med 7:889–894
Higgins JPT, Green S (2008) Cochrane Handbook for Systematic Reviews of Interventions. Wiley, Chichester
Higgins JPT, Thompson SG (2004) Controlling the risk of spurious findings from meta-regression. Stat Med 23:1663–1682
Holmes MV, Perel P, Shah T, Hingorani AD, Casas JP (2011) CYP2C19 genotype, clopidogrel metabolism, platelet function, and cardiovascular events: a systematic review and meta-analysis. J Am Med Assoc 306(24):2704–2714
Jones B, Roger J, Lane PW, Lawton A, Fletcher C, Cappelleri JC, Tate H, Moneuse P (2011) Statistical approaches for conducting network meta-analysis in drug development. Pharm Stat 10:523–531
Kirsch I, Deacon BJ, Huedo-Medina TB, Scoboria A, Moore TJ, Johnson BT (2008) Initial severity and antidepressant benefits: a meta-analysis of data submitted to the food and drug administration. PLoS Med 5(2):e45
L'Abbé KA, Detsky AS, O'Rourke K (1987) Meta-analysis in clinical research. Ann Intern Med 107:224–233

Lane PW (2011) Meta-analysis of incidence of rare events. *Stat Methods Med Res.* (accepted for publication)

Lau J, Ioannidis JP, Schmid CH (1998) Summing up evidence: one answer is not always enough. Lancet 351:123–127

Peters JL, Sutton AJ, Jones DR, Abrams KR, Rushton L (2008) Contour-enhanced funnel plots help distinguish publication bias from other causes of asymmetry. J Clin Epidemiol 61:991–996

Salanti G, Ades AE, Ioannidis JPA (2011) Graphical methods and numerical summaries for presenting results from multi-treatment meta-analysis: an overview and tutorial. J Clin Epidemiol 64:163–171

Sterne JAC, Egger M (2001) Funnel plots for detecting bias in meta-analysis: guidelines on choice of axis. J Clin Epidemiol 54:1046–1055

Sterne JAC, Sutton AJ, Ioannidis JPA, Terrin N, Jones DR, Lau J, Carpenter J, Rücker G, Harbord R, Schmid CH, Tetzlaff J, Deeks JJ, Peters J, Macaskill P, Schwarzer G, Duval S, Altman DG, Moher D, Higgins JPT (2011) Recommendations for examining and interpreting funnel plot asymmetry in meta-analyses of randomised controlled trials. Br Med J 342. doi:10.1136/bmj.d4002

Sutton AJ, Abrams KR, Jones DR, Sheldon TA, Song F (2000) Methods for Meta-analysis in Medical Research. Wiley, Chichester

Thijs V, Lemmens R, Fieuws S (2008) Network meta-analysis: simultaneous meta-analysis of common antiplatelet regimens after transient ischaemic attack or stroke. Eur Heart J 29:1086–1092

Thompson SG (1994) Why sources of heterogeneity in meta-analysis should be investigated. Br Med J 309:1351–1355

Thompson SG, Higgins JPT (2002) How should meta-regression analyses be undertaken and interpreted? Stat Med 21(11):1559–1573

Thompson SG, Smith TC, Sharp SJ (1997) Investigating underlying risk as a source of heterogeneity in meta-analysis. Stat Med 16(23):2741–2758

van Houwelingen HC, Arends LR, Stijnen T (2002) Advanced methods in meta-analysis: multivariate approach and meta-regression. Stat Med 21:589–624

Whitehead A (2002) Meta-analysis of Controlled Clinical Trials. Wiley, Chichester

Chapter 16
Visualization of QT Data for Thorough QT Study Analysis and Review

Christoffer W. Tornøe

Abstract The focus of this chapter is on visualization of QT data for thorough QT (TQT) study analysis and review. The use of graphics is particularly important for QT data due to the high variability and to explore the adequacy of heart rate QT correction, baseline adjustments, choice of positive control to establish assay sensitivity, and the relationship between exposure and QT prolongation. A QT knowledge management system implemented in the R package "QT" standardizes and automates the QT data analyses, graphical representation, and reporting. This allows for easier communication of the results because the analyses are consistent and enables pooled data analysis across TQT studies to address drug development related questions.

16.1 Introduction

The objective of a thorough QT (TQT) study is to determine whether a drug has a pharmacological effect on cardiac repolarization as detected by a QT prolongation on the surface electrocardiogram (ECG) as described in the guideline (ICH E14). Refer to Chap. 18 for more information about ECGs and the QT interval.

The ΔQTc (time-matched baseline adjusted QTc) and $\Delta\Delta$QTc (time-matched change from placebo- and baseline-adjusted QTc) used for the central tendency and concentration–QTc analyses are calculated by

$$\Delta \text{QTc}_{\text{drug}}(t) = \text{QTc}_{\text{drug}}(t) - \text{QTc}_{\text{drug}}(\text{baseline})$$

C.W. Tornøe (✉)
Quantitative Clinical Pharmacology, Novo Nordisk A/S, 108-110 Vandtårnsvej,
Søborg 2860, Denmark
e-mail: cwto@novonordisk.com

$$\Delta\text{QTc}_{placebo}(t) = \text{QTc}_{placebo}(t) - \text{QTc}_{placebo}(\text{baseline})$$

$$\Delta\Delta\text{QTc}(t) = \Delta\text{QTc}_{drug}(t) - \Delta\text{QTc}_{placebo}(t),$$

where the baseline QTc is calculated as the mean of the pre-dose ECG measurements for each treatment.

A key component for assessing the QT prolongation potential of a drug is to create informative graphs that are tailored to address the key questions to be answered. For that purpose, the R package "QT" (requires SAS™) was developed based on the accumulated review experience from the FDA's interdisciplinary review team for QT (IRT-QT) (Tornoe et al. 2011). The purpose was to develop standardized graphics to increase the productivity, consistency, and quality-thereby ensuring faster and easier communication between the team of inter-disciplinary scientists.

While other papers focus on issues related to TQT study design, conduct, and analysis methods, the objective of this chapter is the visualization of QT data for TQT study analysis and review to make informed decisions through the following 10 steps:

1. Data integrity
2. QT correction method
3. Adequacy of sampling times
4. Baseline corrections
5. Assay sensitivity
6. Measure of central tendency
7. Assessing delay between drug and QT effects
8. Relationship between drug exposure and QT prolongation
9. Prediction of QT prolongation at different exposure levels
10. Benefit-risk assessment

16.2 Developing Standardized Graphics for QT Data

The high variability and low signal-to-noise ratio (trying to exclude a 10 ms QT prolongation) in QT measurements makes it difficult to spot trends and relationships based on individual data points. For this purpose, tailored quantile plots were developed.

The quantile plot is generated by binning the independent variable (e.g., RR or concentrations) into quantiles (bins with equal numbers of observations) and plotting the local median or midpoint of the observations in each of the independent variable bins against the corresponding local mean dependent variable (e.g., QT or $\Delta\Delta$QT and associated 90% confidence interval). This is the binning method implemented in the R package.

A potential issue arises when using the local means across the dependent variable quantiles. The different bins can potentially be imbalanced in the number of individuals. Taking the mean of all observations in each bin and calculating the standard deviation for that mean ignores that some of the observations are correlated since they arise from the same subject. This is most likely to be an issue when pooling data from different studies with different variability in PK. One way to account for this imbalance is to use precision weighted averages in each bin by

$$SD^2 = \frac{1}{\sum_{i=1}^{N} \frac{1}{\sigma^2 / n_i}}$$

$$\bar{M} = SD^2 \sum_{i=1}^{N} \frac{1}{\sigma^2 / n_i} \bar{M}_i,$$

where σ^2 is the variance of the measurements in the bin, N is the number of subjects in the bin, and n_i is the number of measurements for the ith subject in the bin. M_i is the mean of the ith subject's measurements that fall into the bin.

Another way to avoid the complex calculations of the mean of the dependent variable is to calculate it for each sampling time thereby ensuring that only one measurement per subject is present in each of the bins. The disadvantage, however, is that the bins are not distributed evenly over the range of measurements.

16.3 QT Knowledge Management System

The QT knowledge management system developed for automatic QT data analyses and reporting is implemented in the R package "QT" (http://www.cran.r-project.org) that performs the data manipulation and graphical presentation of results, executes the linear mixed-effects analyses in SAS, and generates an analysis report with key results and figures.

The data template for the R package is shown in Table 16.1.

The "QT" R package consists of 5 main functions (the R and SAS code can be found at http://qttool.googlecode.com):

- QTcorrections: This function performs the QT-RR analysis on off-drug treatment data and evaluates the ability of the different QT corrections to remove the heart rate effect on on-drug treatment data using PROC MIXED in SAS
- DataCheck: The DataCheck function visualizes key data to check the integrity of the analysis dataset
- MeanData: The MeanData function calculates and plots the mean profiles, and creates the dataset for the QTc-time and concentration–QTc analyses

Table 16.1 Data template for QT package

Column identifier	Description	Units	Type
subjid	Subject identifier		Character
treat	Treatment group		Character
period	Period		Numeric
day	Day relative to first dose of period	Days	Integer
time	Time relative to dose	h	Numeric
rr	RR interval	ms	Numeric
hr	Heart rate	bpm	Numeric
qt/qtc	QT/QTc interval	ms	Numeric
qt.bs/qtc.bs	QT/QTc baseline	ms	Numeric
qt.cfb/qtc.cfb	QTc/QTc change from baseline	ms	Numeric
conc	Parent drug concentration		Numeric
meta	Metabolite concentration		Numeric
moxi	Moxifloxacin concentration		Numeric
wt	Body weight	kg	Numeric
age	Age	years	Numeric
sex	Gender	M = male, F = female	Controlled term

- QTtime: The central tendency analysis of QTc versus time is performed in SAS using PROC MIXED
- QTconc: The concentration–QTc analysis is performed in SAS using PROC MIXED. Three models are estimated, i.e., (1) "with intercept," (2) "no intercept," and (3) "intercept fixed to zero with variability." The results are summarized in tables and graphs

16.4 Visualization of QT Data in Ten Steps for TQT Study Analysis and Review

In the following, the 10 visual steps for TQT study analysis and review are illustrated using simulated data included in the R package "QT." The simulated TQT study is a 4-way crossover design with placebo, moxifloxacin (active control), and therapeutic and supra-therapeutic doses of the drug candidate. The colors used for the different treatment arms are consistent in the following figures with black, orange, blue, and red representing placebo, moxifloxacin, therapeutic dose, and supra-therapeutic dose, respectively. Mean and 90% confidence interval are used throughout.

16.4.1 Data Integrity

Initially, the derived dataset containing QT-, RR-, and concentration–time profiles are inspected by visualizing the data to ensure the merging of data is performed correctly. In Fig. 16.1, the QT, RR, moxifloxacin, and drug X concentrations are plotted for the 4 treatments to assess whether there are outliers, unrealistic values, or unit differences in the measurements.

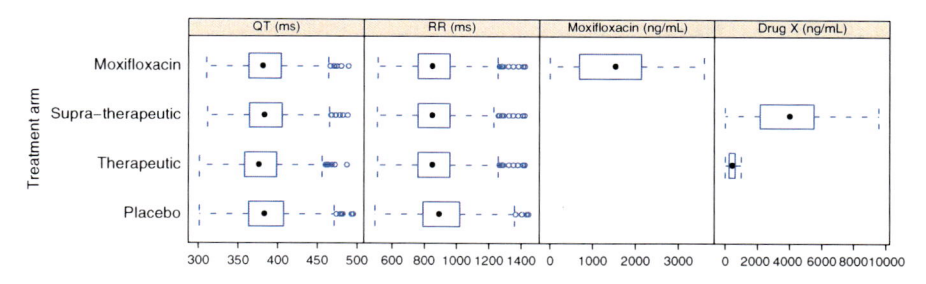

Fig. 16.1 Box plot of QT, RR, moxifloxacin, and drug X for each treatment

16.4.2 QT Correction Method

Before assessing the QT prolongation potential of a drug, the heart rate effect should be removed first since it is known to affect the QT interval. The relationship between QT interval length and heart rate (using the RR interval) is investigated using the available off-drug data only (Fig. 16.2 top). Given the high variability in QT measurements, it is difficult to assess the adequacy of the QT correction method when visualizing the individual data points and comparing it to the estimated regression line (Fig. 16.2 top). However, by using the previously described quantile plot a different intercept (at RR = 1,000 ms) and slope between males and females are clearly seen in Fig. 16.2 (bottom).

In Fig. 16.3, the individual relationship between QT, QTcB (Bazzett's), QTcF (Fridericia), and QTcI (Individual correction) and RR indicates that QTcB overcorrects for heart rate whereas QTcF and QTcI both seem to be appropriate correction methods.

The choice of the heart rate correction method to use for further analyses is further investigated by estimating a linear model using on-drug data to see whether the slope is significantly different from zero indicating whether there still is a relationship between QT and heart rate. The quantile plots in Fig. 16.4 show that the slopes for QTcF and QTcI are both significantly different from zero on a 0.05 α-level but QTcI appears to be the most appropriate QT correction methods.

Furthermore, to ensure that it is not only a global trend shown between the different QT correction methods (QTc) and RR, the average sum of squared individual slopes are calculated and compared to the QTcI having the lowest average (see Table 16.2).

Fig. 16.2 Placebo QT vs. RR relationship. (*Top*) Observed placebo QT-RR for males (*gray squares*) and females (*black circles*) together with the mean predictions (males = *blue*, females = *red*). (*Bottom*) Mean (90% confidence interval) predicted QT-RR curves. The *dots* represent the observed median RR quantiles and associated mean (90% confidence interval) QT. RR quantile ranges for males (*blue*) and females (*red*) are shown along the *x*-axis

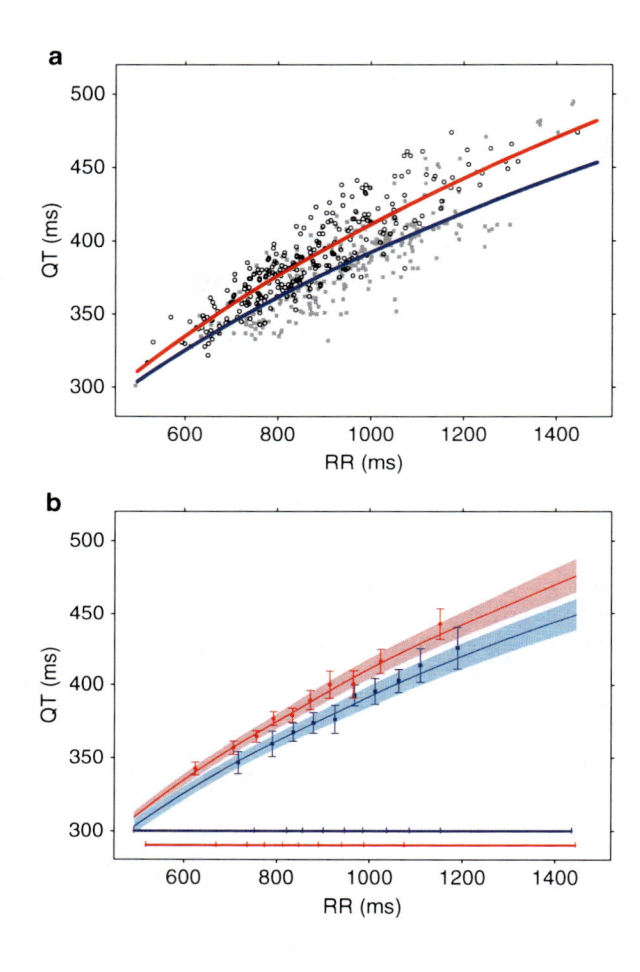

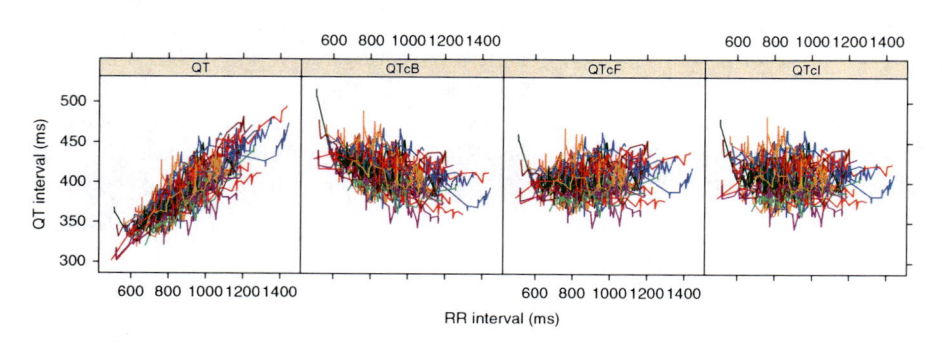

Fig. 16.3 Individual QT, QTcB, QTcF, and QTcI versus RR interval

Fig. 16.4 Mean (90% confidence interval) predicted QTcF and QTcI versus RR relationship (*solid line* and *shaded area*) for male (*blue*) and female (*red*). The *dots* represent the observed median RR quantiles and associated mean (90% confidence interval) QT. RR quantile ranges for males (*blue*) and females (*red*) are shown along the *x*-axis

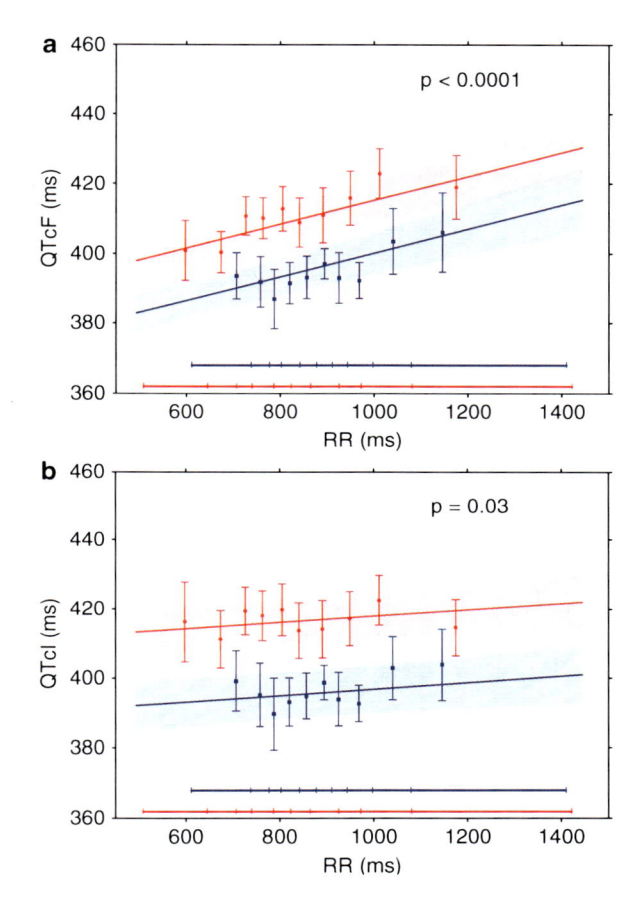

Table 16.2 Average sum of individual squared slopes for different QT-RR correction methods

α-QTcF (where QTcF is subset to α)	0.001816
α-QTcI (where QTcI is subset to α)	0.000448

$\alpha_x = 1{,}000/n\sum b_x(i)^2$, where $b_x(i)$ is the estimated slope for correction x

16.4.3 Adequacy of Sampling Times

The timing of ECG samples should be guided by the available information about the PK properties of the drug candidate and the QT effect should be characterized throughout the anticipated dosing interval (FDA 2005).

ECG and PK should be sampled frequently enough to ensure the peak drug concentration is captured and ECG recordings at time points around the maximum concentration time, t_{max} in case the peak effect on QT does not correspond to peak concentration. The sampling should be for at least 24 h post dose in case the cardiac

Fig. 16.5 Mean (90% confidence interval) concentration–time profiles following a therapeutic dose (*blue line*), supra-therapeutic dose (*red line*), and 400 mg moxifloxacin (*orange line*)

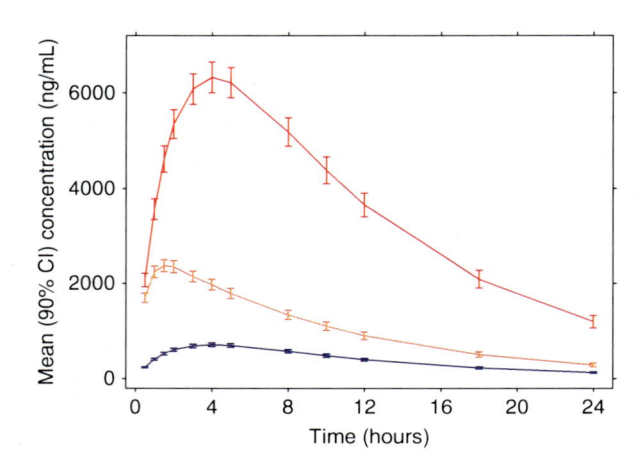

repolarisation effect is delayed. The PK concentration–time profiles for therapeutic, supra-therapeutic doses and the positive control moxifloxacin are shown in Fig. 16.5 with a t_{max} around 4 h for the test drug and around 2 h for moxifloxacin. The sampling is more frequent up until 5 h post dose to ensure the individual C_{max} (maximum concentration) is captured. The time-matched ECGs are sampled just prior to the PK sample to ensure that the venipuncture does not affect the ECG sample.

16.4.4 Baseline Correction

Baseline corrections are necessary in TQT studies due to the large inter-individual variability. For a crossover study design, baseline measurements before each period are used with the diurnal variability in QTc accounted for by each subject receiving all 4 treatments at exactly the same time points. In Fig. 16.6, the mean QTcI and ΔQTcI (change from baseline) are illustrated showing a clear separation of moxifloxacin and supra-therapeutic dose from placebo and the therapeutic dose. The diurnal variation in all treatment arms is clearly visible with peaks around 4 and 18 h post dose and nadirs around 12 and 24 h post dose. This variability can be a result of many factors including activity level, postural changes, circadian patterns, and food intake. Finally, it is noticed that all treatment arms return to baseline 24 h post dosing where the exposure of the drug is negligible (see Fig. 16.6).

16.4.5 Assay Sensitivity

The purpose of including a positive control is to ensure confidence in the ability of the TQT study to detect changes in the mean QT prolongation of around 5 ms

Fig. 16.6 Mean (90% confidence interval) (*top*) QTcI and (*bottom*) ΔQTcI (change from baseline) following placebo (*black line*), 400 mg moxifloxacin (*orange line*), therapeutic dose (*blue line*), and supra-therapeutic dose (*red line*)

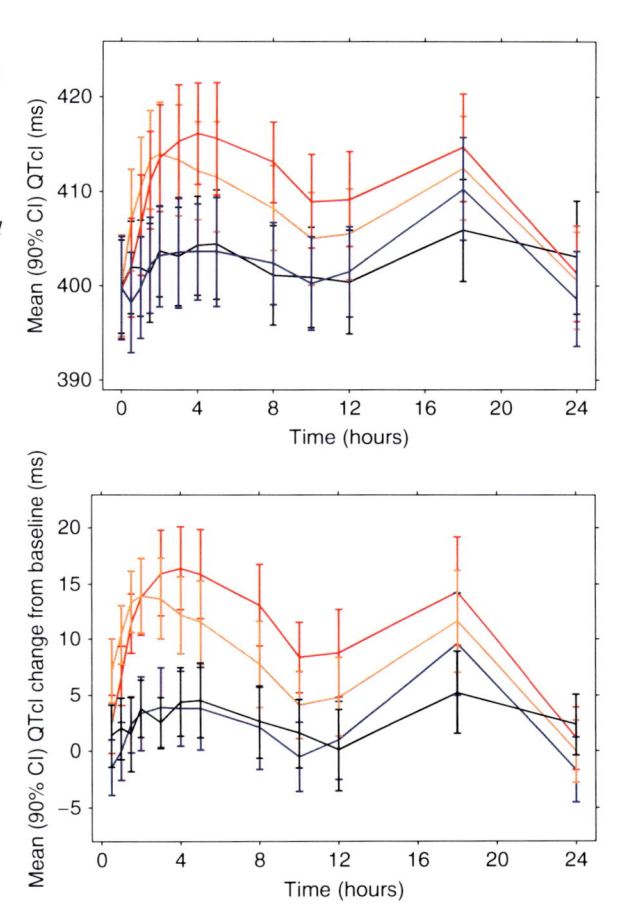

(FDA 2005). Moxifloxacin (a fluoroquinolone used for respiratory infections) is the most commonly used positive control in TQT studies because it reliably prolongs the QT interval with no significant effect on heart rate and is considered relatively benign.

Assay sensitivity is established if at least one time point excludes a 5 ms difference in the mean ΔΔQTc (baseline and placebo adjusted) with a one-sided 95% confidence interval thereby preserving at least 50% of the previously reported ΔΔQTc effect of 10–14 ms following 400 mg moxifloxacin (Bloomfield et al. 2008). The shape of the ΔΔQTcI-time profile is shown in Fig. 16.7 with peak effect excluding 5 ms around 2–4 h post dose and return to baseline at 24 h post dose as expected (Florian et al. 2011). Similarly, the slope of the moxifloxacin concentration–QTc relationship shown in Fig. 16.7 can be used to confirm assay sensitivity in cases with reduced moxifloxacin exposure, e.g., due to over-encapsulation (Florian et al. 2011).

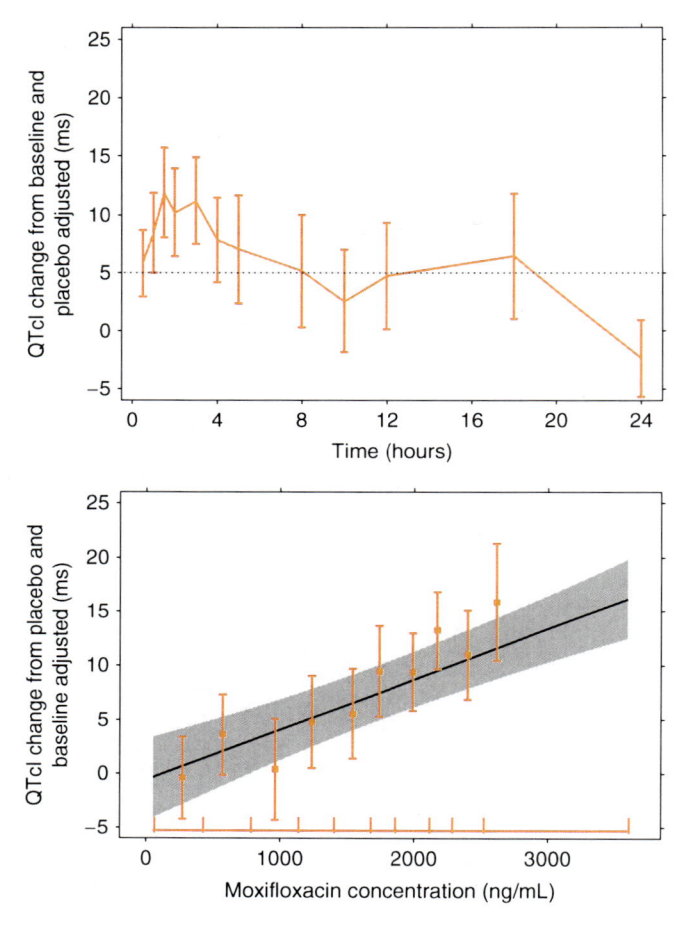

Fig. 16.7 (*Top*) Mean (90% confidence interval) ΔΔQTcI–time profile and (*bottom*) mean (90% confidence interval) moxifloxacin concentration–ΔΔQTcI observations (*dots*), and the estimated relationship (*solid black line* and *shaded gray area*) following 400 mg moxifloxacin with the moxifloxacin concentration quantile range is shown along the x-axis

16.4.6 Measure of Central Tendency

The FDA E14 guidance (FDA 2005) sets the threshold of regulatory concern for the drug candidate around 5 ms evidenced by an upper bound of the one-sided 95% confidence interval around the mean ΔΔQTc (time-matched difference in baseline and placebo adjusted QTc) excluding 10 ms. The TQT study is considered to be negative if the mean excludes 10 ms at all time points. In Fig. 16.8, the therapeutic treatment arm clearly excludes 10 ms at all time points whereas the upper 95% confidence interval includes 10 ms between 1.5 and 18 h post dose for the supra-therapeutic treatment arm.

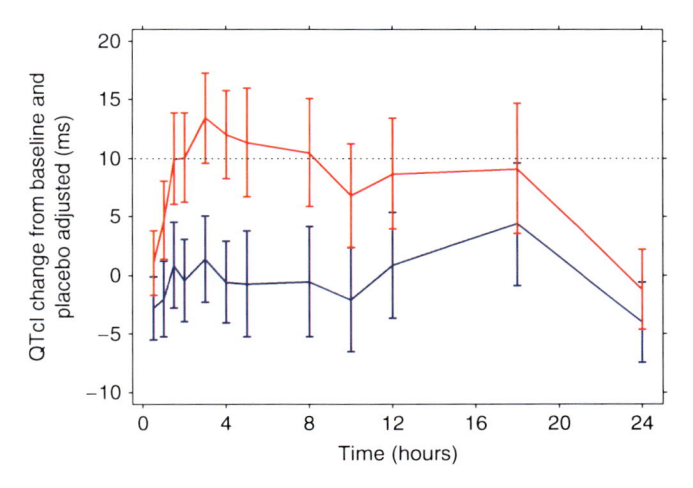

Fig. 16.8 Mean (90% confidence interval) ΔΔQTcI (baseline and placebo adjusted) following a therapeutic (*blue line*) and supra-therapeutic (*red line*) dose

16.4.7 Assessing Delay Between Drug and QT Effects

Before investigating the potential relationship between drug exposure and QT effect, it is important to assess whether there is a temporal delay between drug concentrations and ΔΔQTc. This could potentially indicate that an active metabolite is causing the QT prolongation, and it often requires more advanced modeling to account for the delay if the metabolite is not measured directly. Due to the inter-individual variability in PK and QT, the individual QTc versus drug concentration is expected to show very different slopes (positive and negative) and potentially also hysteresis (the effect on QT lags the change in concentration). Instead, the mean concentrations at each sampling time are plotted against the corresponding mean (90% confidence interval) ΔΔQTcI and connected in chronological order in Fig. 16.9. The slope for the therapeutic and supra-therapeutic treatment arms appear similar and there does not appear to be any clear sign of hysteresis except for the last 2 sampling times at 18 and 24 h post dose. A linear exposure-response model therefore seems adequate to assess the relationship between drug concentration and ΔΔQTcI.

16.4.8 Relationship Between Drug Exposure and QT Prolongation

An adequate TQT study should ensure that the dose- and exposure-response relationship for QT prolongation has been characterized at concentrations covering the worst case clinical exposure scenario in order to support regulatory review (Garnett et al. 2008).

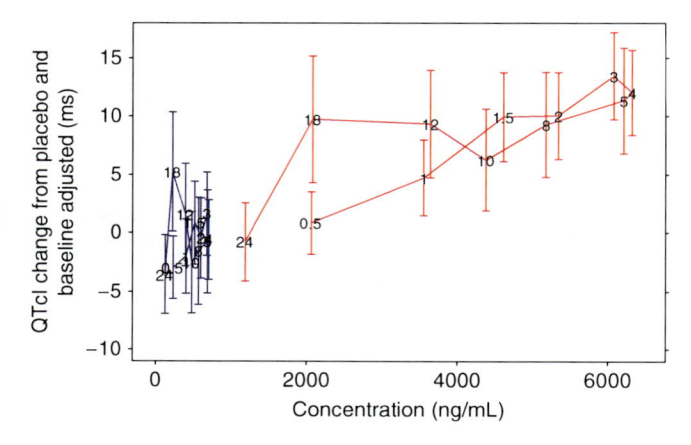

Fig. 16.9 Mean (90% confidence interval) ΔΔQTcI following a therapeutic (*blue line*) and supra-therapeutic dose (*red line*) connected in chronological order with the numbers representing the sampling time

It is important to understand the relationship between drug exposure and QT prolongation when (1) the primary E14 analysis is positive, (2) it is not possible to test doses high enough to cover the worst case clinical exposure scenario, or (3) when the TQT study is positive based on the E14 analysis but there is lack of dose- and exposure-response (Garnett et al. 2008, 2011).

The relationship between drug concentration and ΔΔQTcI is shown in Fig. 16.10 with a clear linear dependency, and the relationship appears similar for therapeutic and supra-therapeutic exposures.

The individually estimated linear mixed-effects regression lines are furthermore shown in Fig. 16.11 (top) to assess whether the relationship between concentration and ΔΔQTcI is more or less pronounced compared to the population mean predictions. All subjects have a positive slope very similar to the estimated population slope while there is some variability in the intercept. The residuals are shown in Fig. 16.11 (bottom) to see whether there are outliers or subjects that are poorly fitted by the linear mixed-effects model.

16.4.9 Prediction of QT Prolongation at Different Exposure Levels

The ΔΔQTc predictions at the geometric mean peak concentration following therapeutic and supra-therapeutic doses are assessed and compared to the primary E14 analysis. Figure 16.11 (top) shows the mean (90% confidence interval) ΔΔQTcI of 0.00 (−2.50; 2.51) and 12.5 (9.49; 15.5) ms at geometric mean peak concentrations of 691 and 6,230 ng/mL, respectively. These predictions based on the concentration–ΔΔQTcI relationship are consistent with the primary E14 analysis mean

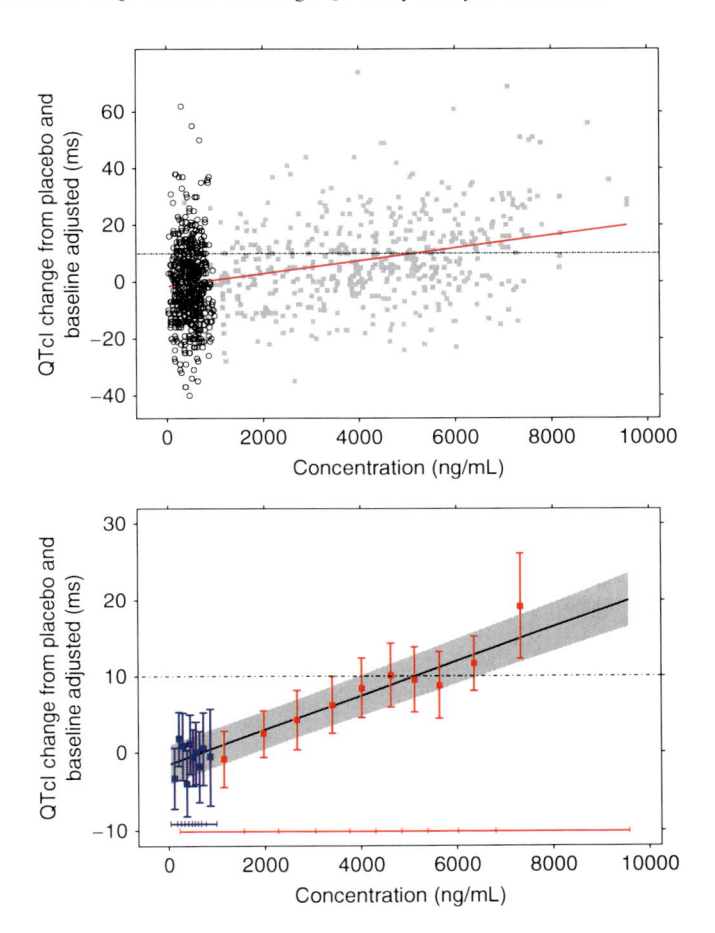

Fig. 16.10 ΔΔQTcI versus drug concentration. (*Top*) Observed data with population mean predictions (*solid red line*) and (*bottom*) observed concentration quantile–ΔΔQTcI plot with population mean and 90% confidence interval (*solid black line* with *shaded gray area*). The drug concentration quantiles following therapeutic (*circles*) and supra-therapeutic (*squares*) doses are shown as *horizontal bars* along the x-axis and the 10 ms threshold is shown as a *dotted line*

(90% confidence interval) estimates of 4.39 (−0.86; 9.65) and 13.4 (9.62; 17.3) ms at 18 h and 3 h post therapeutic and supra-therapeutic doses, respectively.

16.4.10 Benefit-Risk Assessment

The concentration–QTc relationship is used to perform the benefit-risk assessment for drugs that prolong the QT interval. The first question to ask is whether the supra-therapeutic dose covers the highest expected clinical exposure scenario. This is done by adding the fold-change in C_{max} and AUC for identified intrinsic and extrinsic

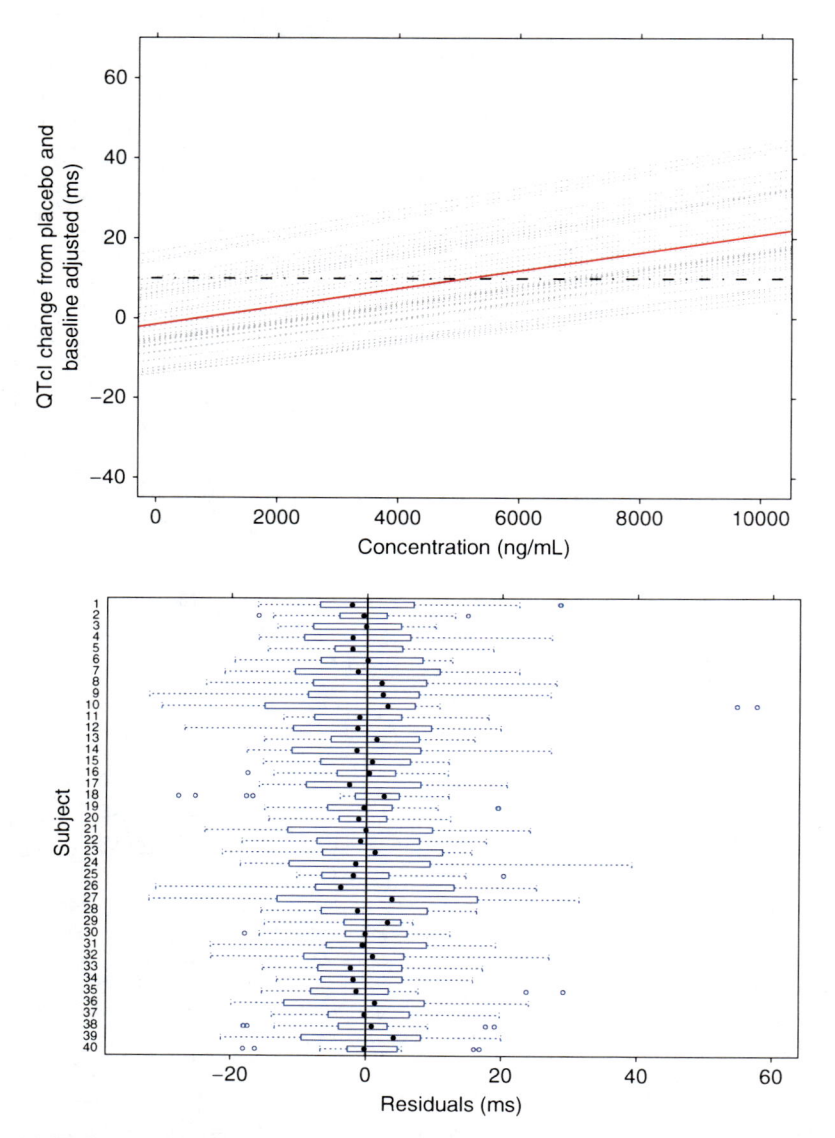

Fig. 16.11 Comparison of individual and population prediction linear mixed-effects regression lines (*top*). Box-plots of residuals for each subject (*bottom*)

factors (e.g., gender, race, organ impairment, drug–drug interactions, or food effects). This way the QT risk in subpopulations can be assessed and appropriate dose adjustments and ECG monitoring can be derived in order to write informative drug labels.

For the example shown in Fig. 16.12 (bottom), the geometric mean peak supra-therapeutic exposure shown in red is 9-fold higher than the therapeutic exposure

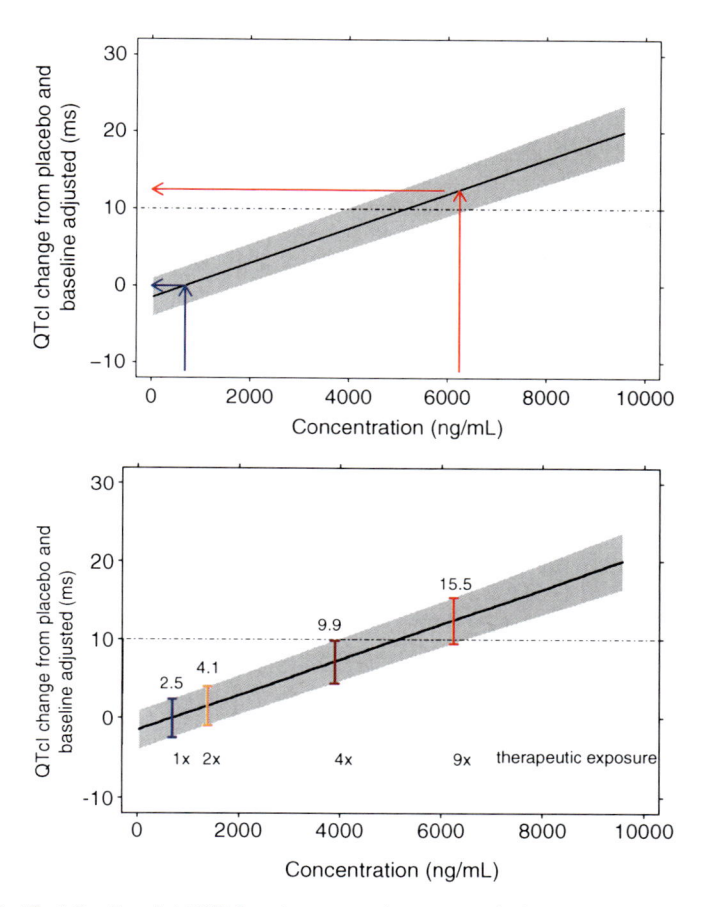

Fig. 16.12 (*Top*) Predicted ΔΔQTcI at the geometric mean peak drug concentrations following therapeutic (*blue*) and supra-therapeutic (*red*) doses. (*Bottom*) Predicted ΔΔQTcI at different multiples of therapeutic exposures

shown in blue and clearly prolongs the mean QT interval by more than 10 ms. However, if the worst case clinical exposure scenario is between 2- and 4-fold, there does not appear a high risk of QT prolongation.

16.5 Concluding Remarks

The purpose of creating a QT knowledge management system was to automate QT data analyses and reporting for consistent and timely review of TQT studies. Furthermore, the developed graphics and reporting standards allow for easier communication (internally and externally) of results.

An important aspect of QT data analyses is to explore the adequacy of the model assumptions (e.g., heart rate QT correction, baseline adjustments, choice of positive control to establish assay sensitivity, linear or nonlinear concentration–QTc relationship, and direct or delayed effects) through model diagnostics and graphical analyses.

The system furthermore enables leveraging prior information to answer drug development related questions through pooled data analysis since the data and results are in a consistent format that easily can be combined. Contributions to improve the science include (1) the use of concentration–QT modeling for regulatory review of new drugs (Garnett et al. 2008) and (2) evaluating TQT study design features on moxifloxacin response (Florian et al. 2011).

References

Bloomfield D, Kost J, Ghosh K et al (2008) The effect of moxifloxacin on QTc and implications for the design of thorough QT studies. Clin Pharmacol Ther 84:475–480

Florian JA, Tornøe CW, Brundage R, Parekh A, Garnett CE (2011) Population pharmacokinetic and concentration–QTc models for moxifloxacin: pooled analysis of 20 thorough QT studies. J Clin Pharmacol 51(8):1152–1162

Food and Drug Administration (2005) Guidance for industry: E14 clinical evaluation of QT/QTc interval prolongation and proarrhythmic potential for non-antiarrhythmic drugs, U.S. Department of Health and Human Services, Food and Drug Administration. http://www.fda.gov/downloads/RegulatoryInformation/Guidances/ucm129357.pdf. Accessed 7 Aug 2011

Garnett CE, Beasley N, Bhattaram VA, Jadhav PR, Madabushi R, Stockbridge N, Tornøe CW, Wang Y, Zhu H, Gobburu JV (2008) Concentration–QT relationship play a key role in the evaluation of proarrhythmic risk during regulatory review. J Clin Pharmacol 48(1):13–18

Garnett CE, Lee JY, Gobburu JVS (2011) Contribution of modeling and simulation in the regulatory review and decision-making: U.S. FDA perspective in clinical trial simulations. AAPS Adv Pharmaceut Sci Ser 1:37–57

International Conference on Harmonisation (2005) Guidance on E14 clinical evaluation of QT/QTc interval prolongation and proarrhythmic potential for non-antiarrhythmic drugs; availability. Notice. Fed Regist 70:61134–61135

Tornoe CW, Garnett CE, Wang Y, Florian J, Li M, Gobburu JVS (2011) Creation of a knowledge management system for QT analyses. J Clin Pharmacol 51(7):1035–1042

Chapter 17
Graphics for Safety Analysis

Peter W. Lane and Ohad Amit

Abstract We present a range of graphics designed for reporting the analysis of safety data. For adverse events (AEs), we show a comparative dot-and-interval plot of all the main AEs in a trial and also comparative plots of cumulative incidence and hazard rate for individual AEs. For laboratory data, we show a scatterplot designed to help identify potential liver toxicity and 2 trellis plots that can show several laboratory measurements in a single graph, showing changes from baseline or the relationship between the measurements. We also give an example of a profile plot for individual patients. Finally, for ECG data we show a comparative cumulative distribution plot and a comparative boxplot profile showing how distributions change over time. We also show a simple comparative profile plot of means of an ECG measurement over time. We produced each graph using one of GenStat™, SAS™ and S-PLUS™ (using code very similar to R), as indicated in the text, and the programs and data are available from the Web site associated with the book.

17.1 Introduction

The analysis of safety data mostly takes the form of simple descriptive statistics, displayed in a tabular or graphical form. For example, the number and percentage of patients experiencing adverse events may be presented, or the means or medians of

P.W. Lane (✉)
Quantitative Sciences, GlaxoSmithKline, Stevenage, UK
e-mail: peter.w.lane@gsk.com; peterwlane@gmail.com

O. Amit
Quantitative Sciences, GlaxoSmithKline, 1250 S. Collegeville Road,
Collegeville, PA 19426-0989, USA
e-mail: ohad.amit@gsk.com

clinical laboratory measurements. Graphs are ideal for communicating this type of information concisely. A particular advantage over tabulation is that the descriptive statistics can often be presented in conjunction with the patient data that have been summarized, to put the statistics in context. Another advantage is that the human eye is able to detect anomalies and patterns in pictures better than in tables of numbers, and a graphical display allows more effective communication.

We consider here 3 types of safety information: adverse events, liver toxicity and cardiac safety. We illustrate graphical methods of displaying this information, based on work by a GlaxoSmithKline team of which we were members, and which was reported in a paper by Amit et al. (2008). We have updated these in the light of recent developments and added some new examples. There are clearly many other types of safety information, but we suggest that many may be displayed graphically using the same approach as we use here for the types on which we concentrate.

For each graph in this chapter we present a version drawn by one package (named in the text). The data and code can be found on the website associated with this book. Most of the data are from a single anonymized clinical trial, which we will refer to as the Safety trial.

17.2 Adverse Events

There are numerous adverse events (AEs) reported in each clinical trial, so displays need to be tailored to highlight important information, such as the most common events and events of special interest. The SPERT (Safety Planning, Evaluation and Reporting Team) have recommended a three-tier approach for signal-detection and analysis of AEs (Crowe et al. 2009). The first tier is made up of AEs of special interest, identified in advance of running trials, and the second and third tiers of other AEs that are considered common and uncommon, respectively: "common" is suggested by Crowe et al. to be more than about 1% incidence in any treatment arm, though this will depend on the size of the trial. The methods in this section are suitable for AEs in Tiers 1 and 2; AEs in Tier 3 are best reported with simple summary statistics.

17.2.1 Dot-and-Interval Plot of AE Incidence

A dotplot is an ideal display to show and compare AE incidence in a randomized clinical trial. This type of display was introduced by Cleveland in the context of showing counts and proportions, and is generally considered superior to barcharts and piecharts (Cleveland 1993). Figure 17.1 shows a two-panel display which enhances the simpler dotplot by adding statistical information comparing the incidence rates of AEs in the Safety trial. In this example, we display all AEs that

17 Graphics for Safety Analysis 327

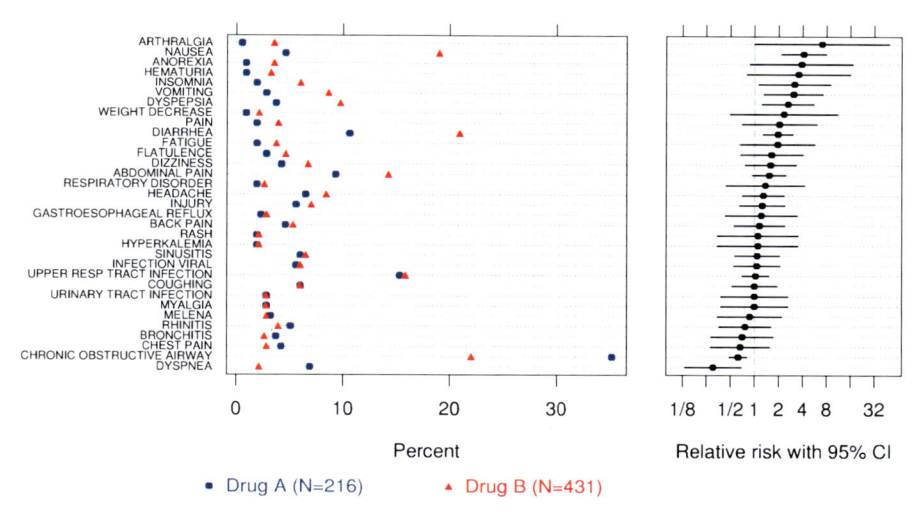

Fig. 17.1 Dot-and-interval plot of AE incidence

had overall incidence greater than 2%, along with relative risks and the asymptotic confidence intervals.

It is clinically valuable to see the actual risk differences as well as relative risks in a single snapshot, to put the statistical ratios into context. Note that the adverse events are ordered by relative risk. Other statistics can also be considered for the right-hand panel, such as risk differences, odds ratios or hazard ratios, depending on the objectives of the display for its audience. It is useful to give further context by adding information about the number of patients in the safety population of the trial: here, we have added that to the key. Colour is used in a modest way to help distinguish the 2 treatments, but note that different symbols are used as well in case the graph is viewed in black and white.

This graph was drawn using S-PLUS™.

17.2.2 Cumulative Incidence of an AE

Cumulative incidence over time is often of interest with AEs, as the time at which such events manifest themselves can be critical in guiding regulators and prescribers regarding monitoring and clinical use of a drug. Figure 17.2 shows a cumulative incidence plot of the gastrointestinal AEs from the same trial as above. It is constructed in much the same way as a Kaplan–Meier (KM) plot, taking account of censored information because many patients withdrew from this trial. It is better to display the information as (1–"survival") against time here, rather than as survival

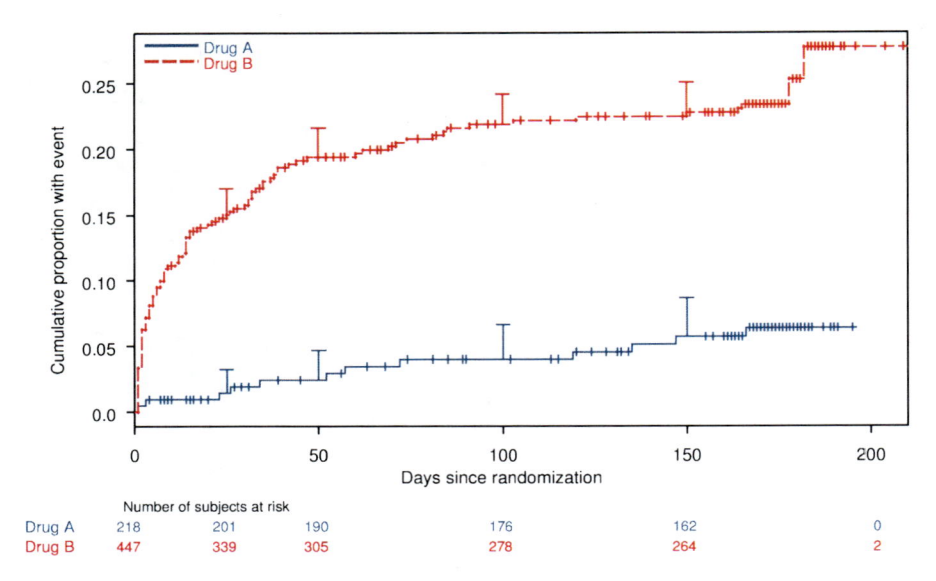

Fig. 17.2 Cumulative incidence of an adverse event, with SEs at selected time-points

against time as in the KM plot because incidence is the focus for AEs. There is also the point that most people are used to seeing survival plots with the *y*-axis ranging fully from 0 to 1, which would cramp the information at the top of the frame (Pocock et al. 2002).

The plot has several enhancements compared to a simple KM plot. First, the numbers of subjects at risk are displayed as strategic points along the *x*-axis in a lower margin, to quantify the steadily decreasing population as subjects withdraw over time. Second, the actual censoring times of subjects on each arm are marked as a "rug-plot" on top of each step function representing the cumulative proportion. Third, the SEs of the estimated proportions are indicated at the same strategic points to show how much precision has been achieved. In this case, we have shown these as positive error bars only, but they could alternatively be negative bars, the usual two-sided bars, or indeed show 95% confidence intervals instead. Note that colour is used as in Fig. 17.1, and that different line styles are used in case the graph is viewed in black and white. In some settings, such as clinical trials without a fixed follow-up, use of competing risks methodology (Pintilie 2007) should be considered in order to estimate the cumulative incidence curves. This methodology would be particularly useful in trials where subjects are treated until the occurrence of a specific event (e.g., disease progression), separate from the safety event of interest.

This graph was drawn using S-PLUS™, which provides an option to add the tricky part, i.e., the rug-plots.

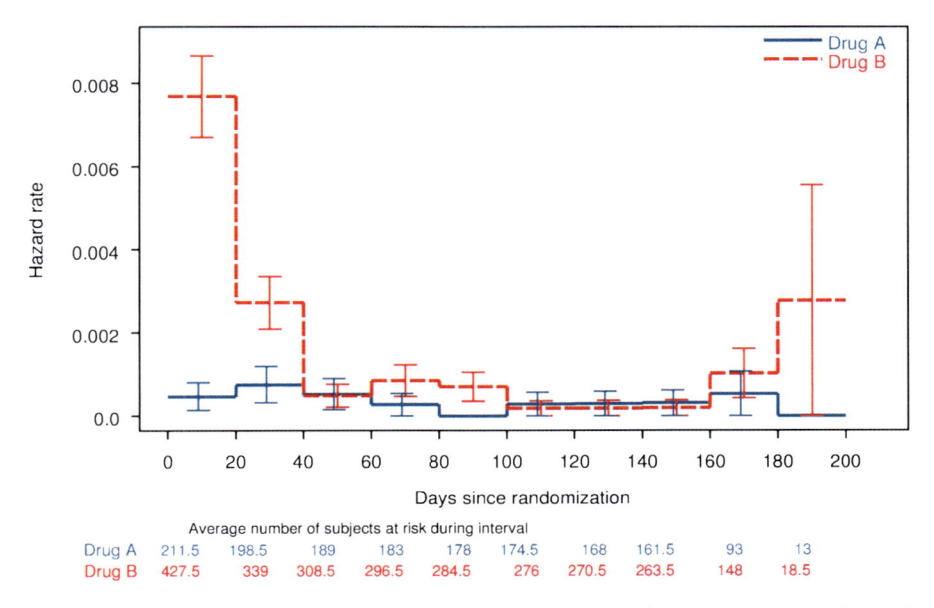

Fig. 17.3 Comparative hazard function for gastrointestinal AEs of concern: nausea, abdominal pain, diarrhoea and vomiting

17.2.3 Hazard Rate for an AE

The information about incidence of an AE can also be displayed as a hazard rate function. Figure 17.3 shows this for the same data, with the hazard rates estimated in successive 20-day intervals (again taking account of censoring) and drawn as a pair of step functions.

As for Fig. 17.2, this has been enhanced with a lower margin giving the average number of subjects at risk during each time period, and SE bars for each hazard estimate. The choice of time periods can be important to illustrate the differences between the two drugs effectively. Note that there is no SE for periods where the hazard was 0, as the estimate of SE is formally 0 for such periods. The lines are again differentiated both by colour and line-style.

This graph was drawn with S-PLUS™.

17.3 Liver Toxicity

Drug-Induced Liver Injury (DILI) is "the single most common adverse effect that can result in failure to obtain regulatory approval to market a new drug, and post-marketing regulatory actions include labelling restrictions and withdrawal from the marketplace" (Watkins 2005). In general, hepatic safety is the second most common reason for termination due to safety during drug development. DILI is the most frequent cause of acute liver failure in patients evaluated for liver transplantation. There are 3 main types of liver toxicity that may be observed: directly destructive, indirect (or metabolic) and cholestatic.

Intrinsic or direct liver injury (e.g., that seen with acetaminophen) is:

- Predictable
- Dose-related
- Similar in animals
- Relatively common
- Observed after a short interval

On the other hand, idiosyncratic liver injury (e.g., that seen with Troglitazone) is:

- Unpredictable
- Often dose-independent (Lammert et al. 2008)
- Not seen in animals
- Relatively rare: 1 in 10,000 to 1 in 100,000
- Usually observed after a longer interval

There are 2 main types: hypersensitivity and metabolic.

Typically in clinical trials, 4 cardinal variables are monitored for liver toxicity using what are described as liver function tests (LFT):

- ALT: alanine aminotransferase
- TBL: total bilirubin
- AST: aspartate aminotransferase
- ALKP: alkaline phosphatase

ALT, AST and TBL are of particular interest because of a criterion that is generally accepted as a surrogate for potential DILI, known as Hy's Law. While there are several clinical aspects to the determination of a Hy's Law case, the laboratory criteria are defined as an elevation of ALT or AST together with simultaneous or subsequent elevation of bilirubin. An occurrence of such a simultaneous elevation indicates the potential for severe liver injury and which in turn could predict for acute liver failure. Andrade et al (2005) have reported a 10% fatality from drug-induced liver injury with jaundice.

17.3.1 Scatterplot to Assess Drug-Induced Liver Injury

The FDA has adopted a criterion for Hy's Law (Wilke et al. 2007), generating a signal when the following conditions are all met:

- ALT or AST ≥ 3xULN (upper limit of normal measurements)
- TBL ≥ 2xULN
- ALKP ≤ 2xULN

17 Graphics for Safety Analysis 331

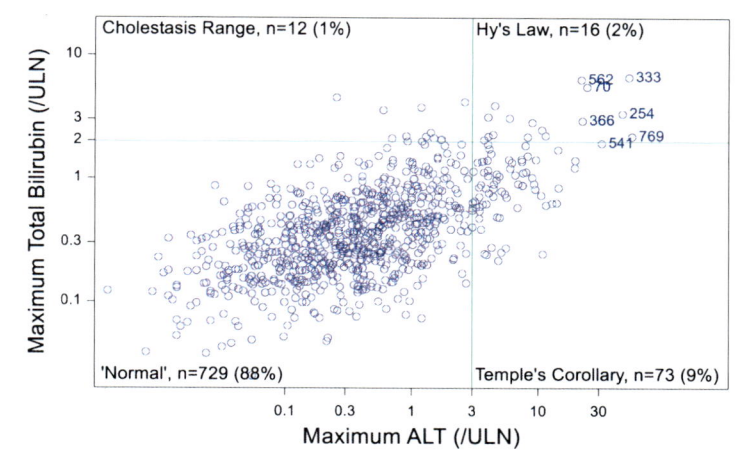

Fig. 17.4 Scatterplot of Total Bilirubin versus ALT used as a signal for DILI

A graphical approach suggested as part of the FDA DILI guidance to evaluate potential Hy's law cases is shown in Fig. 17.4 (using simulated data). This is a simple scatterplot of maximum TBL for subjects during the course of a trial against maximum ALT, with reference lines and annotation associated with the Hy's Law criterion. This concentrates just on ALT and TBL, and a similar graph can be drawn for TBL versus AST.

The figure is split into 4 quadrants with the upper right quadrant indicating the potential for a Hy's Law case. The bottom right quadrant showing subjects with elevated ALT but without elevated TBL is associated with another conjecture called Temple's Corollary. It has been hypothesized that a significant number of patients within this quadrant will predict for the presence of a Hy's Law case at some point in time. The top left quadrant is referred to as the Cholestasis Range, associated with Gilbert's Syndrome (high TBL but normal ALT).

The aspect ratio of this graph is worth noting: it emphasizes the ALT measurements by having a longer axis, and this corresponds to the fact that one can observe far more extreme ALT measurements as multiples of ULN than with TBL. In addition, ALT is typically more predictive of clinical harm in DILI because TBL outliers may be due to Gilbert's Syndrome and therefore not so important.

This graph was drawn using GenStat™.

17.3.2 Scatterplot Trellis of Shifts from Baseline Measurements

A standard tabular summary of LFTs that evaluates shifts in individual LFT measurements is shown in Table 17.1. The number of subjects who "shifted" to a higher LFT value relative to their baseline value is shown in the table.

Table 17.1 A standard tabular summary of LFTs, evaluating shifts in individual LFT measurements

Test	Time		Treatment A				Treatment B		
		n	Any increase	Increase >3xULN	Increase >5xULN	n	Any increase	Increase >3xULN	Increase >5xULN
ALT	Week 4	18	2 (11%)	1 (6%)	0	18	2 (11%)	1 (6%)	0
	Week 6	18	2 (11%)	1 (6%)	0	18	2 (11%)	1 (6%)	0
	Week 8	18	2 (11%)	1 (6%)	0	18	2 (11%)	1 (6%)	0
	Post Rx	18	3 (17%)	1 (6%)	0	18	3 (17%)	1 (6%)	0

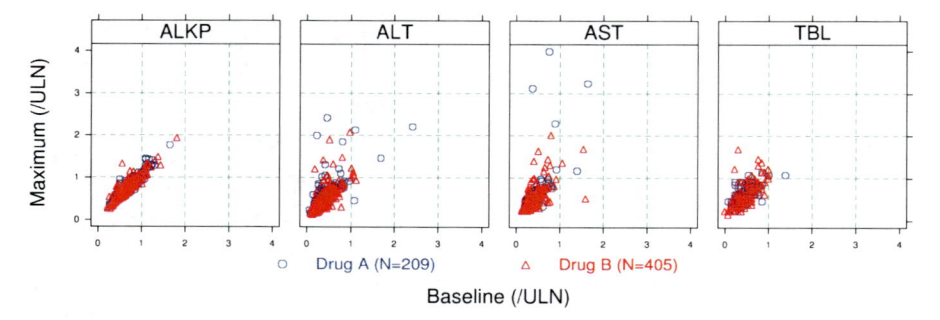

Fig. 17.5 Trellis of scatterplots of maximum LFT measurements versus baseline

A concise graphical summary of the same sort information from the Safety trial is provided in Fig. 17.5.

There are several important elements to this graph. First, when comparing an active drug against a control, as here, it can be more informative to arrange that the control points (blue circles) are drawn last: the distribution of extreme points from the active drug then appears as a "frill" around the central mass of blue points, allowing quick visual appreciation of the potential effect of the drug. However, some software makes this difficult, and we could not find a way to achieve it in Fig. 17.5 using S-PLUS™.

Interpretation of the absolute distribution of the observations for each treatment can be misleading, as the points tend to lie above and to the left of the centre diagonal of each graph. This is an inevitable consequence of using a maximum of several values on the y-axis: in this case there were 8 visits during the trial. Because of natural variation, the distribution of a maximum of 8 observations is inevitably shifted upwards compared to the distribution of a single (baseline) measurement, regardless of any effect of the drugs. Note that the reference lines have been updated since publication of this figure in Amit et al. (2008) to take account of the criteria described in the FDA DILI guidance.

Two other features of this graph can be of importance for interpretation. The distribution of LFT measurements is usually very skewed, particularly when the

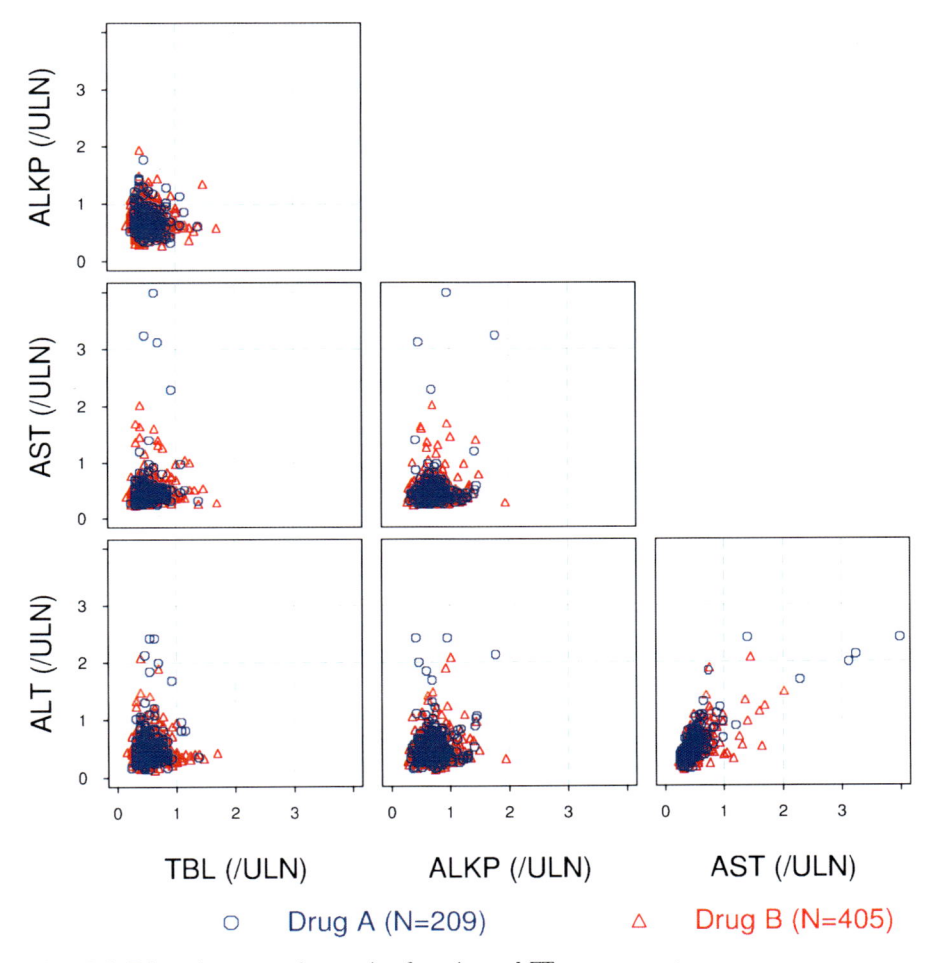

Fig. 17.6 Triangular scatterplot matrix of maximum LFT measurements

patient population has significant elevations, as seen in Fig. 17.4. The graphical display of the relationships can then often be improved by using a log scale as in that figure, which allows display of all the extreme values without overemphasizing those values within the figure.

17.3.3 Scatterplot Matrix of Maximum LFT Measurements

The association between the various LFT measurements can be displayed in the matrix plot as shown in Fig. 17.6. This shows a triangular array of each of 4 LFTs against each other, allowing quick visual assessment of the interrelated information

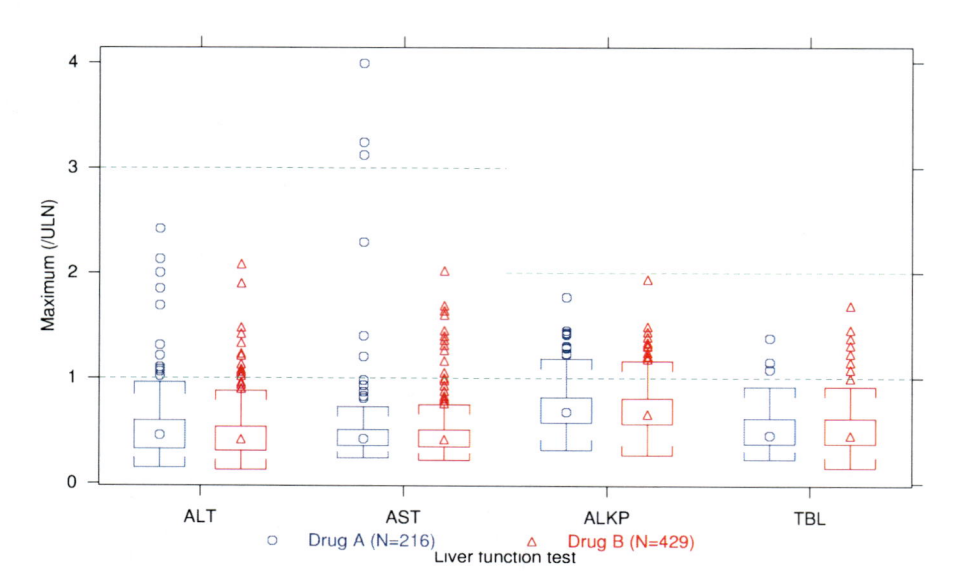

Fig. 17.7 Parallel boxplots of LFT measurements

associated with signals such as Hy's Law. Like Fig. 17.4, Fig. 17.6 can also provide a quick visual assessment of potential Hy's law cases and can also show another important relationship between ALT and AST. The latter can help in further defining the nature of the liver signal of a particular compound.

The individual scatterplots are designed much the same as those in Fig. 17.5. Here, however, we succeeded in arranging for S-PLUS™ to plot the control treatment, Drug A, on top, so that any differences in distribution for Drug B appears as a fringe around that for Drug A.

17.3.4 Parallel Boxplot of LFT Measurements

If the shift information is not of particular interest, and the association between different measures is not to be concentrated on, a simpler graph can give a visual report of the distributions. Figure 17.7 uses boxplots to display the distributions, with an emphasis on the outlying points—a key feature of boxplots.

These boxplots are those defined as "schematic diagrams" by Tukey (1977), with the whiskers extending outside the box no further than 1.5 times the box width. The extreme points are all individually marked, which is ideal for safety measurements of this kind where the interest focuses on them.

This graph was drawn using S-PLUS™, requiring the definition of a transposition function in order to be able to orient the boxes vertically.

17 Graphics for Safety Analysis   335

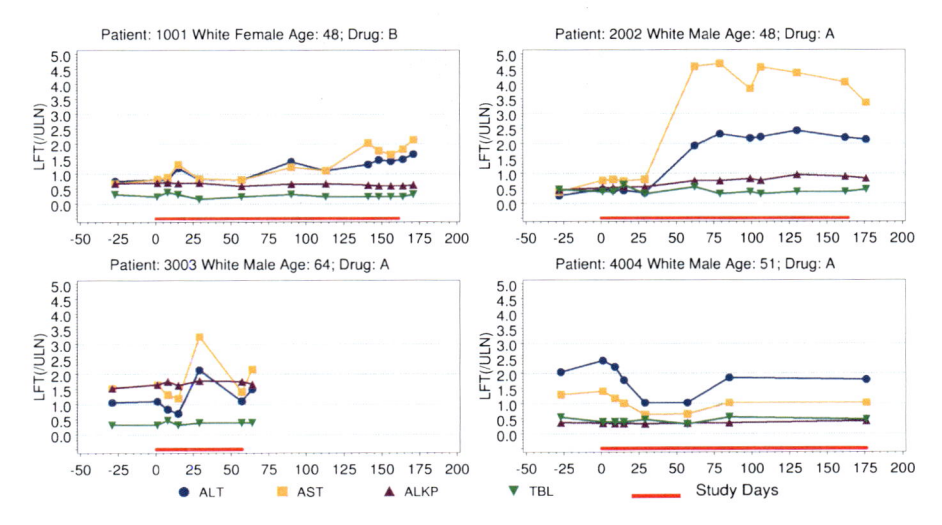

Fig. 17.8 Customized patient profile display of LFT measurements

17.3.5 Patient Profile of LFT Measurements

As previously noted, much of safety analysis is concerned with individual subjects rather than summary data. Figure 17.8 provides a powerful and concise summary of liver function information on selected individual subjects (in this case, on the basis of any LFT exceeding 2xULN during the trial).

All 4 liver function parameters are plotted as a function of time for each individual subject, allowing ready assimilation of several pieces of information. These include the time course of the elevations relative to treatment, the presence or absence of simultaneous elevations, outcomes of dose interruptions, dose reductions as they relate to the elevations and outcomes of a patient subsequent to an interruption or reduction. When data from many individual patients are needed, a series of displays of this kind can be produced. Once the first screen has been viewed and understood, the remainder can be quickly assessed as long as the display style is kept consistent.

This graph was drawn using SAS™.

17.3.6 Other Possibilities

Another aspect of laboratory data that can be of great interest is the way in which AESIs accumulate over time under different treatment regimes. This can be effectively displayed with a cumulative incidence plot, as in Fig. 17.2.

Table 17.2 Other clinically meaningful lab measurements to consider

Category	Clinical interpretation	Measurements
Haematotoxicity	Grades 3 and 4	Red cell count, mean cell volume, platelets, haematocrit, haemoglobin, reticulocyte count, white cell total count and differentials
Nephrotoxicity	Creatinine increase from baseline >0.3 mg/dl (as defined by Acute Kidney Injury Network)	Serum creatinine, blood urea nitrogen and creatinine kinase
Lipids		Cholesterol, thyroxin, LDL, HDL and triglycerides
Rhabdomyolysis/ muscle injury	By profiling	Creatinine phosphokinase (CPK), AST, lactate dehydrogenase (LDH), red cells or myoglobin on urine test
Paediatrics	Growth; CNS functioning; reproductive or endocrine status	Height, weight, BMI, often transformed into z-scores; serial assessments of IQ, as an example, for long-term studies; androgens, estrogens and relevant hypothalamic hormones
Suicidality	Incidence rate (e.g., in a forest plot)	Ideation, attempts, deaths

There are many other lab measurements that could readily be displayed in graphical form. Table 17.2 shows a list of lab measurements to consider. All of the graphical methods described above may be applied to these other lab variables.

17.4 Cardiac Safety

One of the major issues that have led to drug withdrawals has been cardiovascular incidents, so there was an early focus on cardiac safety in safety data analysis. QT prolongation and Torsades de Pointes is of primary concern, but any conduction-interval prolongation (e.g., PR prolongation) could be a potential safety concern and possible showstopper for a new drug. Figure 17.9 shows a stylized ECG trace annotated with the letters from which some of the heart rhythm measurements are associated. Other crucial issues are heart failure (predicted by ejection fraction) and myocardial infarction (predicted by troponin levels).

The main derived endpoints from heart traces are the RR, PR, QRS and QT intervals. The last of these, usually in a corrected form and called the QTc interval, is the key one for general studies. It is a marker for cardiac toxicity: prolonged QTc can lead to increased risk of Torsades de Pointes (TdP)—a rare but life-threatening arrhythmia. This is a Sentinel Event that the FDA require reporting as soon as there is awareness of a case. FDA guidance on evaluation of QTc is as follows:

- Increases to >500 ms are of clinical concern
- Increases to >480 and 450 ms are also of interest

17 Graphics for Safety Analysis

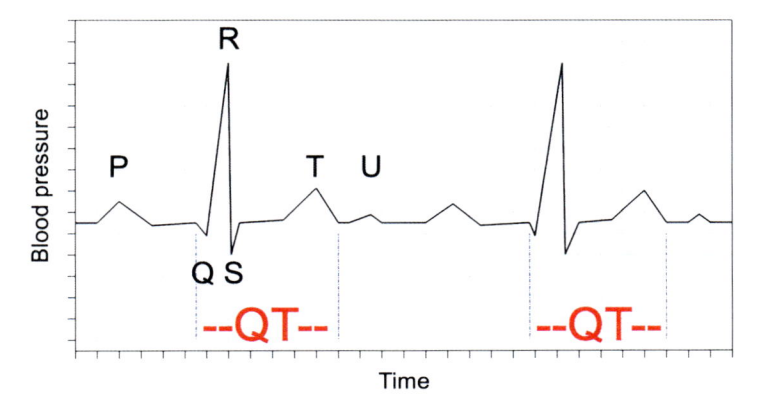

Fig. 17.9 A stylized ECG trace

- Changes from baseline of >60 ms (regardless of absolute value) are of serious clinical concern
- Change from baseline of >30 ms are of clinical concern

Relevant questions to ask based on the guidance are:

- Is there a significant change over time in the distribution of QTc results?
- How many people report a significant shift in QTc values, i.e., an increase of >30 or >60 ms?
- How many subjects report a QTc interval of >450, >480 or >500 ms?

Many of the graphical displays in the previous section are clearly appropriate to cardiac measurements like QTc, such as the scatterplot of shifts in Fig. 17.4.

17.4.1 Cumulative Distribution Plot of QTc

For a comparative trial, a plot showing the detailed distributions of critical measurements like QTc can be invaluable for giving reassurance or highlighting areas of concern. Figure 17.10 shows a cumulative distribution function (CDF) plot from the Safety trial.

This display allows close scrutiny of the distributions, with reference to clinical criteria. Note that the percentage of subjects with a change greater than 0 is just over 10%. This is as expected, as explained before: QTc was measured at baseline and at 8 visits, so the chance that the baseline measure is the smallest of these would be 11% (i.e., 1/9) if the drugs have no effect and successive measurements on a patient

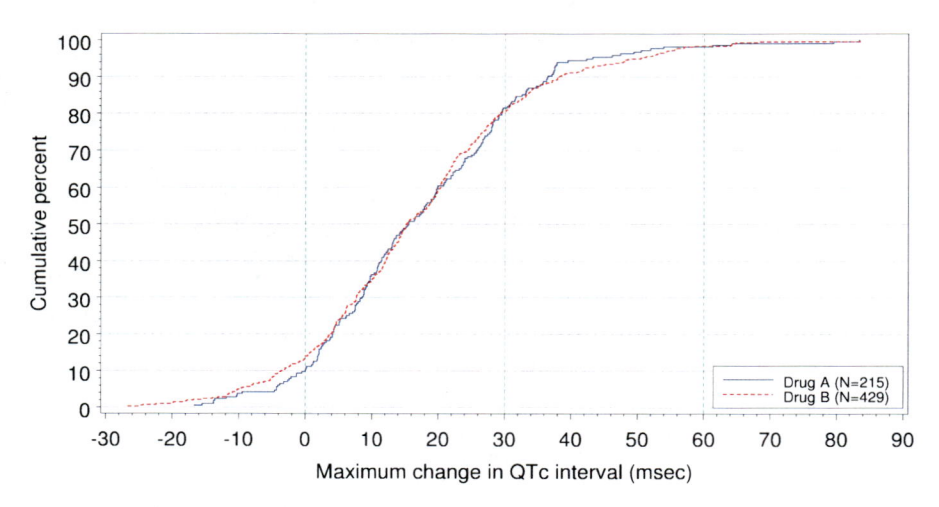

Fig. 17.10 CDF plot of maximum QTc changes

can be taken as independent. Note that it is particularly useful to show a faint grid with this graph, as this helps detailed interpretation of any differences noted between the two step functions.

This graph was drawn using SAS.

17.4.2 Boxplot Profile of QTc

The boxplot, illustrated in Fig. 17.7, can also be used to display the change in the distribution of a variable over time profile. Figure 17.11 compares the distribution of QTc changes from baseline on the 2 treatments arms of the Safety trial, for each of the 8 visits. In addition, a right-hand margin has been added to show the distribution of the maximum change from baseline. The numbers of patients measured at each visit is also displayed in a bottom margin, as before, but here the margin has been brought inside the frame, which can help to emphasize the relationship between the values and the plotted information.

Each individual point representing a change greater than 60 ms has been labelled here with the patient number: this draws attention here to Patient 194, who had an increased level of QTc from Week 12 onwards. Some labels are overwritten, but it would be difficult to arrange to separate them; clearly, the amount of labelling needs to judged carefully if it is to be of use. An alternative to labelling would be to list the values of concern in a separate table.

This graph was drawn using GenStat™.

17 Graphics for Safety Analysis

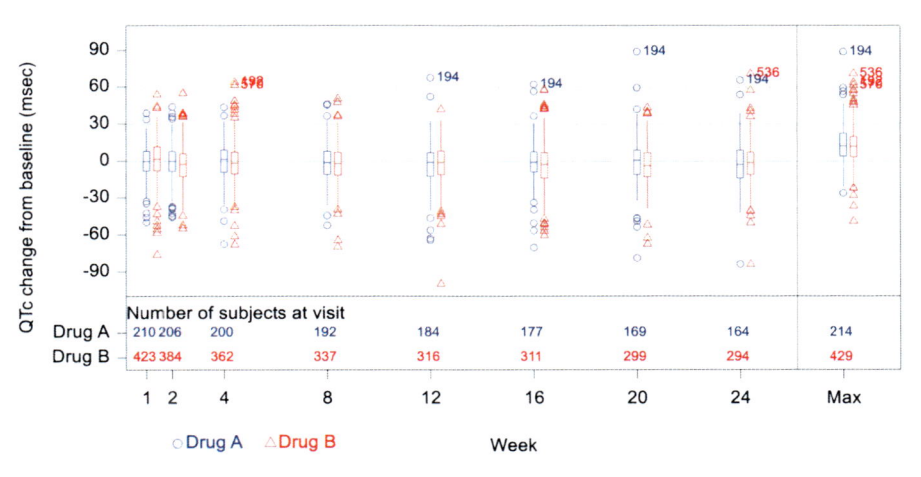

Fig. 17.11 Boxplot profile of QTc changes from baseline

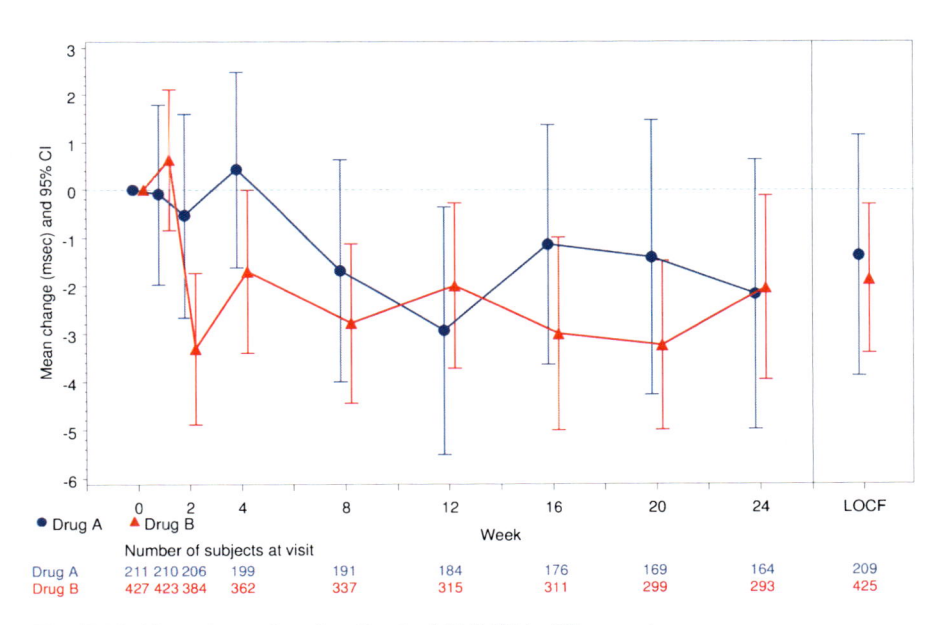

Fig. 17.12 Mean change from baseline (and 95% CI) in QTc over time

17.4.3 Mean Profile of QTc

As well as showing the individual subject data in the above distribution plots, it may be useful to focus on the evidence for systematic difference between the drugs. Figure 17.12 shows the mean QTc changes over time, with confidence limits to put the small differences into context.

This kind of display is commonly seen for reporting efficacy, showing how the effect of treatments compared in a trial change over the course of the trial. Usually, it is the difference at the end of the trial that is of primary interest, and that difference may be adjusted to try to take account of interfering factors, such as drop-out of patients from the trial. One such method is "last observation carried forward," as shown here, but more advanced methods using multiple imputation or mixed modelling are now preferred (Mallinckrodt et al. 2008).

This graph was drawn using SAS™.

17.5 Conclusion

We have described 11 different graphical designs that we recommend for displaying safety information. We used each design to display particular safety outcomes, but many of the designs can of course be used for a wide range of different outcomes, and indeed for efficacy outcomes as well. We have omitted a large class of graphical designs, which are being used increasingly in pharmaceutical companies to monitor safety of drug development programmes. These are interactive designs which allow the viewer to modify the display using a graphical interface or drill down to find further information about aspects of interest in the initial display. Other chapters in this book describe these, in particular Chap. 10.

Acknowledgments Most of these displays were developed in GlaxoSmithKline by a team working on graphics for safety, and several programmers contributed ideas and code. We are indebted in particular to Richard Heiberger (Temple University) and Mike Durante (GSK) for S-PLUS™ code, and Shi-Tao Yeh (GSK) for SAS™ code.

References

Amit O, Heiberger RM, Lane PW (2008) Graphical approaches to the analysis of safety data from clinical trials. Pharm Stat 7:20–35

Andrade RJ, Lucena MI, Fernández MC, Pelaez G, Pachkoria K, García-Ruiz E, García-Muñoz B, González-Grande R, Pizarro A, Durán JA, Jiménez M, Rodrigo L, Romero-Gomez M, Navarrao JM, Planas R, Costa J, Borras A, Soler A, Salmerón J, Martin-Vivaldi R (2005) Drug-induced liver injury: an analysis of 461 incidences submitted to the Spanish registry over a 10-year period. Gastroenterology 129:512–521

Cleveland WS (1993) Visualizing data. Hobart, Summit, NJ

Crowe BJ, Xia HA, Berlin JA, Watson DJ, Shi H, Lin SL, Kuebler J, Schriver RC, Santanello NC, Rochester G, Porter JB, Oster M, Mehrotra DV, Li Z, King EC, Harpur ES, Hall DB (2009) Recommendations for safety planning, data collection, evaluation and reporting during drug, biologic and vaccine development: a report of the safety planning, evaluation, and reporting team. Clin Trials 6:430–440

Lammert C, Einarsson S, Saha C, Niklasson A, Bjornsson E, Chalasani N (2008) Relationship between daily dose of oral medications and idiosyncratic drug-induced liver injury: search for signals. Hepatology 47:2003–2009. doi:10.1002/hep. 22272

Mallinckrodt CH, Lane PW, Schnell D, Peng Y, Mancuso JP (2008) Recommendations for the primary analysis of continuous endpoints in longitudinal clinical trials. Drug Inf J 42:303–319

Pintilie M (2007) Analysing and interpreting competing risk data. Stat Med 26:1360–1367

Pocock SJ, Clayton TC, Altman DG (2002) Survival plots of time-to-event outcomes in clinical trials: good practice and pitfalls. Lancet 359:1686–1689

Tukey JW (1977) Exploratory data analysis. Addison-Wesley, New York, NY

Watkins PB (2005) Idiosyncratic liver injury: challenges and approaches. Toxicol Pathol 33:1–5

Wilke RA, Lin DW, Roden DM, Watkins PB, Flockhart D, Zineh I, Giacomini KM, Krauss RM (2007) Identifying genetic risk factors for serious adverse drug reactions: current progress and challenges. Nat Rev Drug Discov 6:904–916

Chapter 18
Cardiac Safety

Richard J. Anziano

Abstract Cardiac Safety is an important consideration in drug development. This chapter focuses on interval data derived from the 12-lead surface electrocardiogram (ECG) and graphical approaches that enable a better understanding of the data. A motivating example for the link between ECG interval data and cardiac safety is followed by an orientation to the intervals and their relationship to cardiac function. QT interval corrections for heart rate are discussed as are types of data collection. Sample data that are publically available are used in many of the examples from the thorough QT trial presented. Graphics are used to demonstrate some of the attributes of these trial designs such as time matching and use of active controls.

18.1 Introduction

Electrocardiograms (ECGs) are routinely included in clinical trials to assess the cardiac safety of experimental therapies. Prolongation of the QT interval, for example, is associated with some potentially fatal cardiac arrhythmias. Graphical methods presented in this chapter are intended to help interpret ECG interval data and provide a better understanding of the sources of variability in the interval measurements.

A good motivating example related to cardiac safety and ECG data is terfenadine. Terfenadine was the first non-sedating antihistamine and came to the market in 1985. By 1990, after more than 100 million people were exposed, mounting evidence of

---

R.J. Anziano (✉)
Primary Care Statistics, Pfizer Inc., Eastern Point Road,
Groton, CT 06340, USA
e-mail: Richard.J.Anziano@Pfizer.com

serious and sometimes fatal arrhythmias caused the FDA to issue a warning letter. In 1992, the FDA required a black box warning on using terfenadine with CYP3A4 inhibitors, and later in 1997 terfenadine was removed from the market. In 1993, Honig et al. (Honig et al. 1993) published their findings from a rigorous evaluation of terfenadine alone and terfenadine + ketoconazole. In only 6 subjects receiving both regimens they were able to show profound increases in the QT interval.

This example is meant to highlight an important question. If Honig et al. were able to quantify such profound prolongation in so few subjects after coadministration of terfenadine with ketoconazole, could changes have been observed much earlier in the development of this compound? Analysis of QT and concentration are routine now but at that time were not. Perhaps a simple scatter plot of QTc by concentration with an overlay of any smoothed or fitted line would have highlighted prolongations at higher concentration.

Graphical displays are a very efficient and effective way of distilling large quantities of data, examining relationships between variables, and aide in the interpretation of the results of the trials. These displays also allow effective dialog with other scientists.

18.2 Interval Data

Electrocardiogram interval data, used to assess cardiac safety in clinical trials are usually obtained through a standard 12-lead surface electrocardiogram (ECG).

More intensive ambulatory assessments can be obtained through Holter monitoring, whereby interval data are generated continuously. The advantage of Holter monitoring is that more data are generated throughout the day and across a wider range of heart rates. The increase in heart rate range prior to the administration of study drug makes calculation of individual subject QT corrections for heart rate possible. The major disadvantage of Holter monitoring is the lack of control over the sources of variability such as the subject position, environmental conditions, and inter-machine variability. These factors as well as meal timing and time of day are controlled in order to minimize sources of variability in measurement of the different intervals.

It is useful to review a stylized ECG segment in order to understand the intervals, and their relationship to the electrical activity in the heart. In doing so, one can appreciate that the different intervals are really lengths, or durations, and that some lengths are dependent on others.

Figure 18.1 depicts a stylized single beat of the heart in normal sinus rhythm. On the y-axis is the amplitude of the electrical signal and the x-axis represents time. This single set of amplitudes, P through U, is called a complex. The pattern of depolarization and repolarization controls the contraction (depolarization) and relaxation (repolarization) of each of the atria and ventricles. The sequence of contractions and relaxation is what produces the pumping motion of the heart resulting in efficient ejection of blood from the heart into the circulatory system.

18 Cardiac Safety

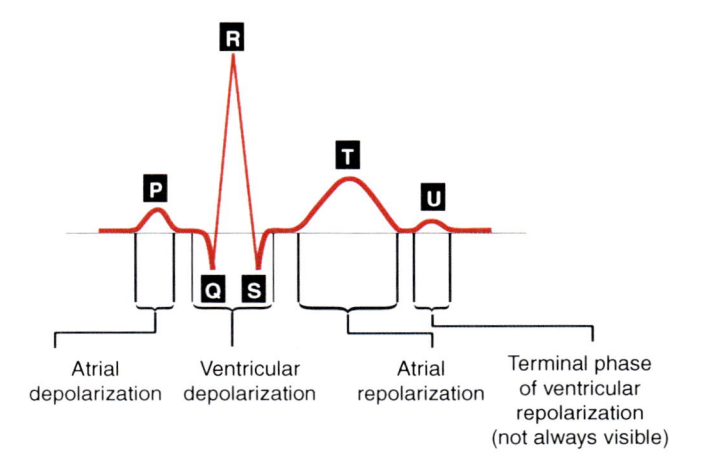

Fig. 18.1 Stylized heartbeat and corresponding ECG segments

Typically, there are 4 interval measures that are returned from a 12-lead ECG. These are the PR, QRS, QT, and RR. These intervals are durations of elapsed time between the different points of the wave-form measured in milliseconds. These durations are often called lengths. The length of the PR is the elapsed time between the beginning of the P-wave to the R-wave; The QT is the length from the beginning of the Q to the end of the T; and QRS is the length from the beginning of the Q to the end of the S. The RR is the distance between the R of one complex and the R of the next complex. The interval data obtained from the ECG machine is usually the average across 3 consecutive beats for each interval.

The more pronounced the change in slope in the amplitude of the electrical signal the more accurate the assessment of the interval. For example, measuring the distance from the R to the R of the next beat is more accurately derived than the distance from the Q to the end of the T-wave. This is because the R, where the amplitude changes rapidly, is very sharply defined and easily identifiable whereas the end of the T-wave is not. As a result, there is more measurement variability in the QT interval than in the RR interval.

Drugs that reduce the amplitude of the T-wave can make it more difficult to determine the end of the T-wave and thus increase the variability of the QT over that seen at baseline. Checking the variability observed in early trials of any new experimental drug against variability of historical data from similar designs may be an indication that the drug under study is flattening the T-wave.

It is useful to note that the RR interval (in ms) is inversely related to the heart rate (beats per minute, bpm): $RR_{ms} = (1{,}000_{ms/s} \times 60_{s/min})/hr_{bpm}$. As such, an increase in heart rate results in less distance between the R of one complex and the R of the next. Compression of the RR results in shorter distances between other parts of the complex such as QT. As such, a shorter RR (high heart rate) relates to shorter QT intervals.

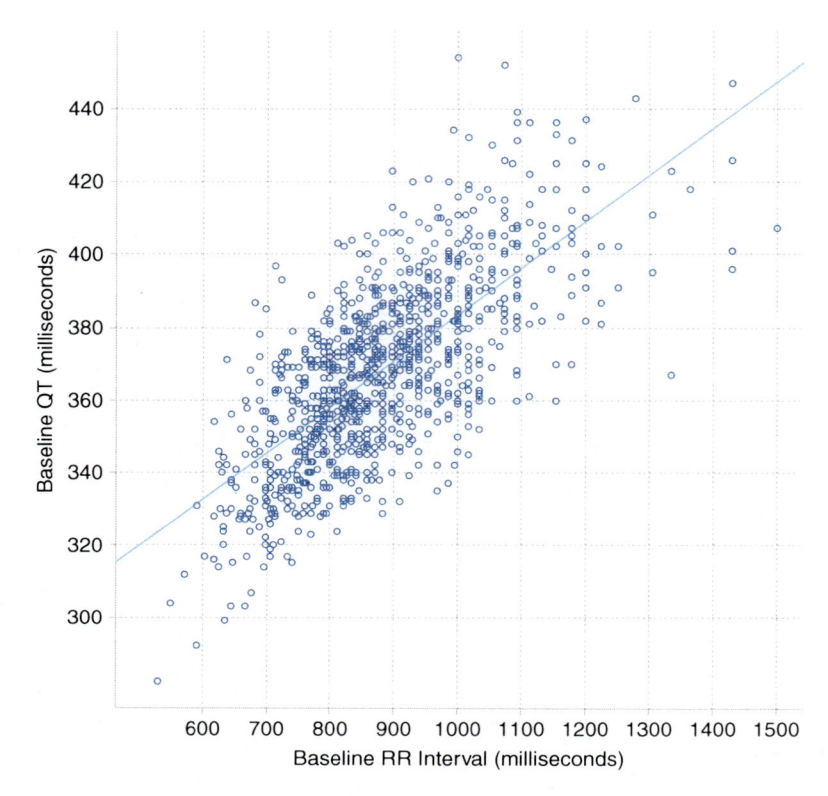

Fig. 18.2 Relationship of QT to RR at baseline

Figure 18.2 shows the baseline ECG interval data from a large phase III program. Each symbol represents an individual subject's QT and corresponding RR interval. A linear regression line is overlaid onto the figure ($f(RR) = 259.2 + 0.134(RR)$; $R^2 = 0.49$) to show the relationship between these 2 intervals.

18.3 QT Correction for RR

One of the derived interval parameters, QTc, is an attempt to "correct" the QT for RR or make the QT independent of the RR interval. Figure 18.2 shows a linear regression for illustration, but most often the corrections are derived from log transformations of both the QT and RR. The form of the log-linear correction is **QTc = QT × RR$^{-\beta}$**, where the exponent β is the magnitude of the rotation of the cloud of points around a pivot point of RR = 1,000 ms; the larger the value of β the more the rotation. With this concept of rotation of QT values around RR = 1,000, common corrections can be

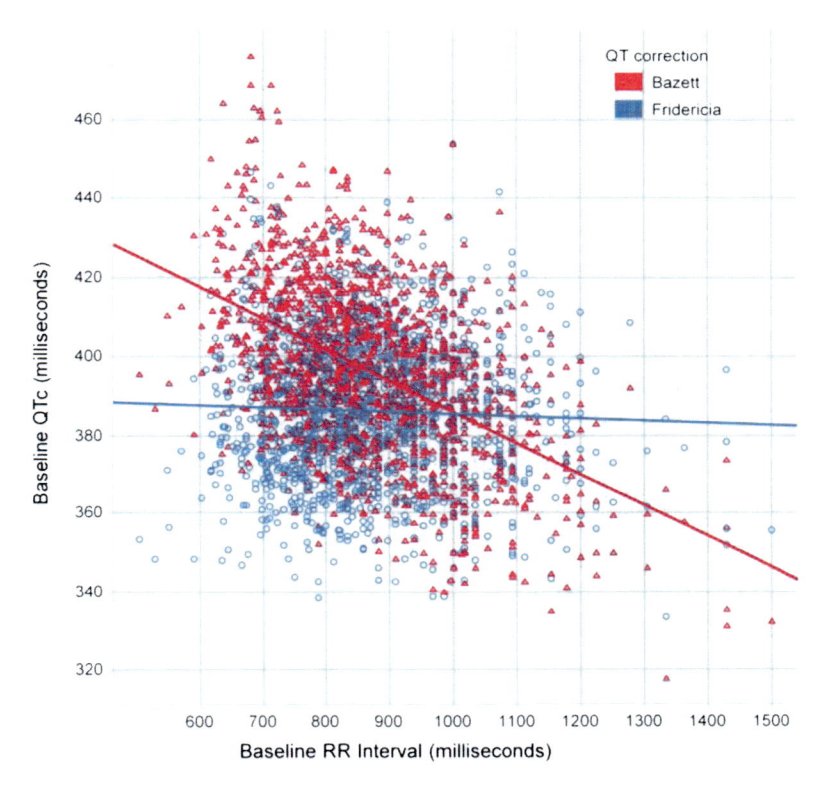

Fig. 18.3 Comparison of QT corrections

better understood. For example, Bazett correction (denoted QTcB) where $\beta=0.5$ has more rotation than Fridericia (denoted QTcF) where $\beta=0.33$. Often Bazett correction is considered to "overcorrect" with respect to the rotation of points described above. This "over-rotation" of the points results in a slightly negative correlation between the resulting QTcB and the RR interval.

Figure 18.3 shows the same set of subjects as in Fig. 18.2, where the QT is replaced with the Bazett and Fridericia corrected QT versus RR. Each subject contributes 2 QTc values for each RR. The over-rotation for the Bazett correction is reflected by the regression line with a negative slope shown in pink. It should also be noted that the differences are more extreme the more the RR departs from 1,000 ms or a heart rate of 60 bpm. At an RR interval of 1,000 ms, QT=QTcB=QTcF or any other β. Contrast that to the subject with the RR of 600, the difference between Bazett and Fridericia is over 30 ms. Similar differences between corrected QT values are also observed for large RR values (low heart rates) such as those often observed in healthy volunteer studies. It is for this reason that careful consideration

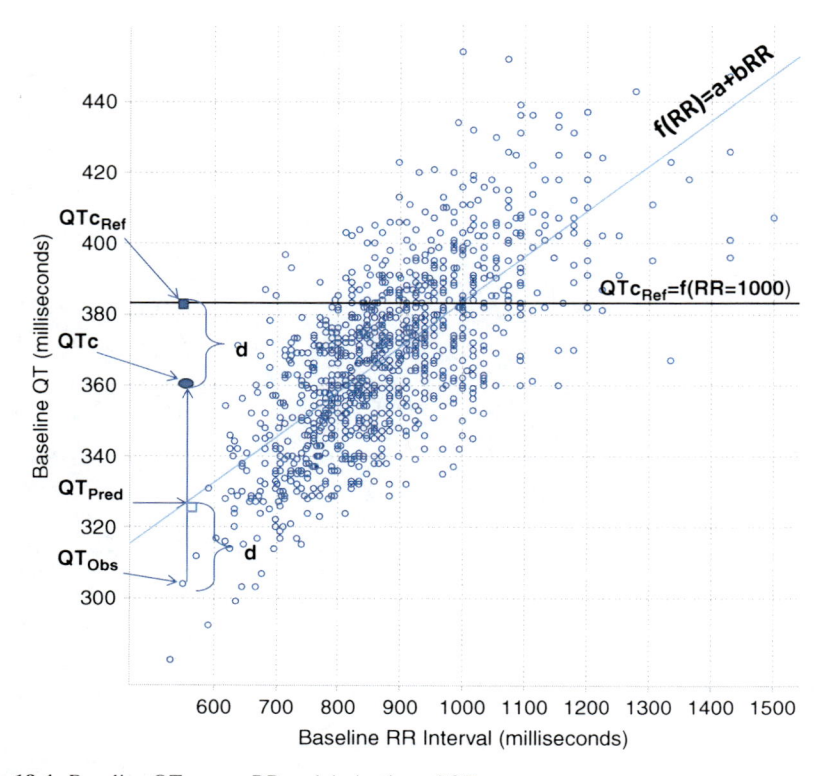

Fig. 18.4 Baseline QT versus RR and derivation of QTc

be given to the QT correction when comparing data to historical data. Having different corrections or a consistent but inappropriate correction among different datasets can create difficulty in interpretation.

Application of a linear model to baseline data is often used to derive a study population correction. Figure 18.4 uses the same data as in Figs. 18.2 and 18.3 and illustrates a case where there is one baseline observation per subject. The application of the least-squares regression line [f(RR) = 0.124 (RR) + 259] produces a predicted QT [QT pred] for each observed QT. The difference between these 2 values [QT obs − QT pred] is equal to the difference between the derived QTc [QTc] and the reference QTc [QTc Ref = QT pred at reference RR interval, here illustrated at 1,000 ms]. Therefore, QTc = QTc Ref + (QT obs − QT pred). For instance, a QT of 304 ms is observed at an RR of 548 ms (HR of 109 bpm). The linear regression model predicts a QT of 326 ms [0.124(548) + 259]. The difference of −22 ms [304 − 326] is added to the reference QTc, 383 ms [0.124(1,000) + 259], yielding a QTc of 361 ms.

A similar linear regression model could be used on log-transformed QT and RR data to arrive at a population value for β. The value of β from this type of model would necessarily have a horizontal or 0 slope in Fig. 18.3. Similar linear regression on the log scale produces a value for β that defines the amount of rotation required to achieve a 0 slope in the regression of the resultant QTc on RR at baseline.

If there is enough pre-dose data within an individual, such as that obtained via Holter monitoring described in Sect. 18.2, it is possible to have individualized corrections that could be different values of β for each subject. Though this type of correction has desirable attributes, these attributes are often outweighed by the limitation of range of the observed RR intervals. This limited range in RR and variability in the measure of QT can lead to nonsensical corrections that are incompatible with physiology of the heart. An example of this is a negative slope value which implies QT shortening with increasing RR.

18.4 QT Prolongation

Disruption of the normal sinus rhythm results in the heart becoming less efficient at ejecting the blood into the circulatory system. These arrhythmias can sometimes be fatal. The example shown in Fig. 18.5 is a stylized depiction of Torsade de Pointes (TdP). Literally translated this means twisting of the points. Depolarization and repolarization of the atria and ventricle are not happening in sequence, and hence the blood is not being effectively ejected from the heart. In effect the heart is in spasm and is no longer a pump, which explains why TdP is often fatal.

It has been shown that QT intervals in excess of 500 ms are associated with TdP and other potentially fatal arrhythmias. In 1993, Bednar et al. (Bednar et al. 2001) reported the results of a comprehensive literature search for all approved non-cardiac medications listed as prolonging the QT interval. QT were reported in 86 cases, and QTc in 116 cases of TdP. The QT and QTc values associated with these cases of TdP are noted in Table 18.1. More than 80% of TdP cases were associated with QT values of at least 500 ms. After correcting the QT interval for heart rate, more than 90% of the TdP cases were associated with QTc values of at least 500 ms.

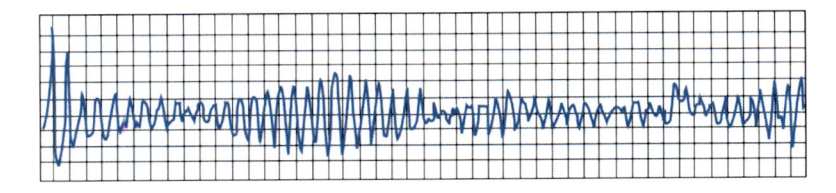

Fig. 18.5 ECG rhythm strip showing TdP

Table 18.1 Relationship between QT and QTc Interval and TdP

QT interval (N=86)		QTc interval (N=116)	
Range (ms)	TdP cases n (%)	Range (ms)	TdP cases n (%)
<500	17 (19.8)	<500	9 (7.8)
500–549	9 (10.5)	500–549	13 (11.2)
550–599	16 (18.6)	550–599	24 (20.7)
600–649	33 (38.4)	600–649	36 (31.0)
650–699	6 (7.0)	650–699	21 (18.1)
≥700	5 (5.8)	≥700	13 (11.2)

18.5 Study Types

For practical purposes, studies are classified into 2 categories depending on whether ECG data are captured as part of routine safety data monitoring or whether estimation of the drug effect on QT is the primary objective as in "Thorough QT Studies" (tQT).

If captured for routine safety, ECG data collection is likely sparse and the patient populations can range from healthy volunteers to special populations such as elderly or renally impaired to the target patient population for the intended therapy under development.

ECG data are often collected once pre-dose and once post dose. In large phase III trials, data collection is likely to be coincidental with routine study visits and steady state dosing. Data of this type, infrequent but on a wide variety of subjects, is useful for exploration of safety signals because of the volume of data and the diversity of subjects under study.

Population pharmacokinetic approaches that relate QTc changes to concentration of parent and/or active metabolite are often used to explore effects in various populations such as elderly subjects, those with impaired renal or liver functions, or different genotypes.

18.6 Thorough QT Studies

Additional information on graphical methods used for tQT studies can be found in the chapter by Tornøe (Chap.16) as well as in the FDA draft guidance (Food and Drug Administration 2005). This section utilizes graphics to demonstrate the importance of different features of the thorough QT such as time matching and use of active control.

The objective of the thorough QT study is to estimate the effects on the QT interval of an experimental therapy compared to placebo. This trial is intended to be a rigorous and intensive evaluation of a therapy's potential to prolong the QT interval.

The study design of the tQT trial attempts to control sources of variability due to time of day, meals, concomitant medications, patient demographics, and medical history. The thorough QT study is conducted in healthy volunteers and where possible with crossover designs. Subjects are randomized to sequences where experimental therapy, positive control, and placebo are assigned in a random order according to Latin Square, often balanced for first order carryover. Crossover designs aide the precision of the estimation because all comparisons are within subject, and matched for time of day.

There are situations where crossover trials are not feasible. For example, if an experimental therapy such as a monoclonal antibody, has an extremely long half-life, the washout required between periods of the crossover makes the trial extremely long. This long duration makes it difficult to retain subjects into the trial, and hence the advantage of the within subject comparisons wanes as a result.

Positive control groups are used to establish assay sensitivity. The positive control is selected based on its ability to reliably produce a mean increase in QTc of at least 5 ms without an appreciable impact on the heart rate. Five milliseconds is typically used in accordance with the regulators' threshold of concern as noted in ICH E14 guidelines.

For crossover studies, each study period is separated by an adequate washout. This washout allows pk and pd to return to baseline to minimize the potential for carry-over. The ECG interval data are collected at the beginning of each study period prior to dose at set times throughout the day. This is usually referred to as the baseline day. The baseline day is replicated each period in order to examine the potential for pharmacodynamic carry over in the interval data. Within each period the interval data are again collected at the same times of the day on a post dosing day. Time-matching of the baseline and post baseline values allows adjustment for diurnal fluctuation of the QT interval. In addition to diurnal fluctuations, there are also large postprandial effects that are addressed in the design of the thorough QT study. Fixing the times of the meals across the baseline and post baseline days and across the study periods allows the analysis to effectively remove the meal effects.

18.6.1 Sample Data from tQT Trial

Some sample ECG interval data from a thorough QT study are publically available on CTSPedia. These data were developed by the author in support of the Industry FDA Academia Safety Working Group (Industry FDA Academia Safety Graphics Working Group 2012). The intent here is to share the fictionalized data to allow the readers to follow along and reproduce the figures presented and extend those figures to their own clinical trial analyses. The dataset is provided as a comma-delimited ASCII file at the following link: http://www.ctspedia.org/wiki/pub/CTSpedia/StatGraphTopic017/tqt.dat.

The design of this sample tQT trial is a placebo and active controlled, 4 treatment, 4 sequence, crossover trial design. This design is presented in Table 18.2. The sequences are selected utilizing a Latin Square balanced for first order carry over.

Table 18.2 Thorough QT crossover design

Example tQT crossover design							
	Period 1	Washout	Period 2	Washout	Period 3	Washout	Period 4
Sequence 1	A		B		C		D
Sequence 2	B		D		A		C
Sequence 3	C		A		D		B
Sequence 4	D		C		B		A

A = placebo; B = active control; C = exp. low dose; D = exp. high dose

The objective of this trial is to examine the effect on the placebo controlled QT interval of 2 doses in a fictional experimental therapy.

The doses of the experimental therapy are noted in the dataset as low and high. In this specific case, we suppose there are no issues with dose limiting toxicity or toleration issues, so the dose is escalated beyond that which was anticipated to have clinical effect. If the dose cannot be escalated beyond the clinically efficacious dose, an alternative might be to give the dose with another drug intended to inhibit the metabolizing enzymes or transporters to produce higher concentrations of experimental drug.

The active control in this study is moxifloxacin. Moxifloxacin is chosen because of its ability to reliably produce mean placebo adjusted changes of at least 5 ms. The half-life of moxifloxacin is approximately 12 h making it suitable for use in the crossover design. By utilizing an active control, we gain information on the trails' assay sensitivity. If no changes are observed in moxifloxacin, there may be a cause to believe that the trial has failed.

ECG interval data are collected at day 0 (pre-dose baseline day) and at day 4 within each of the 4 study periods. The collections are matched based on the time of day. Meal times and content are controlled on the baseline day and day 4 to 8 am (hour 0) and noon (hour 4). For this experiment, interval data were collected at hour 0, 0.5, 1, 2, 3, 4, 6, 9, and 12 for each of day 0 and 4 for each period. Hour 0 corresponds to 8:00 am.

There are 48 subjects in the dataset. If there were no missing data there would be 48 subjects × 9 time points per day × 2 days per period × 4 periods per subject = 3,456 observations.

Variable names are constructed whereby the first part of the variable name denotes the interval, and a second part denotes the day (0 or 4) or change (c = day 4 − day 0; or cc = pbo adjusted change from baseline or sometimes called delta–delta).

For example qt_0 is the variable name associated with the QT interval from day 0; qt_c is the time-matched change from baseline (day 4 − day 0 within the same period); and qt_cc is the placebo adjusted change from baseline. The placebo adjusted change from baseline is computed as the within subject difference between treated and placebo of the changes from baseline. Because this value is a difference (trt − pbo) of differences (day 4 − day 0), it is often referred to as "delta–delta" or "double–delta".

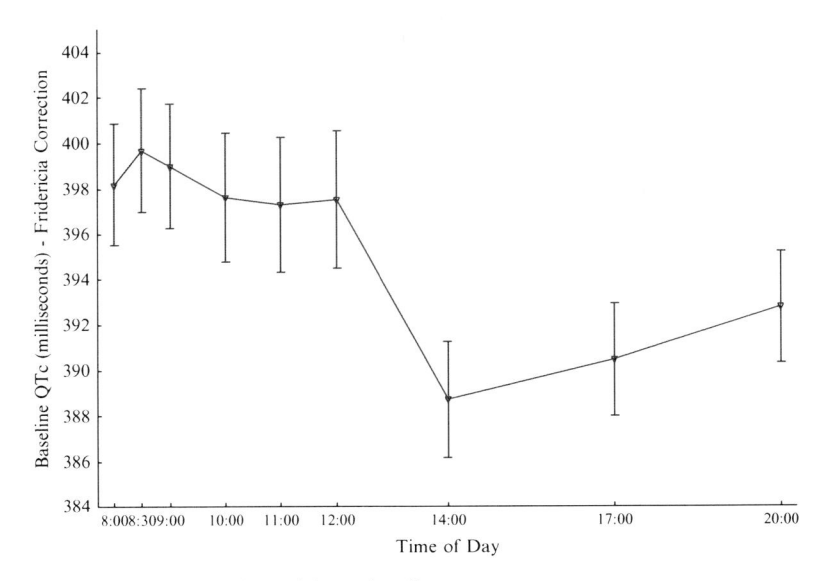

Fig. 18.6 QTc (95% CI) by time of day at baseline

Figure 18.6 illustrates the diurnal variation or circadian patterns and effects of meals. Each subject contributes 4 baseline days of observations, 1 baseline day per period, so we can get a rather precise estimate of the effects of meals or time of the day. The baseline data are free from treatment effect. This assumes, of course, that there was no pharmacodynamic carry-over from period to period. The meal in this case was at noon, note the postprandial change in QTc of nearly 10 ms, illustrating the need to control meal times on the days when ECG data are collected.

By subtracting the day 0 from the day 4 values, we should remove the diurnal and meal effects and be left with effects due to treatment (effects of period or sequence aside). Figure 18.7 plots the difference between day 4 and day 0 by time of day for the placebo period. Ideally, placebo should show no effect and the means by time should be distributed around the horizontal 0 reference line.

As stated above, one of the benefits of the crossover design is that each subject acts as its own control. The interpretation of the thorough QT study is dependent on the placebo controlled changes from baseline profiles. Thus placebo controlled change from baseline is derived from subtracting the placebo changes from baseline from each of the treated changes from baseline within each subject. The resulting difference of differences (delta-delta) or placebo adjusted change from baseline is plotted against time to produce 3 placebo adjusted profiles for each of the experimental drug doses as well as active control (Fig. 18.8).

The active control produced a rather pronounced profile over time indicating our experiment was able to discern the moxifloxacin drug effect on the QTc from

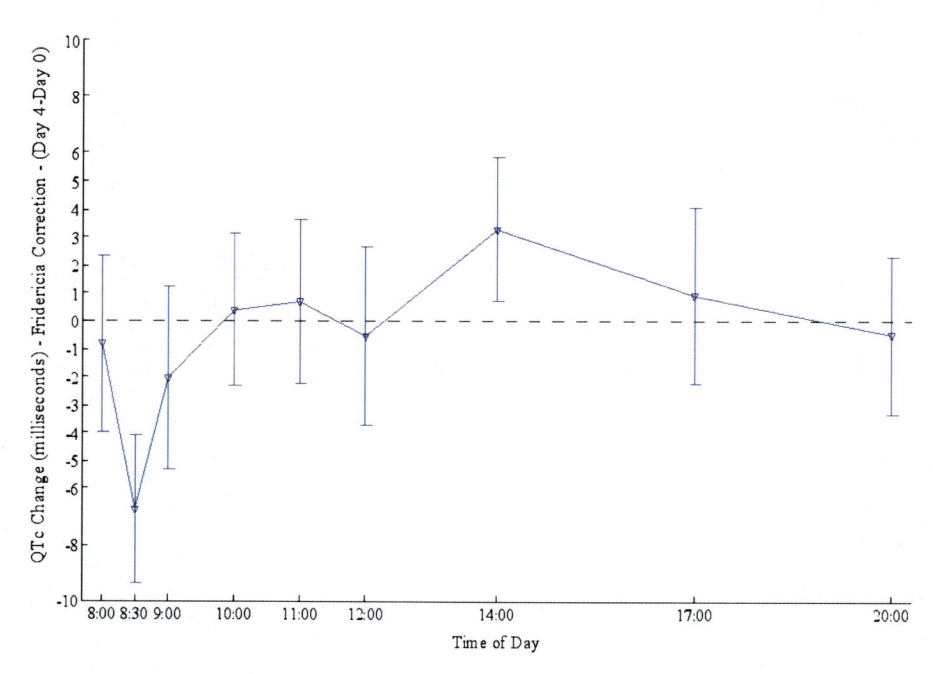

Fig. 18.7 QTc Change from baseline (95% CI) by time of day for placebo

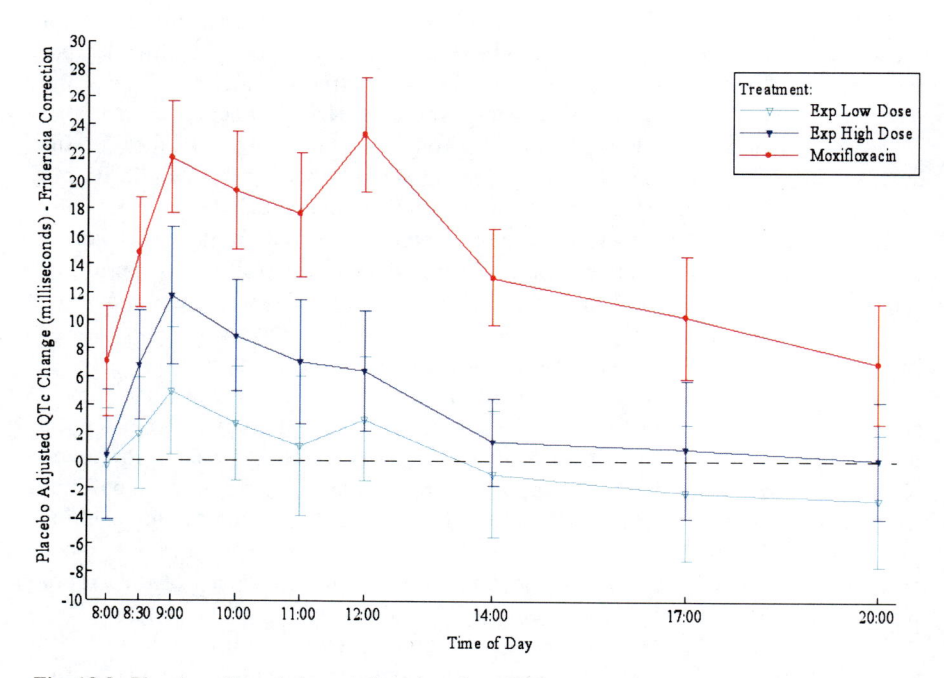

Fig. 18.8 Placebo adjusted changes from baseline (95% CI) in QTc

18 Cardiac Safety

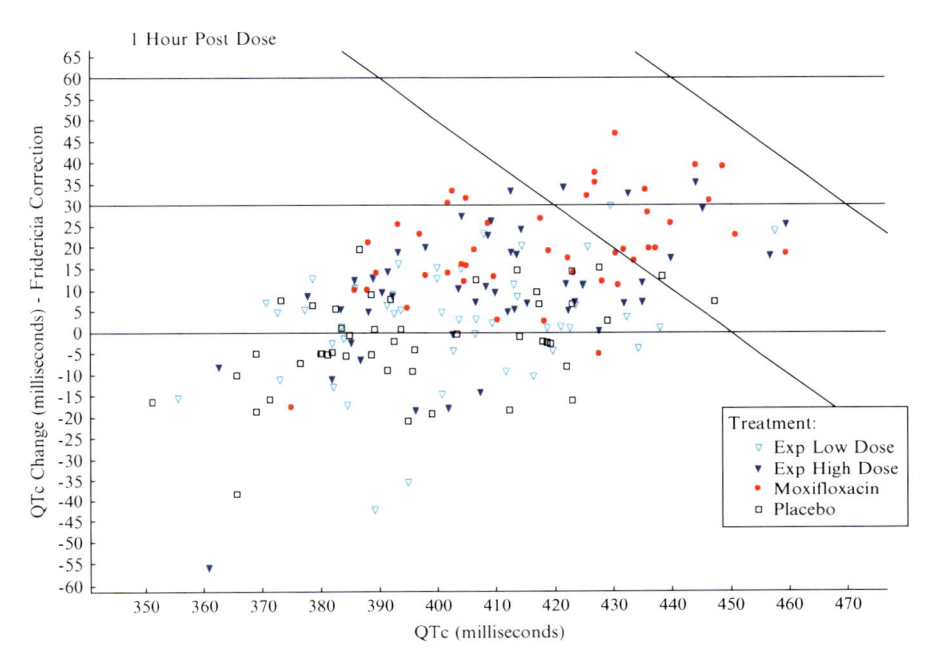

Fig. 18.9 Categorical QTc analysis at t_{max} for experimental drug

background noise. The 2 doses of the experimental drug appear to have changes that are dose related.

In addition to estimating a profile based on a measure of central tendency, the upper tail of the distribution is examined by means of comparing extreme values against predefined thresholds for the raw QTc as well as the changes from baseline in QTc. These analyses are often referred to as categorical analyses. A typical tabular display shows the number and percent of subjects with maximum values exceeding 450, 480, and 500 ms for the raw QTc and 30 and 60 ms for the change from baseline. Unfortunately the tabular display cannot link the changes to the raw values. By plotting the change from baseline values against the baseline values, as in Fig. 18.9, we can examine both the changes from baseline and the raw QTc values at one time. Figure 18.9 was generated at the t_{max} of the experimental drug. The changes from baseline are displayed on the ordinate and the corresponding time matched baseline along the abscissa. Each symbol represents an individual subject and there are 4 values corresponding to the 4 study periods. There are horizontal reference lines that correspond to 0, 30, and 60 ms change. From left to right the 3 diagonal reference lines represent 450, 480, and 500 ms. A quick visual check reveals that from the left, the first diagonal intersects the horizontal 60 ms change reference line at a baseline of 390 ms. This intersection is at 390 ms baseline and 60 ms change, or 450 ms for the raw post baseline value. Plotting the data in this way allows us to examine whether the changes are likely a function of a low baseline value or whether for a given treatment the change from baseline is high in addition to the raw QTc being long. Large QTc changes associated with a high raw value are worse than a large change that is associated with a low baseline resulting in a "typical" raw value.

18.7 Concluding Remarks

This chapter links the cardiac function to the interval data derived from the 12-lead surface electrocardiogram. Increases in QT and QTc are associated with potentially fatal ventricular arrhythmias. Graphical methods show the dependence of QT on RR and the importance of evaluating the QT correction. For the thorough QT study, graphics were used to demonstrate the importance of time matching to address diurnal and postprandial effects as well as the use of active control. Graphical methods used in describing categorical data link changes to the baseline values, which offers a view that not all changes are created equal. This linkage is impossible to see in tabular form.

Graphics are an important tool for the statistician, analyst, or physician both to gain a better understanding of the data at hand, as well as an effective and efficient communication vehicle for discussing and interpreting data with other scientists.

References

Bednar M, Harrigan EP, Anziano RJ, Camm AJ, Ruskin J (2001) The QT interval. Prog Cardiovasc Dis 43(5 Suppl 1):1–45

Food and Drug Administration (2005) Guidance for industry: E14 clinical evaluation of QT/QTc interval prolongation and proarrhythmic potential for non-antiarrhythmic drugs. U.S. Department of Health and Human Services, Food and Drug Administration. http://www.fda.gov/downloads/RegulatoryInformation/Guidances/ucm129357.pdf. Accessed 7 Aug 2011

Honig PK, Wortham DC, Zamani K et al (1993) Terfenadine-ketoconazole interaction: pharmacokinetic and electrocardiographic consequences. JAMA 269:1513–1518

Industry FDA Academia Safety Graphics Working Group. CTSPEDIA URL (2012): http://www.ctspedia.org/do/view/CTSpedia/StatGraphHome

Part IV
Operations, Marketing, and Post-Approval Graphics

Chapter 19
Data Visualization for Clinical Trials Data Management and Operations

Ted Snyder

Abstract Clinical development today requires orchestrating an ever-increasing number of data sources, from trial data in electronic data capture (EDC) systems and third party data transfers to operational data in clinical trial management systems (CTMS). As the number of data sources increases, so too do the headaches caused by data quality issues and the need for timely integration, as well as the operational challenges of orchestrating a cross-functional team and third party clinical research organizations (CRO).

Data visualization is becoming an important tool that allows organizations to better understand and work through many of these issues. It can help identify data and process issues early, reduce the time to lock the database, improve resource planning, and facilitate better communication amongst project teams. The following pages provide some example visualizations that illustrate practical uses of data visualization throughout the lifecycle of a clinical trial.

19.1 Introduction

Clinical development today requires orchestrating an ever-increasing number of data sources, from trial data in electronic data capture[1] (EDC) systems and third party data transfers to operational data in clinical trial management systems[2]

---

[1] Electronic Data Capture (EDC) systems allow data recorded at clinical sites to be entered electronically into a study database. This provides many advantages over traditional paper-based collection, most notably faster entry and data processing.

[2] A Clinical Trial Management System (CTMS) is an electronic system for capturing operational data about a clinical trial such as enrollment, site monitoring visits, and relevant documents. This type of operational data is separate from the actual trial data such as safety and efficacy.

T. Snyder (✉)
Clinical Informatics, Infinity Pharmaceuticals,
780 Memorial Drive, Cambridge, MA 02139, USA
e-mail: ted.snyder@infi.com

(CTMS). As the number of data sources increases, so too do the headaches caused by data quality issues and the need for timely integration, as well as the operational challenges of orchestrating a cross-functional team and third party Clinical Research Organizations (CRO).

Data visualization is becoming an important tool that allows organizations to better understand and work through many of these issues. It can help identify data and process issues early, reduce the time to lock the database, improve resource planning, and facilitate better communication amongst project teams. The following pages provide some example visualizations that illustrate practical uses of data visualization throughout the lifecycle of a clinical trial.

19.2 Data Review and Cleaning

Even with electronic data entry at clinical sites and automated edit checks in EDC, many basic data quality problems still plague trial sponsors, such as incorrect lab units from local labs and transposed fields and numbers. At the same time, the review tool of choice for many users is a stack of printed listings that, while comfortable for many people, are slow to review and provide little insight.

Automated statistical or database programs can be used as a first line of defense to identify potential issues such as outliers and out-of-range values but this can only help so much. Data visualization makes a natural second line of defense by allowing data managers and medical monitors to apply their own knowledge to tease out underlying issues or quickly identify new issues and trends that can be missed by automated programs and textual listings.

The following figures show examples of visualizations for verifying lab values.

Figure 19.1 shows a simple chart of Absolute Neutrophil Count (ANC) lab ranges where each subject is placed along the x-axis and the numerical values are displayed in standardized units on the y-axis. The color indicates whether values are in or out of range, and the horizontal lines indicate a textbook reference range for the lab test. The controls to the left allow the user to configure various parameters of the chart and quickly examine the same data from many angles in order to diagnose potential problems.

- The x-axis may be a categorical grouping such as subject identifier or local laboratory name, or a continuous grouping such as date of collection or time on study
- The y-axis may be numerical values in standardized units or original source units or a derived value such as multiple of upper limit of normal (ULN)
- The coloring of markers may indicate out-of-range status, toxicity grade, or even local laboratory name or source units
- Horizontal or vertical reference lines can add more information by showing normal ranges or reference time points such as date of first dose

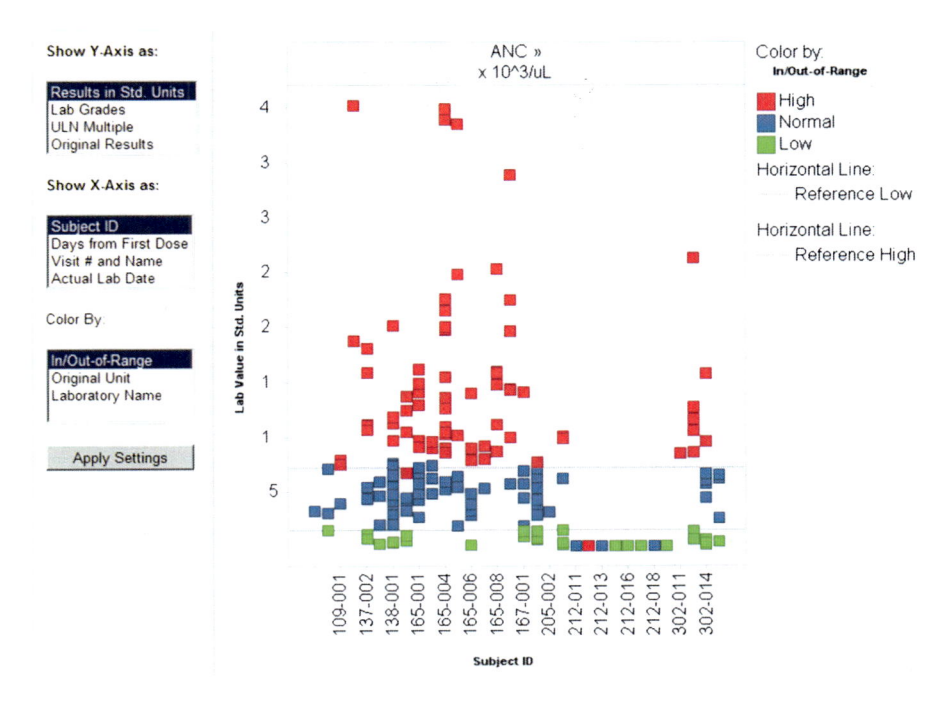

Fig. 19.1 Example visualization for lab value range check. The *y*-axis shows lab results in standardized units, the *x*-axis patient numbers

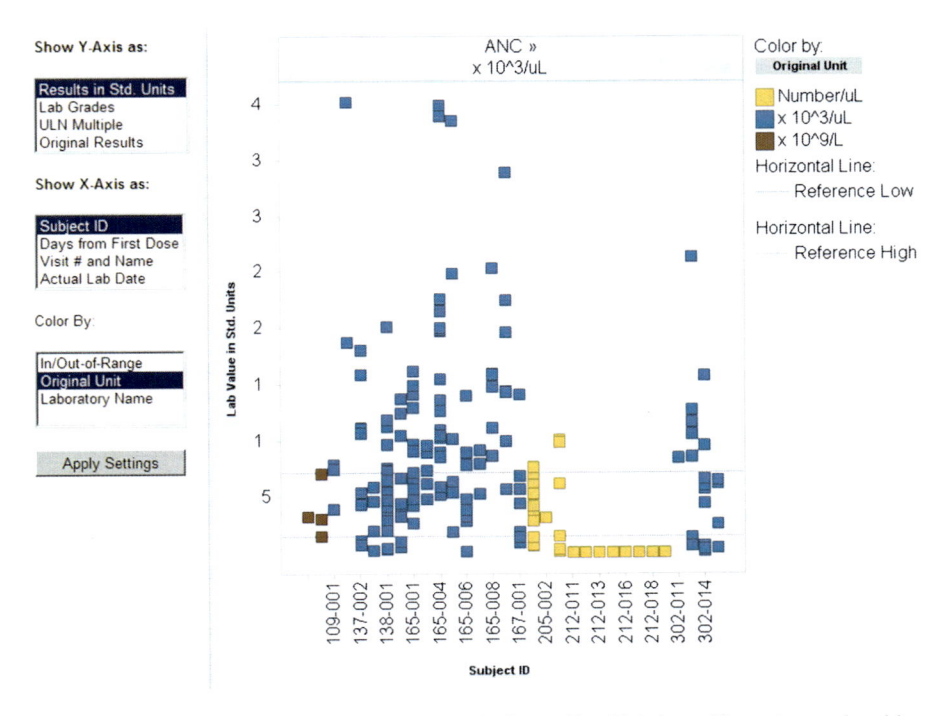

Fig. 19.2 Lab value range check visualization similar to Fig. 19.1, but with markers colored by original lab unit

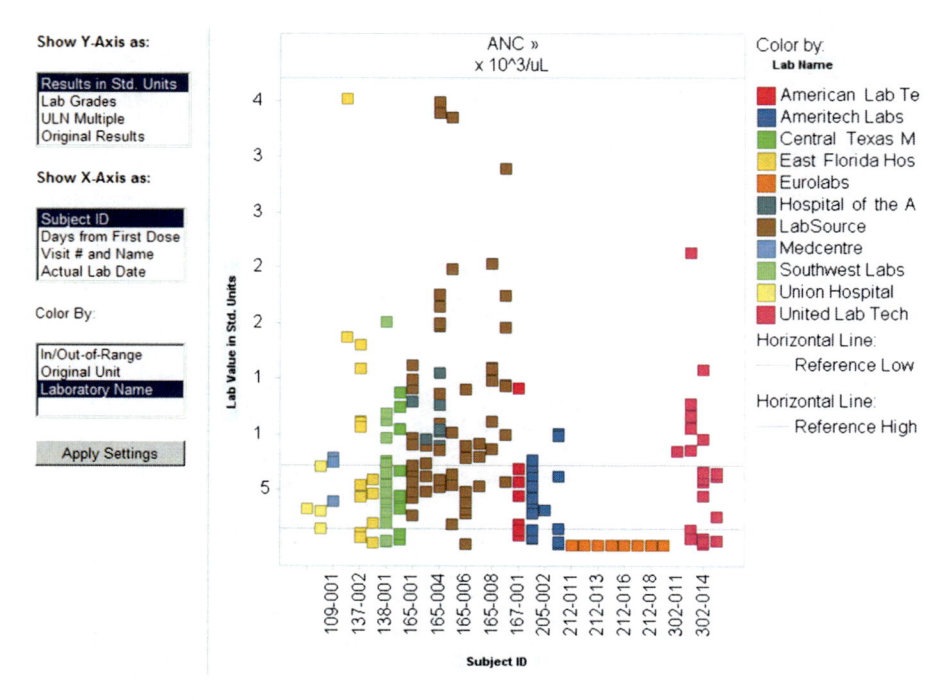

Fig. 19.3 Lab value range visualization with markers colored by lab facility name

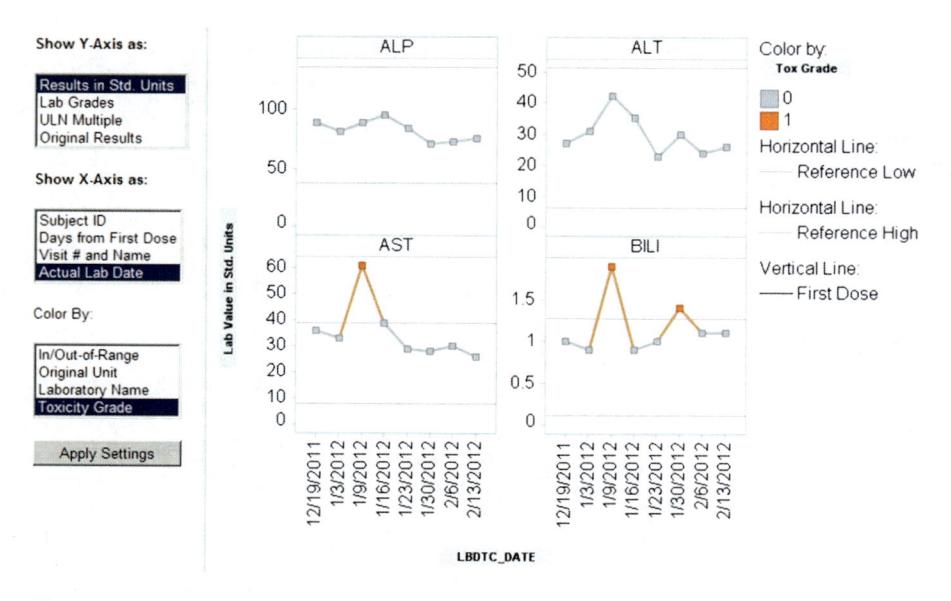

Fig. 19.4 Lab panel visualization displaying 4 lab tests for one subject over time. The *y*-axis shows the lab result in standardized units, scaled for each lab test, the *x*-axis lab date and time by lab test, color indicates toxicity grade

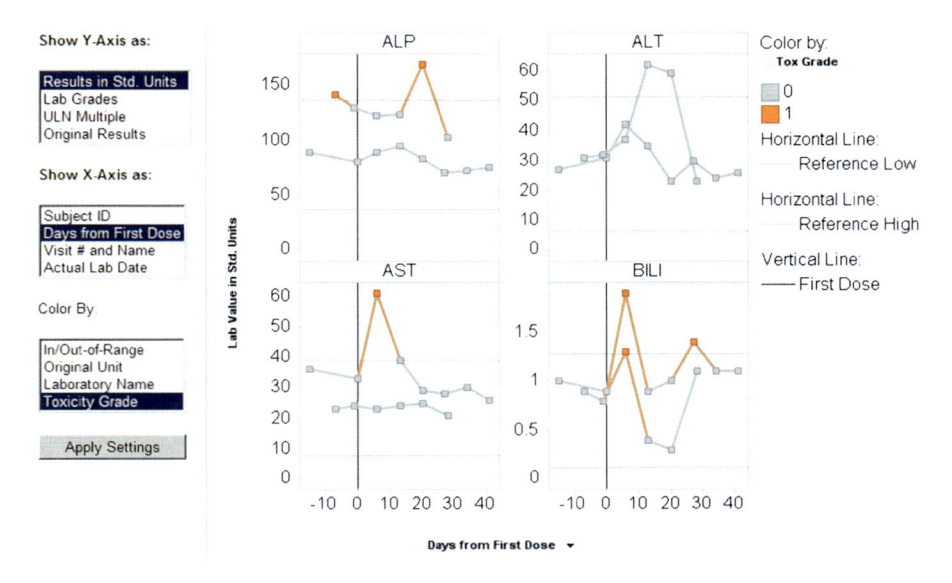

Fig. 19.5 Lab panel visualization showing multiple subjects, each represented by a separate line. The *x*-axis denotes days after first dose for each subject

In addition to displaying values graphically, interactivity is a key capability that allows end users to select any data point of interest and drill down for more information. For example, a data manager can identify the subject number or visit name and check the record for additional information such as whether the data have been verified at the site by a monitor yet. As a result, he can quickly identify potential issues and take necessary actions such as creating a query for the site to verify values. The following example illustrates this with a sample scenario.

An obvious place to start with lab values is to examine whether or not values are in range. Figure 19.1 shows a vertical series of lab values for each subject in the study. Horizontal reference lines indicate standard reference high and low thresholds.

While out-of-range flags are often set automatically based on reference ranges for the lab test, conversion to standardized units often reveals additional issues that become apparent in visualizations such as these. For example, the figure shows a cluster of patients all with very low values, towards the right side. There could be several explanations for this, ranging from clinical reasons to data quality issues. One item of interest is that even though the values are well below the expected range, some are flagged as high and some as in range. It is also interesting that they all appear to come from the same clinical site (i.e., the first 3 digits indicate the site ID). Simple changes to the visualization using the controls on the left side of the screen can help users look closer and determine the problem.

Figure 19.2 depicts the same visualization but the marker coloring has been changed to show the original unit specified with each lab value. Since the values are all converted to the standardized unit, one would not expect to see any difference in range based on the source unit. In this new view one can see that not all values with the source unit of "Number/µL" are affected by the issue, so it does not appear to be a problem with unit conversion factors, however all the affected patients from site 212 do have this unit specified. Further investigation is required.

In Fig. 19.3, the visualization has been changed one more time to color the values by the laboratory name, and it can be seen that all the affected lab values seem to be from the same lab facility. This indicates that the issue is most likely an incorrect unit being provided by this local lab facility and that data management should attempt to sort that out as opposed to querying all the data values in EDC.

A similar but different view using a slightly different configuration is to look at each subject's lab values over time. In Fig. 19.4, a lab panel visualization shows 4 liver function tests for a single subject over time.

This visualization shows the trend of multiple labs at once for a given subject, which again allows a data manager or medical monitor to determine if there are any data quality issues or clinical safety issues.

In Fig. 19.5, the same visualization is again slightly modified to show the same lab tests for multiple subjects, each represented as line within the trellis panel. Here, the x-axis is switched to days from first dose (denoted by a vertical line at $x=0$) to provide a common timeline for subjects.

Visualizing multiple data sets using a common timeline is a powerful technique that can provide a clear picture of a subject's clinical experience and enable rapid identification of more complex cross-CRF data issues.

Figure 19.6 shows a subject timeline with both tabular and scatter plot visualizations of multiple important events from dosing (exposure), efficacy, adverse events, and disposition data sets. Events are arranged with a common timeline of days from first dose, and flow from top left to bottom right. This allows one to do several important things:

- Compare end of dosing with discontinuation to make sure that discontinuation jibes with last dose date
- Compare reason for discontinuation with relevant adverse event and efficacy assessments (here using Response Evaluation Criteria In Solid Tumors or RECIST criteria). In this case the subject had an AE with outcome of "drug withdrawn" that is contemporaneous with the discontinuation reason of "Adverse Event," which makes sense
- Compare follow-up assessments to discontinuation. If subjects are being followed for survival, all assessments after discontinuation should be at regular intervals. If the subject discontinued due to death on study, then follow-up information should not be collected

The scatter plot timeline in this visualization is very powerful because once a user learns how to interpret it, he or she can very quickly identify issues based on the shape of the plot. Additionally, providing a tabular representation side-by-side with a graphic helps the user interpret the plot and allows for display of any records with missing dates.

19 Data Visualization for Clinical Trials Data Management and Operations

Subject ID	Event Date	Event Type	Visit	Days From	Assessment
024-022	12-Jul-2011	First Dose	First Dose	0	First Dose
	30-Aug-2011	RECIST Assessment	Cycle 2	49	Stable disease
	25-Oct-2011	RECIST Assessment	Cycle 4	105	Stable disease
	20-Dec-2011	RECIST Assessment	Cycle 6	161	Stable disease
	09-Jan-2012	Last Dose	Cycle 7 Day 8	181	Last Dose
	11-Jan-2012	Discontinuation	End of Treatment	183	Adverse Event
	15-Jan-2012	Discontinued AE	Ongoing	187	Drug Withdrawn AE
	07-Mar-2012	Follow-Up	Follow-Up	239	Alive at Follow-Up

Fig. 19.6 Subject timeline visualization integrating relevant data from multiple data sets for a single subject

19.3 Clinical Trial Metrics and Operations

Beyond the clinical database, the operational aspects of clinical trials also benefit from visual analysis. These reports can be used by the broader study team, including clinical trial managers, clinical research coordinators, CROs, and data managers.

A simple but effective visualization, to start with, is to look at enrollment over time as well as cumulative trend, as seen in Fig. 19.7. The x-axis has enrollment date at the year and month level and dual y-axes show enrollment in each month (bars) and cumulative enrollment as of the end of each month (line). Months with tall bars (many patients enrolled) correspond to a steeper slope in the cumulative enrollment line.

Data for charts like this may be pulled from a tracking system, such as a CTMS (Clinical Trial Management System), from a third party IVR (Interactive Voice Response) system that handles subject randomization or, in the simplest case, even the trial manager's spreadsheet.

This type of visualization is useful not only to understand and communicate the current status of enrollment but also to compare it to a projected or target enrollment curve (Fig. 19.8).

A logical extension of this is to distinguish and compare subsets of subjects, such as cohorts in dose escalation, domestic versus international sites, or distinct

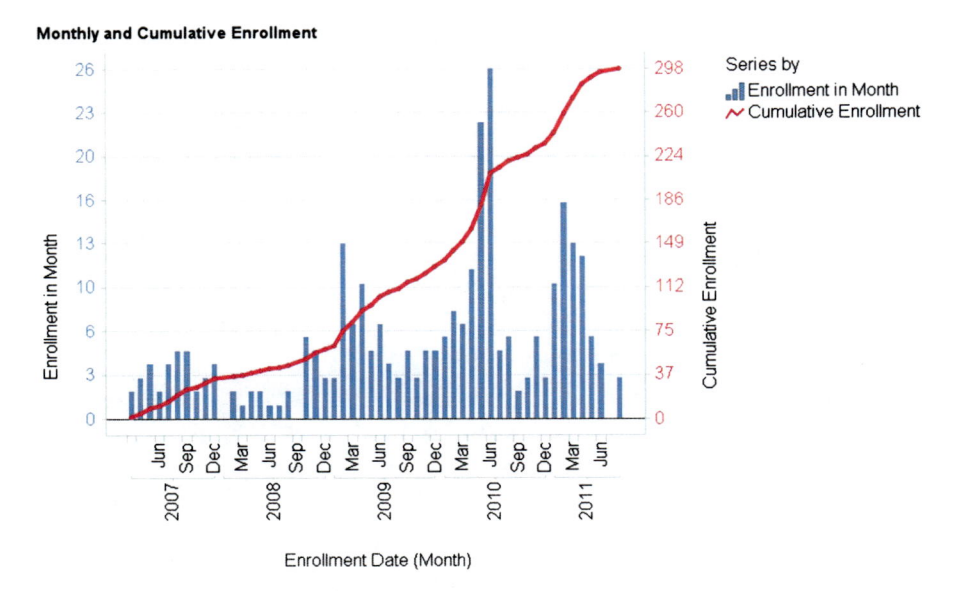

Fig. 19.7 Combination chart showing enrollment by month as bars and cumulative enrollment by month as a line

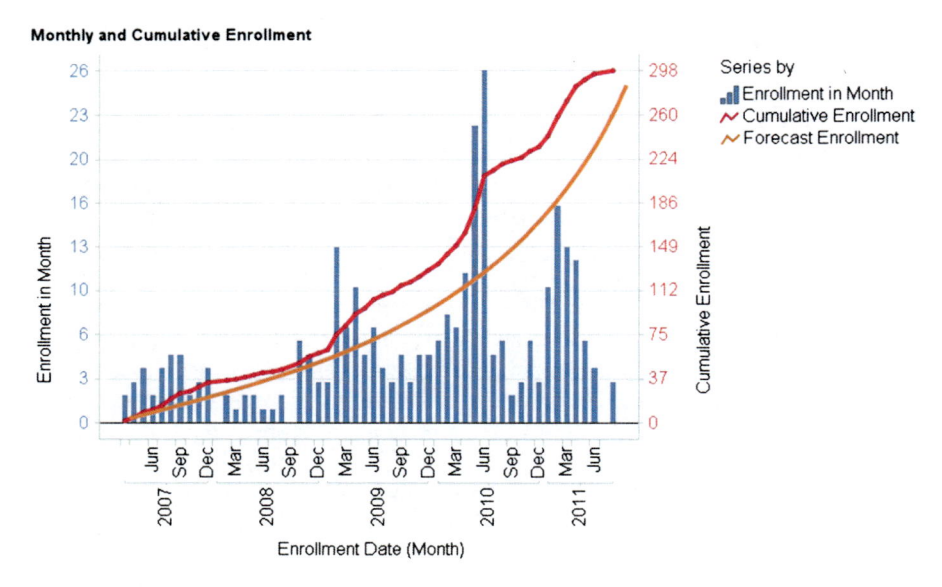

Fig. 19.8 Combination chart showing enrollment and cumulative enrollment compared to a forecast or predicted enrollment curve

19 Data Visualization for Clinical Trials Data Management and Operations   367

Cumulative Enrollment From First Patient In

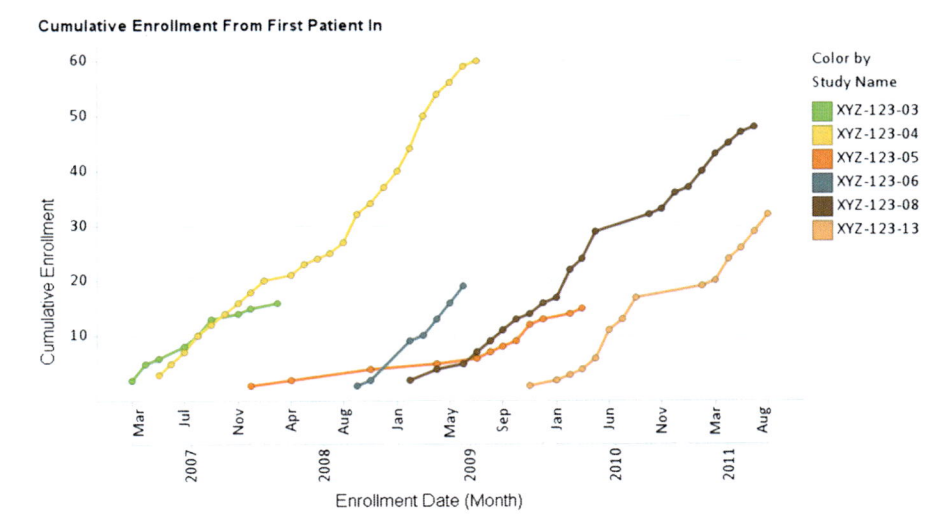

(Line chart with y-axis "Cumulative Enrollment" 0–60, x-axis "Enrollment Date (Month)" spanning Mar 2007 through Aug 2011. Legend "Color by Study Name": XYZ-123-03, XYZ-123-04, XYZ-123-05, XYZ-123-06, XYZ-123-08, XYZ-123-13.)

Fig. 19.9 Line chart comparing cumulative enrollment curves for multiple studies within a program over time

diagnoses. Additionally, at a product development or portfolio management level, enrollment can be compared for multiple studies within a program or even across programs. In Fig. 19.9, cumulative enrollment curves are broken out for several studies within a program.

A further step, as seen in Fig. 19.10, that allows easier comparison of trajectories between studies or subject groups, is to change the *x*-axis from actual date to a normalized timeline of days from "first patient in" (i.e., date that the first patient was dosed). This ability to toggle between an actual and a normalized measure is an invaluable technique in clinical data visualization.

Beyond enrollment, there are a number of other aspects of clinical trial operations that can benefit from the transparency of data visualization, such as metrics around data monitoring and data review. While most EDC systems allow for basic reporting of standard metrics like open query reports and verification status, much more can be done to precisely monitor the progress of the clinical study database.

Figure 19.11 shows a bar chart with the number of data points from the EDC system for several ongoing studies. This type of data can often be extracted at a granular level from the EDC system. The color segments of the bars reflect the verification or monitoring status from the EDC system and the options on the left allow the end user to drill down or slice the data in different ways. For example, within one study, one can visualize the amount of data by subject or CRF, and toggle the "Color by" option between verified and frozen status. When users click on a part of the bar chart they can see the actual data points that need verification or review. Using this type of report can lead to improved monitoring and data review by focusing the team on the most important issues at hand.

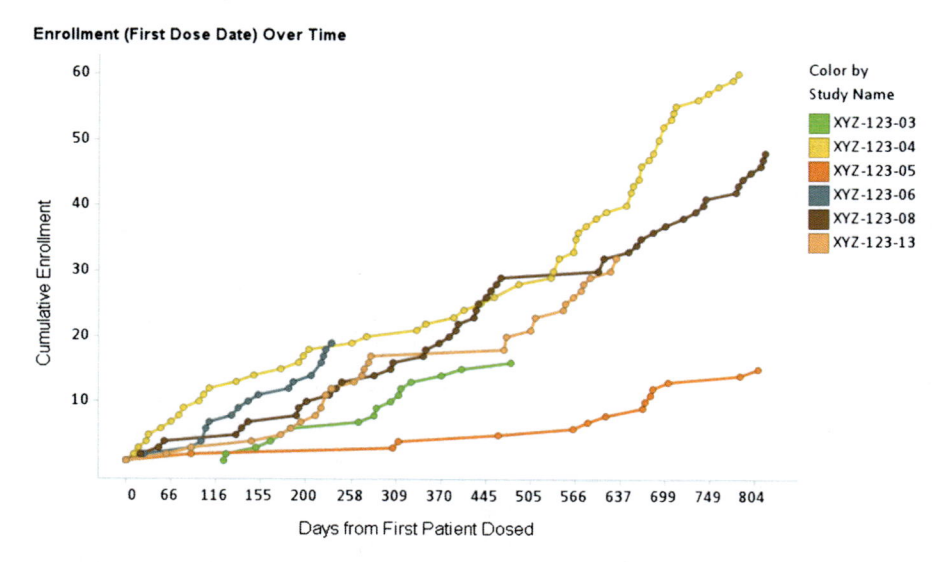

Fig. 19.10 Line chart comparing cumulative enrollment curves for multiple studies over time with *x*-axis as days from "First Patient In" or date of first patient treated

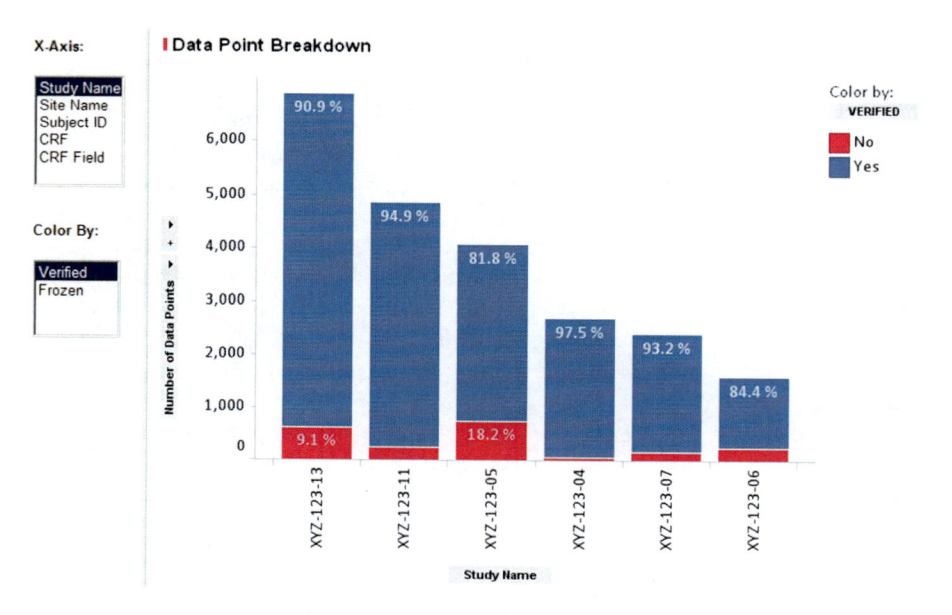

Fig. 19.11 Bar chart showing the current status of data in EDC needing to be source document verified or frozen before database lock

19 Data Visualization for Clinical Trials Data Management and Operations

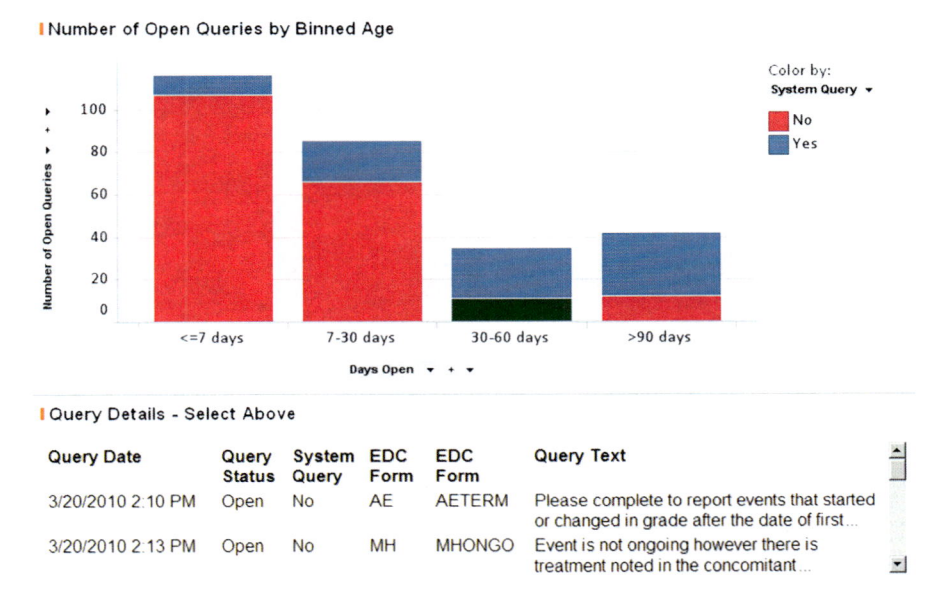

Fig. 19.12 Bar chart showing the number of open queries grouped by relative age and colored by whether or not they are system-generated queries. Details for queries of interest selected in the bar chart are displayed in the table below

The companion to data status is query status, which may similarly be broken out by study, site, subject, CRF, or query age. These visualizations can help not only to track the current status of outstanding queries, but help triage them and identify areas of trouble. For example, in Fig. 19.12 we see queries by relative age segmented by whether they are system queries fired by automated edit checks or by a human user.

The data manager will likely want to address queries outstanding for more than 30 and 90 days first, while queries submitted in the last week can be considered lower priority. Additionally, queries submitted by data managers are usually given higher priority over automated edit check queries fired by the EDC system. Interactivity is again a key feature, because it enables the user simply to click on the segment they want and view the list of high priority queries to be addressed in the table below. The listing of queries can be exported directly from this table into a PDF or spreadsheet and sent to the monitor, CRO, or site coordinator for follow-up.

Additionally, by looking for CRF fields contributing to high numbers of queries, study teams can identify problem areas that may need to be addressed with training for sites or monitors.

Another area with operational impacts that may have clinical consequences is the reporting of adverse events (AE) by sites participating in the study. One of the frequent sponsor challenges with EDC, beyond cleaning the data that are entered, is figuring out which data have not yet been entered. Protocol-scheduled evaluations such as lab values or efficacy assessments can be tracked based on each

Site Profile: Reported AEs by Time on Treatment

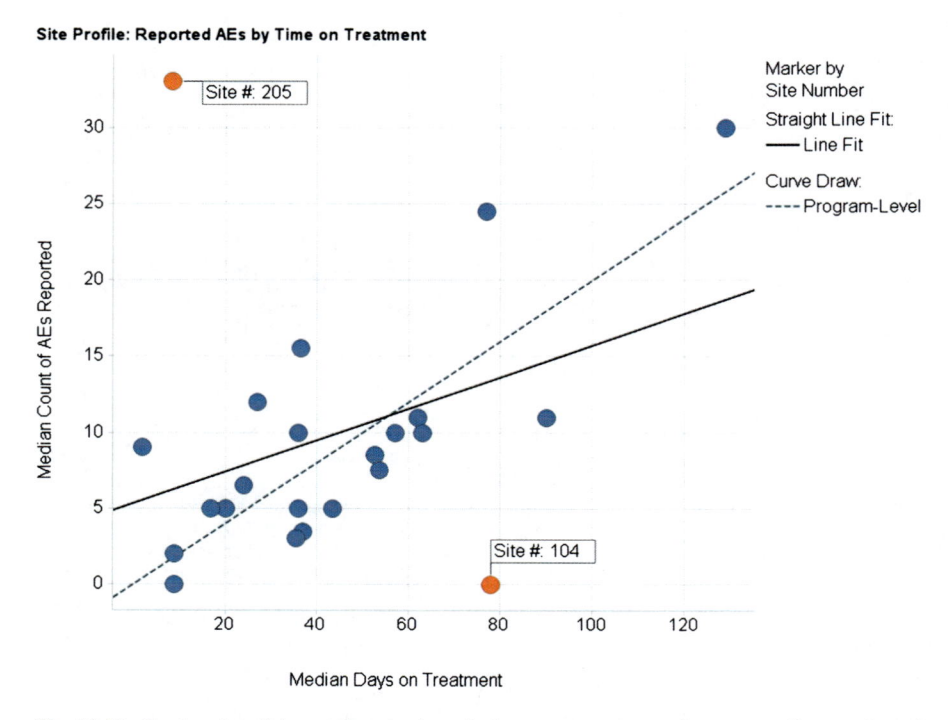

Fig. 19.13 Scatter plot of the median number of adverse events reported versus median number of days on treatment with one marker for each study site

subject's predicted schedule. However, safety data such as adverse events (AE) and concomitant medications (CM) are entered ad hoc by sites based on sponsor guidelines. One way of addressing this problem is to compare typical rates across sites and to other studies.

Figure 19.13 shows a scatter plot with one marker for each site number, where the x-axis indicates the median days on treatment for all patients at the site, and the y-axis the median count of AEs reported at the site. As the linear fit indicates (solid black line), the number of reported AEs should typically increase as time on treatment increases. The second hashed line shows a similar or expected pattern based on historical data from other studies within the program.

While both parameters will vary widely among subjects, we should not expect to see much variation between sites. This visualization is helpful in spotting two potential problems, illustrated with the two selected markers in Fig. 19.13:

- Site #205 (top left) has many more AEs reported for subjects that have been on study a relatively short time. The clinical team may want to review the AEs reported for this site to address potential training issues regarding what severity of adverse events requires reporting

19 Data Visualization for Clinical Trials Data Management and Operations 371

I Days Until Data Entry by Site

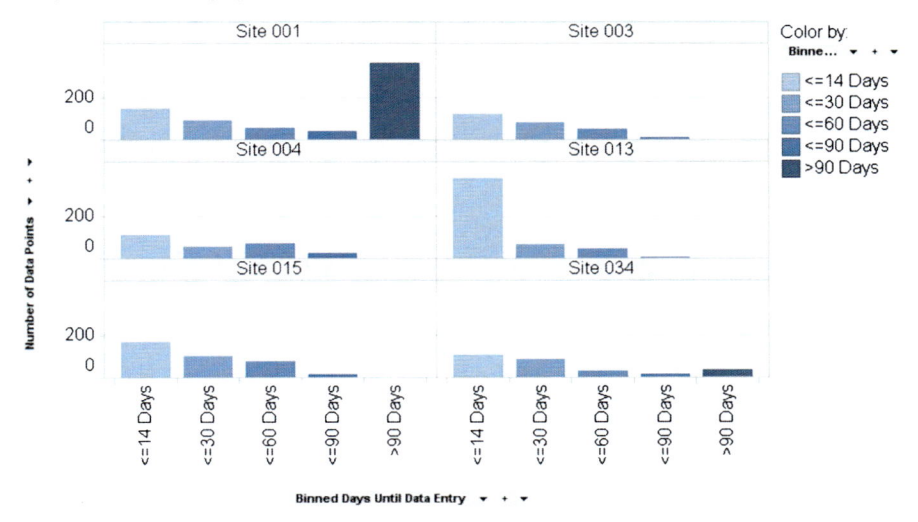

Fig. 19.14 Bar chart showing the amount of data entered by each study site grouped by the relative time until data entry

- Site #104 (bottom center) has no reported AEs for subjects that have been on study for a median of over 2 months. The team may want to work with the sites to address whether data are being reported in a timely manner so as not to get behind with data monitoring and review

One final technique for analyzing data entry behavior at sites is to examine the lag time for sites to enter data after a subject visit. Not all data have an inherent event date, but many data, such as labs and exposure, do have an assessment date that can be compared to the "created" timestamp in EDC. Although this is just a sampling of the data in a study, it can be a good indicator of site performance.

In Fig. 19.14, a bar chart by study site shows the number of sampled data points from EDC (y-axis) binned by the number of days elapsed between the actual date and the created timestamp date in EDC. Although the general trend is that sites enter most data within 14–30 days, some amount of entry lags out over 60 or even 90 days. Also, some sites, such as Site 013, are much better at entering data quickly, but others such as Site 001 are entering a large proportion of data after more than 90 days. Since a backlog in data entry has negative operational implications for a study timeline (if data are not entered they cannot be verified, reviewed, or frozen), this type of visualization can help identify which sites clinical operations or the external CRO need to focus attention on. Since many site contracts are now written with requirements for payment based on timely data entry, this information can be particularly helpful in motivating sites to keep up with data entry.

19.4 Conclusion

The examples illustrated above are a small sampling of the potential for the use of data visualization techniques within clinical data management and operations. Data visualization will not solve all data quality and operational issues facing clinical study teams, but it should be considered an important tool to identify issues faster and to gain a better understanding of clinical data. Finally, beyond the visualization of data, the addition of interactivity and configurability can create tools that allow the broader clinical study team to realize significant operational benefits.

Chapter 20
Post-approval Uses of Clinical Data, Phase IV Data, and Sales and Marketing Data Visualizations

Sam Weerahandi, Birol Emir, and Ed Whalen

Abstract This chapter discusses some useful graphics in Phase IV and Sales and Marketing applications. In the first half of the chapter we show some useful presentations of existing clinical trial data and post-approval Phase IV studies' data. Then, we undertake some applications in Sales and Marketing ranging from various trend analyses to in-depth analysis of field force optimization; in each section, a brief introduction to the application followed by some simple graphics highlighting findings in an easy to interpret manner are presented.

Most illustrations in this chapter are based on simulated data rather than actual data, and yet they are representative of real world situations.

20.1 Introduction

The purpose of this chapter is to provide an introduction to some of the graphics used in a variety of sales and marketing applications and in Phase IV clinical study data. In each of the following sections, we will provide a brief discussion of the objectives, the underlying issues pertaining to the application, and the nature of the data that one typically encounters. Then we present examples of useful graphics that provide insight into the questions and issues one tries to answer, or that provide input to business decision making.

Due to proprietary nature of the data, we use approximations of actual data in most examples. This is accomplished by simulating data from typical ranges of

---

S. Weerahandi (✉) • B. Emir • E. Whalen
Pfizer Inc., 235 East 42nd Street, New York, NY 10017, USA
e-mail: Samaradasa.Weerahandi@pfizer.com

such metrics as the mean, variance, skewness, trend, and so on. Therefore, in our illustration of data analytical methods or examples of typical graphics, such simulated representative data will work as good as real data in each of the applications.

In Sales and Marketing applications, it is important to have simple charts that any senior manger can easily grasp without needing a background in statistics. Basically we need to limit ourselves to such charts as line charts, bar charts, pie charts, and maps. As we illustrate below, simple charts can be presented in a manner that provides many insights while highlighting the value of underlying analyses.

In the Phase IV setting, the audience covers a range from specialized scientists to sales representatives. Therefore, the types of graphics used vary in complexity from simple and quickly understood presentations as described for the Sales and Marketing setting to fairly involved presentations for non-statistician scientists and medical experts. The recipients of the information may use it for decision purposes but more often need it for education and understanding of new findings.

20.2 Post-approval Uses of Clinical Data

Consistent with the ICH general guidelines (ICH 8 1998), we define Phase IV trials as those studies with a valid scientific objective that are initiated after regulators approve of a drug. These include but are not limited to clinical trials of any design, drug–drug interaction trials, and long term safety studies. Phase IV data reach a broader audience compared to a pre-approval, regularity submission environment. Interactions with the users of the analyses are direct and more frequent and include colleagues from marketing and commercial, market access teams, external leading medical experts, and legal representatives (both within and external). Because of these different backgrounds, statisticians need to accommodate people with a wide range of quantitative abilities and communicate at the appropriate level. Hence the dissemination of these results raises considerable challenges.

In this section we summarize graphical data presentations in the clinical communications and marketing settings. These can be separated under 3 titles: internal communications, external communications, and communications to non-statisticians. Internal communications can be of the following types: brainstorming meetings, preparations for congresses, and planning new study design based on accumulated clinical trials data. Examples of the external meetings are public presentations, expert panel meetings, and regulatory interactions (e.g., FDA advisory committees or type C meetings). Communication to non-statisticians includes explanations of statistical methods to legal representatives, marketing colleagues, and clinicians.

Our examples come from epilepsy and neuropathic pain clinical trials settings and will consider both efficacy and safety (adverse events). The primary criterion to establish efficacy in neuropathic pain studies is the endpoint mean pain score, derived from a daily pain diary recorded by the patient using an 11-point numeric (integer) rating scale (NRS). Upon awakening, the patient evaluates the pain experienced during the previous 24 hours by circling the number on the scale that best describes the pain. The integer scale ranges from 0 (no pain) to 10 (worst possible pain). Weekly averages of these pain scores form the dependent variable(s) in analyses. The change from baseline (the average of the values prior to receiving treatment) also typically gets analyzed with more negative changes indicating improvement (reduction) in pain. Finally, what we present here is a small set of examples from a much larger and constantly evolving set of graphical presentations in the post-approval phase.

20.2.1 *Internal Communications and Brain Storming*

Internal communications are integral part of planning and execution of drug development and promotion. In Phase IV we have the following types: brainstorming meetings, preparations for congresses, and designing new studies based on accumulated clinical trials data.

The typical brainstorming meetings may occur twice yearly with a goal such as idea development for new publications based on current and past trials' data. A team of scientists will discuss questions and ideas of interest and statisticians will present background analyses to medical or clinical colleagues. During one of these meetings, we provided a graph to describe the placebo effect, which is statistically quite simple but conceptually useful in one of these cross-disciplinary meetings (Fig. 20.1).

The question of interest behind this figure arose out of an observed increase in pain efficacy among the placebo patients in many trials. This is a big concern since it means that we can easily miss the incremental drug improvements over placebo. So we examined the placebo arms of the 8 completed pain trials. We plotted the weekly averages of the change from baseline NRS pain values against the weeks of the trial duration. Using input from the clinical colleagues, we identified 4 of those studies as the high placebo response (HPR) studies, those that had a greater than 2-point improvement (negative values indicate improvement) from the baseline (shaded as orange). We identified the other 4 studies as low placebo response (LPR) studies (<2-point improvement from the baseline—shaded as green). We also marked the approval status of the studies as whether they are pre- (solid lines) or post-approval (dotted lines). Lastly, we identified the ratio of the active to control arms as 1:1 ratio as red, 2:1 as green and 3:1 as blue colored lines. This presentation easily displays that differences in placebo response to pain were observed for time of the study (pre- or post-approval), ratio of active to control, and study duration. Higher placebo responses on pain were observed in post-approval studies compared to pre-approval studies. Using these initial observations, statisticians could then help the clinical and medical staff hypothesize on reasons for the apparent differences and, in some cases, test those hypotheses. These results were eventually presented at a clinical congress (Freeman et al. 2010).

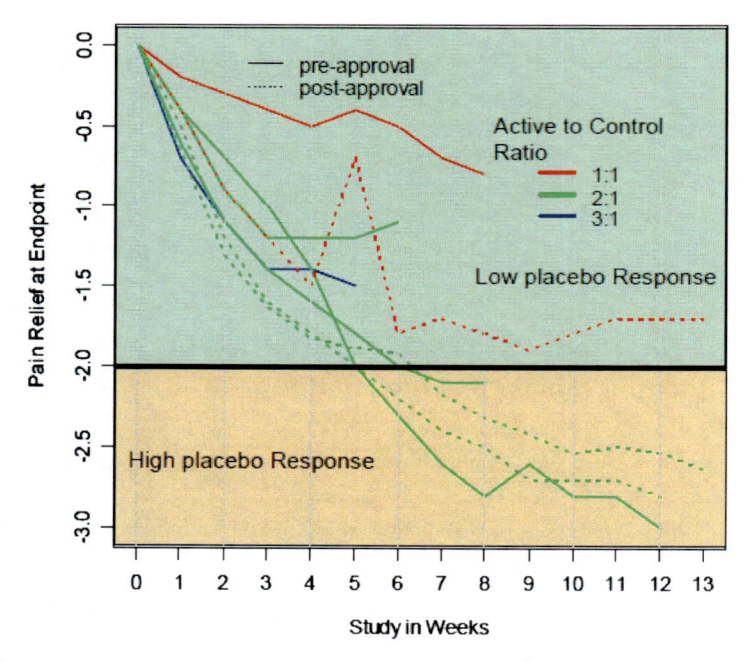

Fig. 20.1 Weekly placebo response over time in mean pain score. This figure uses simple statistics and trial characteristics identification to generate hypotheses on what factors may contribute to increased placebo effect

20.2.1.1 Preparations for Congress Activities

Our clinical team collected patient level pain symptom data using the Neuropathic Pain Symptom Inventory (Bouhassira et al. 2004) scale in 4 separate trials with different disease mechanisms. This 12-item questionnaire was given to the patients at baseline and during treatment. The clinical question of interest was whether patients cluster into different homogenous groups relative to the inventory symptom items. Each NPSI item ranges between 0 (no pain) and 10 (worst imaginable pain) except items 4 and 10, which were excluded from the analysis. The remaining 10 items gather pain symptom information on Burning, Squeezing, Pressure, Electric Shocks, Stabbing, Provoked by Brushing, Provoked by Pressure, Provoked by Cold, Pins and Needles, and Tingling.

Scaling is an important component of the cluster analysis. Because the patient disease populations differ for each of the 4 trials, the cluster analysis will need to account for variation in questionnaire results that arise from population differences. For a given patient, they will rate some items with high scores (lots of that pain sensation) and others relatively low. To factor out disease driven differences we can first normalize the data using a variety of techniques. In Fig. 20.2, each boxplot displays a representative example based on 1 out of every 20 patients with similar

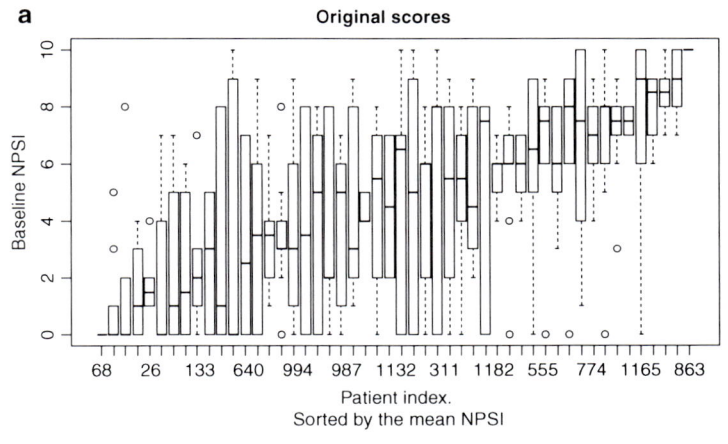

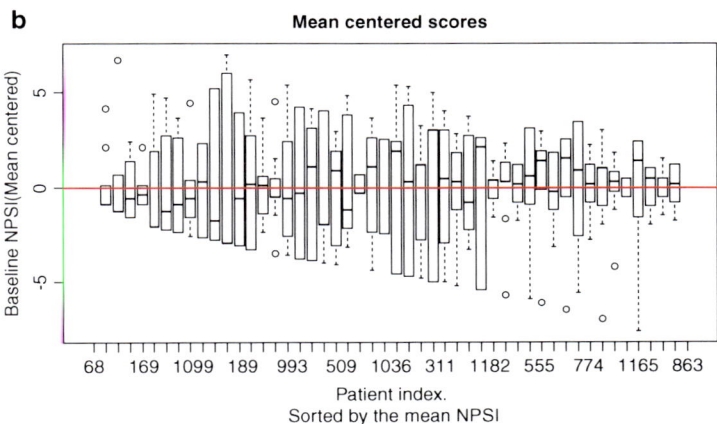

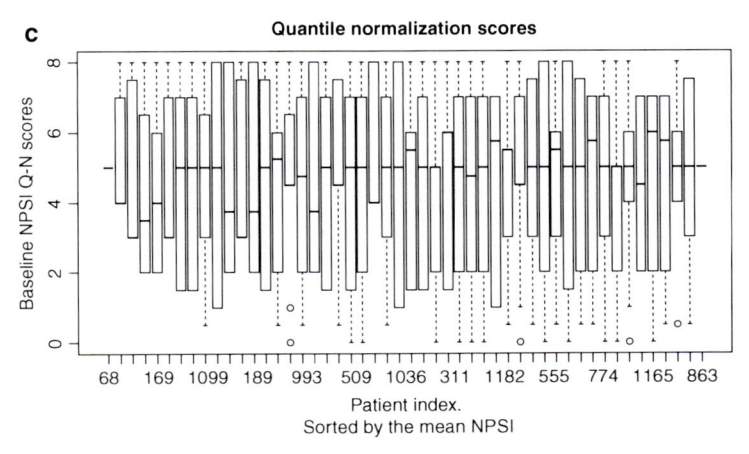

Fig. 20.2 Individual NPSI mean scores versus the patient index (every 20th patient with similar means, sorted from smaller to larger). (**a**) Original NPSI: Lower means have right skewed tails and larger means patients have left skewed tails. (**b**) Mean-subtracted score: Skewness still present after transformation. (**c**) Quantile-normalized data: No more skewness

means. For Fig. 20.2a, we plotted the original scores. Next, for Fig. 20.2b, we subtracted the individual means from each scale and boxplot the same patients. The skewness still exists in these mean-subtracted values. So in Fig. 20.2c, we quantile-normalized (Amaratunga and Cabrera 2001) the data (similar to a Fisher-Yates transformation (Cabrera and Emir 2012)). Using these simple graphics we monitor the effects of the different transformations and decide which to pass through to the cluster analysis step. Subsequent cluster analyses results then were used as part of a presentation at a scientific congress and as part of manuscript.

20.2.2 Planning New Study Design Based on Accumulated Clinical Trials Data

Studies that support a particular concept for drug treatment (proof of concept or POC studies) are an important part of drug development. In most cases, POC studies are the first time that patients get exposed to a new drug to test its efficacy. In the field of epilepsy trials, a leading expert collaborated with us to evaluate a POC design paradigm using our repository of clinical trial epilepsy data (French et al. 2011). The question was how much to shorten the epilepsy patients' exposure in POC trials, which typically required 12 weeks on treatment following a 6–8 week baseline observation period. That is, if the drug is efficacious, we want to take it to next phase of testing as quickly as possible and if it has little or no efficacy then we want to quickly and reliably learn about it. Regardless of the drug's efficacy we also want to minimize each patient's exposure to an experimental treatment for safety reasons.

Our trials include at least 6 weeks of baseline observation during which we collect data on the number of seizures they experience. Once the patient satisfies inclusion and exclusion criteria, he is randomized and takes placebo or one of different doses of drug for at least 12 weeks—the standard epilepsy trial design even for POC. To explore alternatives to these longer exposures we truncate the amount of data used in our analysis and see if we lose much power. We do this by looking at all 6 week by 12 week combinations of baseline and treatment exposures using the original study definitions for an efficacy signal—an odds ratio (OR) of 2 as a cutoff. We reanalyzed the response (a 50% reduction in seizures relative to baseline value) to treatment using a logistic regression with covariates for baseline seizure rate, study, and treatment. We plotted these ORs as a color-coded graph (sometimes called "heat map"), see Fig. 20.3. After consulting with experts, we reached a recommendation to use 4 weeks of baseline and 2 weeks of post-randomization for the POC trial to test a new drug. We also looked at sample size issues but do not present them here.

20.2.3 Sales Representatives

Sales representatives regularly visit physicians to update them about the latest on the safety and efficacy of the medicine that they represent—sometimes called drug detailing. Recent rulings have increased the limits on drug company representa-

Fig. 20.3 Heat map for power in a proof of concept study. This figure shows the ability of a study to detect a difference between an efficacious treatment and placebo based on the amount of baseline observation in weeks (Preweek) and on treatment exposure (Postweek). The lighter colors, yellow or lighter, indicate that the corresponding combination of observation times works well for a given sample size

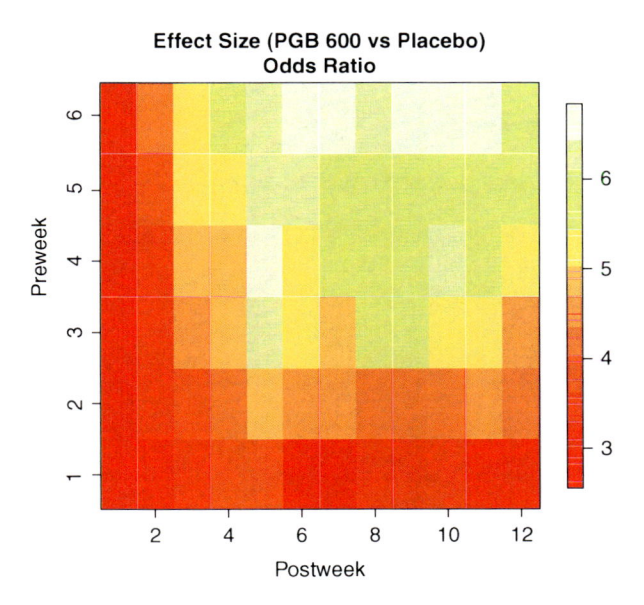

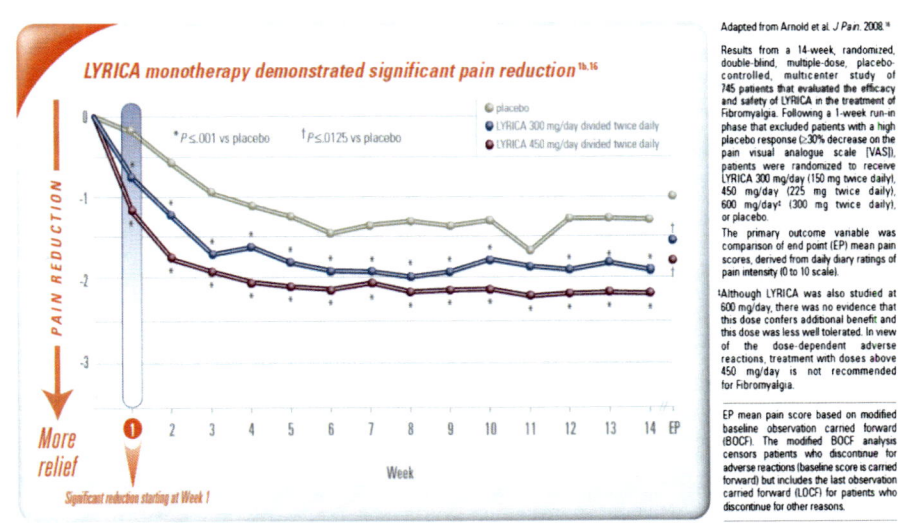

Fig. 20.4 A typical sales promotional graphic showing pain relief maintenance over time and with accompanying fair-balance language for the drug product. Such materials must convey large amounts of statistical information that must be easily understood by non-statisticians in a matter of minutes or less

tives' interactions with the physicians. Hence it is very important to relay the medicine information to the treating physician in a short time, without ambiguity and with a clear "fair balance." We present a typical example of the post-approval use as shown in Fig. 20.4. In this promo piece, we simply graph the LSMEANS mixed model repeated measure (MMRM) values against time (weeks). On the right hand

Odds Ratio (OR) and 95% CIs
Homogeneous Case

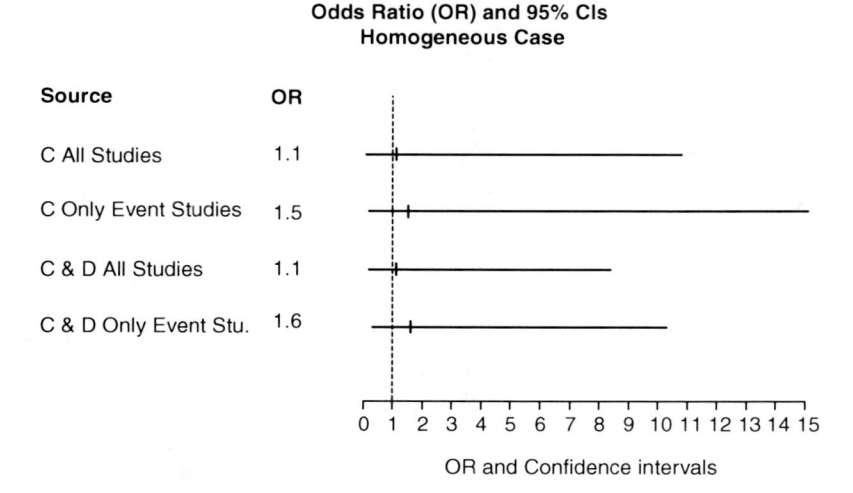

Source	OR
C All Studies	1.1
C Only Event Studies	1.5
C & D All Studies	1.1
C & D Only Event Stu.	1.6

OR and Confidence intervals

Fig. 20.5 Comparison of estimates and intervals for odds ratios when including and excluding studies with no events of interest from estimation process

side, for fair balance the AE information and study details are provided in text. Marketing colleagues test these materials for clarity and ease of understanding. Afterwards, the representatives are trained on how to use them.

With this graph our markers simply wanted to convey the behavior of drug over time. It is the outcome of an MMRM analysis which is not easy to describe in lay language. This graphic displays the p-values, which are always an interest, for each week. The y- and x-axis clearly define the direction of pain improvement.

20.2.4 Regulatory Interactions (Safety—Adverse Drug Effects)

In post-approval settings regulatory bodies such as the FDA or EMEA generate queries to drug companies about marketed products and their safety profile. The following data come from an FDA advisory committee regarding suicide and antiepileptic drugs. The reader can find more detailed information at the FDA website (http://www.fda.gov/ohrms/dockets/ac/08/briefing/2008-4372b1-01-FDA-Katz.pdf).

Figure 20.5 shows a Bayesian analysis of odds ratios for suicide related events for 2 of the drugs in the query. The events are rare so Bayesian methods were applied using Jeffreys prior. The primary FDA analysis uses standard frequentist methods that ignore any studies with no events even though those studies still provide some information on overall exposure. Figure 20.5 shows the effect of excluding these studies for drug C and drugs C and D combined. The "All Studies" intervals are

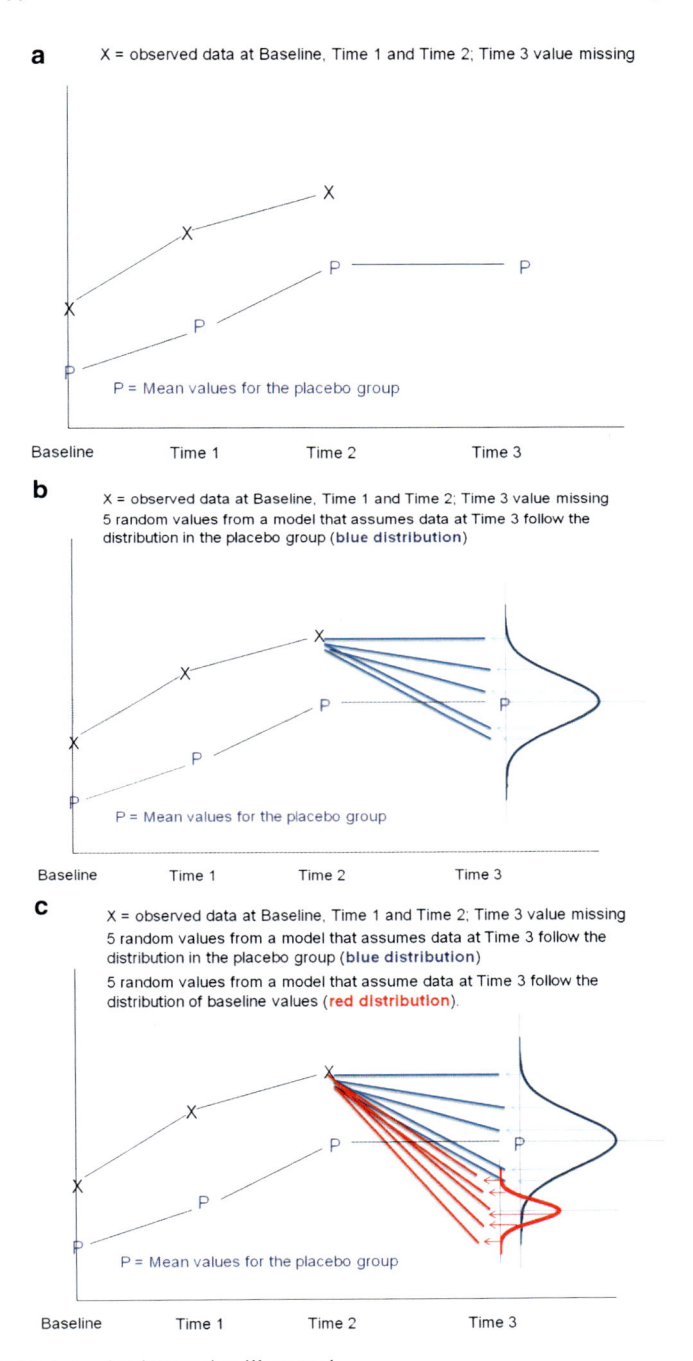

Fig. 20.6 Missing value imputation illustrated

shorter as a result of including the larger body of study information—compared with the "Only Events Studies" interval lengths.

20.2.5 Multiple Imputation Presentation

The statistician's job does not end once the data gathering, analysis and interpretation are completed. We have seen the importance of the dissemination of the results and interpretations to a diverse group of disciplines. Statisticians also get called upon to explain the function and value of methods that non-statisticians may not have previously seen. Figure 20.6 shows a tutorial piece used to explain to company clinicians and other non-statisticians the idea behind multiple imputation (MI) methods and why they may work better than other methods for handling missing data. Note that in presentation the slide illustrates how more than one imputed value arises from the chosen MI distributions.

The audience for these slides wanted to know more about the MI methodology without getting lost in details. In this case, words tended to add confusion but when participants are presented with the build of this slide they then see more clearly the key features of MI. In this MI each patient's missing values will be imputed using one of two ways: (a) using the distribution of the placebo group or (b) using the distribution of the baseline scores. The example patient's observed scores at each time point are indicated by Xs in the slide and the mean of the placebo distribution is indicated by Ps. As the slide builds, each value drawn from the placebo distribution (shown schematically at the right side as a bell curve) appears for each imputation. The bell curve shows the distribution from which these values were drawn. A similar scenario follows in red for imputed values drawn using the baseline scores distribution instead of the post-treatment placebo scores' distribution.

20.3 Sales and Marketing Data

20.3.1 Trend Analysis

Trend breaks of a certain metric is one of the most widely sought information, almost every week, by managers of various sales and marketing organizations. For example, brand managers and sales managers look for this information when they are concerned about possible negative implications due to such factors as competitor promotions, economy, bad publicity, loss of exclusivity of a brand, and so on. They also want to know whether a trend improves following a market trial, a new sales promotion, a

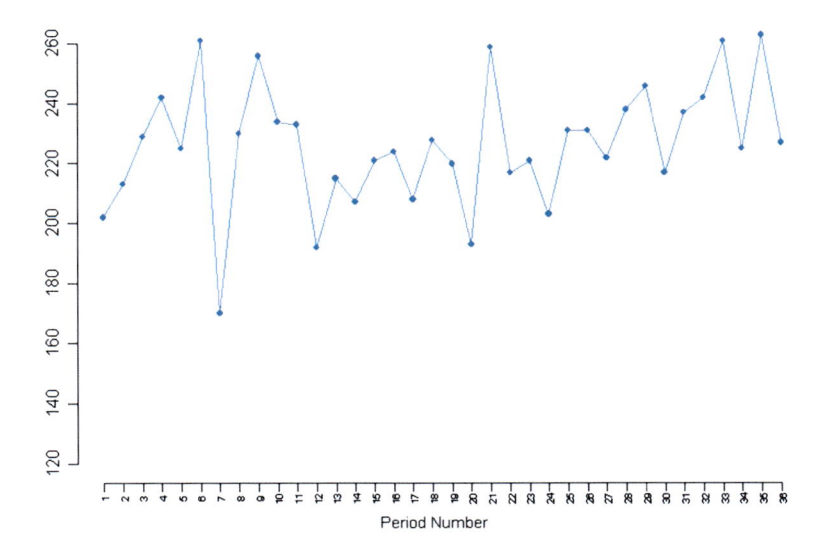

Fig. 20.7 NRx volume by week (period) in district of interest

formulary win, and so on. The metrics they typically look at are NRx (new Rx scripts prescribers write), TRx (total Rx), NRx Market Share, and TRx Market Share. Depending on the state of maturity of the brand they usually know that the current trend is negative or positive, but it is often difficult to tell whether the trend improves or gets worse. Given a set of latest time series data on the metric of interest, they like to know whether a trend break has occurred after an event such as those described above. Since the above metrics are highly noisy, they look for this information week after week following the event to detect as early as possible whether a statistically significant trend break has occurred.

The fact that the detection of trend breaks is not obvious is evident from the time series plot of NRx given in Fig. 20.7 for a hypothetical brand in a certain district. Notice that due to high level of noise in data, it is not very clear whether or not we have sufficient data to declare that a trend break of statistical significance has occurred.

Therefore, it is desirable to provide graphical aid so that a client can see whether the trend is getting better or worse lately. A widely used graphic to accomplish this is the CUSUM (cumulative sum) control chart enabling detection of deviations from a constant trend (Page 1954). To briefly explain how this works in the context of detecting trend breaks, let $Y_1, Y_2, \ldots, Y_T$ be a set of time series data on a metric of interest observed at equally spaced T time periods. The CUSUM control chart is based on recursive residuals from a series of simple regressions of the form

$$Y_t = \alpha + \beta t + \varepsilon_t \tag{20.1}$$

fitted to the increasing window of data points starting with an initial window size of at least 2, starting at time t_0. Let

$$e_t = Y_t - y_t$$

where y_t is the predicted value of Y_t based on the regression fit using data up to time $t-1$. Define

$$v_t = 1 + (t - \text{mean}(X)^2)/\text{SS} \quad \text{and} \quad w_t = e_t / \sqrt{(v_t)},$$

where X is a vector defined as $X = (t_0, t_0+1, t_0+2, \ldots, t-1)$ and $SS = (t-1)\, \text{var}(X)$. Then, the CUSUM plot is based on the vector $U = (U_{t_0}, U_{(t_0+1)}, \ldots, U_{t-1})$, where

$U(t) =$ Cumulative sum of weighted residuals, E_i, up to time t,

where $E_i = w_i/\text{SD}(w)$, where $\text{SD}(w)$ is the standard deviation of the weighted residuals $w = (w_{t_0}, w_{t_0+1}, \ldots, w_T)$. Then, the CUSUM chart for detecting trend breaks would comprise a plot of $U(t)$ as a function t along with confidence bounds,

$$\pm q \left(\sqrt{(T-1)} + 2(t-1)/\sqrt{(T-1)} \right), \tag{20.2}$$

where q is equal to 1.14 for the 99% confidence interval and 0.95 for the 95% confidence interval.

Shown in the chart is CUSUM control chart for trend breaks in the above situation. From the CUSUM plot it is clear that the trend is getting better lately. However, the trend does not come close to crossing the upper confidence bound. This example highlights a major drawback in CUSUM for detecting trend breaks, namely the technique is too slow in detecting trend breaks. Therefore, it is a good idea to carry out a statistical test such as the simple Chow Test (Chow 1960) or the Generalized Chow test under heteroscedasticity (Weerahandi 1987) and present results right on the chart. In applying the Chow test in this situation, we divide the time period into 2 equal halves, estimate the slope in each half and then apply a t-test to conclude whether the change in slopes is statistically significant or not. In the chart below, the finding is reported as an Early Alert by an automated process that generates the chart (Fig. 20.8).

20.3.2 Benchmarking

An equally important notion in performance tracking is benchmarking. In this application, the performance of a certain geographical area, a segment, or any other unit is measured by a certain metric of interest as compared to another geographical area

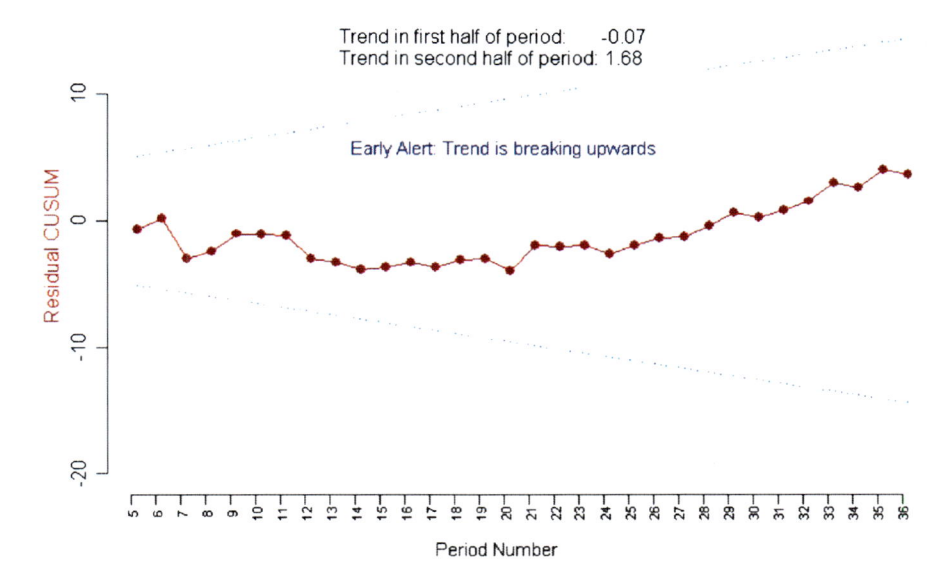

Fig. 20.8 CUSUM control chart for testing trend break in district

of relevance. Here we refer to the former as the *Test Group* and the latter as the *Control Group* or the *Benchmark*. In this application as well metrics such as NRx, TRx, and Market share with respect to NRx or TRx are of interest. For example, if a market trial has been conducted in a region, a market analyst may wish to see whether or not Rx performance in that region is better than in a comparable region. Although eventually one needs to carry out a careful analysis, a simple graphical aid that can be tracked weekly or monthly is desirable before undertaking a time consuming analysis.

With typical data, noise is so high that it is not easy to tell whether the test group is doing better or worse than the benchmark, particularly when the metric of interest is of different magnitude. For example, this is the case when a regional manager wishes to see the sales performance of a certain state when benchmarked against the average performance of the region. As illustrated by the top chart of Fig. 20.9, the issue of different orders of magnitude of metrics can be tackled by presenting each metric on an Index scale as

$$\text{Index at time } t = 100 (\text{Metric at time } t \text{ / Base Metric}), \quad (20.3)$$

thus making the values of each metric vary around 100, where base metric can be set to such quantities as the average of the metric, the value of the metric at the first time period, and so on. In Fig. 20.9, NRx is the metric, and average NRx during the

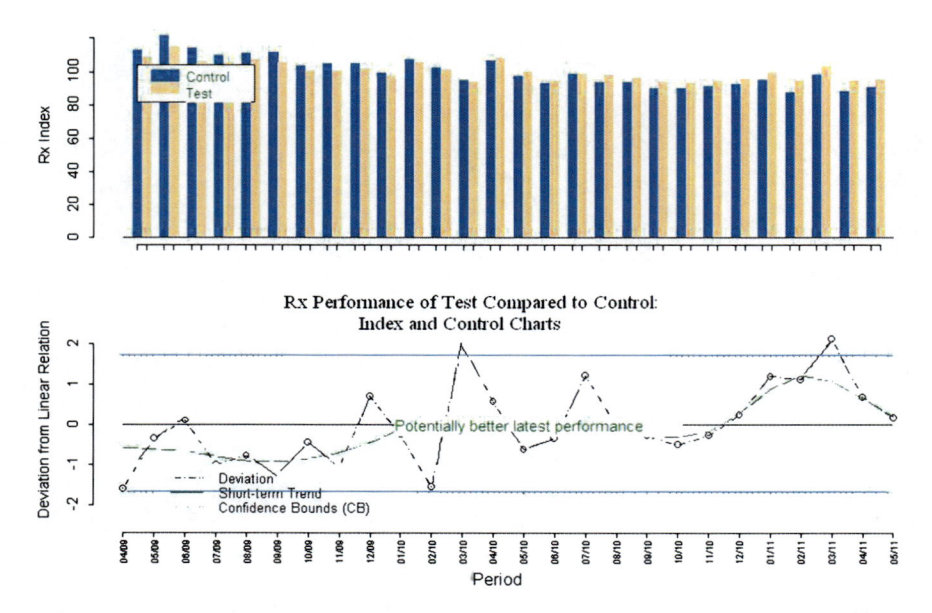

Fig. 20.9 Performance of test group benchmarked against the control group: (1) index chart of 100(Rx/Average Rx), (2) control chart of deviations from constant relationship

study period is taken as the base. The index plot in Fig. 20.9 suggests that the test region might be doing better compared to the control region, but it is not clear whether the improvement is statistically significant or not. Therefore, a benchmarking control chart is presented in the same chart to enable significance testing. The control chart for benchmarking is similar to the CUSUM procedure, except for two modifications: (1) the regressions in (20.1) are run on the observed metric values of the Benchmark in place of time t and (2) weighted residuals E_is are plotted without cumulating.

Deviation of residuals above the 95% upper confidence bound during recent periods would suggest better performance of the test group, and that happening more than once in latter periods would make the improvement highly significant. In the example under study, residuals corresponding to the last 6 months cross the upper confidence only once, but note that all 6 residuals are positive. One can easily reject the hypothesis that this happened by chance at a low level of Type I error, and therefore, we can declare significance and mark this as an early alert of better performance of the test region, as in the bottom chart in Fig. 20.9. In this chart the x-axis, namely the period, is common to each of the two charts. The two charts basically provide the same information, but the index chart is more intuitive to non-statisticians, whereas the control chart helps detecting whether the latest performance of the test compared to the benchmark is statistically significant or not.

20.3.3 Field Force Optimization

One of the primary promotions of most of the pharmaceutical companies is their field representatives calling on doctors. The number of times a field representative going to a doctor's office during a certain period is referred to as the number of details in that period. The field representatives go to doctors' offices and tell the latest on the safety and efficacy of the medications (referred to as detailing) and drop off a certain amount of starters (referred to as samples) since they could positively impact on the way doctors prescribe the brand of interest. Actually field representatives visiting a prescriber's office too often or too in frequently could have negative impact; the former could cause sample cannibalization of sales, and the latter could result in new patients starting on competing brands. Therefore, detailing at a level close to the optimum level is important, a task that requires careful analysis. The prescribing behavior of a doctor is a function of many variables, including details, samples, trend, competition, and so on. Also TV and other type of direct to consumer advertisements affect the way some patients ask for a certain brand by name.

The analysts of many pharmaceutical companies now estimate doctor response to detailing using what is known as marketing mix models, which allow one to control for all drivers of prescribing behavior while estimating the response parameter of interest. Most analysts perform response estimation by a certain set of segments, as opposed to individual prescribers. The segments are based on such attributes as doctors specialty, Rx quintile bucket, region, and so on. Then, the response parameters are estimated by segment in a mixed model setting. A major drawback of response estimation by segment is that individual prescribers in any segment could have nonresponders and high-responders and what we get is the average response for the segment. This approach has limited use and does not allow to optimize the call frequency to individual prescribers. Therefore, some companies such as Pfizer have developed their own method to estimate response for every prescriber, something we will not discuss further due to the proprietary nature of the methodology. When individual prescriber level response estimates are available or segment level estimates are granular enough, one can determine the optimum level of field force by territory of field representative or by district, the lowest level of the management of field representatives' activity.

In this type of application, it is good to present a chart that helps easily understanding of how the optimization works. Figure 20.10 shows how it works using 2 variables of importance in different scales for a given region of interest. When there is no sales cannibalization, typically NRx lift, and hence the revenue lift, we get increases at a diminishing rate, as illustrated by the blue colored curve in the chart. On the other hand the cost incurred by such calls (details) almost linearly increases with the number of calls. Then it is easily seen that the profit as a function of details becomes a concave function with the maximum occurring at a certain point with a reasonable range of details per month. This is illustrated by the red color curve in the chart. In this illustration, the optimal average detail per doctor is slightly above 1.5 details per month in the region of interest. When response estimates for each

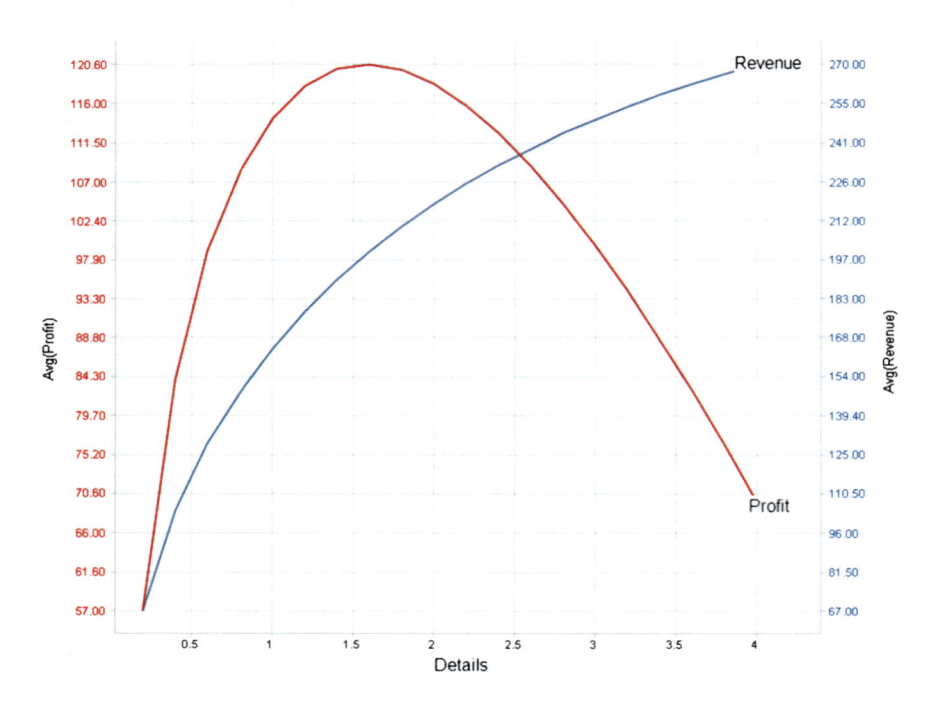

Fig. 20.10 Revenue and profit by number of details per month

region are available by such attributes as prescriber's specialty and quintile, optimum details are computed for each such segment. In the implementation, however, unless one has individual doctor's response estimates, the same calling frequency will have to be maintained for each prescriber in a given segment. If response is estimated by segment rather than by individual prescriber, the low responders in a segment will receive too many details/samples and high responders will receive inadequate details/samples, thus resulting in a sub-optimum solution.

In communicating the benefits of field force optimization with clients, it is important that statistics such as sales lift and profit lift are presented in simple charts such that non-statisticians can easily understand the power and potential of the optimization. In the illustration below, current call frequency (details per month, per prescriber), optimum call frequency, current profit that can be attributed to detailing, and the optimum profit are presented using 2 sets of grouped bar charts. We have implemented this and other graphical displays through an interactive software system that enables metadata display, filtering, and dropdown. In particular, this enables to visualize the annual profit lift potential due to detail optimization when a user moves the mouse from one state into another. For example, for the hypothetical brand under study, the tools highlight that the profit lift potential via detail optimization in New York state is close to a million dollars (Fig. 20.11).

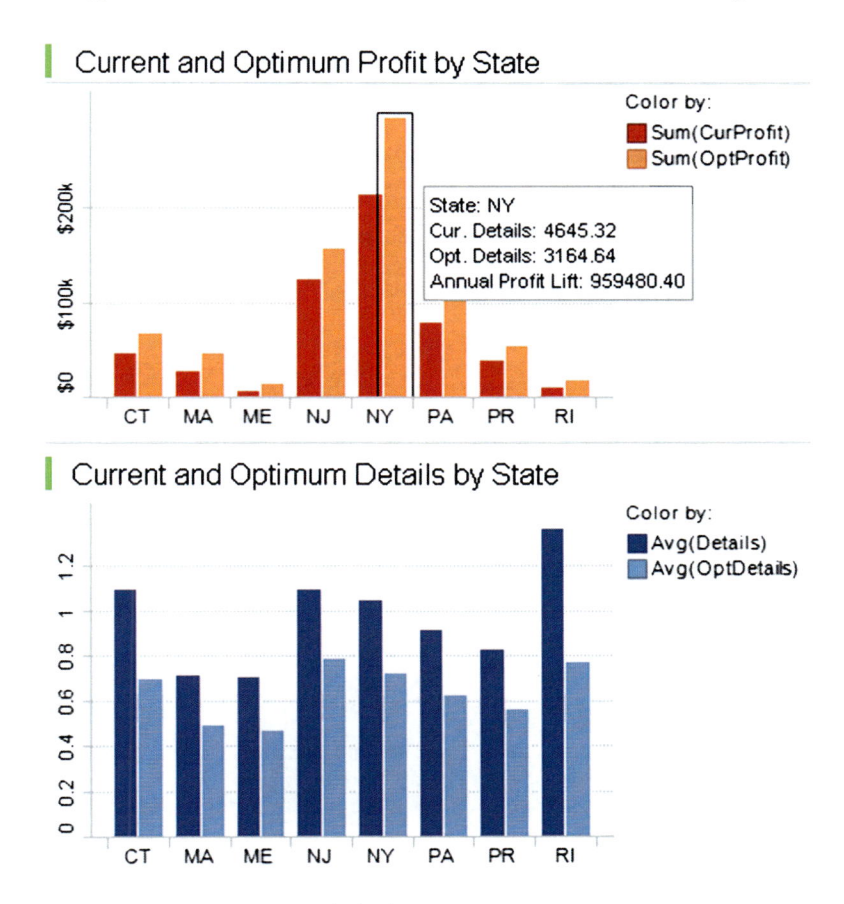

Fig. 20.11 Profit lift due to detail optimization

Further insight concerning the implications of the optimization could be provided with additional charts such as the interactive color-coded map shown below. The interactive software system allows one to look at any regional business unit such as the Northeast business unit in the above example. In Fig. 20.12, we use the optimization carried out for the South business unit using the hypothetical data under study. The interactive color-coded map below shows the states with largest potential with respect to various metrics such as average NRx lift and average profit lift due to optimization in green color, lowest in red color, and average in yellow color, for every state in the South business unit. Those states that fall in between are colored with the corresponding mixed colors. The interactive software system also allows one to move the mouse from one state to another and get a glimpse of important statistics that highlight the profit potential along with related statistics if the optimization is carried out.

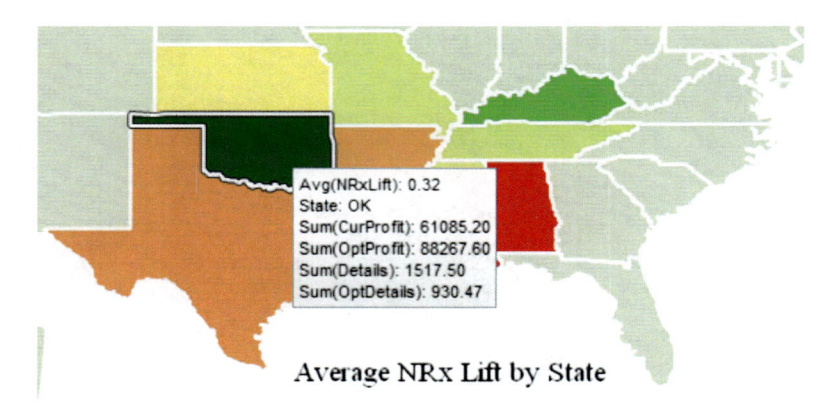

Fig. 20.12 NRx lift and profit lift per prescriber by state

References

Amaratunga D, Cabrera J (2001) Analysis of data from viral DNA microchips. Journal of the American Statistical Association 96:1161–1170
Bouhassira D, Attal N, Fermanian J et al (2004) Development and validation of the neuropathic pain symptom inventory. Pain 108:248–257
Cabrera J, Emir B (2012) Fisher-Yates normalization for questionnaire data. Technical report (2012). Department of Statistic sand Biostatistics, Rutgers University
Chow GC (1960) Tests of equality between subsets of coefficients in two linear regressions. Econometrica 28:591–605
FDA Link http://www.fda.gov/downloads/Drugs/DrugSafety/PostmarketDrugSafetyInformation forPatientsandProviders/UCM192556.pdf
Freeman R, Emir B, Murphy TK, Xu, Y, Whalen E (2010) Placebo response characteristics in eight pregabalin diabetic peripheral neuropathy trials. Presented at third international congress on neuropathic pain, 27–30 May, Athens, Greece
French, J, Cabrera J, Emir B, Whalen E (2011) Designing a new proof-of-principle (PoP) trial for treatment of partial seizures to demonstrate efficacy with minimal sample size and duration—a case study. Pfizer Internal Report (2011). PCBU Statistics, New York, NY
ICH Topic E 8, General Considerations for Clinical Trials (1998) http://www.ema.europa.eu/docs/en_GB/document_library/Scientific_guideline/2009/09/WC500002877.pdf
Page ES (1954) Continuous Inspection Scheme. Biometrika 41 (1/2): 100–115
Weerahandi S (1987) Testing regression equality with unequal variances. Econometrica 55:1211–1215

Chapter 21
Using Exploratory Visualization in the Analysis of Medical Product Safety in Observational Healthcare Data

Patrick Ryan

Abstract Increasing the knowledge about the safety of medical products remains a top priority throughout the pharmaceutical development life cycle and particularly after regulatory approval when the products are used in real-world populations. The increasing availability and use of observational healthcare data, such as administrative claims and electronic health records, provides opportunity for generating better information to support therapeutic decision-making.

Analysis of observational databases is quite different from randomized clinical trials, and offers unique opportunities for exploratory visualization to complement traditional epidemiologic investigations. The Observational Medical Outcomes Partnership was established to conduct methodological research on the appropriate use of observational data to identify and evaluate the effects of medical products in the real world.

Several visualization tools were developed throughout the process, which demonstrate the value in combining standardized analytics with interactive graphics. These tools include a patient profile to study longitudinal patterns within clinical observations for a given person; a cohort profile to evaluate collections of patients; a treemap to assess disease prevalence across a database; a trellis scatterplot to investigate subgroup differences in drug utilization; a heatmap to support evaluation of high-dimensional confounding adjustment; and a multi-method forest plot to enable sensitivity analyses of estimated effects of drug and outcomes. This chapter illustrates these visualizations through a case study exploring the relationship between ACE inhibitor exposure and the subsequent health event angioedema.

P. Ryan (✉)
Epidemiology Analytics, Janssen Research and Development,
1125 Trenton-Harbourton Road, Titusville, NJ 08560, USA
e-mail: ryan@omop.org

21.1 Introduction

Drug safety continues to be a major public health concern in the United States. In order for patients and health care providers to make appropriate therapeutic decisions, they need to be informed of the potential benefits and harms of alternative treatment options. While the efficacy of prescription medicines is generally well-characterized from the series of randomized clinical trials conducted during drug development, the safety profile of medicines is often less certain and more poorly understood (Berlin et al. 2008).

Research suggests that drug safety information is the highest information priority for patients, and that the perception of side effects is influential in many patients' decisions about taking a medicine (Knapp et al. 2004). This patient focus is well-justified. Lazarou et al. estimated that, in 1994, between 76,000 and 137,000 hospital patients died from an adverse drug reaction (ADR), ranking adverse drug reactions as the 4th to 6th leading cause of death (Lazarou et al. 1998), and resulting in health care costs of $3.6 billion annually (Bennett et al. 2007).

Traditional methods of drug safety surveillance involve literature searching and case-by-case analysis of spontaneous adverse event reports, as well as crude frequency counts and calculation of reporting rates (US Department of Health and Human Services 1999). Statistical data mining algorithms are becoming increasingly popular supplementary tools for safety reviewers (Almenoff et al. 2005).

Currently, the FDA conducts spontaneous data mining by applying the Multi-item Gamma Poisson Shrinker (MGPS) method to the Adverse Event Reporting System (AERS) database (DuMouchel 1999; Szarfman et al. 2002). Many groups have recognized the significant limitations in the current system (Berlin et al. 2008; Almenoff et al. 2005; Waller and Evans 2003). As part of the FDA Amendment Act of 2007, Congress mandated the use of observational data (including administrative claims and electronic health records) as part of an active drug safety surveillance system that would supplement current practice (Public Law 110–85 2007).

It is expected that a national active surveillance system will consist of several interrelated processes, including signal detection, signal strengthening, signal validation, and hypothesis testing in a formal pharmacoepidemiologic study (Racoosin 2009). While these observational data sources have been actively studied for pharmacoepidemiologic evaluation studies (Schneeweiss 2009; Schneeweiss and Avorn 2005), appropriate statistical methods for screening observational data to generate and triage hypotheses about potential drug effects are nascent and have not yet been rigorously explored across a network of disparate data sources. Alongside this need for methodological development comes the opportunity to assess the role of exploratory visualization within this new analytical paradigm.

While large-scale clinical trials and pooled meta-analysis results are often desirable to produce the most reliable measure of an effect, they are often logistically infeasible or ethically untenable. Observational studies provide an alternative approach to evaluating drug safety questions that can provide the necessary information about the drug effects to support clinical decision-making. Depending on the questions posed, the primary analysis of an appropriate observational study may provide better information than the analysis of an existing clinical trial data set (Foody 2010). Observational studies provide empiric investigations of exposures and the effects they cause, but differ from experiments in that there is no control of assignment of treatment to subjects (Rosenbaum 2002).

One resource that has provided fertile ground for epidemiologic investigation has been observational healthcare databases. Administrative claims and electronic health record databases have been actively used in pharmacoepidemiology for over 30 years (Strom 2005), but have seen increased use in the past decade due to increased availability at lower costs and technological advances that made computational processing on large-scale data more feasible. Observational healthcare databases, captured as part of the healthcare delivery system, offer researchers the opportunity for a secondary use of data to study effects amongst any observed medical products.

Many such databases contain large numbers of patients that make it possible to examine rare events and specific subpopulations that previously could not be studied with sufficient power (Rodriguez et al. 2001). The large population size makes it possible to estimate absolute incidence rates across a wide array of potential outcomes and to measure amount of exposure in a large population to produce more accurate measures of potential public health impact (Rockhill et al. 1998). Because the data reflect healthcare activity within a real-world population, they offer the potential to complement clinical trial results which suffer from lack of generalizability. Long-term longitudinal capture of data in these sources can enable studies to monitor the performance of risk management programs over time (Weatherby et al. 2002).

Administrative claims databases have been the most actively used observational healthcare data source. Administrative claims databases typically capture data elements used within the reimbursement process. Providers of health care services (i.e., physicians, pharmacies, hospitals, and laboratories) submit encounter information so that they will be paid for services delivered (Hennessy 2006). This commonly includes pharmacy claims for prescription drug fills (providing what drug was dispensed, the dispensing date, and the days supply), and medical (inpatient and outpatient) claims that detail the date and type of service rendered. Medical claims typically contain diagnosis codes used to justify reimbursement for the procedures (also coded). Age and gender can also commonly be inferred from the available data.

In these databases, data are recorded only when a patient has a reimbursable encounter with the health care system that has been properly filed, coded, and adjudicated by the payer (Schneeweiss and Avorn 2005). As a result, many key data elements may not be available. Information on over-the-counter drug use and in-hospital medication is usually unavailable and the patient's actual consumption pattern of the prescription is generally unknown (Suissa and Garbe 2007). Retail pharmacy claims data can be used to study drug utilization pattern, but the completeness of these data can vary by patient age (Polinski et al. 2009) or other unobservable characteristics. Claims can be aggregated by payers, healthcare systems, or data aggregators, though each may have a different perspective on how to define observation periods (whether it be the time insured, the time in the system, or simply the span of time that data was observed). While the databases offer longitudinal coverage, the amount of times that patients persist within a given database can vary significantly because of annual health care coverage choices among employees, eligibility for entitlement programs, or employment status changes.

Electronic health records (EHR) generally contain data captured at the point of care, with the intention of supporting the clinical process. A patient chart may include demographics (birth date, gender, and race), height and weight, and family and medical history. Many EHR systems support provider entry of diagnoses, signs, and symptoms, and also capture of other clinical observations, such as vital signs, laboratory values, and imaging reports. Beyond this, EHRs may often contain findings of physical examinations and the results of diagnostic tests (Schneeweiss and Avorn 2005). EHR systems usually also have the capability to record other important health status indications, such as alcohol use and smoking status (Lewis and Brensinger 2004), but the data may be missing in many patient charts (Hennessy 2006).

Unless integrated across an entire health system, EHR systems are generally maintained independently by physician practices or individual hospitals. The provider and office staff enter information elicited from the patient or generated by the physician, but are also responsible for entering relevant clinical information from services rendered outside the practice, including conditions diagnosed by outpatient specialist physicians or during hospital admissions (Hennessy 2006).

Drug exposure may be inferred from various sources; providers may use the EHR system to capture patient-reported medication history and/or to write prescriptions, but there may be no confirmation that prescription was filled at a pharmacy. As a result of discontinuous care within the US health care system, a patient may have multiple EHRs scattered throughout the providers they have seen, but rarely are those records integrated together, so each reflect a different and incomplete perspective of that person's healthcare experience.

For both administrative claims and electronic health records, drug safety analyses are considered a secondary use of the data. Therefore, the onus is on the researcher to fully understand and assess the relative strengths and limitations of each potential source, prior to conducting an evaluation. Data recorded in either system reflects data used for its primary intent, and therefore, may not necessarily represent the optimal information for study.

For example, diagnoses recorded on medical claims are used to support justification for the payment of a given procedure; this diagnosis could represent the condition that the procedure was used to "rule out" or can be an administrative artifact of being the code used by a medical assistant to maximize reimbursement. Similarly, patients without a diagnosis recorded do not necessarily reflect the absence of a condition, as the code may not be used due to lack of seriousness or convenience to facilitate payment procedures.

A similar limitation exists in EHR systems, where in addition to concerns about incomplete capture, data used for clinical care may not accurately reflect the patient's underlying disease status. For example, physicians may neglect to update conditions that have resolved. Most systems have insufficient processes to evaluate data quality a priori, requiring intensive work on behalf the researcher to prepare the data prior to analysis (Hennessy et al. 2007). Both types of sources require inferences to estimate potential drug exposure. Inferences can be made in administrative claims sources based on pharmacy dispensing records, while inferences for EHR systems rely on the patient self-report and the physician prescribing orders (Hennessy 2006). Neither reflect the timing, dose, or duration of drug ingested, so assumptions are required in interpretation of all study results. Proper investigation of an observational data source prior to analysis is challenging, time-consuming, and susceptible to error. Standardized exploratory analytics and accompanying interactive visualization tools offer the potential to improve both the efficiency and effectiveness of observational research.

21.2 Observational Medical Outcomes Partnership

The Observational Medical Outcomes Partnership (OMOP, http://omop.fnih.org) is a public–private partnership managed by the Foundation for the National Institutes of Health and chaired by the Food and Drug Administration. OMOP is conducting methodological research on the appropriate use of observational healthcare data (administrative claims and electronic health records) for identifying and evaluating the real-world effects of medical products (Stang et al. 2010).

OMOP developed a distributed network of disparate observational data sources, both administrative claims and EHRs, covering de-identified patient-level data for more than 200 million lives. The network of data sources consisted of 5 central datasets and 5 distributed datasets. The central data included one EHR database from GE Centricity (GE) and 4 administrative claims datasets licensed from Thomson Reuters MarketScan® Research Databases: MarketScan Commercial Claims and Encounters (CCAE), Medicare Supplemental Beneficiaries (MDCR), Multistate Medicaid Database (MDCD), and the MarketScan Lab Supplement (MSLR) database.

The 5 distributed datasets were housed by their respective partners—the Department of Veterans Affairs Pharmacy Benefits Management Center for Medication Safety (VA); Humana, Inc. (HUM); Partners Healthcare System (PHCS); the Regenstrief Institute affiliated with Indiana University School of Medicine (RI); and SDI Health (SDI_MID). OMOP also established a community of methodologists to develop and apply statistical models to estimate the strength of association across a wide array of drugs and health outcomes of interest.

These methods include implementations of epidemiologic designs, such as incident user cohort design using high-dimensional propensity score adjustment (HDPS) (Schneeweiss et al. 2009), case-crossover design (CCO) (Schneeweiss et al. 1997), and univariate self-controlled case series (USCCS) (Whitaker et al. 2006), as well as data mining algorithms adapted for longitudinal data, such as disproportionality analysis (DP) (Zorych et al. 2011) and temporal pattern discovery (ICTPD) (Norén et al. 2008).

As part of its research, OMOP has developed a suite of open-source standardized analytical tools for characterizing and exploring patient-level data and aggregate summary results. Such tools include a common data model (CDM) (Overhage et al. 2011) to standardize the structure and terminologies used to represent disparate data sources, and standardized procedures to summarize population-level attributes of each database. Aggregate results were compiled within the OMOP central coordinating center for further analysis and dissemination. The visualizations described in this chapter are the result of this research, and were all generated using TIBCO Spotfire® 3.2.

21.3 Case Study: ACE Inhibitor and Angioedema

Among the test cases used for evaluation in the OMOP experiments was the known positive association between Angiotensin Converting Enzyme (ACE) Inhibitor exposure and angioedema onset. ACE Inhibitors provide a solid basis for methodological research because the class represents a large set of mature products (including lisinopril, benazapril, ramipril, quinapril, and captopril) that are actively used in the broad population.

ACE Inhibitors block the conversion of angiotensin I to angiotensin II within the renin–angiotensin system, which plays an important role in the pathology of hypertension, cardiovascular health, and renal function (Chou et al. 1995; Norris et al. 2010). ACE inhibitors have been found to be effective in the control of hypertension, as well as reduce the risk of acute myocardial infarction among patients with heart failure, left ventricular remodeling after acute myocardial infarction, mortality among patients with severe heart failure and reduced left ventricular ejection fraction, and progression of renal disease among diabetic and non-diabetic patients (Norris et al. 2010).

The Joint National Committee on Prevention, Detection, Evaluation and Treatment of High Blood Pressure (JNC-7) currently recommends ACE inhibitors or Angiotensin Receptor Blockers (ARBs) as first line options for patients with stage 1 hypertension who have diabetes, chronic kidney disease, history of stroke or myocardial infarction, or high cardiovascular risk (Chobanian 2009). While rare in incidence, angioedema has been consistently shown as a potential risk across all ACE inhibitors in clinical trials, and reinforced by observational database studies (Miller et al. 2008).

Angioedema is the rapid swelling of the dermis, subcutaneous tissue, mucosa and submucosal tissues, and can require immediate medical attention to avert airway obstruction and suffocation.

Enalapril was shown to have a 4-fold increase in angioedema risk relative to placebo, from 1 per 1,000 to 4 per 1,000 among all subjects (Kostis et al. 1996). The

ALLHAT study of demonstrated the same incidence and relative effects in lisinopril, with a rate of 4 per 1,000 for lisinopril users, versus less than 1 per 1,000 for the other treatments (ALLHAT 2002).

The HOPE trial showed comparable findings for ramipril (Yusuf et al. 2000). Rates in angioedema were also consistent in trials for captopril (Chalmers et al. 1992) and perindopril (Speirs et al. 1998). The risk of ACE inhibitor-related angioedema is increased in patients of African descent, with an observed 2- (ALLHAT 2002) to 4-fold (Brown et al. Jul 1996) increased risk relative to white Americans. The AASK trial showed the significantly different rates of angioedema among ramipril users over 3.5–6 years of follow-up (6.4%), versus 2.3% and 2.7% for metoprolol and amlodipine, respectively (Chou et al. 1995).

The ACE inhibitor–angioedema association served as one of 9 positive controls within the OMOP experiment. The predictive accuracy of statistical methods was measured on the basis by which analyses could discriminate between the positive controls and 44 negative controls. More generally, the outstanding research question is: to what degree can we learn from observational healthcare data to better understand the effects of medical products? (Madigan and Ryan 2011).

An epidemiological study can be conducted that will estimate the strength of association between ACE inhibitor exposure and angioedema, but without proper context, study results can be easily misinterpreted. Observational research requires a comprehensive understanding of the underlying data at both patient level and population level, and also requires sufficient evaluation of the potential sources of bias that can influence study findings.

Interactive visualization offers opportunities for augmenting traditional epidemiologic practice by enabling efficient exploration of standardized summaries to establish the proper context necessary for interpreting observational study results. As a motivating example, the ACE inhibitor–angioedema relationship is explored using 5 visualizations developed in OMOP as standardized procedures for observational research. These visualizations include:

- *Patient profile*—A longitudinal summary of all clinical observations for a given patient
- *Cohort profile*—A graphical display of all exposed cases within a database to enable identification of patterns across patients
- *Prevalence treemap*—A hierarchical representation of the proportion of patients with drug exposure or disease occurrence that allows for comparisons within and across databases
- *Subgroup trellis scatterplot*—A summary graphic to allow comparisons of drug utilization or disease prevalence within patient subgroups by age, gender, and calendar year
- *Trellis forest plot*—an adaption of typical meta-analysis forest plots to assess heterogeneity across data sources and due to study designs as a means of sensitivity analysis

21.4 Patient Profile

Figure 21.1 is a patient profile constructed using a scatterplot, displaying select clinical observations within the longitudinal record of a single individual. The y-axis identifies the data domains presented in this profile, which includes records from the DRUG_ERA, CONDITION_ERA, and PROCEDURE_OCCURRENCE tables from the OMOP common data model. The x-axis provides the temporal relation in days between each observation, relative to a defined index date. Here, day 0 is the date of the patient's first ACE inhibitor exposure.

Using days to index, rather than calendar dates, allows for complete data display while protecting patient privacy, as observation dates are typically regarded as protected health information. Each point within the graph represents a clinical observation, a record from a specific data table that was recorded on a particular date (transformed into days relative to the index date). The shape of the icon depicts the type of date represented: drug eras represent a span of time that a patient is inferred to have persistent exposure to an active ingredient, so the start date is represented

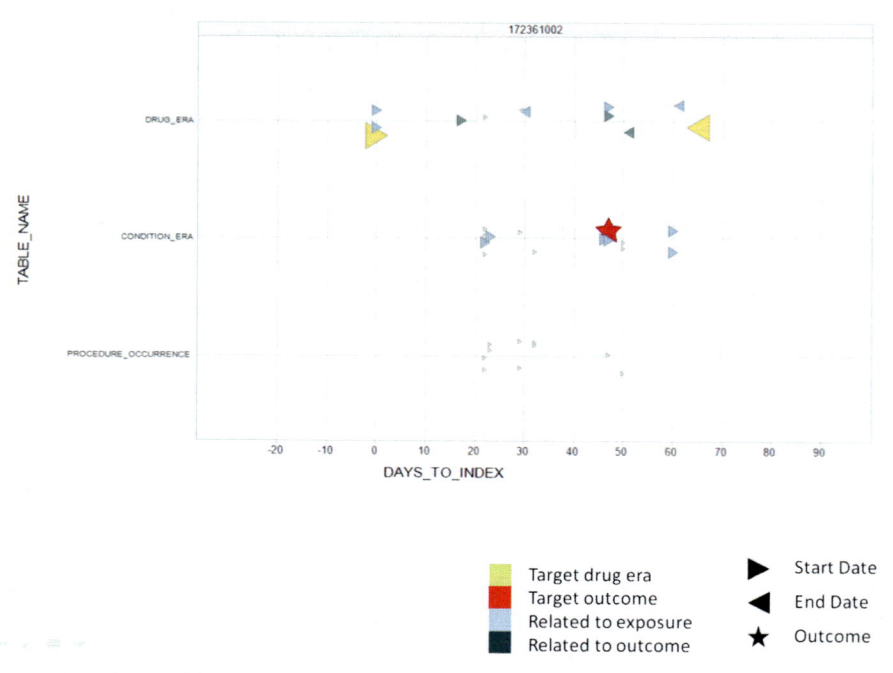

Fig. 21.1 Patient profile: This scatterplot displays select clinical observations within the longitudinal record of a single individual, with days relative to some index date shown in the x-axis and the type of observation categorized on the y-axis. Color, size, and shape are used to illustrate attributes of the observation that is pertinent to exploring the specific drug–outcome relationship within the person's record

by the triangle pointing right, and the end date is represented by the triangle pointing left.

The outcome of interest (here, angioedema) is represented by a star. The color of the points represents that observations' relationship to either the target drug or target outcome; yellow identifies the drug era observations for ACE inhibitor exposure, red highlights the angioedema occurrence; light blue highlights observations that are related to the exposure (e.g., conditions for which the drug is indicated or is used off-label, such as hypertension; and drugs that are alternative treatments for the same indication as the target drug, such as beta-blockers and angiotensin receptor blockers); dark blue highlights observations related to outcome (e.g., drugs indicated as treatments for the outcome, such as prednisone); and gray is used to color observations with no direct relationship to either the target drug or outcome.

The relationships between clinical observations can be derived through automated heuristics using the OMOP standard vocabulary. The size of the icon is used to further focus the user's attention on the target drug and outcome (largest icons), then those observations related to either the drug or outcome (medium size), and to deprioritize focus on the unrelated observations (smallest icons). Within each data domain (table name groups on the y-axis), the points are jittered vertically to minimize data overlap.

As a static graphic, this patient profile serves several purposes. It provides information about the time-to-event relationship by evaluating when the outcome occurs relative to drug initiation; here, the first event occurred 47 days after drug start. The graph also allows for assessing the magnitude of information available for the patient when conduct case adjudication; this patient has no prior medical history before the ACE inhibitor exposure and relatively few comorbidities and concomitant medications observed during the period of ACE inhibitor exposure. However, this figure is limited in its current form in that it does not directly convey what specific drugs, conditions, and procedures are represented at each timepoint.

As an interactive exploratory visualization, the patient profile becomes more powerful. Researchers can hover over each clinical observation to learn more information, such as the specific concept name, the type of relation to either the drug or the outcome, the days relative to index, and the patient's age at the time of the event. Researchers can highlight multiple observations to display labels and begin to posit hypotheses for the clinical circumstances that are observed in the profile.

For example, the researcher could see that specific ACE inhibitor exposure was lisinopril. On day 47, when the angioedema diagnosis was observed, we could see the patient had a pharmacy dispensing for prednisone (a treatment for angioedema) and carvedilol (an alternative antihypertensive treatment). This information strengthens the hypothesis that the angioedema event is true, as it appears the patient received care consistent with a belief in that diagnosis, since they received treatment and were switched to a different antihypertensive not known to be associated with angioedema.

The carvedilol start date and subsequent lisinopril end date underscores one challenge in inferring patterns from administrative claims data. Start dates are typically derived based on the dispensing date recorded on a pharmacy billing claim.

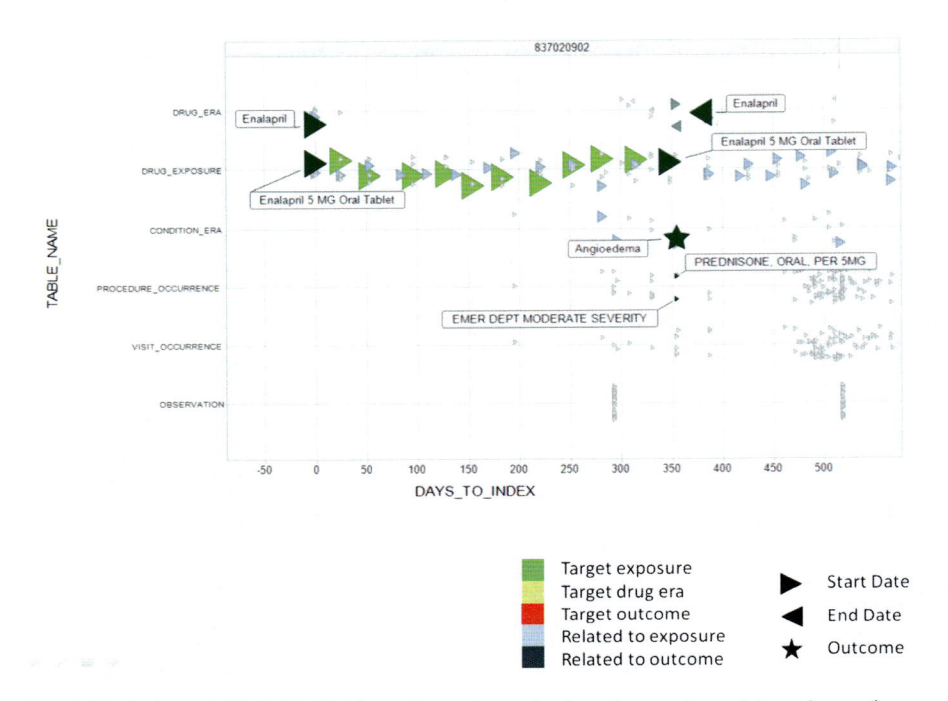

Fig. 21.2 Patient profile with six dimensions, categorized on the *y*-axis, and key observations marked and labeled to facilitate further exploration of the temporal relationship between drug exposure and outcome occurrence, and the plausibility of a causal effect

Pharmacy billing claims also contain other data elements, such as quantity of pills dispensed and number of days supplied, which can be used to infer an exposure end date. However, the true end of exposure is not obtainable from these sources.

In this case, it could be reasonably expected that when the angioedema event occurred, the provider provided a prescription for carvedilol and directed the patient to immediately stop the lisinopril prescription (since lisinopril may have been suspected to have caused the angioedema). If that were the case, the true lisinopril end date would have been day 47, but since this provider–patient interaction is not explicitly recorded, the researcher needs to make assumptions about how to interpret the estimated lisinopril end date that occurs 18 days after the angioedema occurrence represents a potential de-challenge event.

The other value of situating this patient profile within an interactive framework is the ability to rapidly explore patterns across multiple patients and to zoom in and out of longitudinal patterns and data domains. Figure 21.2 shows a patient profile for a different patient with additional data domains shown (DRUG_ERA, DRUG_EXPOSURE, CONDITION_ERA, PROCEDURE_OCCURRENCE, VISIT_OCCURRENCE, OBSERVATION). Unlike the patient shown in Fig. 21.1 who had

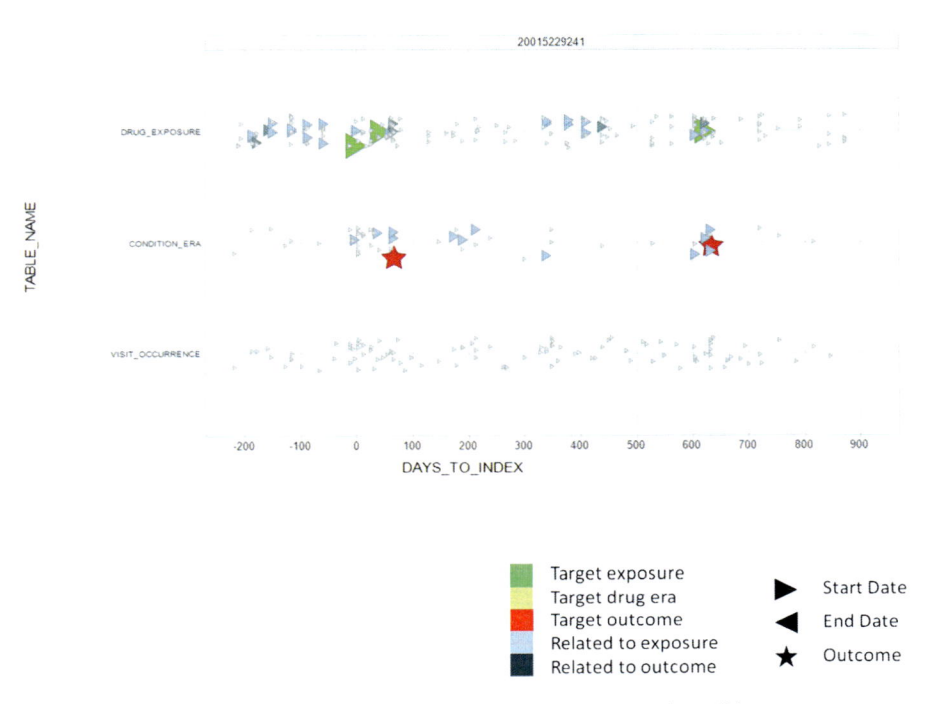

Fig. 21.3 Patient profile with rechallenge pattern. Drug exposure and condition occurrence are categorized on the *y*-axis. The temporal pattern, as displayed by days from index on the *x*-axis, allows users to observe the potential relationship between exposure and outcome, where the outcome was observed during exposure on 2 successive episodes of uses

less than 3 months of time displayed, Fig. 21.3 reveals patient-level data for approximately 2 years.

The DRUG_EXPOSURE table contains all pharmacy dispensing records (displayed on the dispensing date); these records are used to derive the DRUG_ERA periods of inferred persistent use. In this case, the figure shows that the ACE inhibitor era that is over 1 year in length was derived from 12 dispensings which occurred approximately every month, consistent with typical 30-day refill behavior.

The profile reveals that during the first 6 months of ACE inhibitor exposure, this patient had few concomitant medications, and little recorded health service utilization. The patient had greater procedures administered after exposure in the second year. This patient had few observations recorded, with select laboratory values captured at 2 time points.

Interactively exploring this visualization further reveals added insights that may support assessment of the patient's clinical circumstances and events. Filters can be used to select which types of observations to show, so the *y*-axis can be expanded or contracted based on which tables are of interest. A zoom slider can be used on the *x*-axis to allow narrowing focus on particular areas of the longitudinal record (such as the month pre- and post-event) and to zoom out to see the entire 3 years of observation, as shown here. Highlighting select icons, as shown in Fig. 21.2, allows researchers to identify the active ingredient (enalapril) and drill down to the specific dose and form (enalapril 5 mg oral tablet).

To further characterize the angioedema event, researchers can select observations proximal to that event. Here we see the patient had a procedure recorded that indicates the patient was brought to the emergency department with moderate severity, and was administered prednisone on that same date. These observations are consistent with a serious event that required urgent care and resulted in immediate treatment. We note that no additional enalapril prescriptions were dispensed after the event, and the patient began having pharmacy dispensing records for atenolol (a beta blocker indicated for hypertension), indicative of a de-challenge attempt.

Figure 21.3 illustrates a different pattern in exposure and outcome occurrence. This figure provides all records from the DRUG_EXPOSURE, CONDITION_ERA, and VISIT_OCCURRENCE to allow exploration of pharmacy dispensings, diagnoses, and general health service utilization. This patient has over 3 years of longitudinal data capture, including more than 200 days prior to first ACE inhibitor exposure (potentially indicating incident use). We see that the patient had 2 dispensings during the first episode of use, but an angioedema diagnosis was recorded after the second dispensing and no refills were immediately observed.

Initially, this pattern looks consistent with the de-challenge story observed with the prior 2 patient profiles. At day 623, nearly 2 years after the initial exposure, we see the patient received a new dispensing for an ACE inhibitor. An angioedema diagnosis is recorded 9 days after this new ACE inhibitor exposure, and no subsequent ACE dispensings are observed. This pattern is consistent with a rechallenge event, whereby the patient has an exposure, experiences an event, stops the treatment and the event subsides, reinitiates the exposure and the event reemerges.

There is substantial literature in evaluation of rechallenge events from spontaneous adverse event reporting and clinical trials (Hill 1965; Agbabiaka et al. 2008; Arimone et al. 2007; Bandekar et al. 2010; Papay et al. 2009; Perrio et al. 2007), but little research has been carried out to assess the potential for identifying and exploring rechallenge events from administrative claims and electronic health records. To some degree, this may be attributable to the complexities of extracting longitudinal patterns from large observational healthcare datasets that may contain orders of magnitude greater numbers of exposures and exposed cases. Exploratory visualization, through tools like the interactive patient profile, can enable rapid assessment of large numbers of exposed cases and supports generating and evaluating hypotheses of observed patterns across patients.

Patient profiles can be used to identify and characterize expected patterns indicative of potential drug-related outcomes. Equally as valuable, these profiles can support researchers in identifying unexpected patterns and highlighting challenges that may complicate further analyses. Figure 21.4 highlights a 4th patient with ACE inhibitor exposure and subsequent angioedema diagnosis. This patient record

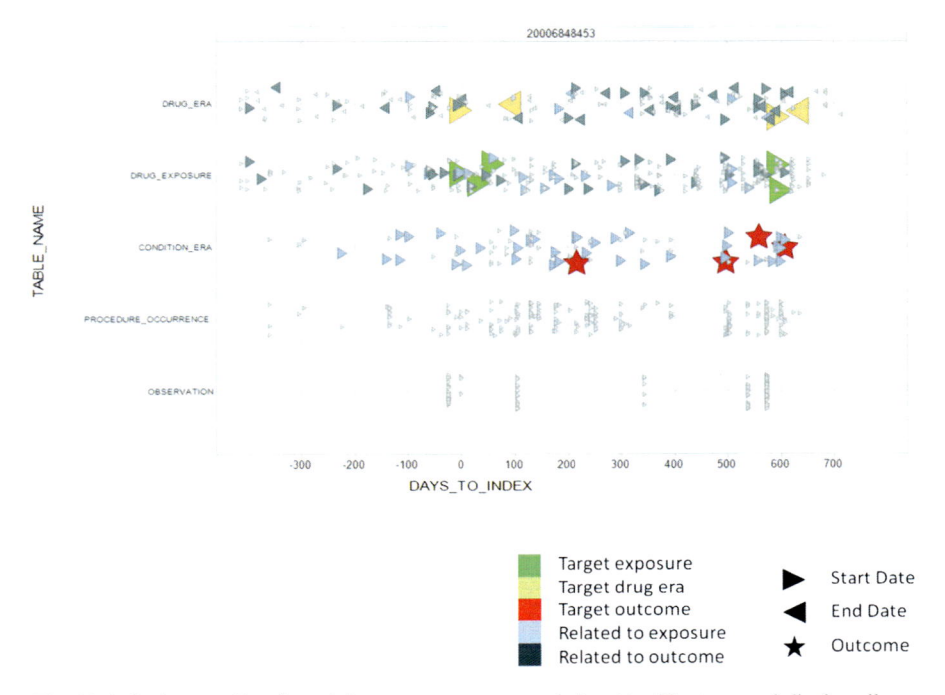

Fig. 21.4 Patient profile of suspicious exposure–event relationship. The temporal display allows the researcher to observe the condition being diagnosed after the first period of drug exposure, and successive diagnoses occurring prior the second exposure

contains clinical observations spanning 3 years of time with active health service utilization throughout.

The patient has a large number of comorbidities and concomitant medications. Moreover, the temporal relationship between ACE inhibitor exposure and Angioedema onset is less clear, as the initial drug era was stopped for 100 days prior to the first angioedema diagnosis, and subsequent diagnoses occurred prior to when the patient reinitiated ACE inhibitor use. The 3 angioedema diagnoses observed between days 500 and 600 may not represent unique events, but instead the same diagnosis being recorded multiple times as part of the follow-up in clinical care.

Evaluating clinical observations in the context of other data in temporal proximity of the event allows for a more critical assessment about the confidence that the researcher should place on each data element, which can be often obscured when looking at each event independently in tabular form or when conducting population-level analyses that do not incorporate patient-level case review.

The patient profile can be powerful tool for researchers to explore patient-level data to develop and evaluate outcome definitions, to refine criteria based on clinical review of quality of cases, and to select specific events that warrant further investigation. Such a tool could be instrumental throughout the lifecycle of a study, from

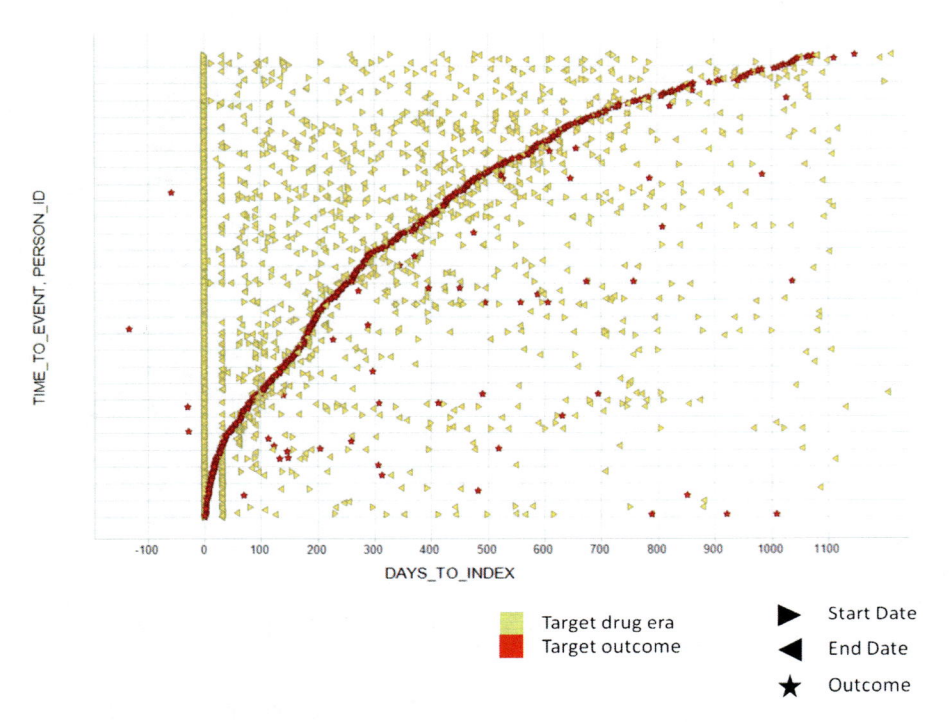

Fig. 21.5 Cohort profile. This scatterplot allows exploration of all patients with co-occurrence of drug exposure and outcome in their longitudinal record. The *x*-axis displays the number of days from an index date, which is the first exposure of ACE inhibitor. The *y*-axis is used to categorize observations by unique person identifier, and sort the patients by the duration from exposure start to subsequent outcome

design to analysis to interpretation of results, and allows researchers to complement the interpretation of population-level effects in the data.

21.5 Cohort Profile

One challenge in reviewing patient profiles is that population-level patterns observable across patients can be difficult to discern. It can be difficult to see the forest when the researcher focuses exclusively on each tree. The cohort profile is a graphical complement to the patient profile, showing the forest by highlighting the key elements from all the trees in the dataset. Figure 21.5 provides a cohort profile representation that allows patient-level information to be evaluated across a population.

The cohort profile is analogous to stacking patient profiles on top of one another, and is similarly constructed as a scatterplot. The *x*-axis is the time relative to an index date, such as the first ACE inhibitor exposure as shown here. The *y*-axis is treated as a categorical variable and is used to order patients by time-to-event (the number of days from the first ACE inhibitor exposure to the next angioedema diagnosis). The y-axis additionally uses PERSON_ID as a further categorization to avoid overlap of patients with the same time-to-event measure. The plot is restricted to displaying the target drug eras, shown as yellow triangles for the start and end dates, and the target outcome, shown as red stars.

The cohort profile can show macro-level patterns across a large set of patients. For example, Fig. 21.5 displays 445 patient profiles, representing all the ACE inhibitor–angioedema events in the Thomson MarketScan Lab Supplemental database. Several types of patterns can be identified in this plot. Notably, the time-to-event distribution becomes apparent, as the plot represents a version of a cumulative distribution function.

We can see that less than 20% of the cases occur within the first 30-day prescription dispensing, suggesting that angioedema is not always an acute onset event. We note that, amongst the entire cohort of events, only 4 patients had an angioedema diagnosis prior to first ACE inhibitor exposure (seen by the red stars with negative days to index). We also note that there is only a modest number of patients with subsequent angioedema events after the initial angioedema diagnosis, based on the number of red stars to the right of the time-to-event distribution.

We observe that ACE inhibitor exposure is often stopped after initial angioedema diagnosis, by noting the proportion of events that are immediately followed by a drug era end date. We can also see a surprisingly high frequency with which patients stop and restart treatment after the angioedema event, based on the number of right-pointing yellow triangles to the right of the time-to-event distribution. Since angioedema is a known side effect of ACE inhibitors and has potential for recurrence, ACE inhibitor use is contraindicated in patients with prior history of Angioedema.

These data suggest that either the angioedema diagnoses are not indicative of true events, or that patients are receiving subsequent exposures that are not recommended. In some instances, these patients demonstrate patterns indicative of a positive rechallenge: (1) the exposure is introduced, (2) the event is observed, (3) the exposure is stopped, (4) the event is no longer observed, (5) the exposure is reintroduced, and (6) the event recurs.

In clinical trials and spontaneous adverse event reports, rechallenge cases often represent strong evidence in support of a causal association. The cohort profile, in conjunction with the patient profile, demonstrate opportunities for how interactive visualization can help researcher navigate to observational data to find patient-level patterns that may be equally valuable in the causal assessment of medical product effects.

21.6 Prevalence Treemap

In most epidemiologic investigations, a specific question is posed and available data are used to yield the least-biased answer. For example, researchers may be interested in estimating the prevalence of hypertension. Often times, additional context

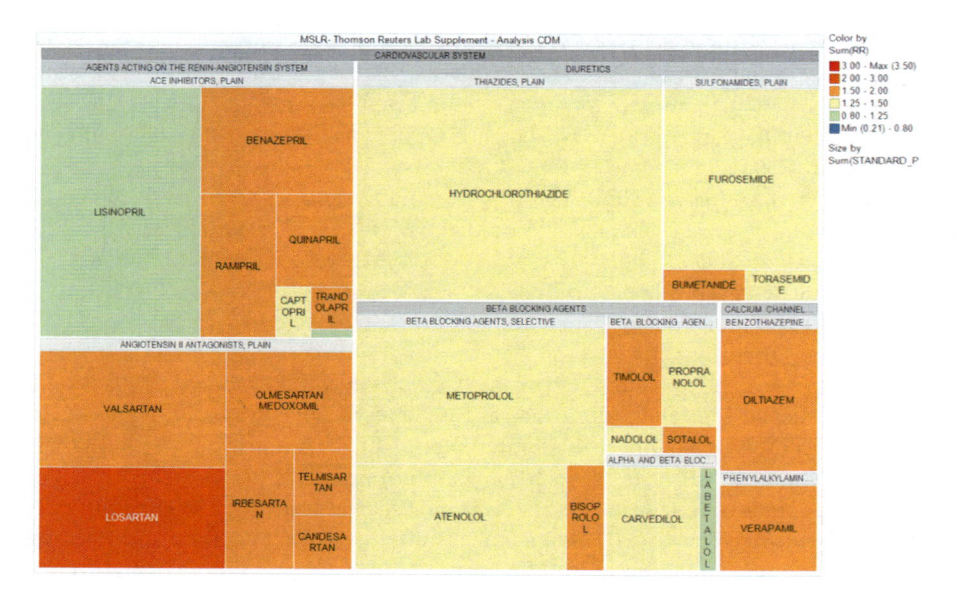

Fig. 21.6 Treemap of drug prevalence. Each *rectangle* represents an active ingredient. The size of the rectangle reflects the standardized prevalence in the MSLR database, and the *color* indicates how different the prevalence in this database is to a referent community of databases. *Rectangles* are organized hierarchically by the ATC classification system

for a given answer can make the response more informative, such as understanding how the prevalence of hypertension compares to other chronic conditions within the same population, or how the prevalence in one database compares to observed hypertension in other data. Access to prevalence information across multiple diseases and across a network of disparate data sources would enable exploratory comparisons.

OMOP developed the Observational Source Characteristics Analysis Report (OSCAR) as an automated routine to estimate standardized prevalence for all drugs and all conditions within a database. OSCAR results from all data partners were compiled in the OMOP central coordinating center, which facilitated cross-source comparisons via treemap visualization.

Treemaps provide a visualization technique to explore relationships within hierarchical entities. The hierarchy is represented through nested rectangles. Each rectangle represents a unique observation, of which 2 attributes for the observation can be displayed based on the rectangle size and color. In Fig. 21.6, each rectangle represents a single active ingredient, and all ingredients are organized by the Anatomical Therapeutic Classification (ATC) hierarchy for medical products. The size of the rectangle indicates the relative prevalence of the drug in the Thomson MarketScan Lab Supplement (MSLR) database, based on the number of persons in the database with at least one exposure to that product. The color of the rectangle indicates how the prevalence in this database compares to the community of data sources.

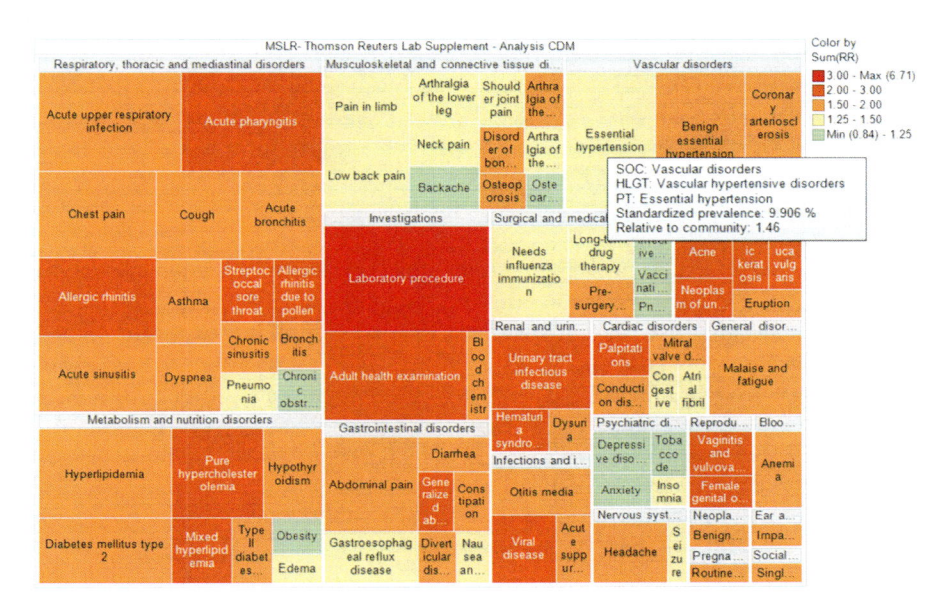

Fig. 21.7 Treemap of disease prevalence. Each *rectangle* represents a condition. The size of the rectangle reflects the standardized prevalence in the MSLR database, and the *color* indicates how different the prevalence in this database is to a referent community of databases. *Rectangles* are organized hierarchically by the MedDRA classification system

The top-left rectangle, "Lisinopril" is classified in the "ACE Inhibitors, plain" class, which is subsumed within the "Agents Acting On the Renin–Angiotensin System," which is one of several classes within the high-level group of "Cardiovascular System." Among drugs within the "ACE inhibitors, plain" class, lisinopril is the most prevalent product in the class, followed by benazepril, ramipril, and quinapril. The green color for the lisinopril rectangle indicates the MSLR prevalence is not substantially different from the average prevalence among the OMOP community, while other ACE inhibitors appear more frequently in MSLR than other sources.

Hovering over "Lisinopril" provides a drill-down summary to display more specific information; the standardized prevalence is 5.22%, which is 9% higher than the average lisinopril prevalence observed across OMOP. The size of the specific products within the ACE inhibitor class, as well as the composite size of the ACE inhibitor class itself, can be compared to products outside the class. These comparisons allow the researcher to observe that other antihypertensive drug classes such as diuretics and beta-blockers have higher prevalence. Such information could be useful for identifying viable comparator drugs to use in cohort studies, or for evaluating preferential prescribing patterns for alternative treatments.

Figure 21.7 provides a treemap that displays disease prevalence for the top 100 conditions across the OMOP community. Conditions are organized hierarchically using the Medical Dictionary for Regulatatory Activities (MedDRA) classification system, with each Preferred Term concept shown within its associated System Organ Class. The graph allows the researcher to immediately identify patterns within the database and relative to data from across the OMOP community. The diagnosis code for "Laboratory procedure" is highly prevalent, and most significantly different in MSLR relative to other sources, which logically makes sense given that MSLR is a population defined as patients with one or more laboratory result.

Other prevalent conditions include: acute upper respiratory infection, hyperlipidemia, and diabetes mellitus type 2. Multiple diagnoses related to hypertension are prevalent, suggesting that study of patients with ACE inhibitor exposure for the primary indication may be viable. In general, since the majority of the rectangles are in orange, we see that the background rate of most conditions is higher in MSLR than in the OMOP community. This insight may be important when considering the impact of comorbidities on estimating the association between ACE inhibitor exposure and angioedema. By hovering over, we see additional details on "Essential Hypertension" to learn that the standardized prevalence is 9.9%, which is 46% higher than the OMOP community average.

21.7 Subgroup Trellis Scatterplot

The prevalence treemaps allow for comparison of population-level effects across concepts and databases. Sometimes, further exploration is needed to characterize subpopulations within a given concept. For example, one may want to assess how the prevalence of hypertension varies by age. Standardized prevalence estimates are composite summaries that are generated from stratified estimates within specific patient subpopulations. Specifically, the standardized prevalence estimates displayed in the treemap are computed based on strata-specific estimates across age groups, gender, and calendar year.

These stratified estimates can provide additional insights if discrepancies between strata can be easily observed. The subgroup trellis scatterplot is one tool that researchers in OMOP have used to try to discern such patterns. Figure 21.8 provides an illustration to explore lisinopril utilization. The plot uses 2-dimensional trellising, with trellis rows displaying different databases and trellis columns partitioning prevalence by year.

Within each trellis plot is a scatterplot with the x-axis used to categorize age in deciles and two series differentiated by color are used to depict gender. The y-axis displays the drug prevalence, as defined by the proportion of patients with the subgroup with at least one exposure.

As expected, the prevalence of lisinopril exposure increases with age, with the highest rates of use in patients over 50 years of age. In MDCR and GE, males have higher prevalence than females; in MSLR, the gender-rates are similar in 2003–2005, but men are observed to have higher prevalence in 2006. MSLR is observed

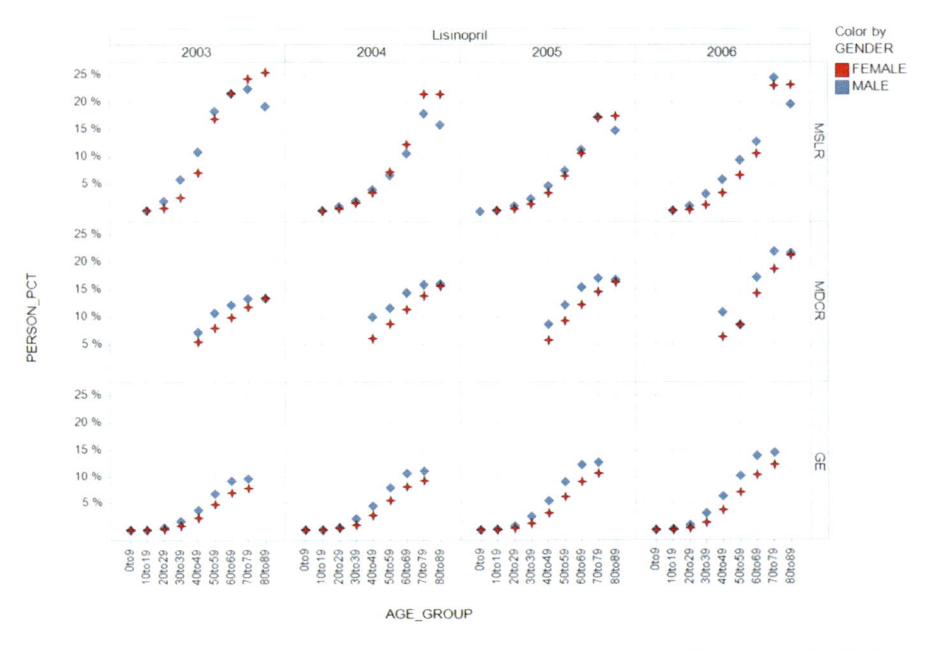

Fig. 21.8 Subgroup trellis scatterplot for lisinopril prevalence. Each *panel* is a scatterplot displaying the relationship between drug prevalance (*y*-axis) and age group (*x*-axis), and panels are trellised by year (*columns*) and database (*rows*)

to have higher lisinopril prevalence across all age groups and genders, consistently across all 4 years, than either MDCR or GE. At least 10% of patients above 50 years of age had at least one lisinopril record in MSLR. Both MDCR and GE exhibit increasing prevalence over time.

One value of the subgroup trellis plot is that it allows rapid exploration of potential patterns in prevalence across multiple dimensions simultaneously. By looking down a column, one can compare databases; by looking across a row, one can examine trends over time; by comparing the slopes of the colors within a given plot, one can study both age and gender differences. Consistencies observed across the trellis plot give indications of macro-level trends that can be expected within subsequent analysis.

21.8 Trellis Forest Plot

A primary output of many observational studies is an effect estimate and a measure of variance. While the relative risk may be the final answer, proper interpretation requires complete understanding of the database, the contributing populations, and the minimization of bias, as has been explored in the prior visualizations. Further

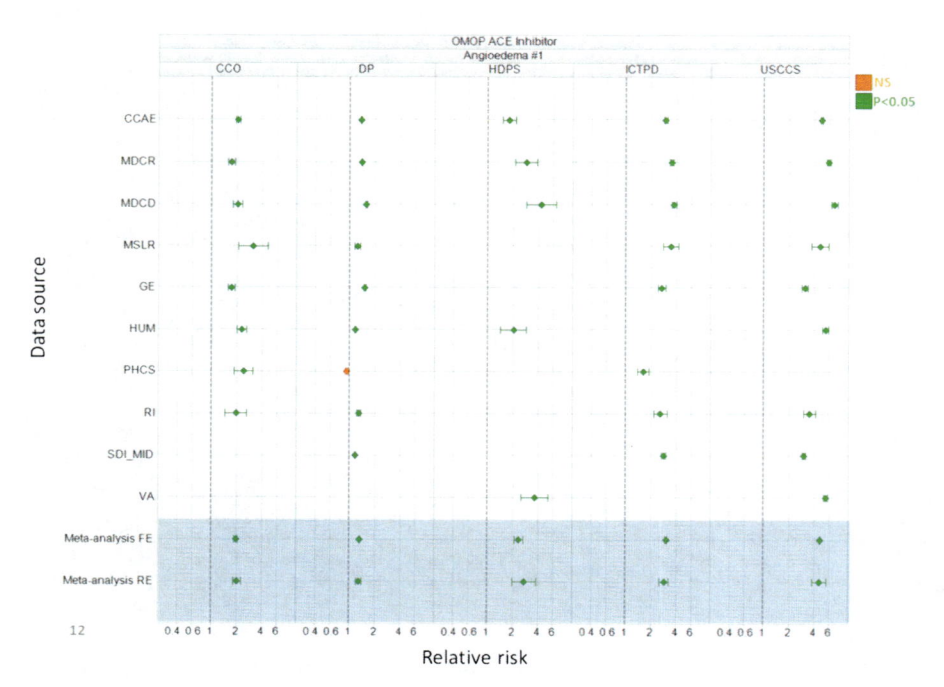

Fig. 21.9 Trellis forest plot for ACE inhibitor-angioedema effect estimates. Each *panel* is a forest plot that displays the effect estimate (*x*-axis) across the network of observational databases (*y*-axis), with the meta-analysis composite estimates provided at the bottom. Panels are trellised by study design in columns to enable evaluation of consistency across data sources and analysis strategies

examination of the final result can also be supported through sensitivity analysis across databases and methodological approaches. In the context of a distributed network of observational databases, effect estimates can be obtained from each contributed data partner and aggregated centrally, and then meta-analytic estimates (both fixed effects and random effects models) can be generated using most statistical software using the source-specific data to evaluate composite summaries.

As shown in Fig. 21.9, the trellis forest plot displays distinct methods within each column. The *y*-axis categorizes each of the contributing databases. Composite estimates using both fixed-effects and random-effects meta-analysis are shown at the bottom, and highlighted in light blue. Within each column, the *x*-axis shows the relative risk plotted on a logarithmic scale, with RR = 1 reference further reinforced with a dashed vertical line. For each database-method combination, the estimated relative risk and associated 95% confidence interval is plotted. The color indicates whether the estimate was statistically significant (green denotes $p < 0.05$, while orange denotes nonsignificance).

This figure demonstrates the consistency in effect estimates observed when studying the ACE inhibitor–angioedema association. With one exception (the disproportionality analysis estimate in the Partners Healthcare database), all effect estimates were positive and statistically significant. The random-effects meta-analysis estimates suggest that the relative risk may exist within a range from RR = 1.5–6, depending on the method employed.

The trellis plot allows examination of heterogeneity within databases by holding method constant and looking down each column. The plot also allows for side-by-side comparison of different methods, which enables examination of how methods vary both in point estimate and standard error. The trellis forest plot provides a multidimensional approach to sensitivity analysis that should allow more comprehensive examination of heterogeneity, a more robust assessment of key factors influencing an observation, and better context for drawing inferences when interpreting effect estimates.

21.9 Conclusions

Observational healthcare data offer tremendous potential for enhancing our understanding of the real-world effects of medical products. These data also present unique challenges for analysis, such as the massive size (millions of patients with billions of observations), irregular data capture, non-constant longitudinal coverage, and a large number of recordable observations.

Standardized analytics with interactive visualizations provide one approach to systematically harness the value of observational data at all stages of analysis, from patient-level case review to population-level pattern detection to design-level confounding assessment to study-level sensitivity analysis and synthesis. As the scope and complexity of observational data continues to grow, exploratory visualization figures are to play an increasingly important role in the analysis of medical product safety.

References

Agbabiaka TB, Savovic J, Ernst E (2008) Methods for causality assessment of adverse drug reactions: a systematic review. Drug Saf 31(1):21–37

ALLHAT (2002) Major outcomes in high-risk hypertensive patients randomized to angiotensin-converting enzyme inhibitor or calcium channel blocker vs diuretic: The Antihypertensive and Lipid-Lowering Treatment to Prevent Heart Attack Trial (ALLHAT). JAMA 288(23): 2981–2997

Almenoff J, Tonning JM, Gould AL et al (2005) Perspectives on the use of data mining in pharmacovigilance. Drug Saf 28(11):981–1007

Arimone Y, Miremont-Salame G, Haramburu F et al (2007) Inter-expert agreement of seven criteria in causality assessment of adverse drug reactions. Br J Clin Pharmacol 64(4):482–488

Bandekar MS, Anwikar SR, Kshirsagar NA (2010) Quality check of spontaneous adverse drug reaction reporting forms of different countries. Pharmacoepidemiol Drug Saf 19(11):1181–1185

Bennett CL, Nebeker JR, Yarnold PR et al (2007) Evaluation of serious adverse drug reactions: a proactive pharmacovigilance program (RADAR) vs safety activities conducted by the Food and Drug Administration and pharmaceutical manufacturers. Arch Intern Med 167(10): 1041–1049

Berlin JA, Glasser SC, Ellenberg SS (2008) Adverse event detection in drug development: recommendations and obligations beyond phase 3. Am J Public Health 98(8):1366–1371

Brown NJ, Ray WA, Snowden M, Griffin MR (1996) Black Americans have an increased rate of angiotensin converting enzyme inhibitor-associated angioedema. Clin Pharmacol Ther 60(1): 8–13

Chalmers D, Whitehead A, Lawson DH (1992) Postmarketing surveillance of captopril for hypertension. Br J Clin Pharmacol 34(3):215–223

Chobanian A (2009) The joint national committee on prevention detection and evaluation of high blood pressure. US Department of Health and Human Services NHLBI, Bethesda, MD

Chou R, Helfand M, Carson S (1995) Drug class review on angiotensin converting enzyme inhibitors. Final report. Portland, OR: Oregon Health & Science University

DuMouchel W (1999) Bayesian data mining in large frequency tables, with an application to the FDA spontaneous reporting system. Am Statist 53(3):177–189

Foody JM, Mendys PM, Liu LZ, Simpson RJ Jr (2010) The utility of observational studies in clinical decision making: lessons learned from statin trials. Postgrad Med 122(3):222–229

Hennessy S (2006) Use of health care databases in pharmacoepidemiology. Basic Clin Pharmacol Toxicol 98(3):311–313

Hennessy S, Leonard CE, Palumbo CM, Newcomb C, Bilker WB (2007) Quality of Medicaid and Medicare data obtained through Centers for Medicare and Medicaid Services (CMS). Med Care 45(12):1216–1220

Hill AB (1965) The environment and disease: association or causation? Proc R Soc Med 58: 295–300

Knapp P, Raynor DK, Berry DC (2004) Comparison of two methods of presenting risk information to patients about the side effects of medicines. Qual Saf Health Care 13(3):176–180

Kostis JB, Shelton B, Gosselin G et al (1996) Adverse effects of enalapril in the Studies of Left Ventricular Dysfunction (SOLVD). SOLVD Investigators. Am Heart J 131(2):350–355

Lazarou J, Pomeranz BH, Corey PN (1998) Incidence of adverse drug reactions in hospitalized patients: a meta-analysis of prospective studies. JAMA 279(15):1200–1205

Lewis JD, Brensinger C (2004) Agreement between GPRD smoking data: a survey of general practitioners and a population-based survey. Pharmacoepidemiol Drug Saf 13(7):437–441

Madigan D, Ryan P (2011) What can we really learn from observational studies?: the need for empirical assessment of methodology for active drug safety surveillance and comparative effectiveness research. Epidemiology 22(5):629–631

Miller DR, Oliveria SA, Berlowitz DR, Fincke BG, Stang P, Lillienfeld DE (2008) Angioedema incidence in US veterans initiating angiotensin-converting enzyme inhibitors. Hypertension 51(6):1624–1630

Norén G, Bate A, Hopstadius J, Star K, Edwards I (2008) Temporal pattern discovery for trends and transient effects: its application to patient records. Paper presented at: Proceeding of the 14th ACM SIGKDD international conference on knowledge discovery and data mining, Las Vegas, Nevada, USA

Norris S, Weinstein J, Peterson K, Thakurta S (2010) Drug class review: direct renin inhibitors, angiotensin converting enzyme inhibitors, and angiotensin II receptor blockers. Accessed on Oct 8, 2012, http://derp.ohsu.edu/about/fi nal-document-display.cfm Accessed Oct 8, 2012.

Overhage JM, Ryan PB, Reich CG, Hartzema AG, Stang PE (2011) Validation of a common data model for active safety surveillance research. J Am Med Inform Assoc 19(1):54–60

Public Law 110–85 (2007) Food and Drug Administration Amendments Act of 2007

Papay JI, Clines D, Rafi R et al (2009) Drug-induced liver injury following positive drug rechallenge. Regul Toxicol Pharmacol 54(1):84–90

Perrio M, Voss S, Shakir SA (2007) Application of the bradford hill criteria to assess the causality of cisapride-induced arrhythmia: a model for assessing causal association in pharmacovigilance. Drug Saf 30(4):333–346

Polinski JM, Schneeweiss S, Levin R, Shrank WH (2009) Completeness of retail pharmacy claims data: implications for pharmacoepidemiologic studies and pharmacy practice in elderly patients. Clin Ther 31(9):2048–2059

Racoosin J (2009) FDA's sentinel initiative—a national strategy for monitoring medical product safety. 2nd Drug Information Association (DIA) Conference on Signal Detection and Data Mining. New York, NY

Rockhill B, Newman B, Weinberg C (1998) Use and misuse of population attributable fractions. Am J Public Health 88(1):15–19

Rodriguez EM, Staffa JA, Graham DJ (2001) The role of databases in drug postmarketing surveillance. Pharmacoepidemiol Drug Saf 10(5):407–410

Rosenbaum P (2002) Observational studies. Springer, New York

Schneeweiss S (2009) On guidelines for comparative effectiveness research using nonrandomized studies in secondary data sources. Value Health 10 Sep 2009

Schneeweiss S, Rassen JA, Glynn RJ, Avorn J, Mogun H, Brookhart MA (2009) High-dimensional propensity score adjustment in studies of treatment effects using health care claims data. Epidemiology 20(4):512–522

Schneeweiss S, Avorn J (2005) A review of uses of health care utilization databases for epidemiologic research on therapeutics. J Clin Epidemiol 58(4):323–337

Schneeweiss S, Sturmer T, Maclure M (1997) Case-crossover and case-time-control designs as alternatives in pharmacoepidemiologic research. Pharmacoepidemiol Drug Saf 6(Suppl 3):S51–S59

Speirs C, Wagniart F, Poggi L (1998) Perindopril postmarketing surveillance: a 12 month study in 47,351 hypertensive patients. Br J Clin Pharmacol 46(1):63–70

Stang PE, Ryan PB, Racoosin JA et al (2010) Advancing the science for active surveillance: rationale and design for the Observational Medical Outcomes Partnership. Ann Intern Med 153(9):600–606

Strom B (2005) Pharmacoepidemiology, 4th edn. Wiley, Chichester

Suissa S, Garbe E (2007) Primer: administrative health databases in observational studies of drug effects—advantages and disadvantages. Nat Clin Pract Rheumatol 3(12):725–732

Szarfman A, Machado SG, O'Neill RT (2002) Use of screening algorithms and computer systems to efficiently signal higher-than-expected combinations of drugs and events in the US FDA's spontaneous reports database. Drug Saf 25(6):381–392

US Department of Health and Human Services, Food and Drug Administration (1999) Managing the risks from medical product use: creating a risk management framework US Department of Health and Human Services, Food and Drug Administration, May 1999

Waller PC, Evans SJ (2003) A model for the future conduct of pharmacovigilance. Pharmacoepidemiol Drug Saf 12(1):17–29

Weatherby LB, Nordstrom BL, Fife D, Walker AM (2002) The impact of wording in "Dear doctor" letters and in black box labels. Clin Pharmacol Ther 72(6):735–742

Whitaker HJ, Farrington CP, Spiessens B, Musonda P (2006) Tutorial in biostatistics: the self-controlled case series method. Stat Med 25(10):1768–1797

Yusuf S, Sleight P, Pogue J, Bosch J, Davies R, Dagenais G (2000) Effects of an angiotensin-converting-enzyme inhibitor, ramipril, on cardiovascular events in high-risk patients. The Heart Outcomes Prevention Evaluation Study Investigators. N Engl J Med 342(3):145–153

Zorych I, Madigan D, Ryan P, Bate A (2011) Disproportionality methods for pharmacovigilance in longitudinal observational databases. Stat Methods Med Res (in print)

About the Editors

Andreas Krause is director and Lead scientist modeling and simulation in the Department of Clinical Pharmacology at Actelion Pharmaceuticals Ltd in Allschwil, Switzerland. He has 20 years of experience in the pharmaceutical industry. Holding a Ph.D. in statistics and econometrics, his areas of expertise include modeling and simulation, also known as pharmacometrics, biostatistics, and bioinformatics.

Graphics and visualization are a key element of his work as evidenced by a multitude of presentations and workshops. Recent publications include *Interactive Visualization of Modeling Results for Effective Team Communication and Decision Making* and *Visualization Concepts for Modeling and Simulation* (the former with Ronald Gieschke, the latter with Kevin Dykstra and Richard Pugh), both in the Journal of Clinical Pharmacology, 50(9), 2010.

Andreas Krause is co-author of *The Basics of S-Plus* (with Melvin Olson, Springer-Verlag, 1997–2005) and co-editor of *Applied Statistics in the Pharmaceutical Industry* (with Steve Millard, Springer-Verlag, 2001).

Michael O'Connell (http://about.me/moconnell) leads the Industry Analytics Group at TIBCO Software, developing analytic solutions across a number of industries including Life Sciences, Financial Services, Energy, Consumer Goods & Retail, and Telco, Media & Networks. He has been working on analytics software applications for the past 20 years, and has published more than 50 papers and several software packages on statistical and graphical methods. Michael has been particularly active in developing statistical graphics applications for review and reporting in clinical development, and has led analytics software and data discovery deployments at most large pharmaceutical companies worldwide. Michael did his Ph.D. work in Statistics at North Carolina State University and is an Adjunct Professor of Statistics in the department.

Index

A
Abdominal pain, 329
Accelerated titration design (ATD), 176
Accuracy, 42, 45, 46, 48, 50, 56, 62
Acetaminophen, 135, 136, 330
Acute coronary syndrome (ACS), 232
Acute myeloid leucemia (AML), 163
ADA. *See* American Diabetes Association (ADA)
Adaptive design, 182
Adaptive study design, 99
Adenosine 5'-triphosphate (ATP) solution, 43
Adverse event(s), 25–28, 30–32, 123, 124, 135–136, 199, 200, 204–206, 209–212, 273, 274, 278, 285, 287–293, 325–328, 364, 369, 370, 374
Adverse Event Reporting System (AERS), 392
Affymetrix, 140, 164–167
Aggregation, 274, 288–292
AIC. *See* Akaike's Information Criteria (AIC)
Akaike's Information Criteria (AIC), 35
Alanine aminotransferase (ALT), 121, 123–125, 127–137, 212–214, 330–332, 334
Alkaline phosphatase (ALKPHS), 121, 124, 125, 128, 130–134, 330
All-cause mortality, 297
Allele, 297
ALT. *See* Alanine aminotransferase (ALT)
American Cancer Society, 165
American Diabetes Association (ADA), 87, 88
American Journal of Epidemiology, 72–74
Amit, O., 26, 204, 300, 325, 326, 332
AML. *See* Acute myeloid leucemia (AML)
Analgesic, 242–243, 246
Analgesic drug, 135

Analysis data model (ADaM), 24, 27, 28, 32
Analysis of covariance (ANCOVA), 96, 97, 219, 225, 229, 232, 233
ANCOVA. *See* Analysis of covariance (ANCOVA)
Andersen, E.W., 71
Andrade, R.J., 330
Angioedema, 396–399, 401, 402, 404, 405, 408, 410
Angiotensin, 396, 398, 407
Angiotensin converting enzyme (ACE) inhibitor, 396–400, 402, 404–408, 410
Angle, 46, 47
Anisimov, V., 34
Annals of Epidemiology, 72, 74
Annals of Internal Medicine, 74
Antibody, monoclonal, 351
Anticholinergics, 218
Antidepressant, 298
Antihistamine, 343
Antiplatelet therapy, 252
Anziano, R.J., 26, 343
Anzures-Cabrera, J., 295
Area, 42–47, 49, 56, 57, 60
Area under the curve (AUC), 42, 43, 57, 58, 60–64
ARIMA. *See* Autoregressive integrated moving average (ARIMA)
Arrhythmia, 336, 343, 344, 349, 356
ASAT. *See* Serum aspartate aminotransferase
Aspartate aminotransferase (AST), 121, 124, 125, 127, 128, 130, 131, 133–135, 212, 213, 330, 331, 334, 336
Aspect ratio, 280, 331
Assay sensitivity, 310, 316–318, 324
AST. *See* Aspartate aminotransferase (AST)

Asthma, 217, 218, 228–230, 236, 238
ATD. *See* Accelerated titration design (ATD)
Atrial fibrillation, 248
AUC. *See* Area under the curve (AUC)
Austin, M., 26, 32, 273
Autoregressive integrated moving average (ARIMA), 35
Axes, 3, 4, 7, 8, 10, 125, 276, 279–282
Axes, multiple, 282
Axis breaks, 281
Axis label, 297
Axis range, 7, 8, 10, 12, 13, 109

B

Background noise, 355
Bar chart, 5, 14, 15, 28, 78, 80, 82, 374, 388
Bar graph, 46
Baseline, 100, 108–110, 114, 258–270
  adjustment, 104, 324
  change from, 175, 183, 186–187, 189, 219, 220, 298, 299, 337–339, 352–356
  correction, 223, 224
Bayes
  empirical, 153
  multi-stage design, 179
Bayesian analysis, 233, 380
Bayesian approach, 268
Bayesian study design, 99
Bazett's correction for QT data, 262, 347
Bednar, M., 349
Benazapril, 396
Benchmarking, 384–386
Benner, A., 139
Bennett, K.B., 50
Berry, D.A., 211
Berry, S.M., 211
Beta-agonist, 233
Beta blocker, 119
Beta distribution, 205, 211
Biclustering, 143, 145–147, 161, 163
Bile system, 119
Bilirubin, 29, 30, 32, 119, 121, 125, 129, 130, 133–135, 200, 201, 206, 212, 213, 330, 331
Binary outcome, 300, 307
Binning, 303
Binomial data, 199
Bioconductor (software), 153
Bioequivalence, 49, 56–58
Biological therapy, 174
Biomarker, 139–167, 176, 218
  classification, 118

data, 17–19, 25
  normalization, 121–122, 125
Black box warning, 344
Blindness, 87
Blood pressure, 117, 119, 120
Blood work, xiv
BMI. *See* Body mass index (BMI)
Bock, J., 12
Body mass index (BMI), 211
Bold typeface, 49
Bolger, F., 44
Bolognese, J.A., 59
Bootstrap, 156–158
Borenstein, M., 301
Box and whisker plot. *See* Boxplot
Boxplot, 17, 94–95, 127, 154, 155, 157, 200–203, 212, 214, 233, 259, 280, 284, 285, 334–335, 338–339, 376, 378
Bradstreet, T.E., 41, 42, 51, 57
Brain storming, 374–378
Brand manager, 382
Breast cancer, 118
Brenner, H., ix
Brewer, C.A., xiii
British Medical Journal (BMJ), 72, 74
Bronchoconstriction, 218
Bronchodilator, 217–239
Brown, H., 51
Brownian motion, 181
Brushing, 376
Bubble plot, 307
Bühlmann, P., 158
Bunionectomy, 242, 243
Burning, 376

C

Calcaterra, J.A., 50
Cancer. *See* Oncology
Cancer, ovarian, 175, 182, 187
Cancer survival rates chart, ix–xii
Captopril, 396
Cardiac effect, 26
Cardiac safety, 326, 336–340, 343–356
Cardiovascular, 255–270
Cardiovascular events, 297, 300
Cardiovascular heart disease (CHD), 297
Carry-over, 351, 353
Carter, L.F., 44
Carvedilol, 399
Categorical data, 11, 199
Censoring, 328, 329
Centering, 142
Center performance, 182

CenterWatch, 174
Central nervous system (CNS), 248, 336
CGH. *See* Comparative genomic hybridization (CGH)
Chambers, J.M., 260
Change from baseline, 4, 8, 12–13, 95–97, 119, 120, 175, 184, 186–187, 189, 279. *See also* Baseline, change from
Chartjunk/chart junk, 47, 48, 51
CHD. *See* Cardiovascular heart disease (CHD)
Chemicals, 174
Chemistry, 288
Chemotherapy, 174
Chen, W., 49
Children, 246, 247
Cholestasis range, 331
Cholesterol, 336
Chromosome, 152, 162–164
Chronic obstructive pulmonary disease (COPD), 217, 218, 220–222, 226–232, 236, 238
Circadian rhythm, 257, 260, 261, 265–267, 270
Circadian variation, 220
Classification, 140, 153–158, 161, 162
Cleveland, W., xiii, 4, 19, 38, 49, 176, 280, 326
Clinical benefit, 175
Clinical Data Acquisition Standards Harmonization (CDASH), 27
Clinical Data Interchange Standards Consortium (CDISC), 24, 27, 28, 32
Clinical pharmacology. *See* Pharmacology
Clinical research. *See* Research, clinical
Clinical Research Asscociate (CRA), 27
Clinical research organization (CRO), 34, 360, 365, 369, 371
Clinical response, 9, 16–18
Clinical trial management system (CTMS), 359–360, 365
Clinical trial metrics, 365–371
Clinicopathological data, 140, 141, 153
Clopidogrel, 297, 300, 302, 303
Cluster (software), 163
Cluster analysis, 376, 378
clusterCons (software), 144
Clustering, consensus, 144, 145
Clustering, k-means, 142–145, 149, 150, 163
Clustering methods, 143
C^{max} (time of maximum concentration), 315, 316, 321
CNS. *See* Central nervous system (CNS)
Cochrane collaboration, 296
Cognitive fit, 45

Cohort profile, 397, 404–405
Cold, 376
Coll, J.H., 44
Color(s), viii, xiii, 3, 4, 6–10, 15, 17, 18, 43, 46, 47, 52, 58, 59, 80, 278, 280–282, 288, 327–329, 360–364, 367, 369
ColorBrewer, xiii
Color-coding, 143, 157
Combination treatment, 175
Common toxicity criteria (CTC), 288, 289
Communication, vii, viii, xiii, 46, 57, 374–378, 382
Comparative genomic hybridization (CGH), 160, 162, 163
Comparison, 4, 6–8, 10–11
Competitor, 382
Complete linkage, 142, 143, 146
Concentration/concentration (drug). *See* Drug concentration
Concentration-QT analysis, 302–324
Concentration-time, 101, 102
Concomitant medication (CM), 123, 124, 135–137, 207, 208, 274–276, 283, 284, 288, 289, 370
Confidence band, 190, 200, 201, 203
Confidence interval, 75–78, 80, 81, 223, 229, 230, 232, 236, 237, 239, 296, 298, 299, 312, 315
Congress activities, 376–378
Consensus clustering, 144, 145
Contours, 302–303
Contraceptive, 41–42
Control, 119, 120, 126, 129, 136, 137
Control chart, 383–386
Control group, 279, 280, 288, 385, 386
Cook, I., 23
Coovert, M.D., 44
COPD. *See* Chronic obstructive pulmonary disease (COPD)
Correlation, 128–132
Country (plot by), 110, 111
Covariates, 100, 104, 108, 110–113
Cox model, 201, 203
Cox proportional hazards model, 252
CRA. *See* Clinical Research Asscociate (CRA)
CRAN, xiii
Creatinine, 202, 336
Creatinine phosphokinase (CPK), 336
CRO. *See* Clinical research organization (CRO)
Cross-over, 312, 316, 351–353
Cross-over studies, 125–12
Cross-over trial, 42, 51
Cross-validation, 155

Crowe, B.J., 326
CTC. *See* Common toxicity criteria (CTC)
CTMS. *See* Clinical trial management system (CTMS)
CTSpedia, xiii, 351
Cumulative incidence, 327–329, 335
CUSUM chart, 384
CYP3A6, 344
CYP2C19, 297
Cytogenetic data, 140

D

Data cleaning, 360–365
Data exploration, 255–270
Data management, 359–372
Data mining, 210–211
Data quality, 104–108
Data review, 360–365
Data status, 369
Data to ink ratio, 7
Decision-making, 392–393
De-escalation, 177, 178
Defibrillator, 257
Delay of effect, 315, 324
Demography, 199
Dendrogram, 141–143, 146, 159, 162, 163
Density of information, 47
Density plot, 285, 286
Depolarization, 344, 349
DerSimonian and Laird method, 298
Descriptive statistics, 325, 326
Design
  Bayesian, 179, 182, 183
  predicted probability, 179, 180
  Simon 4-stage, 179, 180
Deviance, 155, 156, 205
Diabetes, 87–97
Diagnostic plot, 193, 196
Diarrhoea, 329
Diet, 174
Dilevalol, 119
DILI. *See* Drug-induced liver injury (DILI)
3-Dimensional graphs, 14
Dimension reduction, 141, 147–152, 155
Direct to consumer advertising, 387
Discontinuation, 364
Disease
  progressive, 175, 185, 186
  stable, 175, 185, 186
Dispensing, 393, 395, 399, 401_402, 405
Distance metric, Euclidean, 143, 145, 146
Distributional plot, 73, 74, 78–82
Diurnal variation. *See* Circadian rhythm

DLT. *See* Dose-limiting toxicity (DLT)
Dobbins, T.W., 57
Dose escalation with overdose control (EWOC), 176
Dose-limiting toxicity (DLT), 174, 176, 178
Dose optimization, 236, 239
Dose proportionality, 60–63
Dose-ranging trial, 175
Dose-response, 16–17, 91, 123, 127–128, 176, 217–239
Dose selection, 218, 221, 230–236, 239
Dose-toxicity, 176, 177
Dosing regimen evaluation, 50–56
Dot-and-interval plot, 326–327
Dotplot, 205, 300–301, 326
Down-regulation. *See* Regulation
Drill-down, 24, 26, 28, 30, 32, 36, 211, 285
Drop-out, 246–248
Drug concentration, 9, 16, 17, 312, 315, 319–321, 323
Drug-drug interaction, 175
Drug exposure, 303, 319–320
Drug-induced liver injury (DILI), 26, 119, 128–130, 329–332
Drug metabolism and pharmacokinetics (DMPK), 25

E

eCDF. *See* Empirical cumulative density function (eCDF)
ECG. *See* Electrocardiogram (ECG)
ED_{50}, 235
EDA. *See* Exploratory data analysis (EDA)
eDISH. *See* Evaluation of drug-induced serious hepatotoxicity (eDISH)
Edit checks, 360, 369
Efficacy, 4, 117, 218, 231, 232, 236, 238, 242, 243, 246–253, 359, 364, 369, 374, 375, 378, 387, 392
Efficacy data, 4, 27, 183
Egger, M., 301
Ehrenberg, A.S.C., 48, 49
Eisen, M., 141
Electric shocks, 376
Electrocardiogram (ECG), 25, 117, 209, 257, 258, 302, 303, 315, 316, 322, 336–337, 343–346, 349–353
  schema, 257
  time-matched, 316
Electronic case record form (eCRF), 27
Electronic data capture (EDC), 359, 360, 364, 367–369, 371

Electronic health records (EHR), 392–395, 402
EMA. *See* European Medicines Agency (EMA)
E^{max}, 233
E^{max} model, 16
Embolism, 248
Emir, B., 373
Empirical cumulative density function (eCDF), 145
Enalapril, 396, 402
Endocrinology, 336
Enrollment, 182, 187
Enrolment, 359, 365–368
Enrolment data, 26, 33–35, 37
Environmental pollutants, 174
Enzyme, 119, 123–125, 128, 130, 132, 133
    activity, 123, 125
    function, 297
    induction, 130, 132
Epidemiology, 72–74, 79, 80, 82
Epidemiology (journal), 72, 73
Epigenomics, 140
Erroneous data, 105
Escalation of dose with overdose control. *See* Dose escalation with overdose control (EWOC)
Ethinyl, 42
Euclidean distance metric, 142, 143, 146, 148
European Medicines Agency (EMA), 218
Evaluation of drug-induced serious hepatotoxicity (eDISH), 128–130, 133
Event chart, 187–191
EWOC. *See* Dose escalation with overdose control (EWOC)
Exploration, 46
Exploratory data analysis (EDA), 38, 100, 102, 108, 111, 114
Exponential distribution, 252
Exposure to drug. *See* Drug exposure
Expression level (genes), 140
Extreme values, 13–14

F
False discovery rate, 154
Feature screening, 153, 155
Fedorov, V., 34
FEV_1. *See* Forced expiratory volume (FEV1)
Few, S., xiii, 49, 277
Fibrillation, atrial, 248
Field force, 387–390
Finmetrics library, 37
Florian, J.A., 267

Flow diagram, 73, 74, 77, 80, 81
Fluoxetine, 298, 299
Font, 58, 59
Food and Drug Administration (FDA), 24–26, 30, 118, 121, 128, 129, 226, 229–231, 235, 236, 243, 330–332, 336, 344, 350, 351, 392
Food effect, 174
Foot amputation, 87
Forced expiratory volume (FEV_1), 15, 16, 218–230, 232, 236, 238
Forced vital capacity (FVC), 218
Forest plot, 26, 96, 97, 184–188, 296–301, 305, 397, 409–411
Freeman, R., 49
Frequency table visualization, 151
Fridericia's correction for QT data, 262, 347
Frikke-Schmidt, R., 80
Funnel plot, 301–303, 307
Futility, 181, 182
FVC. *See* Forced vital capacity (FVC)

G
Gamma glutamyl transferase (GGT), 121, 124, 130–132
GapMinder, xiii
Garland, K., viii
Gastric inhibitory polypeptide (GIP), 88, 95
Gastroesophageal reflux disease (GERD), 50–56
Gastrointestinal, 327, 329
Gelman, A., 49
Gender effect, 261, 263–265, 267
Gene expression, 140, 141, 145, 146, 148–152, 154, 159, 161, 163
Gene expression data, 140, 146, 150, 159, 161, 163
Gene expression omnibus (GEO), 140
Gene loadings plot, 150, 151
Gene set enrichment analysis (GSEA), 151, 158, 160
Genetic biomarker, 118
Genetics, 117, 118, 121, 140, 153, 162
Genomics, 140, 141, 160
Genotype, 297, 350
GenStat (software), 298, 301–304, 306
GEO. *See* Gene expression omnibus (GEO)
GERD. *See* Gastroesophageal reflux disease (GERD)
ggplot2 (software), 100
GGT. *See* Gamma glutamyl transferase (GGT)
Gilbert's syndrome, 121, 331
Gilder, K., 26, 165

GIP. *See* Gastric inhibitory polypeptide (GIP)
Girgis, I.G., 25, 255, 268
glmnet (software), 155
Glucagon-like peptide (GLP-1), 88, 95
Glucose, 87–97, 118
Goeman, J.J., 158
Goeman's global test, 158
Gough, K., 60
gplots (software), 143
Grable, 41–65
Gradient, 104, 105
Graphical elements, 4, 77–10
Graphical user interface (GUI), 64
Graphics, xi, xv, xiii, vii–ix, 52, 53, 64
graphics (software), 176
Graphs. *See* Graphics
Gray scale, 9, 10, 15, 18
Grey/gray, 99, 102, 105–107
grid (software), 176
Grouping, 141
Growth, 336
GSEA. *See* Gene set enrichment analysis (GSEA)
GUI. *See* Graphical user interface (GUI)

H
Haematocrit, 336
Haematotoxicity, 336
Haemer, K.W., 14
Hamilton scale for depression, 298, 299
Harris, R.L., 49
Harvey, N., 44
Haynes, J.D., 60
Hazard rate, for adverse events, 329
Hazard ratio, 75, 80, 252, 296
HDL, 336
Healthcare data, 391–411
Healthy volunteer, 199
Heart attack, 87
Heart Protection study (HPS), 202
Heart rate, 257, 261, 264, 265, 267, 270
Heat map, 141–147, 163, 378, 379, 391
Heatplus (software), 143
Heiberger, R.M., 4
Hematology, 288
Hemoglobin, 88, 119
Hemoglobin A1c (HbA1c), 87, 88, 93
Henry, G.T., 49
Hepatic function, 119, 129
Herceptin™, 118
Hexagonal binning, 202
Hielscher, T., 139

Hierarchical clustering, 141–143, 145, 162, 163
High-dimensional data, 141, 145, 158–160
Higher dimensions, 4, 14–16
High level group terms (HLGT), 290–292
High level terms (HLT), 290–292
High-througput data, 139, 140
Hink, J.K., 50
Histogram, 124
HLGT. *See* High level group terms (HLGT)
HLT. *See* High level terms (HLT)
Hmisc (software), 176
Holland, B., 4
Holmes, M.V., 297, 303
Holter monitoring, 344, 349
Holt-Winters smoothing, 35
Hormone, 88, 89, 95
Hormone therapy, 174
Hotelling statistic, 151
HTSanalyzeR (software), 158
Hue, 47
Hwang, M.I., 44
Hyperglycemia, 88
Hyperlinking, 209
Hyperparameter, 161
Hypertension, 396, 398, 402, 405, 408
Hypertext markup language (HTML), 208
Hypothesis testing, 137
Hy's law, 121, 128–130, 134, 135, 330, 331, 334

I
IBM Many Eyes, xiii
ICH, 290, 302
ICH 10, 374
ICH E14 guideline, 302
IFPMA. *See* International Federation of Pharmaceutical Manufacturers and Associations (IFPMA)
Ihaka, R., 10
Immune system, 159
Imputation, 252
Imputation, multiple, 382
Incidence rate, 326, 336
Indacaterol, 218–223, 225, 227–236, 239
Individual correction for QTc data, 313–323
Individual patient level, 281–301
Infectious organisms, 174
Inferiority, 180
Infographics, 150–151
INR. *See* International normalized ratio (INR)
Insulin, 88–91, 95

Insurance claim data, 26
Interactive display, 199–215
Interactive graphics, 117–137, 160
Interactive voice response system (IVRS), 365
Interim analysis, 26, 180, 182–183
International Conference on Harmonization of Technical Requirements For Registration of Pharmaceuticals for Human Use, 282. *See also* ICH
International Federation of Pharmaceutical Manufacturers and Associations (IFPMA), 290
International Journal of Epidemiology, 72, 74
International normalized ratio (INR), 248–252
Interpretation, 47, 55
Interquartile distance, 259
Interquartile range, 108, 110–113
Intervals, predictive, 305
Inverse-variance method, 297
Investigational new drug (IND), 25
In-vitro experiments, 25
In-vivo experiments, 25
Iontophoresis, 42–44
IVRS. *See* Interactive voice response system (IVRS)

J
Javascript, 208
Jeffreys prior, 213, 219, 390
Jin, B., 59
Jittering, 58, 59
JMP (software), 176, 285
Jolliffe, I.T., 211
Jones, B., 51, 57
Journal of Clinical Epidemiology, 72, 74
Journal of the American Medical Association (JAMA), 72, 74
Jung, C., xiii
JUPITER study, 202

K
Kaplan-Meier plot(s), 19, 72–77, 80, 81, 182, 183, 189–192, 196, 197, 199, 291, 292, 327
KEGG database, 158–160
Kenward, M.G., 51, 57
Ketoconazole, 344
Key performance indicator (KPI), 33, 36
Kidney failure, 87
Kirsch, I., 298
k-means clustering, 142–145, 149, 150, 163
Kosslyn, S.M., 49

KPI. *See* Key performance indicator (KPI)
Krause, A., 3, 25

L
L'Abbé, K.A., 307
L'Abbé plot, 307
Lab data, 279, 282, 288, 289
Labeling, 281
Laboratory data, 60, 61, 199, 210, 211, 335
Laboratory results, xiii
Lab values, 360–364, 369
Lactate dehydrogenase (LDH), 336
Lalomia, M.J., 44
Lancet, 72, 74
Landscape, 53
Lane, P.W., 26, 295, 300, 325
Lasso estimator, 155
Last patient last visit (LPLV), 33
Late-stage drug development, 241–253
Lattice, 6, 12
lattice (software), 176
Layering, 278–279
Lazarou, J., 392
LDL, 336
Learn and confirm paradigm, 256
Least squares means, 96, 97
Leckart, S., xiii
Legend, 4, 6–10, 17, 18
Length, 43, 47, 54, 60
Lesion, 140–141, 144, 146, 155, 156, 158–160
Lewis, S., 295
LFT. *See* Liver function test (LFT)
Life sciences, vii, xii, xiii
limma (software), 153
Line chart, 364
Line graph, 46
Lines, 4, 7, 9–11, 13, 14, 16–18
Lines (graphical elements), viii
Line types, 282
Linkage, 142, 143, 146
Lin, X., 211
Lipids, 336
Lisinopril, 396, 399, 406–409
Liver function, 350
Liver function test (LFT), 201, 330, 351–355, 364
Liver injury, 119, 121, 125, 128–131, 135, 201
  cholestatic, 130, 131, 133
  hepatocellular, 130, 131
Liver test, 119, 124, 125, 133, 134
Liver toxicity, 326, 329–336
LLN. *See* Lower limit of normal (LLN)
Loess. *See* Lowess

Loess (smoother), 105
Logistic regression, 155, 169, 298
Logrank test, 192, 193, 298, 299
Log scale/logarithmic scale, 75, 78. *See also* Scale, logarithmic
London underground map, viii
Looby, M., 26, 217
Lower limb disorders, 87
Lower limit of normal (LLN), 121, 122
Lowess, 195, 196, 260, 262, 263, 268, 269
LPLV. *See* Last patient last visit (LPLV)
LSMEANS (SAS procedure), 379
Lucas, H.C., 49
Lumley, T., 10
Lung function, 219
Luo, D., 23

M
Madeira, S.C., 147
Magnetic resonance imaging (MRI), 141, 153
Mahalanobis distance, 211–213
MammaPrint® test, 140
Many Eyes. *See* IBM Many Eyes
Maps, 374
Marketing, 24, 26, 27, 35–36, 38, 373–390
Marketing mix model, 387
Marketing trial, 175
Market share, 383, 385
Martingale residuals, 194, 196
Maximum tolerated dose (MTD), 174, 177, 179
MCID. *See* Minimal clinically important dose (MCID)
Mean
 cell volume, 336
 differences, 296, 298
 weighted, 88, 93–97
Mechelen, I.V., 147
Medarov, B.I., 220
Median, 107, 108, 110, 112
Medical claims, 393, 394
Medical Dictionary for Regulatory Activities (MedDRA) coding dictionary, 27, 290, 291, 407
Medical history, 199, 200, 274, 287, 288
Medical product safety, 391–411
Medidata, 36, 37
Meinshausen, N., 159
Merz, M., 26, 30, 117
Meta (software), 298
Meta-analysis, 26, 72, 295–315, 392, 397, 409
Metabolic data, 26

Metabolic disorder, 87
Metabolic hormones, 88
Metabolite, 350
Metabolomics, 140
Meta-regression, 296, 304–313
Metoprolol, 396
Meyer, J., 44, 46
Microarray, 140, 145, 158
MicroArray Quality Control (MAQC-II) project, 140
Microvascular complications, 88
Minimal clinically important dose (MCID), 226, 232, 233, 235
Miskell, A., 26, 87
Mixed model, 379, 387
Model-based analysis, 255
Model-based approach, 232–236, 239
Model-based curves, 73, 74, 77–78, 80, 81
Model building, 255–270
Modeling, 255–257, 260, 268–270. *See also* Pharmacometrics
Model validation, 256
Mohanty, S., 25, 26, 241, 255
Molecular data, 140, 159
Molecular target, 124
Mondrian (software), 285
Monitoring, continuous, 181
Monoclonal antibody. *See* Antibody, monoclonal
Morphine, 246, 249
Moxifloxacin, 257–259, 267–268, 312, 313, 316–318, 352, 353
MRI. *See* Magnetic resonance imaging (MRI)
MSToolkit (software), 100
MTD. *See* Maximum tolerated dose (MTD)
Multi-Item Gamma Poisson Shrinker (MGPS), 392
Multi-panel plot, 95–96
Multiple axes. *See* Axes, multiple
Multiple imputation. *See* Imputation, multiple
Multiple testing, 154, 159
Multi-stage design, 179
Multivariate, 121, 143
Multivariate outlier detection, 214
Muscle injury. *See* Rhabdomyolysis
Myeloma, 140, 142, 150, 152
Myoglobin, 336

N
Nash, D., xi
National Institutes of Health (NIH), 139
Nausea, 329
Nefazodone, 4, 99, 298

Nephrotoxicity, 336
Neurology, 26
Neuropathic pain, 374, 376
Neutrophils, 360
New England Journal of Medicine, 72, 74
Nicholls, A., 99
NIH. *See* National Institutes of Health (NIH)
Nitrogen, 336
Noise, 383, 385
Nonclinical research. *See* Research, nonclinical
Non-responder, 92, 93
Normalization, 121–122, 124, 140, 248, 376–378
Normal range, 121–123, 125, 126, 274, 279, 280, 282, 288, 289, 378

O
Obesity, 174
Observational data, 392, 395, 396, 405, 409–411
Observational medical outcomes partnership (OMOP), 395–397, 399, 405–408
Observed *vs.* predicted. *See* Predicted *vs.* observed
O'Connell, M., 23, 192, 210
Odds ratio, 75, 76, 79, 80, 296, 378, 380
Oliveira, A.L., 147
Omics, 140
Oncology, 26, 173–196
Operating characteristics, 177–182
Operations, 359–372
Operations data, 26, 33–35
Ordering, 141, 143, 146, 298, 313
Orientation (of a graph), 53, 55
Osteoarthritis, 243, 245
Outcome/risk, 73–75
Outlier, 128
Outlier detection, multivariate, 214
Outliers, 199, 201–203, 210, 211, 214. *See also* Extreme values
Out-of-bag error, 156, 157
Out of range values, 123, 125
Ovarian cancer, 182, 187

P
Package insert, xiii
Pain, 42–44, 242–246, 374–376, 379
Pain intensity, 242–244
Pairwise scatterplot. *See* Scatterplot, pairwise
Pancreas, 89
Panebianco, D.L., 42, 57

Panel, 7, 18, 59–61, 101–103, 110, 111, 113, 119, 124, 127–129, 132–140
Parallel coordinate plot, 285
Paroxetine, 298, 299
Partitioning methods, 141, 143–145
Patel, J., 23
Patient, 199, 201–208, 210, 211, 213
Patient profile, 273, 275, 276, 280, 285–289, 291, 397–405
Patient-reported outcome (PRO), 175
Payer, 393
PCA. *See* Principal components analysis (PCA)
PD. *See* Pharmacodynamics (PD)
Penny, K.I., 211
peperr (software), 156
Peptide, 88
Perception, 44, 46, 47, 49–51, 57, 59, 65
Peters, J.L., 303
Pharmaceutical Researchers and Manufacturers of America (PhRMA) organization, 25
Pharmacodynamics (PD), 16, 88, 90, 99–114, 167, 168
Pharmacoepidemiology, 393
Pharmacogenetics. *See* Genetics
Pharmacokinetics (PK), 16, 88, 99–114, 174, 176, 303, 315, 316, 319
  population, 350
Pharmacology, 88, 174
Pharmacometrics, 256
Pharmacy dispensing. *See* Dispensing
Phase 1, 174, 176–179, 282
Phase 2, 175, 179
Phase 3, 175, 179–182
Phase 4, 175, 373–390
Phase 2a, 175
Phase 2b, 175
Phenotype, 153
Physical activity, 174
Pie chart, 374
Pikounis, B., 57
Pins and needles, 376
Pittelkow, Y., 151
PK. *See* Pharmacokinetics (PK)
PK/PD, 16, 99–114
Placebo, 4, 5, 8, 12, 51–57, 59, 88, 90, 94, 124, 125, 127–129, 131, 242, 243, 252, 350–354, 375, 376, 378, 379, 382
  correction, 220, 224
  effect, 257, 261, 265, 267
Plaid model algorithm, 161
Plasma concentration, 42, 57, 60
Platelets, 336
PLATO study, 202

Plotting symbol, 288–291
POC. *See* Proof of concept (POC)
Pocock, S.J., 71
Point estimate, 75, 78, 80, 81
Point symbols, 90
Polygon, 107
Porat, T., 44
Portable document format (PDF), 208
Portrait, 53, 55
Position along a common scale, 46
Positive control, 351
Post-approval trial, 175, 373–390
Post-marketing data, 26
Post-prandial, 88, 90–92
Powers, M., 44, 50
Powsner, S.M., 288
PQ interval, 60, 61
Precision-weighted average, 311
Predicted probability design, 179, 180
Predicted *vs.* observed, 125
Predictive intervals. *See* Intervals, predictive
Preferred terms, 291, 292
Prescott, R., 51
Prescription medicines, 392
Pressure, 376
Prevalence, 397, 405–409
Principal components, 24
Principal components analysis (PCA), 141, 147–150
PRO. *See* Patient-reported outcome (PRO)
Probe set, 140
Profit, 387, 389, 390
Progressive disease (PD), 175, 185, 186
Proof of concept (POC), 175, 378
Propensity, 255
Propensity score, 395
Proportional hazard, 252
Proportional hazard model. *See* Cox model
Proportions, 296
Publication bias, 301–303
Pugh, R., 99
Pulmonology, 26
p-value, 80, 288, 290, 291
P-wave. *See* Electrocardiogram (ECG)

Q
QC. *See* Quality control (QC)
QoL. *See* Quality of life (QoL)
QT, 174, 255–270, 302–324, 336, 343–356
    correction, 310, 313–315, 324, 344–349, 356
    knowledge management system, 311–312, 323
    prolongation, 309, 310, 313, 316, 319–321, 323
QT (software), 303–304
QTc. *See* QT
QTcB. *See* Bazett's correction for QT data; QT
QTcF. *See* Fridericia's correction for QT data; QT
QTcI. *See* Individual correction for QTc data
QTc interval,
Quality control (QC), 104
Quality of life (QoL), 175
Query status, 379
Questionnaire, 376
Quinapril, 396, 407

R
R (software), 298, 302–304, 314
Radiation, 174
Ramipril, 396
Randomized trials, 72
Rare events, 393
Ratkowsky, D.A., 51
Rave. *See* Medidata
RColorBrewer (software), 176
RECIST. *See* Response evaluation criteria in solid tumors (RECIST)
Red cell count, 336
Reference line, 58, 59
Reflux. *See* Gastroesophageal reflux disease (GERD)
Regional business unit, 389
Regularization, 155–156
Regulation, 153, 154, 158
Regulatory interaction, 374, 380–382
Relative influence plot, 212
Relative risk, 204, 211, 327
Renard, D., 217
Repolarization, 344, 349
Reporting, 24, 27, 28, 36, 38
Resampling methods, 155–156
Rescue pain medication, 242–245
Research
    clinical, 174–175, 195, 196
    nonclinical, 166
Responder, 92, 93, 96
Response evaluation criteria in solid tumors (RECIST), 364
    criteria, 364
Reticulocyte count, 336
Revenue, 387, 388
RevMan (software), 300

Index 427

Rhabdomyolysis, 336
Risk
  difference plot, 211, 213
  factor, 140
  ratio, 80, 297, 298, 300, 302
  relative, 300, 313
rms (software), 176
Robbins, N.B., xiii, 4, 49
Rodda, B.E., 57
Roosen, C., 25, 99
RR interval, 312–314. *See also*
  Electrocardiogram (ECG); Heart rate
R-wave. *See* Electrocardiogram (ECG)
Ryan, P., 26, 391

S
S+, 208, 209
Safety, 176–179, 186, 255–270, 374, 378, 380–382, 387, 391–411
  biomarker, 118, 119, 124, 125, 142
  data, 3–5, 16, 17, 20, 25, 26, 199–215, 325, 336
  long term, 175
Salanti, G., 314
Sales, 24, 27, 35–36, 38, 174, 373–390
Sales representative, 374, 378–380
Saline solution, 43
Sampling times, 303, 319, 320
SAS, 6, 88
SAS (software), 298, 300–304, 314
Saturation, 47
Scale, logarithmic, 297, 298
Scaling, 279–280, 282, 376–378
Scatterplot, 5, 6, 8, 17–18, 260, 330–334, 337, 364, 370
  matrix, 333–334
  pairwise, 285, 287
Schenker, N., 42
Schmid, C.F., 49
Schoenfeld residuals, 194, 196
Screening, feature, 153–155
SDV. *See* Source data verification (SDV)
Senior manager, 374
Sensitivity analysis, 236–238
Sentinel event, 336
Seriation methods, 141
Serotonin, 298
Serum aspartate aminotransferase (ASAT), 202
Sethuraman, V., 60
SEURAT (software), 160–163
Sex, 111, 113
SGOT. *See* Aspartate aminotransferase (AST)

SGPANEL (SAS procedure), 6
SGPT. *See* Alanine aminotransferase (ALT)
Shadow, 103, 107, 110–113
Shapes, 282
Shift from baseline, 123, 125–127
Shift plot, 200–203
Shortino, D., 87
Shrinkage, 155
Side effect, 174
Signal to noise ratio, 217–239, 310
Silhouette chart, 144
Sill, M., 139
Simon 4-stage design, 179, 180
Simulation, 255, 256
Single nucleotide polymorphism (SNP), 160, 163
SLD. *See* Sum of the longest diameters (SLD)
Slope, 47, 51, 53–55, 60–62, 104
Small sample size graphics, 87–97
Smith, B.P, 60
Smith, R., 60
Smoking, 113
Smooth curves, 78
Smoothing, 101
Smyth, G.K., 153
Snape, M.D., 75
SNP. *See* Single nucleotide polymorphism (SNP)
Snyder, T., 26, 35, 359
SOC. *See* System organ class (SOC)
Sorting, 142, 275, 278
Source data verification (SDV), 33
Southworth, H., 26, 199, 211
Spaghetti plot, 101, 110, 246
Spatial tasks, 45
Spirometry, 218, 219
S-PLUS, 88, 90
Spotfire, 88, 90, 121, 208, 209, 396
Spotfire (software), 285
Squeezing, 376
Stabbing, 376
Stable disease (SD), 175, 185, 186
Stata (software), 302, 304
Steady state, 219, 221, 225, 229, 258
Stem and leaf plot, 41, 42
Stent thrombosis, 297
Step function, 328, 329, 338
Sterne, J.A.C., 302
Stopping early, 180, 182
STROBE initiative, 72
Stroke, 87, 248, 252, 297, 314
Study data tabulation model (SDTM), 27, 28, 32
Subgroup analysis, 298–300, 397, 408–409

Subgroups, 141, 142, 150, 152
Subjects, 199–203, 206
Subramanian, A., 158
Subsampling, 156
Subsetting, 28
Suicidality, 336
Sum of the longest diameters (SLD), 184–186
Sun, P., 57
Superiority, 180
Surgery, 174
Surrogate endpoint, 88
Surveillance, 392
Survival
    data, 19, 20
    event-free, 175, 189
    overall, 175, 187–192, 194, 195
    progression-free, 175, 183, 189
    relapse-free, 173, 189
survival (software), 176
Sutton, A.J., 302
s4vd (software), 147
S4VD algorithm, 147
Symbolic tasks, 45
Symbols, 4, 5, 7–9, 17
System organ class (SOC), 290–292
Systolic blood pressure, 119, 120

T
Tablejunk/table junk, 48
Tables, 5, 14, 19, 20, 41–65, 199–215
Tables, guidelines for, 47–49
Tabular presentation, 200, 204–206
TBL. *See* Total bilirubin (TBL)
TdP. *See* Torsades de pointes (TdP)
TED conference, xiii
Terfenadine, 343, 344
Text mining, 151
Thompson, S.G., 304
Thorough QT (TQT) study, 175, 257, 258, 302–324, 350–356
Thromboembolic event, 4, 41, 248
Thrombosis, 297
Thyroxin, 336
TIBCO, 121, 396
Tick marks, 51, 52, 56, 58, 59, 281
Time after dose, 101–108, 113
Time course, 121
Time in therapeutic range (TTR), 248–252
Time profile. *See* Time course
Time to event data, 191
Time to event plot, 189–195
Time to progression (TTP), 175, 189

Time to response, 175, 189
Tingling, 376
Tobacco, 174
Tomlinson, J., 295
Tornøe, C.W., 26, 302
Torsades de pointes (TdP), 257, 336, 349, 350
Total bilirubin (TBL). *See* Bilirubin
Total Therapy trials, 140
Toxicology, 118
TQT study. *See* Thorough QT (TQT) study
Transcript (gene), 140, 164
Transcriptomics, 140
Treder, B., 192
Tree map/treemap, 26, 119, 120, 142, 143, 397, 405–408
Trellis, 6, 12, 124, 126, 331–333, 397, 408–410
Trend analysis, 382–384
Trend break, 382–385
Trial cost, 179
Trial design, 176–182
Triglycerides, 336
Troglitazone, 119, 330
TTR. *See* Time in therapeutic range (TTR)
Tufte, E.R., xiii, 4, 6, 19, 20, 38, 49, 50, 176, 288
Tukey, J.W., 19, 334
Tullis, T.S., 49
Tumor, 184
Tylenol™, 141, 242
Type 2 diabetes mellitus (T2DM). *See* Diabetes

U
ULN. *See* Upper limit of normal (ULN)
Units (lab data), 122
Units, standardized, 360–363
Unpowered studies, 87
Unsupervised methods, 141–153
Unwin, A., xiii, 4, 5
Upper limit of normal (ULN), 29, 30, 121, 122, 125, 127–141, 330–332, 335, 360
Up-regulation. *See* Regulation
Urinalysis, 199
Urine test, 336

V
van Houwelingen, H.C., 304
VAS. *See* Visual analog scale (VAS)
Venlafaxine, 298

Index

Vessel revascularization, 297
Vildagliptin, 119
Visual analog scale (VAS), 43
Visualization, 273–293
Visual predictive check,
Vital signs, 199
Volume, 47
Vomiting, 329

W
Wainer, H., 49
Walker, A.M., 87, 141
Wang, J., 26, 241
Warfarin, 248–251
Waterfall plot, 184–186
Web site (for the book), viii, xv
Weerahandi, S., 26, 35, 384
Weighted mean. *See* Mean, weighted
Weiss, A.I., 60
Wenig, T., xiii
Whalen, E., 373
White cell count, 336
Whitehead, A., 302
WHODD, 27
Wikipedia, viii
Wilkinson, L., 49

Williams, P., 77
Wilson, S., 152
Withdrawal, 252
Wong, D.M., 49
Workflow, 118, 121, 124–137

X
Xanomeline (drug), 28–32
x-axis, 7, 10, 13
XML, 208
Xpose (software), 100
XSL, 208

Y
y-axis, 7, 8,10, 12–14
Yuh, L., 58

Z
Zeileis, A., 10
Zhang, A., 281
Zimmerman, H.J., 129
Z score, 246
Zucknick, M., 139

Printed by Publishers' Graphics LLC
SO20130315.19.21.21